DNA *Fingerprinting in Plants*
Principles, Methods, and Applications
Second Edition

Kurt Weising

Hilde Nybom

Kirsten Wolff

Günter Kahl

Taylor & Francis
Taylor & Francis Group

Boca Raton   London   New York   Singapore

A CRC title, part of the Taylor & Francis imprint, a member of the
Taylor & Francis Group, the academic division of T&F Informa plc.

Published in 2005 by
CRC Press
Taylor & Francis Group
6000 Broken Sound Parkway NW
Boca Raton, FL 33487-2742

International Standard Book Number-10: 0-8493-1488-7 (Hardcover)
International Standard Book Number-13: 978-0-8493-1488-9 (Hardcover)
Library of Congress Card Number 2004062869

Library of Congress Cataloging-in-Publication Data

DNA fingerprinting in plants : principles, methods, and applications / by Kurt Weising ...
[et al.]. -- 2nd ed.
   p. cm.
Includes bibliographical references and index.
ISBN 0-8493-1488-7 (alk. paper)
1. DNA fingerprinting of plants. I. Weising, Kurt.

QK981.45.D54 2005
581.3'5--dc22   2004062869

Taylor & Francis Group
is the Academic Division of T&F Informa plc.

Visit the Taylor & Francis Web site at
http://www.taylorandfrancis.com

and the CRC Press Web site at
http://www.crcpress.com

Preface

The new developments in the decade that has passed since the publication of our first edition of *DNA Fingerprinting in Plants and Fungi* have been more impressive than one could ever have imagined at that time. Our first edition encompassed basically all published work that employed DNA fingerprinting in plant or fungal research. In the present edition, we not only had to restrict ourselves to plants, but we also faced the difficult task of extracting a reasonable "core collection" from the tremendous number of scientific articles that had appeared on the topic. We would like to express our apologies to the many authors whose work could not be mentioned because of space limitations, despite the more than 1600 references already listed.

Applications of DNA fingerprinting have blossomed in applied as well as in basic plant sciences. In addition, the diversity of techniques has increased as well. In particular, the balance of hybridization- vs. PCR-based methods has completely been reversed during the last 10 years, with the latter now being the mainstay of most molecular laboratories. We have the strong impression that the publications related to the methodology and applications of PCR-based DNA fingerprinting behave like the DNA in a PCR, i.e., they amplify exponentially. Given that a simple update would never have worked, writing the new edition basically meant writing a completely new book.

The availability of new techniques and new equipment also indicated that we had to write for an even more diverse audience than before. There are still complete novices around, but the starting level of students has generally improved. Although we have still attempted to present the basic protocols and principles, we have also included some background theory as well as numerous references for and descriptions of more sophisticated methodology. The book is therefore intended to serve as a benchtop manual for the beginner as well as a key reference for a wide variety of DNA profiling techniques and applications.

Ten years ago, the average plant molecular marker laboratory employed random amplified polymorphic DNAs (RAPDs) and restriction fragment length polymorphism (RFLP) fingerprinting, whereas the advanced research institutions had already switched to microsatellites and amplified fragment length polymorphisms (AFLPs). Today, the latter two techniques are commonplace in many laboratories, whereas the avant-garde has turned its attention to single-nucleotide polymorphisms (SNPs) and DNA microarrays. It may be assumed that SNPs will become routine markers in the next 10 years or so, but it is quite difficult to predict what kind of exciting novel marker technologies will be on the market in 2014.

If the number of techniques and their applications keep increasing as they have done during the last decade, it is also difficult to imagine what a book like the one you have in your hands will look like another 10 years from now. In any case, we hope that the present book will assist in establishing DNA fingerprinting technology in a broad range of laboratories involved in plant research.

Authors

Prof. Dr. Kurt Weising
University of Kassel
Department of Sciences
Plant Molecular Systematics Group
Kassel, Germany

Prof. Dr. Hilde Nybom
Swedish University for Agricultural
  Sciences
Department of Crop Science Balsgård
Kristianstad, Sweden

Dr. Kirsten Wolff
University of Newcastle upon Tyne
School of Biology
Ridley Building, Claremont Place
Newcastle, United Kingdom

Prof. Dr. Günter Kahl
Plant Molecular Biology
University of Frankfurt
Frankfurt, Germany
and
GenXPro
Frankfurt, Germany

Author Biographies

Kurt Weising, Ph.D. is a professor at the Department of Sciences at the University of Kassel, Germany, where he has led the Plant Molecular Systematics Group since 2000.

Dr. Weising received his Ph.D. degree in 1987 from the Institute of Biology at the University of Frankfurt, Germany. He has done postdoctoral research with Professor Richard Gardner at the School of Biological Sciences, University of Auckland, New Zealand, and with Professor Günter Kahl at the Department of Botany, University of Frankfurt. He also served in an expert mission for International Atomic Energy Agency (IAEA) in Costa Rica.

Dr. Weising is a member of the International Association for Plant Taxonomy (IAPT), International Society for Plant Molecular Biology (ISPMB), Gesellschaft für Biologische Systematik (GfBS), Deutsche Botanische Gesellschaft, Gesellschaft für Züchtungsforschung, and the Senckenberg Research Institute. His research has been funded by the Fritz Thyssen Stiftung (Cologne), the European Commission (Brussels), and the German Research Council (DFG; Bonn).

Dr. Weising has published approximately 70 articles, with topics spanning a wide range from plant chromatin structure and plant transformation via plant genome analysis and DNA profiling methodology to plant molecular systematics and phylogeography. He contributed a number of pioneering studies at the early stage of plant DNA fingerprinting. His current research interests focus on the phylogeny, systematics, and evolution of Bromeliaceae, Chenopodiaceae, and the ant-plant genus *Macaranga* (Euphorbiaceae), with special emphasis on character evolution and speciation processes in mutualistic systems.

Hilde Nybom, Ph.D., is a professor at the Department of Crop Science-Balsgård, at the Swedish University for Agricultural Sciences.

Dr. Nybom obtained her training at the University of Lund, Sweden, receiving an M.Sc. degree in 1977 and the Ph.D. degree in 1987 at the Department of Systematic Botany. She spent a postdoctoral year at the Biology Department at Washington University, St. Louis, MO.

Dr. Nybom is a member of the International Society for Horticultural Science (ISHS), and has edited and/or reviewed numerous manuscripts for the ISHS series *Acta Horticulturae*. She is also a member of the editorial board for *Theoretical and Applied Genetics*, and a member of the review board for *Molecular Ecology*.

Her research is funded mainly by the Swedish Research Council for Environment, Agricultural Sciences and Spatial Planning, but some grants have also been received, e.g., from The Swedish Natural Science Research Council and the European Commission.

Dr. Nybom is the author of approximately 90 articles, mostly in areas related to plant breeding but also branching into ecology, systematics, and population genetics. She has supervised 6 Ph.D. students to their degrees, with several more in the pipeline. Her current research interests are related to the application of various types of DNA markers in population genetics, and to the genetics of plant species with reduced levels of genetic recombination, such as blackberries (*Rubus* subgenus *Rubus*) and dogroses (*Rosa* section *Caninae*). In addition, she is also involved in applied plant breeding, heading a program for breeding high-quality, disease-resistant apple cultivars, which recently resulted in the release of "Frida" and "Fredrik."

Kirsten Wolff, Ph.D., is Reader in Evolutionary Genetics at the University of Newcastle upon Tyne, United Kingdom.

Dr. Wolff received her training at the University of Groningen, the Netherlands, and obtained her M.Sc. in 1982. Her Ph.D., obtained in 1988 at the University of Groningen, was an ecological genetics study of morphological variation in the genus *Plantago*. Her first encounter with DNA fingerprinting was at Washington University with Professor Barbara Schaal (St. Louis, MO). An additional research position with the University of Leiden (the Netherlands) allowed her to improve her molecular skills. Additional Teaching and Research Fellowships at the University of St. Andrews (Scotland) and the University of Neuchâtel (Switzerland) allowed her to apply DNA fingerprinting to a wide range of plant species. In 1999, a permanent position was started at the University of Newcastle.

Dr. Wolff is a member of the Genetical Society, the Steering Committee of the Sheffield Molecular Genetics Facilities, the Scottish Office visiting committee of the Royal Botanical Garden of Edinburgh, and project evaluator of the Research Council of Norway. Research grants have been obtained from the Natural Environment Research Council and the European Commission, one of which is a Marie Curie Training Site specializing in the development of microsatellites.

Dr. Wolff has now published more than 50 articles on the population genetics and evolutionary biology of a wide range of plant (and the odd animal) species. Her interests are in studying population genetic diversity, its distribution and its maintenance. Often molecular tools are used to answer a broad range of ecological or evolutionary questions.

Günter Kahl, Ph.D., is Professor for Plant Molecular Biology at the Biocenter of Johann Wolfgang Goethe-University of Frankfurt am Main, Germany. He currently holds the CSO position with GenXPro, a company for novel technologies in genomics and transcriptomics, located at the Research Innovation Centre (FIZ) in Frankfurt.

After a Ph.D. in plant biochemistry, Dr. Kahl left for two postdoctoral years at Michigan State University (East Lansing, MI), joining Professor Joe Varner and Professor James Bonner at the California Institute of Technology (Pasadena, CA). His main research interests focus on (1) gene technology, in particular the isolation and characterization of plant defense genes and their promoters, and the use of *in vitro* modified defense genes for the improvement of plant crops via gene transfer; (2) plant genome analysis, in particular the development of molecular marker technologies for genomic fingerprinting, the establishment of genetic maps, the use of bacterial artificial chromosome (BAC) libraries, and the map-based cloning of agronomically important genes; and (3) expression profiling of plant tissues with expression microarrays and other high-throughput techniques.

Dr. Kahl is author of more than 250 scientific publications. His work has been supported by the German Research Council (DFG), the European Commission,

Stiftung Volkswagenwerk, Fritz-Thyssen-Stiftung, Gesellschaft für Technische Zusammenarbeit (GTZ), and many others. Given that most of his research is done with crops of subtropical and tropical regions, Dr. Kahl cooperates with a series of international and national research institutions in Japan, the United States, France, United Kingdom, Germany, Syria, India, Mexico, Spain, Colombia, Venezuela, and Chile. He has organized molecular marker courses in many countries and served in expert missions for IAEA, FAO, and UNESCO.

Acknowledgments

We acknowledge the help and support of a large number of people, who were, or still are, part of our laboratories as postdoctoral researchers or Ph.D. students. They have given us the inspiration to not only teach and train them, but also to make our knowledge available to a wider audience. In particular, we want to thank Gudrun Bänfer, Dirk Fischer, Christine Frohmuth, Marie Hale, Bruno Hüttel, Anke Marcinkowski, Carlos Molina, and Martina Rex. However, most of all, we acknowledge the support and patience of those at our home bases — whether they are our partners or pets.

Table of Contents

Chapter 3
Laboratory Equipment

Chapter 7
Linkage Analysis and Genetic Maps

Chapter 8
Which Marker for What Purpose: A Comparison

Chapter 9
Future Prospects: SNiPs and Chips for DNA and RNA Profiling

Appendix 1
Plant DNA Isolation Protocols

Appendix 2
Commercial Companies

Appendix 3
Computer Programs Dealing with the Evaluation of DNA
Sequence Variation and Molecular Marker Data

Repetitive DNA: An Important Source of Variation in Eukaryotic Genomes

In the last few decades, the architecture of the three genomes of a eukaryotic cell (i.e., nuclear and mitochondrial DNA, and chloroplast DNA in plants) has been explored in great detail. These studies have culminated with the determination of complete DNA sequences for the organellar and/or nuclear genomes of a steadily increasing number of species. The first two plant species, for which all three genomes have been sequenced, are the model organisms *Arabidopsis thaliana*[1226,1385,1433] and rice.[503,598,994,1602]

The basic organization of the three genomes present in plant cells is fundamentally different. The **chloroplast DNA (cpDNA)** molecule, typically ranging from 135 to 160 kb in size, is packed with genes and thus resembles the streamlined configuration of its cyanobacterial ancestral genome.[259,1226] In contrast, the **nuclear genome** of plants (and other eukaryotes) can be viewed as a huge ocean of largely nongenic DNA, with some tens of thousands of genes and gene clusters scattered around like small islands and archipelagos. A high proportion of this apparently nonfunctional DNA consists of repeated motifs and may be considered as junk DNA or selfish DNA.[357,525,1015]

The **plant mitochondrial DNA (mtDNA)** shares a number of features with both the nuclear and the chloroplast genome. Thus, plant mtDNA genes have prokaryotic properties just like cpDNA genes, but introns are more common.[756] With about 370 to 490 kb, the three higher plant mtDNAs sequenced so far are about 20 times larger than their animal counterparts, but only about 10% of these sequences represent genes.[756,994,1433] Another 10 to 26% were found to be made up of repetitive DNA, including retrotransposons.[727,994] Thus, the majority of plant mtDNA sequences lack any obvious features of information. The accumulating sequence data also revealed an extensive and ongoing horizontal exchange of DNA between the three different genomes, resulting in a net lateral transfer of genes from the organelles to the nucleus.[756,881,892,994]

Repeated DNA elements comprise the largest space of the nuclear genome in most eukaryotic organisms, and various types of repetitive DNA are also found in the organelles. It is therefore not surprising that a considerable fraction of the currently employed DNA profiling techniques treated in this book relies on mutations of repetitive DNA elements, in one way or the other. In this introductory chapter, we give a brief survey of the types of mutations encountered in eukaryotic genomes in general, and summarize important characteristics of the major classes of repetitive DNA found in plants.

1.1 CATEGORIES OF DNA SEQUENCE MUTATIONS

Mutations in genomic DNA can be classified into several categories (Figure 1.1; for a detailed treatment, see Graur and Li[525]). The simplest and most frequent type of mutation is a **base substitution**, i.e., the substitution of one nucleotide residue in the DNA sequence by another one (Figure 1.1B and C). Base substitutions occur at various rates (see below) and are thought to arise mainly from mispairing during DNA replication.[525] Base substitutions are the molecular basis of single-nucleotide polymorphism (SNP) markers (see Chapter 9). If the exchange involves nucleotides carrying the same type of base (i.e., purine against purine, or pyrimidine against pyrimidine), the mutation is called a **transition**. If nucleotides carrying a different type of base are exchanged (i.e., purine against pyrimidine or vice versa), the mutation is called a **transversion**. Eight different possibilities for transversions but

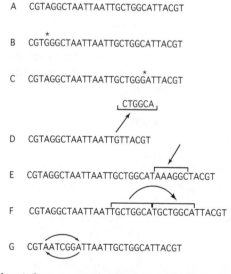

Figure 1.1   Types of mutations commonly encountered in DNA. (A) **Wildtype sequence**; (B) **transition**: A ↔ G exchange at the position marked with an asterisk; (C) **transversion**: C ↔ G exchange at the position marked with an asterisk; (D) **deletion** of the sequence CTGGCA; (E) **insertion** of the sequence AAAGGC; (F) **duplication** of the sequence TGCTGGCA; (G) **inversion** of the sequence AGGCTAA. See text for details.

only four for transitions exist. Nevertheless, transitions are more frequent in nature than transversions. If the mutation affects the first or second codon position within a protein-coding region, the encoded amino acid is often substituted by a different one. Such mutations are said to be nonsynonymous. In contrast, the mutation is called synonymous or silent if the amino acid sequence in the encoded protein remains unchanged.

Insertions and **deletions** refer to the gain and loss, respectively, of a piece of DNA sequence at a particular site (Figure 1.1D and E). These two types of mutations are often collectively referred to as **indels**.[131] They can be of any size between one base pair (bp) and several kilobases. Various mechanisms may be responsible for the generation of indels, including the activity of transposable elements[1208] (see Chapter 1.3), slippage of the DNA polymerase during replication[806] (see Chapter 1.2.2.3), and unequal crossing over between sister chromatids or between two homologous chromosomes. The latter two mechanisms initially produce a **duplication** in one daughter molecule (Figure 1.1F), and a deletion in the other one. Duplications of short sequence motifs are responsible for the majority of cpDNA variation at the population and species level.[1009] For example, Van Ham et al.[1451] found a total of 50 small indels (partly due to mononucleotide repeat variation) in the intergenic *trnL-trnF* spacer of 15 species belonging to the Crassulaceae, Saxifragaceae, and Solanaceae families. Because the presence of a duplicated motif increases the chances for further duplication, long tracts of tandemly repeated DNA sequences may eventually be produced, which are a common element of nuclear genomes (e.g., micro- and minisatellites; see Chapter 1.2).

The exchange of one DNA sequence tract by another one is known as **recombination**. A well-known example is the reciprocal exchange of DNA sequence elements between homologous chromosomes during the meiotic prophase in eukaryotes. **Gene conversion** is a nonreciprocal form of homologous recombination; i.e., one of the two DNA sequence variants involved in the recombination process is lost. **Site-specific recombination** involves the exchange of nonhomologous sequences between two DNA molecules. Because the mutated sequences usually differ in length from those of the wildtype, this type of mutation may also be grouped in the indel category described above. Finally, **inversions** are generated when a piece of DNA is excised and reintegrated in an opposite orientation (Figure 1.1G). Large inversions occur at low frequency in cpDNA, and are reliable markers at deep taxonomic levels (see Graham and Olmstead[519] and references cited therein). Kelchner and Wendel[707] noted that minute inversions also occur in the chloroplast genome, but may often remain unrecognized.

The rates at which the above-described mutations occur can be vastly different, depending on the biology of the organism, the genome under consideration (nuclear, chloroplast, or mitochondrial), and the type of mutation. On the low side of the spectrum, the average rate of silent nucleotide substitution in plant mtDNA was calculated to equal about one third of the neutral rate in cpDNA, and about 1/12 of that in nuclear DNA.[1559] For example, Yang et al.[1587] determined an average rate of 0.16 to 0.23×10^{-9} nucleotide substitutions per site per year in the first intron of the mitochondrial *nad*4 gene from 10 Brassicaceae species. This is about 1/23 of the substitution rate in the nuclear ribosomal internal transcribed spacer region. On the upper end of the scale, probably the highest mutation rates are found in some hypervariable human minisatellites, with germline mutation rates exceeding 10^{-2} changes per generation (reviewed by Bois and Jeffreys[146] and Vergnaud and Denoeud[1465]; see Chapter 1.2.1). It is obvious that the rate of polymorphism detected with a particular marker technique will depend on the type of sequence and genome targeted by the respective marker. The highest mutation rates are often associated with particular classes of repetitive DNA, which are introduced in some detail below.

1.2 TANDEM-REPETITIVE DNA: THE BIOLOGY OF MINI- AND MICROSATELLITES

Depending on their genomic organization, repetitive DNA elements may be classified as either interspersed or tandemly repeated. Interspersed repeats, exemplified by transposable elements, are present at multiple sites throughout the genome (see Chapter 1.3). Tandem repeats, on the other hand, are restricted to fewer loci and consist of arrays of two to several thousand sequence units arranged in a head-to-tail fashion. This kind of organization is also exhibited by some genes, such as the transcription units for histone mRNA and ribosomal RNA. Tandem-repetitive DNA may be further classified according to the length and copy number of the basic repeat units as well as its genomic localization (see Tautz[1368] for a review of the nomenclature):

1. **Satellite** DNA, originally described in the early 1970s,[1294] was named after its separability from bulk DNA by buoyant density gradient centrifugation. Typical satellites consist of very high numbers of repetitions (usually between 1000 and more than 100,000 copies) of a basic sequence motif. Monomer sizes may range from two to several thousand bp, but 100 to 300 bp are most common. Satellites are generally heterochromatic in nature, and are often located in subtelomeric or centromeric regions. Satellite DNAs are only rarely used as molecular markers (e.g., for species identification[1058,1257]) and will not be treated further in this book.
2. The term **minisatellites** was invented in 1985 to describe another family of tandemly reiterated repeats.[662,663] Minisatellites consist of intermediate-sized DNA motifs (about 10 to 60 bp), and show a lower degree of repetition at a given locus compared with satellites. Often, minisatellites form families of related sequences that occur at many hundred loci in the nuclear genome.
3. Tandem repeats made up from very short (i.e., about 1 to 6 bp) motifs were called **simple sequences** by Tautz and Renz.[1369] Later, this class of DNA was coined **microsatellites** (in continuation of the above nomenclature[829]), simple repetitive sequences (SRS), simple sequence repeats (SSRs), or simple tandem repeats (STRs). Microsatellites are usually characterized by a low degree of repetition at a particular locus, but microsatellites consisting of identical motifs may be found at many thousand genomic loci.

Given that mini- and microsatellite tandem arrays occur at multiple sites in the genome, they share some properties of both tandemly repeated and interspersed DNA. Moreover, different mini- and microsatellites often occur intermingled with each other in a particular stretch. If point mutations accumulate within such intermingled repeats, their repeat structure will become more or less obscured, resulting in cryptically simple DNA.[1371] Tandem repeats in general, and mini- and microsatellites in particular, are characterized by highly variable copy numbers of identical or closely related basic motifs. Therefore, this class of DNA polymorphism was also coined **variable number of tandem repeats (VNTRs)**.[973] The evolutionary biology and/or functional significance of mini- and microsatellites have been reviewed by Ellegren,[389] Epplen,[405] Epplen et al.,[407–409] Goldstein and Pollock,[506] Goldstein and Schlötterer,[507] Jarne and Lagoda,[659] Kashi and Soller,[697] Kashi et al.,[698] Li et al.,[811] Moxon and Wills,[958] Powell et al.,[1094] Sutherland and Richards,[1347] Tautz and Schlötterer,[1370] and Vergnaud and Denoeud.[1465]

1.2.1 Minisatellites

Highly polymorphic loci based on tandem repeats were first detected in the human genome in the early eighties.[105,1577] In 1985, Jeffreys et al.[663] demonstrated that radioactive probes specific for such repeats detect multiple hypervariable DNA loci on Southern blots carrying restriction-digested human DNA, resulting in individual-specific fingerprints. The term minisatellites coined by Jeffreys et al.[662] was initially applied to tandem repeats of 10 to 50 bp units, carrying a common GC-rich core sequence of 10 to 15 bp, but repeats with longer unit size and higher AT content were also identified. Since these pioneering studies, minisatellite loci have been cloned and sequenced from numerous organisms, including humans,[41,973,1466,1566] cattle,[484] mouse,[147,737] birds,[548] and plants.[171,641,1413,1414,1551] Examples of single minisatellite repeat units cloned from various organisms and genomes are compiled in Figure 1.2.

...ACAGGGGTGTGGGG...	human	Bell et al. [105]
...AGGAATAGAAAGGCGGGYGGTGTGGGCAGGGAGRGGC..	human	Wong et al. [1566]
...GGAGGTGGGCAGGAXG...	human	Jeffreys et al. [662]
...CTGGGCAGGGAGGA...	mouse	Kominami et al.[737]
...AGGGAAGGGCTC...	willow warbler	Gyllensten et al. [548]
...GGGGACAGGGGACACCC...	willow warbler	Gyllensten et al. [548]
...CTATACAGGGCTGGTT...	salmon	Bentzen & Wright[119]
...GCCTTTCCCGAG...	yeast	Andersen & Nilsson-Tillgren [29]
...GAGGGTGGXGGXTCT...	M13 phage	Vassart et al. [1456]
...GGAGGAGGAAGGGGAGAGGAAGGAGGT...	rice	Winberg et al.[1553]
...AGGATGGCATGGAGGTGGAGGAGGACATGGCGG...	*Arabidopsis*	Tourmente et al. [1413]
...TATTATTATTAGTATA...	*Orchis* chloroplast	Cafasso et al. [206]
...TATTTAATTGCGTTGCTCGACCAACGGGAGAGG...	*Beta* mitochondrion	Nishizawa et al. [991]

Figure 1.2 Examples of (mostly GC-rich) minisatellite repeat units cloned from various organisms. The last two lines exemplify AT-rich minisatellite repeat units detected in plant cpDNA and mtDNA, respectively.

1.2.1.1 Chromosomal Localization and Association with Other Repeats

In most species examined to date, minisatellites were distributed unevenly across the nuclear genome. Early *in situ* hybridization experiments revealed a prevalent localization of human minisatellites in subtelomeric regions.[1195] A significant increase of minisatellite frequency toward the telomeres was also found by Vergnaud and Denoeud,[1465] based on the analysis of 34.6 Mb of human chromosome 22. In other mammals, a subtelomeric location of minisatellites is less obvious,[26] and plant minisatellites have a tendency to cluster around the centromeres.[171,1413,1465] In any case, the uneven coverage of chromosomes should be considered when analyzing minisatellite-derived marker data. Minisatellites are frequently associated with other types of repeats, including microsatellites[40] and transposons.[600,641,888] For example, the interior of a transposable element belonging to the *Basho* family of rice (see Chapter 1.3.2) was found to be associated with an AT-rich 80-bp minisatellite, which exhibited a variable number of tandem repeats.[641]

1.2.1.2 Mutability and Evolution

Since the first discovery of minisatellites, numerous mechanisms have been discussed as possible causes for tandem repeat variability, including replication slippage, transposition, extrachromosomal rolling circle replication, and a variety of recombinational events.[146,658,1372,1465,1570] The currently accepted view holds that at least two different types of mutational mechanisms need to be distinguished. The vast majority of minisatellites are assumed to display moderate mutation rates in both mitosis and meiosis. In this group, mutations presumably originate via DNA replication errors.[549,732,1372] However, a few minisatellites in the human genome display extraordinary high mutation rates only during meiosis (5×10^{-2} per cell per generation, and higher). In this type of minisatellite, mutant alleles were shown to contain segments from both parental alleles, providing evidence for interallelic exchange. Moreover, a strong bias of mutational events toward the 5′-end of the tandem array was observed.[942] It was therefore suggested that the major mutational process is based on a complex gene conversion mechanism, involving the nonreciprocal transfer of repeat units from a donor allele into the 5′-end of a recipient allele.[668]

Extensive research on hypervariable minisatellites in humans and transgenic systems finally revealed that meiotic hypervariability is caused by the physical proximity between a minisatellite and a hot spot for double-strand breaks (reviewed by Armour et al.,[43] Bois and Jeffreys,[146] and Vergnaud and Denoeud[1465]). Following the induction of such a double-strand break during the meiotic prophase, complex recombinational processes are initiated that eventually lead to (1) a variation in the copy number and (2) internal rearrangements of the minisatellite alleles on both homologous chromosomes. The resulting heterogeneity in the arrangement of distinguishable repeat units was exploited for a specific molecular marker technique targeted at hypervariable minisatellites, called minisatellite variant repeat mapping[665,667,941] (see Chapter 2.3.10.1).

1.2.1.3 Minisatellites in Organellar Genomes

Minisatellites are not restricted to nuclear DNA. For example, they are also regularly present in the control DNA region of animal mtDNAs (reviewed by Lunt et al.[855]). More recently, minisatellites were also reported from plant mtDNA.[991,1318] For example, Nishizawa et al.[991] identified four unrelated minisatellites in sugar beet mtDNA (Figure 1.2). One array of 32-bp units (*rrn26*) varied in copy number between 2 and 13 among seven beet accessions. Sperisen et al.[1318] found two minisatellites consisting of basic repeat units of 32 and 34 bp, respectively, in the intron of the *nad*1 gene of *Picea abies* (Norway spruce). The repeat region was polymorphic at the intraspecific level, exhibiting 18 size variants. A database search demonstrated the presence of minisatellites in the mtDNA of many other plant species.[1318] Finally, minisatellite-like sequences were also identified in the chloroplast genomes of several plant taxa, including *Sorbus aucuparia*[720] and various orchids[107,206,280,281] (Figure 1.2). At present, minisatellites in plant mtDNA and cpDNA represent a largely untapped source of molecular markers at the intraspecific level (but see Cozzolino et al.[281] and Sperisen et al.[1318]).

1.2.1.4 Potential Functions of Minisatellites

The functional significance of minisatellites for eukaryotic genomes is still a matter of debate. Indications for potential functions have been obtained in a number of studies. For example, nuclear proteins were identified that specifically interact with certain minisatellites.[264,708,1416,1487,1584] Such interactions were postulated to serve regulatory purposes in, for example, recombination,[1486,1487] transcriptional activation,[708,1417] and/or splicing,[1426] to name a few. Moreover, minisatellites may constitute fragile chromosome sites[1348] and could thus be involved in chromosomal translocations. Finally, minisatellites are sometimes present in genes as, for example, in human genes encoding an epithelial mucin[1350] and an involucrin.[380]

1.2.1.5 Minisatellites as Molecular Markers

Minisatellites have been exploited as molecular markers in various ways, but two techniques clearly prevail. In one method, minisatellite-complementary probes are hybridized to restriction-digested genomic DNA to produce highly variable restriction fragment length polymorphism (RFLP) fingerprints[295,663,1456] (see Chapter 2.2.3.1). This technique has been used extensively in the past, but is not applied so frequently anymore. Alternatively, minisatellites are used as single primers in a polymerase chain reaction (PCR; e.g., in direct amplification of minisatellite DNA [DAMD][1309]; see Chapter 2.3.5.2). A more sophisticated approach is the minisatellite repeat variant mapping technique described by Jeffreys et al.[665,667,941] (see Chapter 2.3.10.1).

1.2.2 Microsatellites

The existence of tandem repeats consisting of very short (i.e., 1 to 6 bp) sequence motifs in eukaryotic genomes was first recognized in the early 1970s, when $(TAGG)_n$

repeats were found in the satellite DNA of a hermit crab.[1294] Since then, a large
number of studies based on Southern hybridization, molecular cloning, and/or database screening have documented the ubiquitous presence of these so-called microsatellites in bacterial, fungal, plant, animal, and human genomes (see surveys of
Beckmann and Weber,[97] Cardle et al.,[214] Chin et al.,[246] Depeiges et al.,[329] Dieringer
and Schlötterer,[344] Echt and May-Marquardt,[377] Field and Wills,[441] Gur-Arie et al.,[546]
Jurka and Pethiyagoda,[680] Katti et al.,[699] Morgante et al.,[951] Panaud et al.,[1036] Sharma
et al.,[1277] Tautz and Renz,[1369] Toth et al.,[1412] Van Belkum et al.,[1438] and Wang et al.[1499]).

In plants, the presence of microsatellites was first demonstrated by RFLP fingerprinting with oligonucleotide probes.[129,1521–1523] Plant microsatellites were first
cloned in 1991,[269] and PCR-generated, locus-specific plant microsatellite markers
(see Chapters 2.3.4 and 4.8) were first reported in 1992.[14] Initial studies suggested
a lower abundance of microsatellites in plants as compared with animals.[769,949,1499]
However, more recent surveys based on large data sets from the *Arabidopsis*, rice,
maize, soybean, and wheat genome demonstrated that microsatellite frequencies in
plants are higher than previously anticipated.[214,951] For example, Cardle et al.[214]
searched 27,000 kb of genomic DNA sequences from *A. thaliana* for the presence
of all possible mono- to pentanucleotide repeats. They found an average frequency
of one microsatellite per 6.3 kb, which is equivalent to the situation in mammals.[97,680]

1.2.2.1   Categories of Microsatellites

If all self-complementary and overlapping motifs are merged into single motifs,
there are 501 possibilities of nonredundant mono- to hexameric repeats; i.e., two
monomeric, four dimeric, 10 trimeric, 33 tetrameric, 102 pentameric, and 350
hexameric patterns (compiled by Jurka and Pethiyagoda,[680] see Figure 1.3A for
examples). The most abundant motifs found in mammalian genomes proved to be
$(A)_n$ and $(CA)_n$ as well as their complements,[11,97,680,1412] whereas $(A)_n$, $(AT)_n$, $(GA)_n$,
and $(GAA)_n$ repeats are the most frequent motifs in plants.[214,951,1412,1499] Mononucleotide repeats consisting of A/T tracts are also present in chloroplast genomes[1092,1093]
(see Chapter 1.2.2.4).

Microsatellites composed of tri-, tetra- and pentanucleotide motifs are generally
less common than mono- and dinucleotide repeats. Estimates are extremely variable,
depending on the motif, the genomic localization (introns vs. exons vs. 5'- and
3'-untranslated regions vs. intergenic regions), and the species under consideration
(for details, see Toth et al.[1412]). As a general rule, trinucleotide repeats are the
predominant type of microsatellites found in exons, whereas repeats consisting of
multiples of one, two, four, and five base pairs are rare in genes.[152,214,680,951,1412,1597]
This is not surprising, considering the fact that slippage of one or more trinucleotide
units does not affect the triplet periodicity imposed by the open reading frame,
whereas frameshift mutations resulting from the insertion/deletion of other types of
repeat units will completely change the amino acid sequence downstream of the
mutated site.

Another way to categorize microsatellites relates to the degree of perfectness of
the arrays. Weber[1512] recognized three classes, comprising (1) perfect repeats, which
consist of a single, uninterrupted array of a particular motif; (2) imperfect repeats,

A

Mononucleotide repeats:	...AAAAAAAAAAAAAAAAAAAAAAAA...
Dinucleotide repeats:	...CACACACACACACACACACACACACA...
Trinucleotide repeats:	...CGTCGTCGTCGTCGTCGTCGTCGT...
Tetranucleotide repeats:	...CAGACAGACAGACAGACAGACAGA...
Pentanucleotide repeats:	...AAATTAAATTAAATTAAATTAAATT...
Hexanucleotide repeats:	...CTTTAACTTTAACTTTAACTTTAA...

B

Perfect repeats:	...$(AG)_{32}$...	*Cicer*
	...$(TAT)_{25}$...	*Cicer*
	...$(CAA)_{7}$...	*Cicer*
Imperfect repeats:	...$(TC)_{6}A(TC)_{13}$...	*Cicer*
	...$(AG)_{12}GG(AG)_{3}$...	*Cicer*
Compound repeats:	...$(AT)_{6}(GT)_{42}AT(GT)_{5}(GT)_{10}$...	*Cicer*
	...$(AT)_{14}(AG)_{8}$...	*Cicer*
	...$(GAA)_{21}...(TA)_{23}$...	*Cicer*
Chloroplast:	...$(T)_{5}C(T)_{17}$ ..	*Nicotiana*
	...$(T)_{14}$...	*Nicotiana*
	..$(CT)_{8}TTTC(T)_{12}$...	*Macaranga*
Mitochondrion:	...$(G)_{11}$...	*Pinus*

Figure 1.3 (A) Examples of perfect microsatellites made up from mono-, di-, tri, tetra-, penta-, and hexanucleotide repeats, respectively. (B) Examples of perfect, imperfect, and compound microsatellites cloned from different genomic compartments. Sequences of nuclear microsatellites from chickpea (*Cicer arietinum*) are derived from the work of Hüttel et al.[633] The cpDNA microsatellite sequences from tobacco (*Nicotiana tabacum*) and *Macaranga indistincta* are derived from the articles by Weising and Gardner,[1520] and Vogel et al.,[1476] respectively. The *Pinus* mtDNA microsatellite was identified by Soranzo et al.[1313]

in which the array is interrupted by one or several out-of-frame bases; and (3) compound repeats, with intermingled perfect or imperfect arrays of several motifs. Examples for these different categories are given in Figure 1.3B. Weber[1512] also showed that the level of polymorphism exhibited by PCR-amplified $(CA)_n$ microsatellites in humans is positively correlated with the number of uninterrupted, perfect repeats at a given locus. These findings were later supported by numerous studies in animals (e.g., Blanquer-Maumont and Crouau-Roy[141]) and plants (e.g., Bryan et al.,[79] Saghai-Maroof et al.,[1210] Smulders et al.[1302]).

1.2.2.2 Chromosomal Localization and Association with Other Repeats

Extensive genetic mapping projects in humans,[342] animals,[346,1287] and plants[808,836,1136,1181,1362,1378] indicated that short tracts of microsatellites (i.e., total size <100 base pairs) are quite evenly dispersed throughout the genome, albeit some local clustering occurs. In plants, the extent of clustering obviously depends on the species. Thus, Temnykh et al.[1378] observed a relatively uniform distribution of microsatellite markers in rice, whereas Ramsay et al.[1136] and Li et al.[808] found dense clusters of microsatellite markers around centromeric regions in barley. A predominant association with centromeres was also reported for exceptionally long microsatellites (i.e., repeats consisting of more than 20 units of GA, AT, CA, and/or GATA) cloned from tomato.[38] Microsatellites cloned from undermethylated, presumably low-copy tomato DNA sequences showed the same type of association,[39] whereas microsatellites derived from expressed sequence tag (EST) databases (see Chapters 4.8.4.2 and 4.8.4.3) mapped to euchromatic regions.[39]

In part, pericentromeric clustering of markers on genetic maps may be explained by reduced recombination rates in heterochromatic regions. This is obviously the case for wheat chromosomes 2A, 2B, and 2D, where microsatellite markers clustered around centromeres on a genetic map,[1181] but were evenly distributed on a physical map.[1180] Nevertheless, fluorescent *in situ* hybridization studies using metaphase chromosomes[288,513,1244] and molecular sequence data[36,172,1482] demonstrated that long tracts (>1 kb) of microsatellites indeed exist, and that some of these are clustered at centromeric and other heterochromatic locations.

In primate genomes, A/T-rich microsatellite repeats are frequently associated with *Alu* sequences, which are the major type of retroelement found in humans[34,381,1623] (see also Chapter 1.3.1). It was hypothesized that A/T-rich microsatellites evolve from the poly(A) tail of the *Alu* elements.[34,972] An intimate association between microsatellites and retroelements was also reported from other mammals[701] and plants,[957,1135] suggesting a common evolutionary history of both types of repeats. For instance, Ramsay et al.[1135] found that 41% of 290 microsatellite-containing clones from barley genomic libraries enriched for dinucleotide repeats also harbored other repetitive DNA elements. In rice, AT-rich microsatellites are frequently associated with a transposon of the miniature inverted-repeat transposable element (MITE) superfamily[13,1379] (see also Chapter 1.3.3).

Morgante et al.[951] screened a large data set from five plant genomes for the presence and distribution of microsatellites. In contrast to the above findings, these authors found a preferential association of microsatellites with the unique, non-repetitive DNA fraction. A similar preference is also suggested by the often unexpectedly high frequencies of microsatellites in 5'- and 3'-untranslated regions of genes, as detected in EST and cDNA libraries.[460,609,690] For example, Fraser et al.[460] recorded on average 3% of microsatellite-containing clones in cDNA libraries of *Actinidia* species, which favorably contrasts with the 1% of positive clones in an unenriched genomic library of *Actinidia chinensis*.[1525] Obviously, this topic deserves further study. In any case, the physical proximity of at least some microsatellites and retroelements has been exploited by a number of marker techniques, using PCR primer pair combinations specific for these two repeat classes (e.g., *copia*-SSR,[1113] REMAP[682]; see Chapter 2.3.5).

1.2.2.3 Mutability and Evolution

The mutability of microsatellites has been the subject of a vast number of studies, and only a few aspects can be mentioned here. For more comprehensive treatments of the topic, the reader should consult the reviews by Ellegren,[389] Estoup et al.,[416] Goldstein and Pollock,[506] Goldstein and Schlötterer,[507] Jarne and Lagoda,[659] and Li et al.[811] Studies of microsatellite mutation processes can be divided into three categories[1098]: (1) theoretical modeling under various assumptions[338,1437]; (2) direct analysis and characterization of *de novo* mutations in the germ line[1099,1101,1400]; and (3) the indirect analysis of past mutations by comparative sequencing of alleles from orthologous loci, both within species and between species with known phylogenetic relationships.[81,694,1454]

Microsatellite mutation rates proved to vary considerably depending on the locus, the length of the repeat motif, the organism, and sometimes the allele. Values reported from humans[224] and various animals including mouse,[296] pigs,[388] birds,[167,1101] fish,[1287] and flies[1240,1250,1251] range from about 5×10^{-6} to 1.5×10^{-2} mutations per locus per gamete per generation. Particularly low rates were observed in *Drosophila*,[1250,1251] which could perhaps be explained by a downward mutation bias of microsatellite length in this organism.[574] On the upper end of the scale, germline mutation rates at the percentage level were reported for an $(AAAG)_n$ and an $(AAGAG)_n$ repeat in barn swallow.[167]

In plants, only few direct measurements of microsatellite mutation rates are yet available. Diwan and Cregan[349] reported the formation of new alleles in a soybean mapping population at a rate of 2×10^{-4}. Thuillet et al.[1400] determined the same average rate of 2×10^{-4} for 10 microsatellite loci from durum wheat, but rates at the individual loci varied between zero and 10^{-3}. Higher rates were reported for long $(TAA)_n$ repeats (with n = 19 to 51) in inbred populations of chickpea.[1429] Averaged over 15 loci, values of 1.0×10^{-2} and 3.9×10^{-3} were calculated for a long-lived and a short-lived chickpea cultivar, respectively.[1429] Vigouroux et al.[1469] investigated rates and patterns of mutations at a large number of microsatellite loci in six maize inbred lines. An average rate of 7.7×10^{-4} mutations per generation was estimated for dinucleotide repeat loci, whereas no single mutation was detected in micro-satellites with repeat motifs longer than 2 bp. Mutation rates have also been studied for chloroplast $(A)_n$ repeats in *Pinus torreyana*.[1111] No variation was present at 17 cpSSR loci. A maximum mutation rate of <3.2 to 7.9×10^{-5} was therefore calculated from the number of individuals investigated (n = 64) and the number of generations from a presumed bottleneck 3500 to 8500 years ago.[1111] It should be noted that mutation rates obtained by direct measurements generally suffer from the relatively small numbers of mutants and allele generations (e.g., one single mutation in 157,680 allele generations of *Drosophila*[1250]), and therefore represent only rough estimates.

Results from a vast number of studies dealing with *de novo* mutations and/or comparative sequencing of microsatellite alleles from orthologous loci can be summarized as follows:

1. Within species, changes in allele length are most often caused by size alteration of the repeat itself. Small-scale variation in repeat copy number is generally thought to result from a mutational process called replication slippage or slipped strand mispairing.[525,806] Slippage implicates mispairing of the newly replicated strand during the replication process and most often involves gain or loss of a single repeat unit.[167,1469] A length-independent slippage process was proposed as an additional mechanism to create very long microsatellites.[344] *In vitro* experiments showed that replication slippage may in fact result in considerable amplification of a given microsatellite.[1238]

2. Repeat number is usually highest in the organism from which the repeat has been cloned. This phenomenon was first observed when primers designed for human microsatellite loci were applied to other primates.[141,1198] Amos and Rubinsztein[27] interpreted these observations as an indication for a directional bias toward repeat elongation in microsatellite evolution. However, opposite trends were also reported (i.e., directional bias toward a loss of repeats in *Drosophila*[574]). It is now generally accepted that the observed phenomena are better explained by ascertainment bias rather than directional evolution (see Chapter 4.8.4.3).

3. Mutability of repeat number is considerably reduced when the repeat sequence has been interrupted by, e.g., point mutations[387] or variant repeats.[1074] This is probably explained by the greater opportunity for slippage provided by longer repeats. Early studies on microsatellite variability[1512] had already shown a positive correlation between the level of polymorphism and the total size of a perfect array (see Chapter 1.2.2.1). Consequently, microsatellite mutation rates are often allele-specific rather than locus-specific, with higher mutation rates observed in longer repeat arrays.[167,671,1240] Allele size-dependent instability is especially pronounced in certain kinds of trinucleotide repeats such as $(CAG)_n$, $(CTG)_n$, and $(CCG)_n$, which can adopt unusual DNA conformations during DNA replication (concept of dynamic mutation[1160–1162]).

4. A minimum number of repeats are needed to initiate the elongation of a repeat by slipped strand mispairing.[920] Longer repeats may also suddenly be created if a base substitution bridges an interruption between two repeats. By analyzing a microsatellite locus in a human globin pseudogene, Messier et al.[920] found the threshold for this so-called birth of a microsatellite to be ~5 to 6 GT repeats. Primmer and Ellegren[1098] showed that size expansion over evolutionary timescales may already start with repeats as short as $(AG)_2$. Conversely, the so-called death of a microsatellite is thought to be initiated by the formation of an interruption.[1074,1373] In a model of microsatellite evolution suggested by Kruglyak et al.,[753] the average genome-wide microsatellite repeat length in an organism results from a balance between slippage and point mutations.

5. Insertions, deletions, and point mutations are also frequent in microsatellite-flanking regions,[938] which apparently is the main cause for the limited transferability of microsatellite markers across species (see also Chapter 4.8.4.3).

6. Size homoplasy is common; i.e., alleles of identical size do not necessarily share an identical sequence, and even if they do, they need not be identical by descent — an inevitable consequence of the fast, forward–backward, stepwise mutation

process.[30,415,480,526,671,1017] Size homoplasy has also been observed in chloroplast microsatellites.[366,555,1476]

7. Patterns of microsatellite evolution may differ substantially among loci, with some repeats being relatively stable and others highly unstable (see Primmer and Ellegren[1098] for a discussion).

Taken together, the mutational processes governing microsatellite evolution are very complex, and care needs to be taken when microsatellite markers are used in population genetics (see also Estoup et al.[416]; Chapters 5.6 and 6.3.2.1).

1.2.2.4 Microsatellites in Organellar Genomes

Poly(A/T) repeats are the only type of microsatellites that are regularly present in the chloroplast genome, mainly in introns and intergenic regions[1092,1093,1114] (Figure 1.3B). Some chloroplast microsatellites appear to be associated with mutational hotspots in the cpDNA molecule. One example relates to the spacer between the *rpl*2 and *rps*19 genes, in which Goulding et al.[517] identified a polymorphic poly(A) tract in tobacco cpDNA. The position of this microsatellite was found to be conserved in many other angiosperms.[51,180,528,1436,1520] Microsatellites appear to be rare in plant mtDNA, with one single explicit report of a $(G)_n$ repeat from several conifer species[1313] (Figure 1.3B).

1.2.2.5 Potential Functions of Microsatellites

The functional impact of microsatellites for eukaryotic genomes is still incompletely understood. The majority of copies are probably selfish, because they amplify and propagate in the absence of counterselective pressure.[1015] However, a wide variety of possible roles of microsatellites at particular genomic locations have been discussed in countless papers, involving almost any process taking place in eukaryotic nuclei. Only three of these are mentioned here:

1. Microsatellite-like repeats are structural elements of both telomeres and centromeres. Telomeric repeats, which are found at the extreme ends of eukaryotic chromosomes, actually represent a "special version" of microsatellites.
2. Some microsatellites bind nuclear proteins (see Epplen et al.[408] and references cited therein) and may, for example, serve as a landing pad for transcription factors that enhance or reduce the expression of neighboring genes (e.g., the GAGA factor[498,521]).
3. Some microsatellites (especially trinucleotide repeats) are transcribed and then often encode tracts of identical amino acids.[152,1597] For example, CAG repeats are translated into glutamine repeats, which are integral sequence components of various transcription factors.[485] Expansion of CAG, GAA, and GCG/GCA repeats in human genes were found to be associated with an ever-increasing number of neurodegenerative diseases (reviewed by Brown and Brown[173] and Cummings and Zoghbi[292]).

For comprehensive surveys of putative functions of microsatellites, see Epplen,[405] Epplen et al.,[407–409] Li et al.,[811] Moxon and Wills,[958] and Richards and Sutherland.[1160–1162]

1.2.2.6   Microsatellites as Molecular Markers

Numerous methods have been developed that exploit microsatellites as molecular markers, in one way or the other. The most important variant is the locus-specific PCR amplification of nuclear and organellar microsatellites with flanking primers (Chapters 2.3.4 and 4.8). Other methods use microsatellite motifs (instead of flanking regions) as single PCR primers (Chapters 2.3.5.3 and 4.6), as PCR primers in combination with other primer types (Chapters 2.3.7.3 and 2.3.7.4), or as hybridization probes (Chapters 2.2.3 and 2.3.6).

1.3 TRANSPOSABLE ELEMENTS

It is safe to assume that most interspersed repeats in the eukaryotic nucleus have acquired their current genomic location by transposition. Mobile genetic elements that are able to change their position within the genome were first discovered by Barbara McClintock in maize more than 50 years ago. Since then, a vast number of such mobile DNA elements have been detected and characterized, and it soon became apparent that these so-called transposons are regular and ubiquitous constituents of eukaryotic genomes.[1208] According to the mechanism of transposition, mobile genetic elements of eukaryotes can be divided into two classes. **Class I transposons** disperse via an RNA intermediate. Given that reverse transcription of RNA into DNA is involved in this process, they are more commonly called **retrotransposons**. In contrast, **class II transposons** propagate (or jump) via a DNA intermediate.

1.3.1   Class I Transposons

Class I transposons propagate via an **RNA intermediate**, which is reverse-transcribed into a cDNA. According to their genomic organization and gene content, retrotransposons (also called retroelements in a more general term) may be further divided into:

1. Retroviruses
2. Long terminal repeat (LTR) retrotransposons
3. Long interspersed elements (LINEs)
4. Short interspersed elements (SINEs)

LINEs and SINEs are also referred to as non-LTR retrotransposons.[1243] For each type of retrotransposons, active as well as defective copies have been found. In general, inactive elements outnumber active copies by a factor of several thousand. The occurrence, biology, and evolution of retrotransposons have been reviewed by Bennetzen,[113] Flavell et al.,[452] Grandbastien,[520] Graur and Li,[525] Kumar and Bennetzen,[759] Saedler and Gierl,[1208] and Schmidt.[1243]

Retroviruses are distinguished from other types of retroelements by the presence of an *env* gene in their genome. The protein encoded by this gene allows retroviruses to enter and leave their host cell. Retroviruses are therefore the only infectious type

A   LTR Retrotransposons

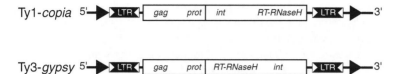

B   Non-LTR Retrotransposons

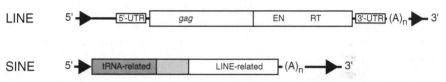

Figure 1.4   Structure of major types of LTR and non-LTR retrotransposons found in plant genomes (not to scale). Both types of elements are bordered by short target site duplications (black triangles). (A) **LTR retrotransposons** are characterized by long terminal repeats (LTRs) and encode a capsid protein (*gag*), an RNase H, a reverse transcriptase (RT), a protease (*prot*), and an endonuclease (*int*). Ty1-*copia* and Ty3-*gypsy* elements differ from each other by the relative arrangement of *int* and RT-RNaseH domains. (B) **Non-LTR retrotransposons** have no LTRs, but harbor an (A)$_n$ tract at their 3′-end. They are subdivided into LINEs and SINEs. Only LINEs carry genes involved in their own transposition (encoding a capsid protein, an endonuclease and a reverse transcriptase). See text for further details.

of retroelement: they spread from cell to cell, and also from organism to organism.[520] It was hypothesized that retroviruses evolved from LTR retrotransposons, possibly in close conjunction with their typical host organisms, the vertebrates. Indications for retroviral activities were also found in invertebrates. The possible existence of retroviruses in plants is discussed controversially.[1073]

Both **LTR retrotransposons** and retroviruses are characterized by the presence of about 300- to 500-bp-long direct repeats at both ends of the element. These so-called LTRs contain control sequences for the initiation and termination of transcription as well as for polyadenylation (Figure 1.4A). The DNA between the LTRs is normally between 3 and 5 kb long, but can exceed 10 kb. It encodes (1) a capsid protein, which packages the viral RNA into a virus-like particle; (2) an RNase (RNase H); (3) a reverse transcriptase, which generates a cDNA from the full-sized message; (4) a protease, which is needed for processing the polyprotein; and (5) an endonuclease, which serves as an integrase (Figure 1.4A). The mechanism of transposition is quite well understood. In the first step, an active element is transcribed by host-encoded RNA polymerase II and translated on host ribosomes. Once the retrotransposon-encoded gene products are available, the retrotransposon RNA is packaged into a virus-like particle, reverse-transcribed into a cDNA, and integrated at another site of the genome.

LTR retrotransposons can be classified into two groups that are distinguished by the arrangement of the integrase and reverse transcriptase genes along the element. These groups were named **Ty1-*copia*** and **Ty3-*gypsy***, respectively (Figure 1.4A), following the nomenclature given to the initial representatives of both groups, which were first described for yeast (transposon yeast: Ty1 and Ty3) and *Drosophila* (*copia* and *gypsy*). Both types of retroelements are ubiquitous in many plant species (see Flavell et al.,[450,451] Hirochika et al.,[599] Manninen and Schulman,[874] Suoniemi et al.,[1346] and Voytas et al.[1483]).

Transposition via RNA leaves the original copy in the genome, and therefore results in a replicative increase in the copy number of retrotransposons (the so-called copy-and-paste mechanism of transposition). The extent of this increase can be quite different, depending on the type of element and the species. For Ty1-*copia*-like elements, reported copy numbers range from a few in *Arabidopsis*[739] up to several millions in *Vicia faba*.[1055] Moreover, LTR retrotransposons often provide landing pads for other retrotransposons, resulting in complex arrangements of retroelements nested within each other.[1223]

Several sequence variants of the same element are frequently found in the same genome, especially in plants.[450,451,739,1055] For example, Konieczny et al.[739] sequenced 20 clones with homology to Ty1-*copia* transposons from *Arabidopsis thaliana* and found that these belonged to 10 different families. Together, these families were estimated to account for about 100 kb, which is 0.1% of the *Arabidopsis* genome. Flavell et al.[451] compared Ty1-*copia* transposons from eight different plant species. These elements proved to be extremely heterogeneous, both intra- and interspecifically. For example, all 31 investigated potato clones had different sequences, and homologies within the reverse transcriptase gene region ranged from 5 to 75% at the amino acid level. The 27 clones analyzed by Pearce et al.[1055] in *Vicia faba* were also all different, and phylogenetic analysis of sequences placed some elements closer to retrotransposons found in other species than to *V. faba*. These results indicate that the evolution of LTR retroelements does not necessarily parallel that of their hosts, suggesting that horizontal transfer between evolutionary distant species may play a significant role.[449]

Retrotransposons lacking the terminal repeats are known as **non-LTR retrotransposons** (Figure 1.4B). They can be further subdivided into **LINEs** and **SINEs**. LINEs and SINEs are the predominant types of retrotransposons in the vicinity of mammalian genes, but also occur in other eukaryotes. LINEs are several kilobases long, have a characteristic poly(A) tract at their 3′-end, and are flanked by short direct repeats (3 to16 bp) that result from the repair of the staggered breaks generated by the integration process. Two open reading frames are usually present, one encoding a capsid protein, and the other encoding an endonuclease and a reverse transcriptase domain.[444] LINEs are particularly well characterized in *Drosophila* and mammals. The human genome contains about 100,000 LINEs of the so-called L1 type, but the vast majority of copies are defective.

LINEs are also commonly found in plants.[755,787,993,1245,1569] For example, Wright et al.[1569] identified 17 different non-LTR transposons in the small genome of *A. thaliana*, all with low copy number. One element resided in the mitochondrial genome, as did several LTR retrotransposons of the Ty1-*copia* and Ty3-*gypsy* type.[727,1433] Noma et al.[993] identified non-LTR retrotransposons in 27 of 33 plant species investigated, including monocots, dicots, gymnosperms, and horsetail (*Equisetum*). Neighbor-joining analyses showed a relatively good congruence between gene trees and host trees, indicating a predominantly vertical transmission of non-LTR retrotransposons. Plant LINEs generally exhibit high levels of sequence heterogeneity, which may be the consequence of the accumulation of mutations in these mostly defective elements.[1243]

With about 100 to 500 bp, SINEs are much smaller than LINEs. They are not able to transpose on their own. but require the activity of a reverse transcriptase *in trans*. Probably the best known representatives of SINEs are the *Alu* repeats, which occur in about 1,000,000 copies in the genome of humans and other primates.[525] SINEs are derived from processed pseudogenes. Their intact ancestors are host genes encoding small cytoplasmic RNAs such as tRNAs and 7SL-RNA. Like LINEs, SINEs are flanked by short target site duplications and harbor an A-rich tract at their 3′-end. Like their tRNA progenitors, SINEs carry two internal promoters recognized by host RNA polymerase III. Upstream of the poly(A) region, some SINEs share limited sequence homology with LINEs.[1008] It was hypothesized that these regions are recognized by the transpositional machinery of LINEs.[1008,1243] SINEs were found in several plant taxa,[796,957,1589,1595] but do not seem to play a predominant role in plant genomes.

1.3.2 Class II Transposons

Class II transposons disperse via a DNA intermediate and are characterized by short terminal inverted repeats (TIRs). The internal regions encode one or two genes responsible for transposition. As in the case of retrotransposons, there are autonomous as well as defective elements. The latter can only transpose when active elements are present in the same genome. Transposition usually follows a nonreplicative cut-and-paste mechanism. Hence, copy numbers are small to intermediate (usually less than a few hundred), and class II transposons therefore comprise only a small part of the genome. They often integrate in gene-rich regions, which makes them useful tools for gene isolation by transposon tagging.[492,567,1443] The occurrence, biology and evolution of class II elements have been reviewed by Graur and Li[525] and Saedler and Gierl.[1208]

In plants, class II transposons can be grouped into at least four superfamilies, three of which (*Ac*, CACTA, *Mu*) were first characterized in maize. Transposons of the ***Ac* family** (e.g., *Ac* in maize; P-elements in *Drosophila*) code for a single gene (a transposase). Transposons of the **CACTA family** (e.g., *En/Spm* in maize, *Tam1* from *Antirrhinum majus*) carry two genes, encoding a transposase and a DNA-binding protein, respectively. ***Mutator*-like elements** also encode two genes and are characterized by much longer TIRs than the other two families.

Another superfamily known as **Tc1-*Mariner*-like elements** (**MLEs**) probably is particularly widespread in nature (reviewed by Plasterk et al.[1083]). Tc1 was initially detected in the nematode *Caenorhabditis elegans*, and *Mariner* in the fly *Drosophila mauritiana*, but related elements were identified in other animals as well as in the nuclear genomes of humans, ciliates, fungi, and plants (e.g., in *A. thaliana*[787]). Like members of the *Ac* family, MLEs contain a single gene coding for a transposase that is flanked by TIRs. Expression of the transposase appears to be necessary and sufficient for transposition, facilitating the spread to new hosts. The ubiquitous occurrence of MLEs across the tree of life may well be caused by horizontal gene transfer.[1083]

1.3.3 Unclassified Transposons

Miniature inverted-repeat transposable elements (MITEs) are a superfamily of transposons that are characterized by small size (<500 bp), short TIRs, AT-richness, high levels of internal sequence divergence, the potential to form secondary structures, relatively large copy numbers (typically >1000 per haploid genome), and a preference for sequence-defined integration sites such as TA or TAA (reviewed by Wessler et al.[1532]). The MITE families *Tourist*, *Stowaway*, and *Heartbreaker* have initially been detected in the Poaceae,[184–187,1532] but more recent data show that MITE-like transposons are also present in other plant species[217,787] as well as in eukaryotes outside of the plant kingdom (Casa et al.[216] and references cited therein). More than 10 MITE families have since been characterized.[13,187,217,1424,1611] All MITEs share most or all of the common structural features outlined above, but there is no sequence homology between the various families.

Akagi et al.[13] reported a short (393 bp) MITE-like element from the rice genome. This novel element, which was called *Micron*, occurs in about 100 to 200 copies and lacks TIRs. Sequence analysis of 19 homologues revealed a high degree of sequence conservation (> 90%), and a consistent association with $(AT)_n$ microsatellites on both flanks. This could be explained by specific targeting of AT-rich microsatellites by *Micron* elements.[13]

Given that MITEs share features of both class I and class II elements, their classification remains elusive. MITEs probably move via a DNA intermediate, although direct evidence is lacking.[1532,1611] Like class II elements, MITEs show a preference for genic regions.[1611] However, copy numbers far exceed those found in other DNA transposons, suggesting that MITEs could transpose by a mechanism that leaves the donor copy intact (e.g., by a gap repair mechanism also known from *Drosophila* P elements). MITEs are the most abundant type of transposons associated with plant genes. Their small size suggests that transposase functions are normally provided *in trans*. However, Le et al.[787] also identified unusually large members of the MITE family in *A. thaliana* that potentially encode a transposase.

A novel family of DNA transposons with unknown transposition mechanism called *Basho* was first identified in *A. thaliana*[787] and later in rice.[1424] Some of the rice *Basho* elements harbor an internal polymorphic AT-rich minisatellite.[641] This again demonstrates the often close association between tandem and interspersed repeats in the plant genome.

1.3.4 Transposons and Genome Evolution

Transposons have long been considered as a prototype of selfish DNA,[357,1015] whose presence can have deleterious consequences for the host. For example, genes around a transposon insertion site can be blocked, rearranged, or their regulatory patterns changed by altering the specificity of promoters. In the case of retrotransposons, the potential to spread rapidly by a copy-and-paste mechanism may lead to genomic obesity associated with a waste of energy. In some plant species, retrotransposons occupy more than half of the genome.[1223] It therefore seems logical that the host genome has evolved a number of mechanisms to keep the activity of transposable elements under tight control.

One important control mechanism appears to be epigenetic silencing by DNA methylation. Under normal circumstances, the vast majority of transposable elements are heavily methylated, which presumably inhibits the expression of the genes needed for transposition. Interestingly, transcriptional control is relaxed under certain stress conditions. Several studies have shown that, for example, tissue culture stress or treatment with microbial elicitors leads to an activation of retrotransposon expression and a concomitant increase in the transposition rate (reviewed by Capy et al.[213]). Such a response to stress could make sense. Transposition mutagenizes the genome, thereby increasing the genetic plasticity and diversity. This may provide an increased chance of responding properly to changed environmental conditions. Stress-induced transposition of previously dormant elements would also explain the long-known phenomenon of somaclonal variation; i.e., the observation that plants regenerated from tissue culture often show a high frequency of mutations (reviewed by Karp[695]; see also Chapter 6.2.3). The observation that transposons may also act as so-called useful parasites[213] has challenged the earlier concept of junk or egoistic DNA, and perhaps represents just one among several potentially beneficial effects of transposon activity on host genome evolution. For more comprehensive treatments of transposable element contributions to plant genome evolution, see the reviews by Bennetzen,[113] Fedoroff,[432] and Kidwell and Lisch.[712]

1.3.5 Transposons as Molecular Markers

A wide variety of molecular marker techniques use PCR primers directed toward transposable elements, either alone[227,383,461,1116] (see Chapter 2.3.8.2) or in combination with other types of primers. Thus, LTR retrotransposon-specific primers have been combined with microsatellite-specific primers in *copia*-SSR[1113] and REMAP[681] (see Chapter 2.3.5), and with amplified fragment length polymorphism (AFLP) primers in sequence-specific amplification polymorphism (S-SAP)[1510] (see Chapter 2.3.8.2). AFLP primers were also used together with primers specific for DNA transposons, such as the petunia *Ac*-like element *dTph1*[1443] and the *Heartbreaker* element belonging to the MITE superfamily[216,1044] (see Chapter 2.3.8.2).

2

Detecting DNA Variation
by Molecular Markers

In this chapter, we review current strategies used to visualize DNA polymorphisms by electrophoresis-based methods. Each section starts with a short introduction to the principles and the historical development of the respective technique, followed by a summary of the properties, advantages, disadvantages, and application areas of the markers generated. This chapter also provides a survey of the plethora of acronyms for the various DNA profiling techniques at hand. Experimental protocols of commonly used polymerase chain reaction (PCR)-based marker systems, comments on technical aspects, reaction parameters, reproducibility, robustness, and transferability of markers, as well as on modifications of the standard techniques, are given in Chapter 4.

2.1 PROPERTIES OF MOLECULAR MARKERS

The analysis of genetic diversity and relatedness between or within different populations, species, and individuals is a central task for many disciplines of biological science. During the last three decades, classical strategies for the evaluation of genetic variability, such as comparative anatomy, morphology, embryology, and physiology, have increasingly been complemented by molecular techniques. These include, for example, the analysis of chemical constituents (so-called metabolomics), but most importantly relate to the development of molecular markers. Marker technology based on polymorphisms in proteins or DNA has catalyzed research in a variety of disciplines such as phylogeny, taxonomy, ecology, genetics, and plant and animal breeding.

The following properties would generally be desirable for a molecular marker:

1. Moderately to highly polymorphic
2. Codominant inheritance (which allows the discrimination of homo- and heterozygous states in diploid organisms)
3. Unambiguous assignment of alleles

4. Frequent occurrence in the genome
5. Even distribution throughout the genome
6. Selectively neutral behavior (i.e., no pleiotropic effects)
7. Easy access (i.e., by purchasing or fast procedures)
8. Easy and fast assay (e.g., by automated procedures)
9. High reproducibility
10. Easy exchange of data between laboratories
11. Low cost for both marker development and assay

No single type of molecular marker fulfills all of these criteria. However, one can choose between a variety of marker systems, each of which combines some — or even most — of the above-mentioned characteristics. Properties of molecular markers and their application in various areas of research have been reviewed by Avise,[57] Bachmann,[65] Baker,[71] Caetano-Anollés and Gresshoff,[200] Epplen and Lubjuhn,[406] Gupta et al.,[544] Henry,[593] Hillis et al.,[597] Hoelzel,[603] Karp et al.,[696] Lee,[791] Nybom,[1000] and Winter and Kahl.[1554]

2.2 TRADITIONAL MARKER SYSTEMS

2.2.1 Protein Markers and Allozymes

For the generation of molecular markers based on protein polymorphisms, the most frequently used technique is the electrophoretic separation of proteins, followed by specific staining of a distinct protein subclass. Less commonly, specific proteins are detected by monoclonal antibodies with an attached fluorescent label. Although some earlier studies focused on seed storage protein patterns, the majority of protein markers are derived from allozymes.

Allozyme analysis is relatively straightforward and easy to carry out.[963] A tissue extract is prepared and electrophoresed on a nondenaturing starch or polyacrylamide gel. The proteins of this extract are separated by their net charge and size. After electrophoresis, the position of a particular enzyme in the gel is detected by adding a colorless substrate that is converted into a dye under appropriate reaction conditions. Depending on the number of loci, their state of homo- or heterozygosity, and the enzyme configuration (i.e., the number of separable subunits), from one to several bands are visualized. The positions of these bands can be polymorphic and thus informative.

Sometimes the terms isozyme and allozyme, incorrectly, are treated as interchangeable. **Isozymes** are enzymes that convert the same substrate, but are not necessarily products of the same gene. Isozymes may be active at different life stages or in different cell compartments. **Allozymes** are isozymes that are encoded by orthologous genes, but differ by one or more amino acids due to allelic differences. The main advantages of allozyme markers are their codominant inheritance and the technical simplicity and low cost of the assay. Disadvantages include the restricted number of suitable allozyme loci in the genome, the requirement of fresh tissue, and the sometimes limited variation. Advantages and drawbacks of allozyme-based as compared with DNA-based marker analyses are discussed in more detail in Chapter 8.1.

Allozyme electrophoresis has successfully been applied to many organisms from bacteria to numerous fungal, plant, and animal species since the 1960s.[575] These studies encompassed various fields (e.g., physiology, biochemistry, systematics, genetics, and breeding) and purposes (e.g., evaluation of mating systems, ploidy levels, and hybrid origins). Methodology and applications of allozyme analysis have been reviewed by Baker,[71] Hamrick and Godt,[562] May,[900] and Murphy et al.[963]

2.2.2 DNA Sequencing

Polymorphisms at the DNA level can be studied by numerous approaches. Certainly the most direct strategy is the determination of the nucleotide sequence of a defined region,[899,1219] and the alignment of this sequence to an orthologous region in the genome of another, more or less related organism (see also Chapter 4.3.3 and reviews by Alphey[25] and Hillis et al.[597]). The extent of homology between various sequences can be deduced from the alignment, and phylogenies can be reconstructed by a variety of approaches and algorithms (for reviews, see Archibald et al.,[33] Felsenstein,[435] Hall,[556] Huelsenbeck and Crandall,[628] Huelsenbeck et al.,[629] Page and Holmes,[1027] and Swofford et al.[1352]).

DNA sequencing provides highly robust, reproducible, and informative data sets, and can be adapted to different levels of discriminatory potential by choosing appropriate genomic target regions. On the negative side, DNA sequencing can be prohibitively tedious and expensive when very large numbers of individuals have to be assayed (e.g., in population genetics, phylogeography, and marker-assisted plant breeding programs). Another disadvantage, at least for certain areas of research, is the highly specific sampling of only a small part of the genome. For example, phylogeny reconstructions based on DNA sequence data generally result in gene trees, which do not necessarily reflect the species tree.[363] Many of the PCR-based molecular markers described in Chapter 2.3 instead provide a measure of genome-wide genetic variation.

DNA sequencing has been greatly facilitated by the advent of the PCR,[960,1212] which made it possible to isolate orthologous DNA regions from any organism of interest with unprecedented speed. Universal primer pairs were designed on the basis of sequence information for conserved parts of the DNA, and the PCR-amplified target regions were either sequenced directly or sequenced after cloning.[597] The popularity of DNA sequencing was further enhanced by the development of fluorescence-labeled primers and nucleotides that could be used for the automated detection of DNA molecules in gel- or capillary-based sequencing instruments.[1299] With readings of up to 1200 base pairs, fluorescence sequencing provides much higher resolution than the traditional approach using radioisotopes. Moreover, it is easier to perform, and sequence data are directly transferred to a computer. The fact that the technical equipment is more expensive than traditional sequencing facilities is not a real problem because custom sequencing services have become widespread and relatively inexpensive.

The efficiency and speed of fluorescence-based DNA sequencing technology paved the way for several huge sequencing projects, which have resulted in the completion of drafts of whole genome sequences from several eukaryotic model organisms, including yeast,[504] the nematode *Caenorhabditis elegans*,[1386] two ecotypes of *Arabidopsis thaliana*,[1385] two varieties of rice,[503,1602] *Drosophila melanogaster*,[4] mouse,[1389] rat,[1390] and man.[1388]

Molecular systematics is another important application area for DNA sequencing, especially for evaluating medium- and long-distance relationships. In plants, higher order taxonomic studies are mostly based on slowly evolving DNA regions, such as the chloroplast *rbc*L gene,[259,1009] the nuclear 18S[1306] and 26S ribosomal RNA genes,[763] and various mitochondrial genes.[234] For studies at the infrafamilial and infrageneric level, the internal transcribed spacers (ITS1 and ITS2) within the nuclear ribosomal gene clusters,[73,74] and a wide range of introns and intergenic spacers in plant chloroplast and mitochondrial DNA have become most popular.[657,1358] Sets of universal PCR primer pairs have been developed that allow PCR amplification and subsequent sequencing of certain DNA regions from almost any plant species of interest.[73,326,371,372,529,558,1216,1358] To solve particularly difficult problems such as the origin of angiosperms, multigene analyses were performed that combined sequence data from three to 17 genes derived from all three genomes.[234,762]

In recent years, DNA sequencing also has become popular for population genetic studies. In an approach coined phylogeography by Avise et al.,[59] intra- and interspecific phylogenies are reconstructed from DNA sequence haplotypes derived from the chloroplast, mitochondrial, or nuclear genome. These phylogenies are compared with the current geographical distribution of the respective lineages, allowing important conclusions on historical population processes (reviewed by Avise[58] and Schaal et al.[1229]; see Chapter 6.5).

The methodology of DNA sequencing has been reviewed by Alphey,[25] Ausubel et al.,[56] and Sambrook and Russell.[1217] The various applications of DNA sequence analyses for molecular systematics have been reviewed by Hillis et al.[597] and Soltis et al.[1308]

2.2.3 Restriction Fragment Length Polymorphism (RFLP) Analysis

Molecular marker methods usually evaluate DNA sequence variation without sequencing. The first DNA marker generation exploited so-called restriction fragment length polymorphisms (RFLPs).[154] Restriction enzymes are endonucleases produced by a variety of prokaryotes. Their natural function is to destroy invading, foreign DNA molecules by recognizing and cutting specific DNA sequence motifs, mostly consisting of four, five, or six bases. Each enzyme has a specific, typically palindromic recognition sequence, and the bacteria usually protect their own DNA from being cut by methylating the cytosine or adenine residues within this sequence.[904]

Digestion of a particular DNA molecule with a particular restriction enzyme results in a reproducible set of fragments of well-defined lengths. Point mutations within the recognition sequence as well as insertions or deletions between two recognition sites result in an altered pattern of restriction fragments, and may thus bring about a screenable polymorphism between different genotypes (Figure 2.1). A list

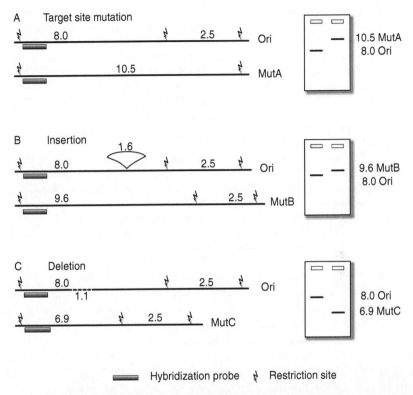

Figure 2.1   Molecular basis of RFLPs. An RFLP can originate from the mutation of (A) a restriction enzyme target site as well as (B) from the insertion or (C) deletion of a piece of DNA between two target sites. RFLPs are typically visualized by electrophoresis of restriction-digested genomic DNA on agarose gels, followed by Southern blotting and hybridization with a sequence-specific probe (see text for details). Hybridization signals derived from the original wildtype allele (Ori) and the mutated alleles (MutA, B, and C) are indicated in the right panel of each figure. Numbers refer to the length of a particular restriction fragment in kilobase pairs.

of all known restriction enzymes, their recognition sequences, methylation sensitivity, commercial availability, and other useful information is compiled in the REBASE database,[1177] which is available at http://rebase.neb.com/rebase/rebase.html.

2.2.3.1 Nuclear RFLPs and DNA Fingerprinting

RFLPs can be derived from the nuclear, chloroplast, and mitochondrial genome. The analysis of nuclear RFLPs involves several experimental steps. First, high molecular weight genomic DNA is extracted from the organism of interest, and digested with one or more restriction enzymes. The resulting fragments are separated according to size by gel electrophoresis. The gel is Southern-blotted onto a membrane, and one or more specific fragments are visualized by blot hybridization with a labeled probe. Two categories of probes are usually chosen:

1. **Locus-specific probes** recognize one or a few specific regions of the genomic DNA, resulting in easy-to-screen codominant markers. Hybridization probes are either anonymous in nature (i.e., obtained from a cDNA or genomic library of the investigated species) or specific for certain genes. Ribosomal RNA genes (i.e., the coding region of 18S, 5.8S, and 25S rRNA, also referred to as rDNA) have been popular sources for RFLPs in plants, because the same probes can be applied to a wide range of species, and polymorphisms are easy to detect due to the high abundance of these sequences.[49,717,1209]

2. **Multilocus probes** are usually designed to recognize tandemly repeated DNA motifs such as mini- or microsatellites (see Chapter 1.2). These probes create complex banding patterns, which were coined DNA fingerprints by Jeffreys et al.[663] Because of a variable number of tandem repeat-type polymorphism,[973] DNA profiles generated by mini- and microsatellite-specific hybridization probes are highly variable, and often individual-specific. In a variant coined oligonucle-otide fingerprinting, radiolabeled oligonucleotides specific for microsatellite motifs such as (GATA)₄ are used as probes, which are hybridized to genomic DNA immobilized in dried agarose gels.[20,1519,1522]

The main advantages of RFLP markers are their codominance and high reproduc-ibility. Drawbacks as compared with PCR-based techniques are the tedious experi-mental procedures, and the requirement of microgram amounts of relatively pure and intact DNA. For quite some time, locus-specific RFLP markers served as stan-dard tools for the construction of genetic maps (reviewed by Tanksley et al.[1364]), starting points for map-based cloning of genes,[891] cultivar identification,[1131] and phylogenetic studies[315] (reviewed by Dowling et al.[362]). Multilocus RFLP markers, on the other hand, were mostly used for forensic purposes, parentage analyses, and genotype identification (for reviews see Burke et al.,[188] Epplen and Lubjuhn,[406] Pena et al.,[1062] Weising and Kahl,[1519] and the first edition of this book). Typical multilocus RFLP fingerprints generated with microsatellite-specific hybridization probes are shown in Figure 2.2.

2.2.3.2 *RFLPs in Chloroplast and Mitochondrial DNA*

The chloroplast DNA (cpDNA) molecule is approximately 150 kb in size and consists of an inverted repeat separating one large single copy (LSC) and one small single copy (SSC) region. Recombination in cpDNA is absent or very rare (but see Marshall et al.[886]). Plant mitochondrial DNA (mtDNA) is much larger, and plant mtDNA sequences are thought to evolve relatively slowly (Wolfe et al.[1559]; see Chapter 1). As a consequence of intragenomic recombination, the general architec-ture of the plant mtDNA molecule is highly variable, and different forms and sizes of plant mtDNA can be found within a single mitochondrion, a cell, or an individ-ual.[66,1033,1034] Both cpDNA and mtDNA are present in hundreds of copies per cell, and each acts as a single heritable unit. Inheritance is uniparental, in contrast to the biparentally transmitted nuclear DNA (reviewed by Birky[134] and Reboud and Zeyl[1146]). In most cases, transmission is through the female parent. The best-known exception to this rule is the paternal transmission of cpDNA in most but not all gymnosperms.

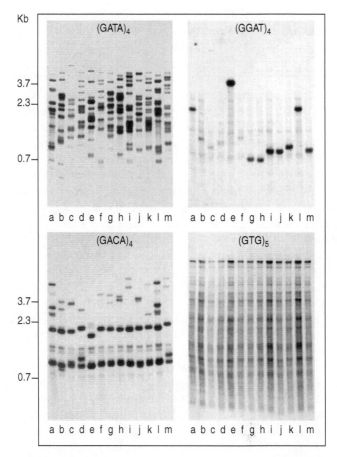

Figure 2.2 Oligonucleotide fingerprints of chickpea (*Cicer arietinum*). Five-microgram aliquots of genomic DNA from one individual plant each of 13 landraces (lanes a to m) were digested with the restriction enzyme *Taq*I, separated on a 1.4% agarose gel, in-gel hybridized with ^{32}P-labeled probes, and autoradiographed. The same gel was rehybridized with four different probes. Positions of size markers are indicated (Kb, kilobase pairs). The patterns range from highly variable [e.g., (GATA)$_4$] to completely monomorphic [e.g., (GTG)$_5$]. Fingerprints observed in lanes (a) and (l) are identical with each probe, suggesting a close relationship of these two samples.

RFLPs in organellar DNA can be visualized by one of three experimental strategies. In the classical approach, mtDNA and/or cpDNA are extracted separately from the nuclear DNA, which can be achieved by ultracentrifugation in density gradients or, alternatively, by differential extraction procedures.[880,933,1418] The organellar DNA is then digested with one or more restriction enzymes, electrophoresed on agarose or polyacrylamide gels, and RFLPs are detected by ethidium bromide or silver staining. The second approach relies on Southern blot hybridization. It starts with the isolation of total genomic DNA (see Chapter 4.2), followed by digestion with restriction enzymes (see Chapter 4.3.1). The myriad of resulting restriction fragments are electrophoresed on agarose gels and blotted to a membrane. Organellar DNA fragments are then visualized by hybridization with a labeled total cpDNA or mtDNA sequence, or a specific cpDNA or mtDNA sequence. The third strategy involves the amplification of a defined region of the organellar genome by PCR. The resulting PCR products are digested with restriction enzymes, and fragments are separated by gel electrophoresis and stained with ethidium bromide. This latter technique, commonly referred to as PCR-RFLP or cleaved amplified polymorphic sequences (CAPS),[738] is treated in more detail in Chapters 2.3.2 and 4.9.

RFLPs of cpDNA have been studied extensively in plants, and have proven to be valuable for molecular systematic studies above the species level,[259,657,1009,1178] as well as for phylogeographic analyses within species (reviewed by Newton et al.[988] and Schaal et al.[1229]; see Chapter 6.5). Hybridization-based approaches have now largely been replaced by direct sequencing and CAPS technology.

With mtDNA, the situation is different in plants and animals. Animal mtDNA is relatively small (approximately 15 to 20 kb), its gene order is highly conserved, and the rate of sequence divergence is higher than that in nuclear DNA. These properties made mtDNA a valuable source for RFLPs in population studies, especially for the analysis of maternal lineages and population history (reviewed by Avise[57,58]). However, mtDNA RFLP analysis in animals is now being replaced successively by direct sequencing. In plants, the analysis of mtDNA RFLPs has not been very attractive, mainly because of the high incidence of intramolecular recombination (see above). Traditional RFLP analyses of plant mtDNA have only been performed for a few purposes; e.g., to analyze cytoplasmic male sterility[736] or to follow seed migration routes in gymnosperms.[355] More recently, interest in revealing plant mtDNA polymorphisms has increased considerably. The availability of complete mtDNA sequences[756,994,1433] and other sequence information has allowed the design of consensus PCR primers[326,371] which greatly facilitated direct sequencing (Chapter 2.2.2) and PCR-RFLP (CAPS; Chapter 2.3.2) studies.

2.3 THE PCR GENERATION: MOLECULAR MARKERS BASED ON *IN VITRO* DNA AMPLIFICATION

The invention of the polymerase chain reaction (PCR) by Mullis and coworkers[960,1212] revolutionized the methodological repertoire of molecular biology. This technique allows us to amplify any DNA sequence of interest to high copy numbers *in vitro*, thereby circumventing the need for molecular cloning. To amplify a particular DNA sequence, two single-stranded oligonucleotide primers are designed, which are complementary to motifs on the template DNA. The primer sequences are chosen to allow base-specific binding to the two template strands in reverse orientation. Addition of a thermostable DNA polymerase in a suitable buffer system and cyclic programming of primer annealing, primer extension, and denaturation steps result in the exponential amplification of the sequence between the primer-binding sites, including the primer sequences.

Now that RFLP markers are about to celebrate their 25th birthday, they have already been largely replaced by more sensitive and convenient PCR-based marker technologies. PCR assays were not only developed to screen for restriction site variation (as manifested in the CAPS [Chapter 2.3.2] and amplified fragment length polymorphism [AFLP; Chapter 2.3.8] approach), but also for other types of polymorphisms outlined in the forthcoming sections. PCR technology has the ability to create large numbers of markers in short periods of time, requires little experimental effort, and works with nanogram amounts of DNA. Moreover, PCR markers are amenable to automation, which is an important requisite for the high-throughput assays needed in molecular breeding programs. Whereas the traditional locus-specific, hybridization-based RFLPs are still in use in some laboratories, mainly because of their robustness and reliability, RFLP analysis with multilocus hybridization probes is now found on the Red List of Threatened Marker Technologies.

2.3.1 Principle of the PCR

PCR is based on the enzymatic *in vitro* amplification of DNA. Since the introduction of thermostable DNA polymerases in 1988,[1212] the use of PCR in research and clinical laboratories has increased tremendously, and tens of thousands of publications as well as numerous books document the success of the technique (e.g., Innis et al.[639] and Mullis et al.[960]). In a typical PCR assay, three temperature-controlled steps can be discerned, which are repeated in a series of 25 to 50 cycles. A reaction mix consists of:

1. A buffer, usually containing Tris-HCl, KCl, and $MgCl_2$
2. A thermostable DNA-polymerase, which adds nucleotides to the 3′-end of a primer annealed to single-stranded DNA (ssDNA)
3. Four deoxyribonucleotide triphosphates [dNTPs]: dATP, dCTP, dGTP, dTTP
4. Two oligonucleotide primers
5. Template DNA

The selectivity of the reaction is determined by the choice of the primer(s). Primers are single-stranded pieces of DNA (oligonucleotides) with sequence complementarity to template sequences flanking the targeted region. To allow for exponential amplification, the primers must anneal in opposite directions, so that their 3′-ends face the target amplicon. Amplification is most efficient when the two primer binding sites are not further apart than approximately 4 kb. However, amplification products of more than 10 kb can be obtained under optimal conditions.[241]

The principle of the cycling reaction is outlined in Figure 2.3. In the first step of the first cycle, the original template DNA is made single-stranded by raising the temperature to about 94°C (denaturing step). In the second step, lowering the temperature to about 35 to 65°C (depending on primer sequence and experimental strategy) results in primers annealing to their target sequences on the template DNA (annealing step). The primers will preferably hybridize to binding sites that are identical or highly homologous to their nucleotide sequence, although some mismatches (especially at the 5′-end) are allowed. For the third step, a temperature is chosen at which the activity of the thermostable polymerase is optimal; i.e., usually 65 to 72°C (elongation step). The polymerase now extends the 3′-ends of the DNA–primer hybrids toward the other primer binding site. Because this happens at

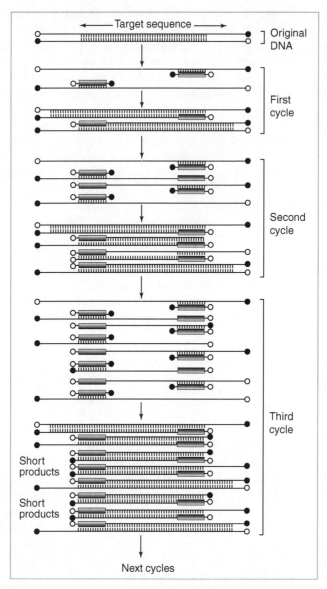

Figure 2.3   Principle of the polymerase chain reaction. A target DNA sequence is exponentially amplified with the help of flanking primers and a thermostable DNA polymerase. The reaction involves repeated cycles, each consisting of a denaturation, a primer annealing, and an elongation step. Primers are represented by shaded boxes. The 5'- and 3'-ends of DNA single strands are indicated by open and closed circles, respectively. In the initial stage of the reaction, both shorter and longer products are generated. Only the shortest possible fragments are amplified exponentially, and therefore predominate the final product almost exclusively. See text for details.

both primer-annealing sites on both DNA strands, the target fragment is completely replicated (cycle 1).

In the second cycle, the two resulting double-stranded DNAs are again denatured, and both the original strand and the product strand now act as a template. Repeating these three-step cycles 25 to 50 times results in the exponential amplification of the target amplicon between the 5′-ends of the two primer binding sites (short products in Figure 2.3). Other, longer fragments are also generated, but these are only linearly amplified and their proportion in the final product is negligible. See Chapter 4.3.2 for general strategies to optimize the outcome of a PCR.

One of the main reasons for the versatility of the PCR technique is that any kind of primers can be chosen, depending on the purpose of the study. For example, any particular DNA sequence of interest can be amplified by a pair of **specific primers**, which are designed on the basis of DNA sequence information. Such a strategy is chosen for gene isolation or for the analysis of transferred genes in transgenic organisms. Specific primers based on unique flanking sequences are also used to analyze nuclear or organellar microsatellites, which are among the most important PCR-based marker systems (see Chapters 2.3.4 and 4.8). On the other side of the spectrum, **arbitrary primers** can be constructed that amplify anonymous genomic DNA sequences under appropriate experimental conditions (see Chapters 2.3.3 and 4.4). Between the two extremes, there are numerous possibilities to construct **semi-specific primers**. These are directed toward sequence elements that belong to a more or less well-defined subset of the genome. Most semispecific primers target repetitive DNA motifs, amplifying sequences that reside between two consecutive elements of the repeat (see Chapters 2.3.5 and 4.5). Specific, semispecific, and arbitrary primers can be used in various combinations. Their potential for different experimental purposes is almost endless.

2.3.2 Cleaved Amplified Polymorphic Sequences

The idea to create molecular markers by digesting PCR products with restriction enzymes dates back to Williams et al.[1548] and Arnold et al.[50] Several acronyms have been created for this marker technique, of which **PCR-RFLP** and **CAPS**[738] are most frequently used. CAPS markers are generated in two steps. In the first step, a defined DNA sequence is amplified using a sequence-specific primer pair. This may already result in differently sized and hence informative PCR fragments.[1548] In the second step, the PCR product is digested with a restriction enzyme, usually with a four-base recognition specificity. The digested amplification products may or may not reveal polymorphisms after separation on agarose gels.

As opposed to conventional RFLP analysis (see Chapter 2.2.3), the CAPS approach does not require radioactivity or blotting steps, but instead exhibits all the attractive attributes of PCR-based techniques. Because *in vitro*-amplified DNA remains unmethylated, CAPS markers are also insensitive to DNA methylation. Like RFLPs, CAPS markers are codominant. The possibility to distinguish homo- and heterozygous states makes the procedure particularly attractive for mapping purposes (reviewed by Drenkard et al.[368]). Because only a subset of base substitutions is targeted, and small insertion–deletion events may escape detection, CAPS assays yield less information than direct sequence analysis of a PCR product. With decreasing costs for DNA sequencing, CAPS markers are therefore expected to be replaced continuously by direct sequencing (e.g., in phylogeography; see below and Chapter 6.5).

In principle, CAPS markers can be generated from either nuclear or organellar DNA. For example, Konieczny and Ausubel[738] analyzed nuclear DNA fragments that had already been mapped to specific chromosome arms of *Arabidopsis*, whereas Purugganan and Wessler[1116] amplified and digested DNA fragments of the maize transposable element *magellan* in a CAPS variant coined **transposon signatures**. In plants, defined regions of the chloroplast genome have been major targets of the CAPS approach. Thus, restriction site variation of PCR-amplified cpDNA has been applied extensively for phylogenetic reconstruction at various taxonomic levels (reviewed by Jansen et al.[657] and Olmstead and Palmer[1009]). CAPS assays also facilitated the screening for intraspecific cpDNA RFLPs, which have rarely been detected with traditional methods (see review by Soltis et al.[1307]). Consequently, chloroplast CAPS markers became standard tools for phylogeographic analyses below the species level[373,1229] (see Chapter 6.5). In these studies, noncoding cpDNA regions are amplified by PCR with sets of universal primers that bind to conserved coding regions[326,372,529,558,1216] (see Chapter 2.2.2). Aliquots of the resulting PCR products are digested with one of a set of restriction enzymes, and the identified polymorphisms are combined into nonrecombinant cpDNA haplotypes. Statistical parsimony networks can be reconstructed that reflect the genetic distances among these haplotypes.[1090] Comparing genetic relationships with geographical distribution patterns has yielded important insights into, e.g., the postglacial recolonization routes of tree species into central Europe[2,988,1359] (see Chapter 6.5.1.2).

2.3.3 PCR with Arbitrary Primers: RAPD and Its Variants

The methods described in the following section use primers of arbitrary nucleotide sequence to amplify anonymous PCR fragments from genomic template DNA. Typically, single PCR primers are used under relaxed stringency conditions, and no prior knowledge of DNA sequence is required. The basic principles of the technology have been presented by three independent groups in the early 1990s, each suggesting a different protocol.[201,1527,1546] Since then, numerous modifications in primer design, cycling conditions, separation and visualization of PCR products, and overall strategy have been suggested. Three main streams of PCR with arbitrary primers can still be distinguished:

1. The **random amplified polymorphic DNA (RAPD)** procedure introduced by Williams et al.[1546] is technically the simplest version. It employs single primers with 10 nucleotides and a GC content of at least 50%. PCR products are separated on agarose gels and detected by staining with ethidium bromide.

2. The **DNA amplification fingerprinting (DAF)** protocol was suggested by Caetano-Anollés et al.[201,202] DAF makes use of very short primers (often only five to eight

nucleotides long) at relatively high concentrations (~3 μM), with either low- or high-stringency annealing steps and two- instead of three-temperature cycles in the PCR. The resulting fragments are resolved in polyacrylamide gels and visualized by silver staining. Descendants of DAF include DNA profiling with **mini-hairpin primers**[198] and the generation of **arbitrary signatures from amplified profiles (ASAP)**[199] (see Chapter 4.4.2.1). For a review of DAF and its variants, see Caetano-Anollés.[197]

3. **Arbitrarily primed PCR (AP-PCR)**, introduced by Welsh and McClelland,[1527] is the most complicated variant. In this technique, oligonucleotides of 20 or more bases, originally designed for other purposes, are used as primers. Two cycles with low stringency (allowing for mismatches) are followed by 30 to 40 cycles with high stringency. Radiolabeled nucleotides are included in the last 20 to 30 cycles only. PCR products are separated by polyacrylamide gel electrophoresis and made visible by autoradiography. AP-PCR can be simplified by silver staining of polyacrylamide gels[1236] or by separating the fragments on agarose gels and staining with ethidium bromide.

The term **multiple arbitrary amplicon profiling (MAAP)** encircles the characteristics of all three families of techniques adequately[203] but has received little attention. A more recent term is **arbitrarily amplified DNA (AAD)**.[197] For convenience, we will use the term RAPD for all types of PCR with arbitrary primers, as is done by most authors.

All arbitrarily-primed PCR techniques have in common that (1) fingerprint-like multilocus banding patterns are produced, (2) no prior knowledge of genomic DNA sequences is needed, and (3) the primers can be universally used for most pro- and eukaryotes. Although *a priori* nothing is known about the identity and the sequence context of a particular PCR product, its presence or absence in different organisms can serve as an informative character. A flow sheet of an RAPD experiment is depicted in Figure 2.4, typical RAPD gel patterns are exemplified in Figure 2.5.

2.3.3.1 The Molecular Basis of RAPDs: Significance of Mispriming and Competition among Priming Sites

To obtain an amplification product with only one primer, there must be two identical (or at least highly similar) target sequences in close vicinity to each other: one site on one strand and the other site on the other strand, in an opposite orientation. RAPD polymorphisms can theoretically result from several types of events: (1) insertion of a **large** piece of DNA between the primer binding sites may exceed the capacity of PCR, resulting in fragment loss; (2) insertion or deletion of a **small** piece of DNA will lead to a change in size of the amplified fragment; (3) the deletion of one of the two primer annealing sites results in either the loss of a fragment or an increase in size; (4) a nucleotide substitution within one or both primer target sites may affect the annealing process, which can lead to a presence versus absence polymorphism or to a change in fragment size (Figure 2.4).

The number of fragments that can be expected theoretically from one primer, annealing with 100% homology, can be calculated from primer length and the complexity of the target genome, assuming that the nucleotides are present in equal

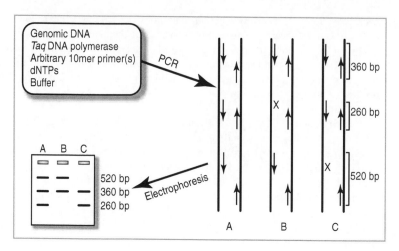

Figure 2.4   Strategy of PCR with arbitrary primers. Genomic DNA, a thermostable DNA polymerase, one (or two) primer(s) of arbitrary sequence, the four deoxyribonucleotide triphosphates (dNTPs) and a suitable buffer are combined into a reaction tube and subjected to PCR. The primers anneal to anonymous target sequences of the template DNA. If two primers (depicted as arrows, not drawn to scale) anneal in an opposite direction and at a suitable distance from each other, the DNA sequence between the two primers is amplified. PCR products are separated by gel electrophoresis and visualized by, e.g., ethidium bromide staining. Various mechanisms may result in presence versus absence polymorphisms (see text). For example, a base substitution within a primer target site (indicated by x) may interfere with primer annealing, and thus prevent the amplification of the respective fragment.

proportions. Williams et al.[1547] gave the equation: $b = (2000 \times 4^{-2n}) \times C$, where b is the expected number of fragments per primer, n is the primer length in nucleotides, and C is the genome size in base pairs per haploid genome. For example, in a plant species such as maize (genome size of 6×10^6 kb), 10.9 fragments with a 100% homology between primer and template are expected per 10-nucleotide primer, whereas only 0.029 fragments figure for yeast (1.6×10^4 kb). The results of many investigations, however, suggest that the number of fragments per primer is largely independent of genome complexity. Thus, plants with large genomes such as onion[1543] do not exhibit more complex RAPD fragment patterns than plants with comparatively small genomes, such as *Arabidopsis*.[1151] Similarly, the ploidy level of a plant does not seem to influence the number of RAPD fragments per primer (see Figure 2.5). A positive correlation of ploidy level and band number has nevertheless been observed under the more stringent conditions of AFLP analysis[8,693] (see Chapter 6.3.3.5).

The independence of RAPD fragment number from genome size and ploidy state may be explained by mismatch and primer competition. To allow for mismatch (which is especially desired when species with low genome complexity are analyzed), RAPD-PCR is usually performed at low stringency, i.e., at annealing temperatures of 35 to 45°C (see Chapter 4.4). To investigate the competition phenomenon more closely, Williams et al.[1547] performed RAPD experiments in which DNA samples of two organisms were pooled at different ratios. When DNAs of two individuals from the same species were examined separately, both exhibited a characteristic

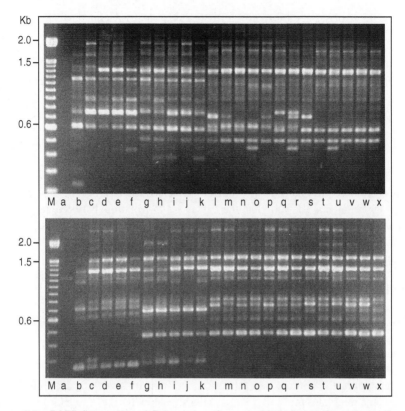

Figure 2.5   RAPD fingerprints of *Pelargonium*. Genomic DNA aliquots from nine cultivars of *Pelargonium peltatum* (lanes c to k), 13 cultivars of *P.* × *hortorum* (lanes l to *x*), and one individual of the wild species *P. peltatum* ssp. *dibrachya* (lane b) were amplified with the arbitrary 10-mer primers OPG-4 (upper panel) or OPG-6 (lower panel; Operon, Alameda). RAPD products were separated on 1.5% agarose gels and stained with ethidium bromide. The two species as well as certain groups of cultivars are easily distinguished from each other. The cultivars analyzed on lanes (b) to (f) and (s) to (x) are diploid, all others are tetraploid. Apparently, the different ploidy levels have no influence on the intensity of banding patterns. A weak ghost band (see Chapter 4.4.2.7) is visible in lane (a) in the lower panel, loaded with a negative control without template DNA. Positions of size markers (lane M) are indicated (Kb, kilobase pairs).

banding pattern. If they were pooled prior to amplification, the bands derived from each individual were amplified in proportion to the input of their genomic DNAs. However, some bands were only poorly amplified, and detected only if the specific DNA was added in excess. In a second experiment, DNA from soybean (having a high complexity genome) was mixed with DNA of a cyanobacterium (having a low complexity genome). All amplified RAPD fragments originated from the soybean genome, even if the cyanobacterial DNA was added in excess. The conclusion from these experiments was that the amplification reaction is determined in part by competition for genomic priming sites. Primers will preferably bind to target sites with a higher degree of homology. These are more likely available in a more complex genome.

The issue of competition was further addressed by Halldén et al.,[557] Hansen et al.,[568] Heun and Helentjaris,[596] Reineke et al.,[1150] and Staub et al.[1326] Heun and Helentjaris[596] observed only few aberrant RAPD fragments in maize F_1 hybrids. In a set of experiments conducted by Halldén et al.,[557] DNA samples of doubled haploid *Brassica napus* lines were mixed in a 1:1 ratio, producing artificial heterozygotes. In 84 of 613 cases, one of the expected parental bands failed to amplify in the mixture. The overall error rate per heterozygous situation was thus 14%. Similar results were obtained under different reaction conditions (14% error rate), in experiments using template DNA of *Bacillus cereus*, which has a much simpler genome (16% error rate), and in experiments using 1:1 mixtures of sugar beet inbred lines (18% error rate).[568] In contrast, only low frequencies (0.2%) of mixture-specific extra bands attributable to heteroduplex molecules were observed. Southern blot hybridization experiments showed that there was no correlation between target DNA sequence copy number and competitive strength of a fragment.[557] This indicates that the amplification success is determined by the actual DNA sequence rather than by copy number. Finally, serial dilution experiments of one genome into another showed that in fact all polymorphic RAPD fragments were subject to competition, which proved to be a quantitative rather than a qualitative phenomenon.

In a series of experiments on aberrantly inherited RAPD fragments from the gypsy moth, *Lymantria dispar*, Reineke et al.[1150] demonstrated that the synthesis of one particular RAPD product was suppressed by the presence of the template for another product in the same reaction. Both products had closely related sequences that were highly repetitive in the *Lymantria* genome. To explain this extreme form of competition, the authors developed a model that involves physical interactions between the templates for both fragments during the PCR. Taken together, the results summarized above show that the genetic background is a strong determinant of whether a particular RAPD fragment is amplified or not. Competition for priming sites appears to be a general feature of RAPD reactions and probably occurs with all primers and in all kinds of organisms.[557]

Caetano-Anollés et al.[204] showed that the eight nucleotides closest to the 3'-end of a primer are crucial for the generation of a particular band. If they were identical, primers with eight, nine, or 10 nucleotides resulted in identical or highly similar banding patterns. In contrast, patterns generated by related primers of five to eight nucleotides were different in complexity and length distribution. Interestingly, decreasing primer length also decreased the number of products, whereas the mean size of the amplified bands was increased. A model proposed to explain these findings illustrates the competitive nature of PCR with arbitrary primers.[204] According to this model, DNA amplification is modulated at two levels. First, primer target sites are selected in a template screening phase. The selectivity at this stage is determined by primer sequences, and influenced by reaction conditions. Bona fide as well as mismatch annealing may occur, resulting in a complex family of primary amplification products. In subsequent rounds of amplification, the newly formed molecules may interact in diverse ways. Given that an amplified ssDNA molecule generated by a single RAPD primer has palindromic ends, it can self-anneal to form a hairpin loop. The model suggests that competition occurs among single-stranded template

DNA, primers, and the terminal palindromic sequences to form double-stranded DNA, a primer-target DNA complex, or an intramolecular hairpin loop in the ssDNA, respectively (see Caetano-Anollés[197] for illustrations). In addition, the model suggests that the different types of molecules tend to reach an equilibrium, and only a subset of potential target sites is amplified to high copy numbers.

The hairpin symmetry may not be confined to the primer sequence itself, but may also extend into internal regions of the fragment, thereby further stabilizing the hairpin loop. For an efficient amplification of a given fragment, hairpin loop formation must be out-competed by the primer–template duplex. It appeared that very short primers, five or six nucleotides long, form less stable hybrid molecules with ssDNA than do longer primers. A higher frequency of hairpin loop formation would thus explain the lower complexity of banding patterns obtained with shorter primers. Conversely, it was found that large hairpin loops (formed by long fragments) are less stable than shorter loops. Consequently, large hairpin loops are probably less effective as competitors, which would explain why the size distribution of amplification products from very short primers is biased toward longer fragments.

The working model presented by Caetano-Anollés et al.[204] was expanded by Rabouam et al.[1124] on the basis of Southern blot hybridization experiments and DNA sequencing of cloned RAPD fragments of a bird and a nematode. The latter authors found that several additional types of artefactual inter- and intrastrand interactions can take place, including nested primer annealing to internal regions of RAPD fragments. For a more detailed description of the early stages of primer–template interactions in arbitrarily-primed PCR, see Caetano-Anollés.[197]

2.3.3.2 Properties of RAPD Markers

Given that RAPD primer sequences are arbitrarily chosen, the genome is expected to be sampled randomly. Most RAPD fragments are inherited as dominant markers, i.e., they are either present or absent. A fragment is seen in the homozygous (AA) as well as in the heterozygous (Aa) situation, and only the absence of the fragment clearly reveals the underlying genotype (aa). Williams et al.[1546] and Fritsch and Rieseberg[463] found that at least 95% of RAPD fragments were dominant markers, whereas the remaining behaved codominantly, i.e., as two alleles with a different size. Echt et al.[378] found no codominant RAPD fragments using 19 different primers. For many applications, the dominant nature of RAPD fragments is a disadvantage (e.g., in population genetics; see Chapters 5.6 and 6.3.2.1 and Lynch and Milligan[863]). Dominant inheritance is not problematic in haploid situations, which are encountered in the megagametophytes of gymnosperms.[644,1423]

The use of RAPD fragments as molecular markers is further complicated by variation in band intensity. The brightness of a given band will depend on several factors, including the degree of repetitiveness of the targeted DNA region, the extent of primer–template mismatch, and the presence or absence of competing target regions in the genome. Variation in the intensity of comigrating bands is one of the annoyances encountered during the conversion of an RAPD banding pattern into a binary character matrix (see Chapter 5.1.2).

There are also reports of RAPD fragments that are not inherited according to Mendelian expectations. In general, three different types of observations were made:

1. Nonparental bands are present in the progeny.[632,1149,1167,1262] Most of these bands probably represent artefactual heteroduplex molecules.[61,995] Such heteroduplex formation can occur when two allelic DNA segments differing by one or more base substitutions, insertions, and/or deletions are amplified during the PCR. Reannealing of two identical alleles results in a homoduplex, whereas reannealing of two different alleles produces a heteroduplex molecule. Alternatively, hetero-duplexes may also result from the interaction of PCR products from different (i.e., paralogous) loci. Because of conformational changes caused by nucleotide divergence between the two alleles, heteroduplexes generally migrate at different rates from homoduplexes in gel electrophoresis.[1539] Heteroduplexes can introduce error in various types of analysis, including testing for paternity, estimating genetic relatedness, and studying pedigree. Their impact is expected to be more pronounced in outcrossers (allogamous plants) than in selfers (autogamous plants), due to a higher degree of heterozygosity in the former.[995]
2. Some parental bands are completely absent from the progeny.[557,596] These observations are probably a consequence of competition for target sequences (see Chapter 2.3.3.1 for a discussion of this phenomenon).
3. Some parental bands are inherited in a strictly uniparental manner.[1,1467] Such bands are thought to originate from organellar DNA rather than from nuclear DNA.

Considering the small size and complexity of organellar genomes, only few RAPD fragments are expected to result from cytoplasmic DNA. Lorenz et al.[843] compared RAPD patterns derived from separately isolated *Beta vulgaris* DNA of the nuclear genome, chloroplast genome, mitochondrial genome, and total genome, respectively. Reproducible RAPD profiles could be obtained from both organellar DNAs using various primers, and the organellar origin of RAPD fragments was confirmed by Southern hybridization. Four of five mtDNA-specific RAPD fragments proved to be unique for either male-fertile or male-sterile sugar beet plants.[844] In experiments on Douglas fir, an unexpected 45% of all RAPD bands scored were inherited in a strict maternal manner, and were thought to be derived from mtDNA.[1] Taken together, these experiments demonstrate that a variable and sometimes large portion of RAPD fragments may be of organellar origin, and therefore exhibit aberrant inheritance patterns. In this context, it is noteworthy that mitochondrial and chloroplast sequences became constituents of the nuclear DNA by horizontal gene transfer[892] (see Chapter 1), and can therefore be detected in nuclear DNA purified by the most stringent criteria.

2.3.3.3 Advantages, Limitations, and Applications of RAPD Markers

The greatest advantage of the RAPD approach is its technical simplicity, paired with the independence of any prior DNA sequence information. Many researchers were enthusiastic about the novel marker technique, and myriad RAPD studies were initiated in the 1990s. Thus, a literature search in 1996 already revealed 3000 references to RAPDs.[1125] Despite a number of drawbacks (see below), RAPDs are still widely used. Main application areas include the identification of cultivars and clones, genetic mapping, marker-assisted selection, population genetics, and molecular systematics at the species level, to name just a few (see Chapter 6).

One obvious disadvantage that RAPDs share with other multilocus markers is their dominant nature, which limits their use for population genetics and mapping studies.[863] RAPDs also turned out to be sensitive to slight changes in reaction conditions, which interfere with the reproducibility of banding patterns between separate experiments, PCR instrumentation, and laboratories[397,962,1064] (see Chapter 4.4.2). This high sensitivity is at least in part a consequence of the nonstringent PCR conditions, which are needed to allow for mismatch priming. More recently, there has been a shift in the relative ratio of published multilocus marker studies, with RAPDs continuously being replaced by the more stringent (but also more complicated) AFLP technology and its modifications (see Chapters 2.3.8 and 4.7).

2.3.3.4 Sequence-Characterized Amplified Regions

In 1993, Michelmore et al[1040] introduced a new type of RAPD-derived molecular marker, which circumvented several of the drawbacks inherent to RAPDs. The new markers were generated by cloning and sequencing RAPD fragments of interest, and designing long (24-mer) oligonucleotide primers complementary to the ends of the original RAPD fragment. When these primers were used in a PCR with the original template DNA, single loci called **sequence characterized amplified regions (SCARs)**[1040] were specifically amplified. These SCARs either retained the dominant segregation behavior of the original RAPD fragment or were converted into codominant markers. Whereas the generation of SCAR markers is somewhat laborious, the SCAR concept exhibits several advantages over RAPD markers, especially for genetic mapping:

1. Stringent PCR conditions can be applied that exclude competition between primer binding sites. This results in reliable and reproducible bands that are less sensitive to reaction conditions.
2. SCAR markers are locus-specific. Codominant inheritance — if present — can therefore easily be identified. Codominant SCARs are more informative for genetic mapping than dominant RAPDs.
3. RAPD fragments often contain interspersed repetitive DNA, and can thus not be used as hybridization probes for identifying a clone of interest in map-based cloning programs. In contrast, SCAR primers can be used to screen pooled genomic libraries by PCR.
4. The reproducible amplification of defined genomic regions allows comparative mapping (as has been done with RFLPs) and synteny studies between related species.

The concept of generating locus-specific SCARs from anonymous PCR fragments is not restricted to RAPDs, but was applied to other multilocus marker techniques such as AFLPs.[1272,1582] Fragments of interest are physically isolated from a multilocus banding pattern, and either reamplified or cloned prior to sequencing.

Methodology and applications of PCR with arbitrary primers have been the subject of numerous reviews, including Bowditch et al.,[156] Caetano-Anollés,[197] Hadrys et al.,[551] Newbury and Ford-Lloyd,[987] Rafalski,[1125] Tingey and del Tufo,[1402] Waugh and Powell,[1509] Williams et al.,[1547] Wolfe and Liston,[1558] and the first edition of this book.

2.3.3.5 Expression Profiling with Arbitrary Primers

In the early 1990s, PCR-based fingerprinting techniques with arbitrary primers were also developed for expression profiling. These strategies, coined **differential display** by Liang and Pardee[817] and **RNA arbitrarily primed PCR (RAP-PCR)** by Welsh et al.,[1530] produce expression profiles by a two-step procedure. In the first step, partial cDNAs are generated by reverse transcription of a subset of the mRNA population investigated. In the second step, these cDNAs are amplified by PCR and separated by gel electrophoresis.

Two different kinds of primers were applied for reverse transcription. Liang and Pardee[817] used an oligo(dT) primer anchored by two selective bases at its 3'-end, e.g., oligo(dT)$_{11}$CA. For statistical reasons, such a primer will bind to 1/16 of all polyadenylated RNA species present in the sample [i.e., all those that have a TG motif just upstream of their poly(A) tail], provided that mismatches between primer and template are prevented by stringent annealing conditions. The cDNA formation then starts from the poly(A) tail. Welsh et al.[1530] used an arbitrary primer of 10 or 18 bases to initiate cDNA strand synthesis under low-stringency conditions. Such a primer has the potential to anneal anywhere within any RNA molecule. The latter strategy can therefore also be applied to produce partial cDNAs from nonpolyadenylated RNAs, e.g., of bacterial origin. The RAP-PCR variant was also considered more reproducible by some authors.[917]

All subsequent steps are similar for both procedures. First, an arbitrary primer is used to start second cDNA strand formation. Primers of about 10 bases proved to be optimal for this purpose; shorter primers revealed much less amplification products than expected from statistical arguments.[817] The same effect was also observed in RAPD and DAF analysis[201,202,1546] (see Chapter 2.3.3). Double-stranded cDNA products are further amplified by standard high-stringency PCR, with radio-labeled nucleotides or primers being included in the amplification reaction. Finally, the PCR products are electrophoresed on a sequencing gel and visualized by auto-radiography. Tissue-specific RNA composition is reflected by the occurrence of characteristic banding patterns.

One limitation of the technique is that rare mRNAs are likely to be underrepresented: the probability of observing a product is dependent on both priming efficiency and abundance of the target RNA. Thus, highly abundant RNA molecules have a much higher chance of being arbitrarily primed during the first PCR cycles. Another disadvantage is the frequent appearance of false positives. Therefore, any candidate products showing differential expression need to be verified by, e.g., traditional Northern blot analysis.[1217]

Within the first decade after its invention, numerous variations of differential display and related methods appeared in the literature (for reviews, see Liang[816] and

McClelland et al.[905]). Despite the recent availability of much more sophisticated methods of expression profiling (see Chapters 2.3.8.6 and 9), differential display still provides an effective and easy-to-perform screening method for differentially expressed genes in different tissues and/or under different environmental conditions. Moreover, PCR products can be cloned from the gel and sequenced, offering a direct approach for the isolation of the underlying genes.

2.3.4 Microsatellites

Microsatellites, also known as simple sequence repeats (SSRs), consist of tandemly reiterated, short DNA sequence motifs. They frequently are size-polymorphic in a population, due to a variable number of tandem repeats (see Chapter 1.2.2). Microsatellites are ubiquitous components of all eukaryotic genomes, and were also found in prokaryotes.[441,546,1438]

Numerous molecular marker strategies have been developed that exploit the variation of microsatellites and their immediate vicinity (for reviews, see Powell et al.[1094] and Weising et al.[1526]). In the most commonly used approach, sequence information of repeat-flanking regions is employed to design locus-specific PCR primer pairs. Amplification products are then separated on denaturing polyacrylamide gels and visualized by autoradiography, fluorometry, or staining with silver or ethidium bromide (see Chapter 4.8 for details). The principle of the assay is shown in Figure 2.6, and typical gel patterns are exemplified in Figure 2.7. Allele size differences of a single base pair can be revealed by this technique. The resulting markers were variously called **simple sequence length polymorphisms (SSLPs)**,[250]

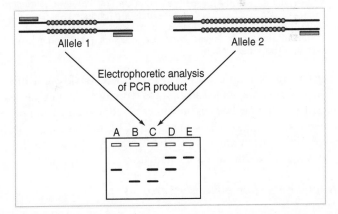

Figure 2.6   PCR amplification of microsatellite DNA. Primer pairs (depicted as shaded bars) are designed to specifically target the 5′- and 3′-flanking region of a microsatellite (symbolized by a row of circles; each circle represents a single repeat unit). PCR products are typically resolved by denaturing polyacrylamide gel electrophoresis that provides single-base pair resolution. Size polymorphisms most commonly result from a variable number of tandem repeats (VNTR), and multiple alleles are usually found in a population or species. Like RFLPs, microsatellite markers are codominant, i.e., both alleles of a diploid organism are detected (lanes [C] and [D]), and homo- and heterozygotes can therefore be distinguished.

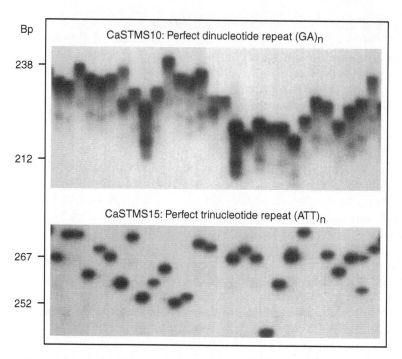

Figure 2.7   Microsatellite alleles revealed at two loci (CaSTMS10 and CaSTMS15) of different accessions of chickpea (*Cicer arietinum*). Radiolabeled PCR products were resolved on sequencing gels and visualized by autoradiography. Single bands are present in the majority of lanes, indicating a low frequency of heterozygotes as expected for a highly inbred species. A typical stutter pattern (see Chapter 4.8.3.2) is observed for the $(GA)_n$ repeat at locus CaSTMS10, which detected 25 alleles among 63 accessions.[633] No stutter bands are produced upon amplification of the $(ATT)_n$ repeat at locus CaSTMS15, which revealed 16 alleles among the same set of accessions.[633] Positions of size markers (M13 sequencing ladder) are indicated (Bp = base pairs).

sequence-tagged microsatellite sites (STMS),[96] **SSR markers**, or **microsatellite markers**.[829] In this book, we will mostly use the term microsatellite and occasionally also refer to SSRs. Other methods use the microsatellite motifs themselves (instead of flanking regions) as single PCR primers (see Chapter 2.3.5.3), as PCR primers in combination with other primer types (see Chapters 2.3.8.3 and 2.3.8.4), or as hybridization probes (see Chapters 2.2.3 and 4.6).

2.3.4.1 Nuclear Microsatellite Markers

The successful application of flanking PCR primers to amplify polymorphic tandem-repetitive DNA regions was actually first demonstrated for minisatellites, which consist of repeat units between 15 and 50 bp.[615,664] In 1989, four groups independently applied the same approach for shorter tandem repeats of the $(CA)_n$-type, i.e., Litt and Luty,[829] Smeets et al.,[1298] Tautz,[1367] and Weber and May.[1513] Litt and Luty[829] also added the term microsatellites to the already polymorphic nomenclature of repetitive DNA. These initial studies already showed that:

1. Single loci are typically amplified, resulting in one or two bands depending on the homo- or heterozygous state in diploid organisms, i.e., microsatellite markers are locus-specific and codominant (Figure 2.6).
2. Many differently sized alleles may exist in a population (Figure 2.7), and the level of heterozygosity can be extremely high.
3. The allelic polymorphism at microsatellite loci is mainly caused by a variable number of repeat units. For example, different alleles of a $(CA)_n$-type microsatellite usually differ by 2, 4, 6, 8, 10, bases etc.
4. PCR-amplified microsatellite markers are inherited in a Mendelian fashion.

It was soon realized that microsatellites are more useful than minisatellites for this kind of analysis, because they are shorter, easier to amplify, more abundant, and more evenly distributed throughout the genome (see Chapter 1.2.2.2). The large number of alleles and high levels of variability among closely related organisms made PCR-amplified microsatellites the marker system of choice for a wide variety of applications (see Chapter 6). To date, more than 1000 research articles have been published that report on the development and/or use of microsatellite markers in plants (see also Chapter 4.8.4.1).

Microsatellites can be subdivided into three classes, comprising (1) perfect repeats, (2) imperfect repeats, and (3) compound repeats[1512] (see Chapter 1.2.2 and Figure 1.3). A direct correlation was often observed between the number of perfect repeats and the level of polymorphism exhibited by PCR amplification.[1210,1302,1512] Although most early studies focused on dinucleotide repeats, other types of microsatellites also proved to be useful. Mononucleotide repeats represent the most frequent type of microsatellites, but often show a strong stuttering effect (see Chapter 4.8.3.2), which may render the detection of single base differences difficult. Microsatellites composed of tri-, tetra- and pentanucleotide motifs[382,715,1264,1311] are much easier to score (less stuttering and greater size difference between alleles) and have a higher chance to be conserved among taxa (see Chapter 4.8.4.3), but are also less abundant than A-, CA- and GA-repeats.

Microsatellites are codominant markers, as are allozymes and RFLPs. However, nonamplifiying alleles (so-called null alleles) are commonly observed.[208,645,676,715,1026,1061,1202] In nonamplifiying alleles, mutations in one or both primer binding sites prevent PCR amplification. Individuals homozygous for a null allele do not show any band at all, whereas heterozygotes have only one band and therefore mimick a homozygote on a gel. Undetected null alleles can give the erroneous impression of an apparent homozygote excess in population studies.[1536] Null alleles can also interfere with the interpretation of inheritance data.[715] The problem may be solved by redesigning primer pairs for the locus, avoiding the mutated primer binding site.[208,645,676,1026] Moreover, multiple microsatellite loci should be examined in population studies to reduce the influence of null alleles.

The popularity of nuclear microsatellites stems from a unique combination of several important advantages, namely their codominant inheritance, high abundance, enormous extent of allelic diversity, and the ease of assessing size variation by PCR with pairs of flanking primers. In addition to some technical problems (stuttering effect, see Chapter 4.8.3.2), the most serious disadvantage is the necessity of sequence information for primer design. The primers used for the first generation of SSR markers were generally deduced from flanking sequences of known micro-satellites found in DNA databases. In model organisms such as man, mouse, and *Arabidopsis*, and for economically important species such as cow, sheep, rice, maize, and tomato, computer-assisted cloning still provides a valuable source for marker generation (see Chapter 4.8.4.2). For the majority of species, however, database entries are still limited or even nonexistent. In some instances, heterologous primers (derived from microsatellite markers in related taxa) may be available and informa-tive (see Chapter 4.8.4.3). In most cases, molecular cloning will be required to retrieve the sequence data needed for the design of microsatellite-flanking primers (see Chapter 4.8.5).

The methodology and applications of nuclear microsatellite markers in plants and other organisms have been the subject of numerous reviews, including Cregan and Quigley,[284] Goldstein and Schlötterer,[507] Gupta and Varshney,[543] Holton,[608] Jarne and Lagoda,[659] Nybom,[1000] Powell et al.,[1094] and Weising et al.[1526]

2.3.4.2 Chloroplast Microsatellite Markers

Microsatellites are also regular constituents of organellar genomes, but occur at much lower frequencies than in the nucleus (see Chapter 1.2.2.4). Powell et al.[1092,1093] first recognized the potential of size-variable mononucleotide repeats to uncover intraspecific variation within the otherwise conserved chloroplast genome. They demonstrated that PCR amplification of these so-called **chloroplast microsatellites** by flanking primer pairs reveals intra- and interspecific length variation, reminiscent of nuclear microsatellites.

Early studies on chloroplast microsatellites (in the following text, also referred to as **chloroplast simple sequence repeats [cpSSRs]**) were mainly performed on conifers, in which sets of five[218] and 20 primer pairs[1462] were constructed that flanked mononucleotide repeats in the fully sequenced *Pinus thunbergii* cpDNA. These primers amplified polymorphic fragments from various species of the pine genus (e.g., *Pinus halepensis*,[181] *Pinus resinosa*[379]), other genera of the pine family (e.g., *Abies*,[1461,1464] *Pseudotsuga*[1467]), and other gymnosperms.[218]

Chloroplast SSR markers share a number of properties that distinguish them from nuclear microsatellites. First, the vast majority of cpSSRs are relatively short (10 to 20 bp) tracts of poly(A/T), whereas tracts of poly(C/G) or dinucleotide repeats are rare (see Vogel et al.[1476] for an exception). Second, initial studies in *Pinus torreyana* suggested lower mutation rates of cpSSRs compared with nuclear micro-satellites.[1111] However, the large number of allele sizes observed at some cpSSR loci (e.g., up to 18 size variants in *Abies alba*[1464]) indicates that microsatellite mutation rates can vary considerably also in the chloroplast genomes, and additional studies in this direction will have to be performed. Third, the lack of recombination in the chloroplast genome makes cpSSR markers not only individually informative, but they also can be combined to form specific cpDNA haplotypes. Genetic diversity measures and phylogeographic studies can then be based on haplotype frequencies and distributions (see Chapter 6.5.1).

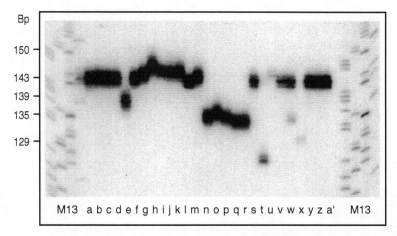

Figure 2.8   Chloroplast microsatellite alleles detected with the ccmp1 primer pair targeted at a (T)$_n$ repeat in the *trn*K intron.[1520] Radiolabeled PCR products from a set of angiosperm species were resolved on sequencing gels and visualized by auto-radiography. Lanes (a) to (m): different genera and species of Solanaceae; lanes (n) to (r): different species of the genus *Actinidia* (Actinidiaceae); lanes (s) to (u): different genera and species of Brassicaceae; lanes (v) to (a): different genera and species from miscellaneous plant families (for details, see Weising and Gardner[1520]). Single fragments are generally accompanied by a set of stutter bands. Positions of size markers (M13 sequencing ladder) are indicated (Bp, base pairs).

cpSSR markers served various purposes so far, including the assessment of paternal vs. maternal plastid inheritance,[218] the detection of hybridization and introgression,[181,422] and especially the analysis of the genetic structure and phylogeography of plant populations[527,528,1032,1041,1476] (see Chapter 6.5.1). If used in conjunction with nuclear markers, cpSSRs can also provide valuable information on the relative importance of gene flow via seeds and pollen, respectively.[524] cpSSR marker technology and applications have been reviewed by Provan et al.[1112,1114] A typical gel pattern showing cpSSR variation across a range of angiosperms is exemplified in Figure 2.8.

2.3.5   Inter-Repeat PCR

2.3.5.1   *From Alu Repeats to Zinc Fingers: Repetitive DNA as a Primer Target*

The family of strategies treated in this section exploits primers complementary to known repetitive DNA sequence elements. To obtain a set of PCR products, closely spaced, inversely oriented copies of the target motifs need to be present at reasonable frequencies. Codominant polymorphisms are expected to result from insertion–deletion events in the inter-repeat region, whereas dominant polymorphisms will be caused by sequence alterations within the primer binding sites. Methods have been proposed that use primers specific for the following DNA sequences:

- Minisatellites[588]
- Microsatellites[542,922,1621]
- (Retro)transposons and other interspersed repeats, e.g., short interspersed element (SINE)-PCR and *Alu*-PCR,[88,985] retrotransposon-based insertion polymorphisms (RBIP),[453] inter-retrotransposon amplified polymorphism (IRAP),[681] and inter-MITE PCR[227]
- Intron–exon splice junctions[1516]
- 5S RNA genes[734]
- tRNA genes[1528]
- Families of zinc finger protein genes[1432]
- Families of plant pathogen resistance genes[240]

There are also methods that use various combinations of repeat-specific primers (e.g., microsatellites + retrotransposon LTRs in retrotransposon-microsatellite amplified polymorphism (REMAP)[681] and *copia*-SSR;[1113] see Chapter 2.3.5.4). This list is not necessarily comprehensive, especially because many additional combinations are possible. In the present survey, we focus on techniques using primers specific for minisatellites (2.3.5.2), microsatellites (2.3.5.3), and transposable elements (2.3.5.4). Because of their ubiquity and interspersed genomic organization, these three repeat types have been most frequently exploited as genetic markers.

2.3.5.2 Primers Directed toward Minisatellites

The high abundance of minisatellites in eukaryotic genomes allows the use of minisatellite-complementary oligonucleotides as PCR primers to generate numerous polymorphic amplification products. Early studies in this direction were performed with fungi. Thus, Meyer et al.[922] showed that the M13 minisatellite repeat unit (see Figure 1.2 in Chapter 1) distinguished and identified different isolates of the human fungal pathogen *Cryptococcus neoformans*, when used as a PCR primer. Heath et al.[588] employed various minisatellite core sequences as primers (including M13), to study fish, bird, and human genomes. After electrophoretic separation of the PCR products, differences between species were found, but intraspecific variation was not observed.

Gustafson and colleagues were the first to apply minisatellite-primed PCR to plants.[93,1309,1617] In an approach coined **direct amplification of minisatellite DNA (DAMD)-PCR**, they amplified genomic DNA of various, mainly diploid *Triticum* species under high-stringency conditions, using known plant and animal minisatellite core sequences as single primers.[1309] When DAMD-PCR products were separated on 2% agarose and stained with ethidium bromide, a unique, moderately complex, polymorphic DAMD profile for each primer was generated, that either constituted a discrete, RAPD-like banding pattern, or a continuous smear. DAMD-PCR products were cloned and used as conventional RFLP probes against Southern blots with genomic DNA. The majority of clones showed some degree of genome specificity, i.e., gave a strong signal with certain species, but not with others.[1309] Other probes hybridized to single or moderately dispersed target sequences in all *Triticum* species, producing polymorphic single- or multilocus RFLP fingerprints, respectively. The

DAMD-PCR technique was later extended to tetra- and hexaploid *Triticum* species and cultivars[93] and to species and cultivars of the genus *Oryza*[1617] with similar results.

In all the above-described studies, sequence analysis showed that tandem repeats were absent from DAMD-PCR products, suggesting that only minisatellite-intervening sequences are amplified. Consistent with these observations, the extent of polymorphism of the DAMD products themselves is mostly limited.[1617] However, if the same products are used as RFLP probes, flanking minisatellites are codetected, resulting in highly variable fingerprints.[588]

2.3.5.3 Primers Directed toward Microsatellites

The successful application of microsatellite-specific oligonucleotides as PCR primers was also first described by Meyer et al.,[922] who amplified DNA from different strains of the human fungal pathogen *C. neoformans* with the primers $(CA)_8$, $(CT)_8$, $(CAC)_5$, $(GTG)_5$, $(GACA)_4$, and $(GATA)_4$. The PCR products were separated on agarose gels and stained with ethidium bromide. Each fungal strain exhibited a specific banding pattern, and serotypes could easily be distinguished. The technique was subsequently applied to numerous other organisms, and several acronyms were proposed, including **single primer amplification reactions (SPAR)**,[542] **inter-simple sequence repeat PCR (ISSR-PCR)**,[1621] and **microsatellite-primed PCR (MP-PCR)**.[1524]

The principle of MP-PCR is shown in Figure 2.9A, and typical banding patterns obtained by agarose gel electrophoresis are exemplified in Figure 2.10. Amplification of inter-repeat sequences will take place if inversely repeated microsatellites are present within an amplifiable distance from each other. Whereas initial priming may occur in different registers within the microsatellite target region, the average product size is continuously reduced by internal priming in successive cycles, and the final product is expected to be primed from the extreme 3'-end of each flanking microsatellite.

Gupta et al.[542] used 23 primers complementary to di-, tri-, tetra-, and pentanucleotide repeats to amplify genomic DNA across a panel of eukaryotes. They found that tetranucleotide repeat primers were most efficient in amplifying polymorphic patterns. GC- as well AT-rich primers worked equally well. Primers representing a combination of two tetranucleotide repeats, or compound microsatellites, were also effective. Single base permutations produced different PCR fingerprints. Banding patterns of higher complexity were observed when radiolabeled PCR products were separated on denaturing polyacrylamide gels and detected by autoradiography. Bands mapped as dominant markers in a segregating maize population.

These results were in part confirmed by Weising et al.,[1524] who used a variety of di-, tri-, and tetranucleotide repeats as PCR primers for the analysis of plant species. Distinct and polymorphic banding patterns were only obtained with tri- and tetranucleotide repeat-specific primers containing a minimum of 25% GC (Figure 2.10). Meyer et al.[922] stressed that MP-PCR combined some advantages of RAPD analysis (i.e., no need for sequence information) and microsatellite analysis (i.e., use of high-stringency annealing conditions, leading to more reproducible banding patterns). The validity of this statement was challenged by the results obtained by Weising et al.,[1524] who found no significant advantage of MP-PCR compared with RAPD analysis in terms of reproducibility and sensitivity to reaction

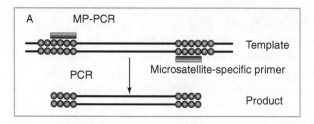

A   MP-PCR

Template

PCR

Microsatellite-specific primer

Product

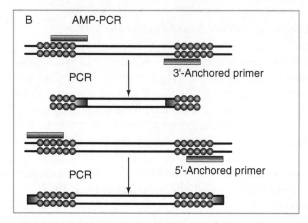

B   AMP-PCR

PCR

3'-Anchored primer

PCR

5'-Anchored primer

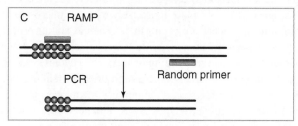

C   RAMP

PCR

Random primer

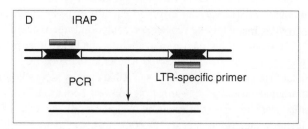

D   IRAP

PCR

LTR-specific primer

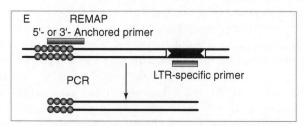

E   REMAP

5'- or 3'- Anchored primer

PCR

LTR-specific primer

conditions (see Chapter 4.5.3). If carefully optimized, both RAPD and MP-PCR are nonetheless expected to yield reliable and reproducible results within the same laboratory.[1097,1139]

In the more sophisticated ISSR variant developed by Zietkiewicz et al.[1621] (also coined **anchored microsatellite-primed PCR [AMP-PCR]**), 5'- or 3'-anchored di- or trinucleotide repeats serve as single PCR primers (Figure 2.9B), and the amplification products are separated on polyacrylamide gels. The anchor is composed of nonrepeat bases and ensures that the amplification is initiated at the same nucleotide position in each cycle. Fingerprints obtained with this technology revealed inter- and intraspecific polymorphisms in a wide variety of eukaryotic taxa.[743,1276,1495,1563,1621]

AMP-PCR has several advantages over unanchored variants of microsatellite-primed PCR. First, primer design ensures annealing of the primer only to the ends of a microsatellite, thus circumventing internal priming and smear formation. Second, the anchor allows only a subset of the targeted inter-repeat regions to be amplified, thereby reducing the high number of PCR products expected from dinucleotide inter-repeat regions to a set of about 10 to 50 easily resolvable bands. Third, functional 5'-anchors ensure that the targeted microsatellite is part of the product (but see Chapter 4.5.3 and Fisher et al.[446]). Potential variable number of tandem repeat (VNTR) polymorphisms within the microsatellite will then contribute to the inter-repeat variation, which could considerably increase the chance of observing a polymorphism.

In a technique designated as **random amplified microsatellite polymorphism (RAMP)**, Wu et al.[1574] combined 5'-anchored mono-, di-, or trinucleotide repeat-specific primers (e.g., $CCGGT_{10}$) and arbitrary 10-mer primers to obtain codominant microsatellite polymorphisms without cloning. The principle of RAMP is shown in Figure 2.9C. To compensate for the different annealing temperatures of the two types of primers and to ensure that microsatellite loci are preferentially amplified, a PCR program is used that switches between high and low annealing temperatures (thermally asymmetric PCR profile[835]), and PCR products are separated on denaturing polyacrylamide gels. Because only the microsatellite primer is end-labeled with ^{33}P, fragments flanked by two RAPD primers remain undetected and only microsatellite-derived bands show up on the autoradiograms. Amplification of different *Arabidopsis* strains and ecotypes resulted in fragment patterns of moderate complexity useful as

---

Figure 2.9 (opposite page) A schematic survey of commonly used DNA profiling techniques based on inter-repeat PCR. (A) **Microsatellite-primed PCR (MP-PCR)**[542,922,1524]: Unanchored tri- or tetranucleotide repeats such as $(GATA)_4$ serve as single PCR primers. (B) **Anchored microsatellite-primed PCR (AMP-PCR)**, also called **inter-simple sequence repeat (ISSR) PCR**[1621]: 5'- or 3'-anchored di- or trinucleotide repeats such as $CG(CT)_6$ or $(CAA)_4CG$ serve as single PCR primers. (C) **Random amplified microsatellite polymorphism (RAMP)**[1574]: 5'-anchored mono-, di-, or trinucleotide repeat-specific primers such as $CCGGT_{10}$) are combined with arbitrary 10-mer primers. (D) **Inter-retrotransposon amplified polymorphism (IRAP)**[681]: Oligonucleotides complementary to long terminal repeats (LTRs) of retrotransposons are used as single PCR primers. (E) **Retrotransposon-microsatellite amplified polymorphism (REMAP)**[681]: An LTR-specific primer is combined with a 5'- or 3'-anchored microsatellite primer. The same strategy is used in ***copia*-SSR**[1113] Microsatellite arrays are symbolized by rows of circles. Retrotransposon LTRs are painted black.

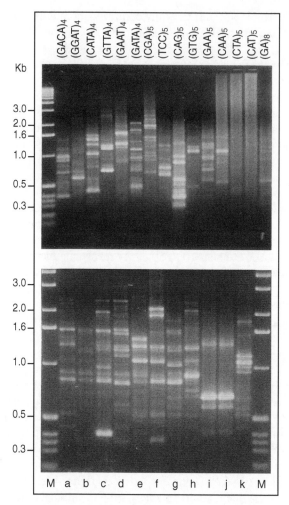

Figure 2.10   Examples of microsatellite-primed PCR (MP-PCR) profiles. Upper panel: Banding patterns obtained by PCR amplification of genomic tomato DNA with different unanchored microsatellite primers. Lower panel: Intra- and interspecific polymorphism within the genus *Actinidia* (kiwifruit) revealed by (GTTA)$_4$ as single PCR primer. Lanes (a) to (d): *Actinidia chinensis*; lanes (e) to (g): *A. deliciosa*; lane (h): *A. setosa*; lanes (i) to (j): *A. chrysantha*; lane (k): *A. arguta*. MP-PCR products were separated on 1.4% agarose gels and stained with ethidium bromide. Positions of size markers (lane M) are indicated (Kb, kilobase pairs).

molecular markers. About two thirds of these fragments appeared to be codominant, and mapped at apparently random loci in the *Arabidopsis* genome.[1574]

Another variant of the same principle, coined **double-stringency PCR (DS PCR)**, was proposed by Matioli and de Brito.[894] *Drosophila* DNA was amplified by a single unanchored (GACA)$_4$ primer in combination with a four times more concentrated 10-mer RAPD primer. The initial 15 PCR cycles were performed at high stringency (e.g., 53°C), which allows only the microsatellite primer to anneal and amplify. During this stage, a population of DNA molecules was generated that is enriched for inter-microsatellite repeats. The annealing temperature was then dropped to low stringency (e.g., 35°C), and PCR was continued for another 25 cycles. During this stage, both primers are expected to anneal, and RAPD primer binding sites internal to the inter-microsatellite repeat will result in the production of subfragments. Banding patterns obtained by the use of microsatellite and RAPD primers in separate assays were very different from those obtained by the combination of both. The fragments were inherited in a Mendelian fashion.[894]

Theoretically, the combination of different 5'-anchored microsatellite primers with different RAPD primers allows for the generation of an almost unlimited number of unique markers, at least a subset of which is expected to be inherited codominantly. Moreover, no cloning and sequencing is required. Despite these attractive features, RAMP and DS PCR were only used in a few plant taxa so far (i.e., *Arabidopsis*,[1574] barley,[95,300-302,1218] and soybean[1495]). In soybean, the inclusion of 10-mer or 14-mer RAPD primers failed to alter the RAMP banding patterns obtained with the anchored microsatellite primers alone, although the same PCR conditions were applied as in *Arabidopsis*.[1495] Surprisingly, novel products were only observed in combination with longer (18 to 20 bp) arbitrary primers.

Several RAMP studies were performed in barley. Becker and Heun[95] combined 5'-anchored, labeled $(GA)_n$ primers with 10-mer, 16-mer, and 20-mer RAPD primers. These longer-than-usual RAPD primers were chosen to ensure comparable annealing temperatures for both primers, thus circumventing the necessity for the thermally asymmetric PCR profile of Wu et al.[1574] To obtain additional polymorphisms, aliquots of the amplification products were digested with the restriction enzyme *Mse*I, resulting in so-called **dRAMPs**. A total of 10 primer combinations resulted in 43 RAMPs and 17 dRAMPs, which identified 40 new loci on a barley RFLP map. Mapping demonstrated that some of the dRAMPs were derived from RAMPs, and that only seven loci defined by dRAMPs were actually unique. This showed that the digestion of RAMP products was of no considerable advantage.[95] Sánchez de la Hoz et al.[1218] used silver-staining to visualize the electrophoresed PCR products from 14 barley cultivars. Bands derived from single RAPD primers, single microsatellite primers, and combinations of both were evaluated separately. Interestingly, phenograms based solely on RAMP markers reflected the known pedigrees of cultivars more faithfully than dendrograms based on RAPDs. Conversely, RAMP-based genetic similarity values were only poorly correlated with coefficients of parentage calculated for 29 spring and 20 winter barley lines.[300] Ten of 35 RAMP markers mapped in barley were scored as codominant.[301]

In summary, genetic marker systems employing anchored or unanchored microsatellite-specific primers, either singly or in combination, became well established and were used for various applications. These include the identification of cultivars,[44,228,1097,1130,1563] genetic mapping,[1222] the assessment of genetic diversity,[689,1215] biogeographical studies,[910] detection of somaclonal variation,[799-801] and molecular systematics.[643,955,1130,1132] For *Pinus radiata*[446] and *Brassica oleracea*,[151] MP-PCR fragments were shown to be enriched for internal repeats, suggesting the presence of microsatellite clusters in the genome. In species containing such clusters, MP-PCR products may serve as a source for the generation of codominant, locus-specific microsatellite markers[446] (see Chapter 4.8.6.3). Methodology and applications of the various microsatellite-primed PCR techniques have been reviewed by Reddy et al.[1147] and Vogel and Scolnik.[1475]

2.3.5.4 Primers Directed toward Interspersed Repeats

Interspersed repetitive elements are fundamental components of eukaryotic genomes. The most important representatives of interspersed repeats are transposons that often contain a set of genes encoding enzymes required for transposition (see Chapter 1.3). Whereas DNA transposons move via a DNA intermediate in a cut-and-paste mechanism, retrotransposons move via RNA, and frequently attain a high copy number through a copy-and-paste mechanism.

One class of retrotransposons is bounded by so-called LTRs that act as transcriptional enhancers and also play a role in the insertion process (see Chapter 1.3.1). LTRs are relatively conserved and offer themselves as primer targets. If two or more retrotransposons reside in close vicinity to each other, outward-facing, LTR-specific primers (used either alone or as a pair) should be able to amplify the intervening DNA sequences. This principle is employed for many PCR strategies relying on interspersed repeats (see below). Kalendar et al.[681] first demonstrated the feasibility of this approach for barley (*Hordeum vulgare*) and related species, using primers directed against LTRs of *BARE*-1, a retrotransposon belonging to the widespread Ty1-*copia* family. The acronym **inter-retrotransposon amplified polymorphism** (**IRAP**) circumscribes the technique (see Figure 2.9D).

Retrotransposons are sometimes accompanied by microsatellites, located at the 5′- or 3′-end, or even within the transposon.[13,1135,1379] The frequent colocalization of both elements led to speculations that microsatellites could provide integration points for homology-driven insertion of retrotransposons into genomic DNA[972,1135] (see Chapter 1.2.2.2). The **REMAP** technique, also introduced by Kalendar et al.,[681] exploits the often close association of microsatellites and retrotransposons by combining outward-facing LTR-specific primers with 3′- or 5′-anchored di- or trinucleotide microsatellite primers (Figure 2.9E). Amplification and separation of products in 2% agarose gels followed by ethidium bromide staining revealed 15 to 30 bands, depending on the species. Banding patterns were completely different if the same anchored microsatellite primer was used alone, indicating that the majority of REMAP bands were derived from sequences bordered by a microsatellite on one side, and by an LTR on the other. Interestingly, the REMAP pattern was considerably more variable than the corresponding ISSR pattern. As was the case with IRAP, the extent of polymorphism was too high to permit the use of REMAP for interspecies comparisons. Below the species level, however, bands generated by REMAP were able to distinguish between closely related cultivars.[681] In a subsequent study, Kalendar et al.[682] were able to demonstrate by REMAP that *BARE*-1 insertion patterns in *Hordeum spontaneum* plants varied on a microgeographical scale and in a way consonant with the ecogeographical distribution of the plants.

Basically the same strategy, coined *copia*-**SSR**, was simultaneously developed by Provan et al.[1113] and also tested on barley. In this study, combination of a radiolabeled *BARE*-1 LTR-specific primer with an unlabeled, anchored microsatellite primer

[BDB(CA)$_7$] revealed multiple polymorphic products on a sequencing gel. As was also observed by Kalendar et al.,[681] more products (especially in the low molecular weight range) and more polymorphisms were obtained by combining both types of primers. However, hardly any bands were amplified by the *BARE*-1 primer alone, as opposed to the IRAP results presented by Kalendar et al.[681] Finally, *BARE*-1 specific primers have also been used in conjunction with AFLP primers in the **sequence-specific amplification polymorphism (S-SAP)** technique[1510] (see Chapter 2.3.8.2).

SINEs represent a nonautonomous class of retroelements (see Chapter 1.3.1). The best known SINEs are the *Alu* repeats, which make up 5 to 10% of primate genomes. *Alu*-directed PCR primers have been used for the generation of human DNA fingerprints in a technique called **Alu-PCR** as early as 1989.[985] Interestingly, primers designed from human *Alu* sequences were reported to generate RAPD-like amplification profiles in various plants, such as banana[88] and sugarcane.[21] Considering that *Alu* repeats are assumed to be fast-evolving and specific for primate genomes, these findings are somewhat surprising. They could indicate the presence of *Alu*-like sequences in plants as assumed by the authors, but could also be explained by unspecific primer binding to relatively unrelated repeats. In our experience, primers of any sequence and length will be able to generate an RAPD-like banding pattern from genomic DNA, provided that (1) PCR conditions are relatively relaxed (as was the case in the touch-up PCR used in the sugarcane study),[21] and (2) there is no specific primer target sequence in the template. Whatever its molecular basis may be, the method has nevertheless been used to clone genus-specific DNA sequences from sugarcane and two related genera. These sequences were then successfully applied to the characterization of interspecific hybrids.[21]

Chang et al.[227] developed a DNA profiling strategy that was based on **inter-MITE polymorphisms (IMPs)**. MITEs constitute a superfamily of plant transposable elements that are characterized by small size (usually <500 bp), ubiquitous occurrence in plants, moderate to high copy numbers, and a preference for the insertion near or within genes (see Chapter 1.3.3). Outwardly facing, degenerated IMP primers were designed from the terminal inverted repeats of the barley MITEs *Stowaway*[186] and *Barfly*,[227] and the utility of these primers was demonstrated by fingerprinting genomic DNA from 26 barley cultivars and a double-haploid mapping population.[186] Separation of fluorescent PCR products on an automated LI-COR sequencer revealed highly complex, polymorphic patterns of up to 120 bands. Eighty-eight loci were mapped, with no significant clustering. A dendrogram based on presence vs. absence data of 89 bands correlated well with a previous RFLP analysis. Finally, the barley MITE-based primers were effectively used in wheat, oat, and maize, demonstrating the transferability of the method within plant families.[186]

Taken together, the transposon-based PCR methods described in this section generate multilocus banding patterns that are often highly variable. In general, the individual bands behave as dominant markers. Although all of these techniques are principally applicable to any plant species, none of them has become as popular as RAPD, microsatellite-primed PCR, and AFLP technology. The main disadvantage of any transposon-based DNA profiling technique (also including S-SAP and related techniques; see Chapter 2.3.8.2) is the need of sequence information for designing element-specific primers.[1056]

2.3.6   DNA Profiling of Genic Regions: RGAP, SRAP, and TRAP

The rapidly growing amount of plant gene and expressed sequence tag (EST) sequence information in databases opened the door to the development of DNA profiling strategies that are directed at specific coding DNA regions of interest. These could be particular genes or groups of genes with known function,[240,341] or protein-coding regions in general.[621,833]

One class of gene-specific genome scanning methods targets plant resistance genes that encode proteins counteracting pathogen attacks. In recent years, several large classes of pathogen resistance genes have been identified in plants, comprising hundreds or even thousands of individual genes and pseudogenes within a single plant genome (reviewed in Ellis et al.[393] and Young[1599]). Whereas overall sequence similarity among these genes proved to be low, several conserved domains have been identified that are presumably involved in the defense response. These include a so-called leucine-rich repeat (LRR) region important for protein-protein interactions, a protein kinase (PK) domain involved in signal transduction, and a nucleotide binding site (NBS) domain, among others. Degenerate PCR primers directed toward nucleotide sequences encoding these conserved domains allowed the cloning of candidate resistance genes in heterologous species.[685,794,1604]

Chen et al.[240] used the same type of primers to amplify resistance gene analogs from rice, wheat, and barley. Fragments were separated in denaturing polyacrylamide gels. Silver staining revealed complex DNA profiles that consisted of 30 to 130 bands. The level of intraspecific polymorphism was high, with, for example, 188 polymorphic bands generated from eight primer pairs in 12 near-isogenic lines of rice. This novel genome scanning technique became known as **resistance gene analog polymorphism (RGAP)** analysis, and proved to be a useful tool for detecting genetic variation associated with disease resistance.[240] Degenerate primer pairs have since been used to develop molecular markers specific for candidate resistance genes in many organisms, including wheat and barley,[240] grapevine,[336] chickpea,[634] (see also Chapter 7.1.4) and cocoa.[757] Most investigations aimed at marker-assisted selection and/or map-based cloning of resistance genes, but RGAP analysis was also applied to population genetics in pine (*Pinus oocarpa* in Nicaragua[341]). In this study, the pattern of genetic diversity detected with RGAP proved to be similar to that obtained with RAPD and AFLP.

Liu and Quiros[833] described a marker technique that specifically amplifies polymorphic junction fragments between exons and flanking DNA. This method was called **sequence-related amplified polymorphism (SRAP)**. Two primers are used, each of which consists of the following three elements (from 5′ to 3′; see Figure 2.11 for examples):

1. An arbitrary filler sequence of 10 to 11 bases at the 5′-end (filler sequences of forward and reverse primers need to be different from each other)
2. The sequence motifs CCGG and AATT in the forward and reverse primer, respectively
3. Three selective bases at the 3′-end

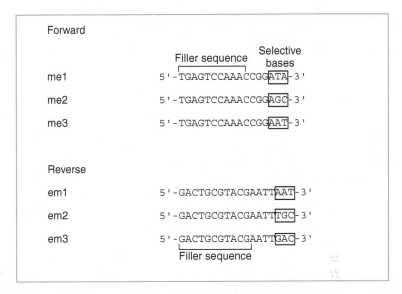

Figure 2.11 Examples of forward and reverse primers used in the sequence-related amplified polymorphism (SRAP) approach.[833] See text for details and explanation of primer elements.

Variation in the selective nucleotides generates a set of primers sharing the same core sequence (i.e., filler1+CCGG and filler2+AATT, respectively).[833]

The rationale behind the primer architecture is the following. Exon sequences are known to be more GC-rich than other regions of the genome, and a random screen of 20 *Arabidopsis* BAC clones from the GenBank database supported this view.[833] Hence, the CCGG core sequence of the forward primer is used to specifically target exonic gene sequences. In contrast, the core sequence of the second primer (AATT) is designed to bind to AT-rich sequences, which are preferentially found in noncoding regions. A PCR using both types of primers in combination is assumed to amplify preferentially junction fragments between genes (i.e., the targets of interest) and flanking noncoding DNA (providing the level of polymorphism needed for an efficient marker system).

Using radiolabeled primer pairs, Liu and Quiros[833] amplified genomic DNA from *Brassica oleracea* with an annealing temperature of 35°C for five cycles, followed by 50°C annealing temperature for another 35 cycles. Low initial PCR stringency allowed for some mismatch binding (as in the AP-PCR technique described by Welsh and McClelland[1527]; see Chapter 2.3.3). In later cycles, high stringency was applied to ensure reproducible amplification of the initial products to high copy numbers. Moderately complex banding patterns resulted after separation on denaturing poly-acrylamide gels and autoradiography. Primer size (17 to 18 bases), GC content (40 to 50%), and the use of a two-primer rather than one-primer system proved to be essential to successfully amplify SRAP bands.[833]

Testing the marker system with recombinant inbred and doubled haploid lines of *B. oleracea* showed that (1) a large proportion of the fragments corresponded to exons in open reading frames, (2) 20% of the fragments were codominantly inherited, and (3) SRAP and AFLP markers were evenly distributed across a genetic map, with no major differences between the two marker techniques.[833] The SRAP technique also worked with other crops, including potato, celery, and lettuce. The preferential amplification of coding regions by SRAP primer pairs was further supported by a more recent study on *Cucurbita pepo*, in which six of six sequenced SRAP fragments showed significant homology to either genes or ESTs from the databases.[440]

A related approach, called **target region amplification polymorphism (TRAP)**, was suggested by Hu and Vick.[621] Common features of SRAP and TRAP include the use of two primers of about 18 nucleotides (one of which targets at a coding region), and nonstringent PCR conditions during the first five cycles. The two techniques mainly differ in the architecture of one of the two primers. In TRAP, the so-called fixed primer is complementary to a particular EST from the database. The second primer was adapted from Liu and Quiros[833] and contains an arbitrary AT- or GC-rich core sequence to anneal with either an intron or exon sequence, respectively. Separation of PCR products obtained from sunflower genomic DNA on denaturing polyacrylamide gels in a LI-COR DNA sequencer resulted in 30 to 50 bands of 50 to 900 bp, depending on the primer pair. Multiplex PCRs were performed, combining one unlabeled fixed primer with two arbitrary primers, each of which was labeled with a different fluorochrome. In this way, two independent banding patterns were produced in the same run.[621]

Taken together, the methods described in this section exhibit a number of advantages over other DNA marker techniques, including speed, simplicity, and their preference for potentially useful genes. The TRAP variant owns the additional advantage of exploiting the huge amounts of EST sequence information in databases. However, the low PCR stringency during the first five cycles may have a negative impact on reproducibility, as is the case for RAPDs and related techniques. Liu and Quiros[833] demonstrated a high reproducibility of SRAP in *B. oleracea*. It remains to be seen whether this also holds for other crops, and for comparisons across laboratories.

2.3.7   Hybridization of Microsatellites to RAPD and MP-PCR Products

This strategy combines arbitrarily or microsatellite-primed PCR with microsatellite hybridization.[255,401,1139,1140,1163] Genomic DNA is first amplified with either a single arbitrary 10-mer primer (as employed in RAPD analysis), or a microsatellite-complementary 15- or 16-mer primer. PCR products are then electrophoresed, blotted, and hybridized to a radiolabeled mono-, di-, tri-, or tetranucleotide repeat probe such as $(GA)_8$ or $(CAA)_5$. Subsequent autoradiography reveals reproducible, probe-dependent fingerprints that are completely different from the ethidium bromide staining patterns, and which are polymorphic at an intraspecific level. This method was called **random amplified microsatellite polymorphisms (RAMPOs)**; please note the difference to RAMP, see Chapter 2.3.5.3,[1163] **random amplified hybridization microsatellites (RAHMs)**,[255] or **randomly amplified microsatellites (RAMS)**.[401] Results from a typical RAMPO experiment are shown in Figure 2.12.

The occurrence of RAMPO bands may be explained as follows. Any RAPD reaction or MP-PCR probably creates many thousand different products of various abundance. The majority of less frequently occurring fragments will remain below

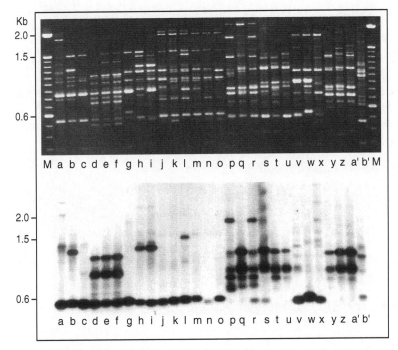

Figure 2.12   Random amplified microsatellite polymorphism (RAMPO)[1163] profiles of *Dioscorea* (yam) species. Genomic DNA aliquots from three accessions each of a set of different *Dioscorea* species were amplified with the arbitrary 10-mer primer OPG-13 (Operon, Alameda). The resulting RAPD products were separated on a 1.4% agarose gel and stained with ethidium bromide (upper panel). The gel was blotted onto a nylon membrane, hybridized with a ^{32}P-labeled (GTGA)$_4$ probe and autoradiographed (lower panel). The pattern of (GTGA)$_4$-detected fragments is less complex than the underlying RAPD pattern and phenetic relationships between samples were easy to evaluate (see text for details of the procedure). Lanes (a) to (c): *Dioscorea rotundata*; lanes (d) to (f): *Dioscorea cayenensis*; lanes (g) to (i): *Dioscorea abyssinica*; lanes (j) to (l): *Dioscorea praehensilis*; lanes (m) to (o): *Dioscorea liebrechtsiana*, lanes (p) to (r): *Dioscorea smilacifolia*, lanes (s) to (u): *Dioscorea minutiflora*; lanes (v) to (x): *Dioscorea togoensis*; lanes (y) to (a′): *Dioscorea burkilliana*; lane (b′): *Dioscorea bulbifera*. Positions of size markers (lane M) are indicated (Kb, kilobase pairs).

the detection level of ethidium bromide staining. However, the ubiquitous presence of microsatellites in eukaryotic genomes provides a means of visualizing a subset of such minor amplification products by hybridization. The signal intensity of fragments harboring a certain microsatellite motif will depend both on the length of this motif and the abundance of the fragment.

For genetic relatedness studies, RAMPO bands appear to be less sensitive to misinterpretation than RAPD bands, because not only the size but also the hybridization signal intensity of two bands (i.e., the presence and copy number of a certain microsatellite) are criteria for homology. In other words, the RAMPO technique contains a built-in hybridization control (see also Chapters 4.6.2 and 5.1.2). Another advantage of the RAMPO technique is the low complexity of banding patterns, facilitating detection of species-specific bands considerably.[1139,1140]

The RAHM,[255] RAMPO,[1163] and RAMS approach[401] were extensively tested with different plant species, but also *Daphnia* (Crustacea).[401] As RAMPO fragments are enriched for the presence of microsatellites, they may be cloned and exploited for the generation of locus-specific microsatellite-flanking primer pairs[401,1430] (see Chapter 4.8.5.4). The methodology of RAMPO and related techniques have been reviewed by Weising and Kahl.[1519] See Chapter 4.6 for some technical aspects. RAMPOs share their fate with other marker technologies that are partly or totally based on blot hybridization. They are only rarely used anymore, because more convenient marker systems are available for most purposes.

2.3.8   AFLP Analysis and Its Variants

Amplified fragment length polymorphism (AFLP) technology was introduced by Zabeau, Vos, and coworkers[1481,1605] and represents an ingenious combination of RFLP analysis and PCR. AFLP technology is applicable to all organisms without previous sequence information, and generally results in highly informative fingerprints. It rapidly became one of the most popular and powerful approaches to detect DNA polymorphisms.

2.3.8.1   The AFLP Technique: Principle, Advantages, and Limitations

The AFLP reaction comprises two principal steps (Figure 2.13 and 2.14A). In the first step, genomic DNA is digested with two different restriction enzymes producing sticky ends, and double-stranded synthetic adapters of a defined sequence are ligated to both ends of all restriction fragments. Adapter and restriction site sequences then provide universal primer binding sites for subsequent PCR reactions that comprise the second step (see below). Given that the restriction sites are not restored by the adapters, restriction and ligation can also be performed in a single step[37,314] (see Chapter 4.7).

Typically, two successive PCRs are performed on the restricted template, using specifically designed primers that allow only a subset of the restriction fragments to be amplified. To achieve this, the 5′-portions of the primers are made complementary to the adapters, whereas the 3′-ends extend by a few, arbitrarily chosen nucleotides (so-called **selective bases** or **selective nucleotides**) into the restriction fragment (Figure 2.13). Exact matching of the 3′-end of a primer is essential for amplification. Therefore, only those restriction fragments are amplified in which the 3′-primer extensions match the sequences flanking the restriction sites. Statistically, each selective base added to one of the primers reduces the complexity of banding patterns fourfold. Thus, only 1/16 of the total set of fragments are amplified if there is one selective base on each side (+1 primers), 1/256 in case of two (+2 primers), and 1/4096 in case of three (+3 primers). A touchdown PCR program[354] is used to maximize specificity. In the standard procedure described by Vos et al.,[1481] one of the selective primers is radioactively labeled, the amplification products are separated on highly resolving sequencing gels, and banding patterns are visualized by autoradiography (see Chapter 4.7.4.3 for variants of the detection procedure). A typical AFLP pattern is shown in Figure 2.15.

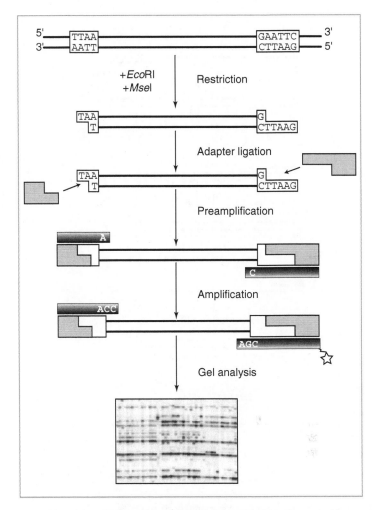

Figure 2.13 Principle of the AFLP strategy.[1481] Genomic DNA is digested with two restriction enzymes (here *Eco*RI and *Mse*I) and specific adapters are ligated to both ends of all resulting fragments. Two successive PCRs are then performed using specific primer pairs, of which the 5′-portions are complementary to the adapters and the restriction site, and the 3′-ends extend by one or a few selective bases into the interior of the restriction fragment. Usually one of the primers is labeled by a radioisotope or a fluorochrome (indicated by a star). Amplification products obtained by the second, selective PCR are separated on sequencing gels. See text for details.

Polymorphisms between two or more genotypes may arise from three sources: (1) sequence variation in one or both restriction sites flanking a particular fragment (such as in RFLPs), (2) insertions or deletions within an amplified fragment (such as in RFLPs), and (3) differences in the nucleotide sequences immediately adjacent to the restriction sites (not assayed in RFLP analysis). AFLPs therefore detect higher levels of polymorphism than RFLPs. AFLP marker bands are mainly dominant, but codominant inheritance can sometimes be evaluated.

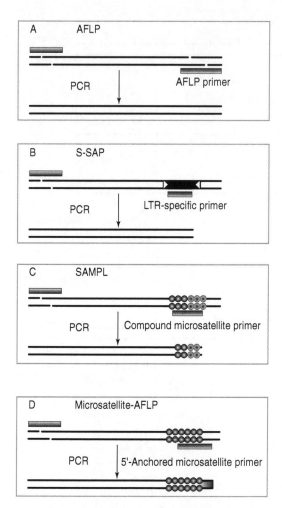

Figure 2.14 A schematic survey of commonly used AFLP variants. (A) **Amplified fragment length polymorphism (AFLP)**[1481,1605]: Genomic DNA fragments are amplified by a pair of AFLP primers, each carrying one to four selective bases at its 3′-end. (B) **Sequence-specific amplification polymorphism (S-SAP)**[1510]: Oligonucleotides complementary to long terminal repeats (LTRs) of retrotransposons or other specific target sequences are used in combination with an AFLP primer. (C) **Selective amplification of microsatellite polymorphic loci (SAMPL)**[950]: Compound microsatellite primers such as $(CT)_4(GT)_4$ are used in combination with an AFLP primer. (D) **Microsatellite-AFLP**[16,1585]: 5′-anchored microsatellite primers such as $TG(CT)_8$ are used in combination with an AFLP primer. Microsatellite arrays are symbolized by rows of circles. Retrotransposon LTRs are painted black.

AFLP technology is characterized by a unique combination of important advantages. First, it is as versatile as PCR with arbitrary primers, because no *a priori* sequence information is needed. Second, high stringency can be applied during PCR, ensuring robustness and high reproducibility of the method[569] (see Chapter 4.7.5). Third, bands of interest may be excised, cloned, sequenced, and transformed into

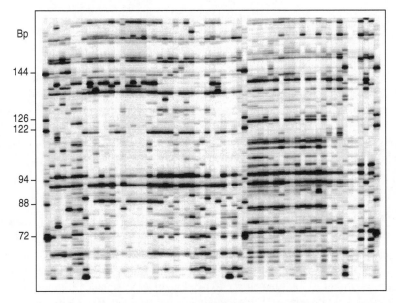

Figure 2.15 Amplified fragment length polymorphism (AFLP)[1481,1605] banding patterns of *Fosterella* species (Bromeliaceae). Genomic DNA aliquots from 60 accessions belonging to different *Fosterella* species were digested with *Hind*III and *Mse*I, ligated to the corresponding adapters, and subjected to two rounds of PCR. Selective primers were *Mse*+CAA and *Hind*+ACA (the latter labeled with the fluorochrome IRDye700). PCR products were separated on a 6% denaturing polyacrylamide gel in an automated LI-COR sequencer. Positions of size markers are indicated (Bp, base pairs).

SCARs by designing specific primers,[1272,1582] as was described for RAPDs[1040] (see Chapter 2.3.3.4). Fourth, a limited set of AFLP primers can be combined to yield a large set of primer combinations, each producing a unique set of amplified fragments. Fifth, the multiplex ratio is not only high (i.e., many potentially informative bands are generated in a single experiment), it can also be tailored by using variable lengths of 3′-primer extensions and/or the choice of enzyme. Thus, only one or two selective nucleotides are required for templates with low complexity such as bacterial genomes,[819] whereas up to eight selective nucleotides may be required for species with very large genomes[565] (for a discussion, see Chapter 4.7.4.2). Pattern complexity is usually chosen to range between 20 and 60 bands (see Figure 2.15). Typical fragment sizes span 50 to 500 bp.

The advantages summarized above make AFLPs a versatile tool for numerous applications, including molecular taxonomy,[77,170,331,693,740] population genetics,[211,481,882,1622] characterization of germplasm collections,[127,602,908] the identification of clones,[1088,1349] sports,[316] and cultivars,[1427] the construction of genetic linkage maps,[314,706,1601] marker saturation at specific genomic regions,[261,1291] building of contiguous sets of overlapping bacterial artificial chromosome (BAC) clones for physical mapping,[722,1291] and many more (see also Chapter 6).

Although it is a very powerful approach, the basic version of AFLP analysis also has a number of limitations, such as dominance of markers, clustering of markers that are generated with a certain pair of restriction enzymes in distinct genomic regions,[706,1601] limited levels of polymorphism in some cultivated species,[69] the requirement for both good quality DNA (to ensure complete restriction), and medium quantities of DNA (as compared with, for example, microsatellite analysis). A number of variations on the basic AFLP theme have therefore been explored, most of which use a combination of one unlabeled AFLP primer with a second, labeled, sequence-specific or semi-sequence-specific primer (see Figure 2.14). Several of these modifications are described in the following sections. Methodology and applications of AFLP and related techniques have been reviewed by Mueller and Wolfenbarger,[959] Vogel and Scolnik,[1475] and Vos and Kuiper.[1480]

2.3.8.2 S-SAP Analysis

One important modification of the basic AFLP technology introduced by Waugh et al.[1510] became known as **S-SAP analysis**. The initial steps of DNA digestion and adapter ligation are identical to the standard AFLP assay. However, only one restriction site-specific AFLP primer is employed in the final amplification step, whereas the second primer is complementary to a defined DNA sequence (Figure 2.14B). In the original S-SAP version,[1510] the specific primer targeted the LTR sequence of the barley retrotransposon *BARE*-1. *BARE*-1 elements are highly abundant in barley (70,000 to 100,000 copies per haploid genome[874]), but the use of three selective bases at the 3′-end of the AFLP primer sufficiently reduced the complexity of banding patterns. The level of polymorphism was higher than that revealed by AFLP primers alone. Segregation data for 54 such fragments were analyzed alongside an existing framework of some 400 other barley markers. Forty-eight of these *BARE-1*-specific S-SAP fragments could be mapped and showed a dispersed distribution across all seven barley linkage groups. All fragments behaved as dominant markers. The even marker distribution across chromosomes was corroborated by *in situ* hybridization.[1510]

The concept of combining gene-specific primers with AFLP primers is not restricted to transposons, but is potentially applicable for targeting and mapping any gene, gene family member, or other DNA sequence of interest. The only requirement is that sufficient sequence information is available for primer design. Thus, an S-SAP variant coined **gene-anchored amplification polymorphism (GAAP)** was designed to amplify DNA sequences in the immediate vicinity of mtDNA coding regions.[845] Gene-specific primer sequences were derived from the fully sequenced mtDNA of *Arabidopsis thaliana*[1433] and combined with adapter-specific primers in a ligation-mediated PCR. The resulting GAAP markers were easily transferable across the family Brassicaceae, but revealed only low sequence variation below the species level.[845]

So far, amplification of transposon insertion sites has remained the main application area of S-SAP. For example, S-SAP targeted at **Ty1-*copia* retrotransposons** (see Chapter 1.3.1) has been applied to pea,[394,1056,1057] sweet potato,[121] and alfalfa.[1086] In all three species, S-SAP markers were more polymorphic than other markers, including AFLP,[121,394,1086] RAPD,[121] and SAMPL.[1086] Moreover, linkage studies revealed evenly dispersed map positions in pea[394] and alfalfa,[1086] as was demonstrated for the *BARE-1* element of barley.[1510]

Related approaches were developed for other transposons, and several acronyms describe the different variants of the technique. An experimental strategy coined **transposon display**[1443] targets *dTph1* elements, a group of class II (DNA) transposons of the *Ac* family present in the genome of *Petunia hybrida*. Ligation-mediated PCR using an AFLP primer in combination with a *dTph1*-specific primer allowed the simultaneous detection of individual transposons as well as the isolation of *dTph1*-tagged genes.[1443] Frey et al.[461] presented a general approach to amplify the insertion sites of the maize *Mutator* (*Mu*) family of transposable elements (see Chapter 1.3.2). In this method, coined **amplification of insertion mutagenized sites** (**AIMS**) by Frey et al.[461] and ***Mu*AFLP** by Edwards et al.,[383] genomic sequences flanking the insertion site are amplified by *Mu*-specific primers in combination with an *Mse*I or *Bfa*I AFLP primer. The *Mu* elements have become the transposons of choice for tagging maize genes, mainly because they appear to preferentially insert within and around genes.[492] Hence, the rationale behind AIMS was to facilitate and accelerate the identification and isolation of tagged genes in transposon tagging programs. The successful use of AIMS for the identification of hundreds of transposon-tagged maize genes has been demonstrated by Hanley et al.[567]

Casa et al.[216] designed a modification of the S-SAP technique that combines an *Mse*I or *Bfa*I-specific AFLP primer with a primer specific for the maize MITE family *Heartbreaker* (*Hbr*; see Chapter 1.3.3). The technique was called ***Hbr* display** by Casa et al.[216] and **MITE-AFLP** by Park et al.[1044] As is the case for *Mu*, *Hbr* is known to preferentially integrate into genic regions. To increase the specificity of the approach, the selective PCR step was performed with a nested *Hbr*-specific primer. Mapping of 213 *Hbr* markers in 100 recombinant inbred lines of maize revealed an even distribution across the 10 maize chromosomes. Park et al.[1044] used MITE-AFLP to study phenetic relationships among diploid rice species carrying the AA genome.

In conclusion, several variants of the S-SAP approach have been described, most of which target transposons. A major disadvantage of the technique is the need for sequence information to design element-specific primers. Although rapid transposon isolation methods based on PCR with adapter primers have been designed,[1056] it may still be necessary to clone and sequence hundreds of candidates to obtain a few functional primer sequences.[121] These shortcomings interfere with a more widespread use of the approach. Nevertheless, S-SAP markers exhibit important advantages for a number of application areas. First, the preference of many transposable elements to integrate into gene-rich regions is certainly beneficial for the purposes of gene isolation by transposon tagging, marker-assisted selection, and map-based cloning. Second, the relatively even distribution of S-SAP markers across genetic maps favorably contrasts with the behavior of AFLP markers, which are often clustered in certain genomic areas.[706] Third, the high level of S-SAP polymorphism, exceeding that of most other PCR-based multilocus marker systems, may be helpful in discriminating closely related accessions. Fourth, as insertional polymorphisms are generated by transpositional events, S-SAP markers can be used to monitor the transpositional activity of the element at an evolutionary timescale.

S-SAP markers have also been advertised as useful tools for phylogeny reconstruction (e.g., in the genus *Pisum*[1057]). However, this application area should be met with caution for at least two reasons. First, the very high extent of polymorphism will normally prevent the use of these markers across species boundaries. Second, genetic distances deduced from insertional polymorphisms do not necessarily reflect genetic distances in terms of sequence divergence. For example, Ellis et al.[394] found AFLP variation, but almost no S-SAP polymorphisms among accessions of *Pisum abyssinicum*. This could be explained by the selective inactivation of the investigated Ty1-*copia* elements in this particular pea lineage.

2.3.8.3 Selective Amplification of Microsatellite Polymorphic Loci

The **SAMPL** technique was introduced by Morgante and Vogel.[950] It combines the high and controllable multiplexing rate of the AFLP technique with the high levels of microsatellite polymorphism by using AFLP-type primers together with compound microsatellite primers (Figure 2.14C).

As in the case of S-SAP, the first steps of the SAMPL procedure are more or less identical to a standard AFLP analysis. Thus, DNA is digested with one rare and one frequently cutting restriction enzyme, and suitable adapters are ligated to the resulting restriction fragments. A preamplification step is then performed with two primers that are complementary to the adapters, plus one specific base extending from the 3'-end into the ligated genomic DNA fragment. In the second (selective) amplification, a single AFLP primer (targeting one of the restriction sites at the fragment ends) is combined with a microsatellite-specific primer, annealing to a compound dinucleotide repeat, which can be anywhere internal to the restriction fragment. Compound microsatellites consist of two or more different simple sequence motifs directly adjacent to each other, and are common elements of plant genomes (see Figure 1.3). SAMPL primers anneal to the junction site of the two microsatellite motifs, providing a perfect 5'-anchor in both directions. TA and GC repeats must be of limited length to prevent self-annealing. SAMPL primers are optimally designed to have similar annealing temperatures as their corresponding AFLP primers (see also Chapter 4.7.3). Successful amplification of AT-rich microsatellites with the SAMPL technique requires relatively short AFLP primers.[1031] One of the two primers is labeled with a fluorochrome or radioisotope, and products are separated on sequencing gels. For SAMPL analysis, theoretically it would be sufficient to digest the DNA with a single restriction enzyme. However, using two enzymes provides more flexibility because both AFLP and SAMPL can then be applied to the same batch of preamplification products (see also Figure 4.1 in Chapter 4.7.3).

Given that the number of possible combinations of restriction enzymes and compound microsatellite motifs is high and each combination can produce from about 10 to 50 bands, the amount of potentially detectable polymorphisms is nearly unlimited. A possible drawback of the method is that the 5'-anchor may not always be functional (see Chapter 2.3.5.3), resulting in blurred banding patterns and stutter bands. To overcome this problem, hot-start PCR conditions may be required (see Chapter 4.3.2.3).

Mendelian inheritance of SAMPL bands has been demonstrated, and relatively high percentages (10 to 20%) of SAMPL markers were scored as codominant.[1028,1556]

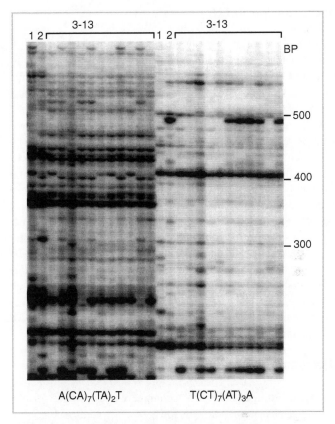

Figure 2.16 Selective amplification of microsatellite polymorphic loci (SAMPL)[950] banding patterns from chickpea. Genomic DNA aliquots from one individual each of 11 recombinant inbred lines (lanes [3] to [13]), resulting from an interspecific cross between *Cicer arietinum* (lane [1]) and *Cicer reticulatum* (lane [2]), were digested with *Eco*RI, and a specific adapter was ligated to both sides of the resulting fragments. Each of the two compound microsatellite primers indicated below the figure were end-labeled with ^{32}P and used for selective PCR in combination with an AFLP primer complementary to the *Eco*RI adapter. PCR products were separated on a sequencing gel and autoradiographed. Positions of size markers are indicated in base pairs. The inbred lines segregate for a number of polymorphic bands.

As could be expected from the incorporation of a microsatellite in the PCR product, SAMPL bands were shown to be more polymorphic than AFLPs in the same species.[852,1194,1292,1556] For example, up to six putative alleles of a single SAMPL locus were detected in *Lactuca sativa*,[1556] and SAMPLs easily discriminated between soybean cultivars and species,[1475] lettuce cultivars,[1556] megagametophytes of Norway spruce,[1028] and 55 genotypes of bread wheat.[1194] The high level of variation exhibited by SAMPL patterns indicates that the technique will be more useful at the intraspecific rather than the interspecific level. A typical SAMPL banding pattern is shown in Figure 2.16.

2.3.8.4   Microsatellite-AFLP

Whereas compound microsatellite primers are employed in SAMPL, 5′-anchored microsatellite primers are used in conjunction with AFLP primers in a method coined **microsatellite-AFLP (MFLP)** by Yang et al.[1585] (Figure 2.14D). Genomic DNA from various plant species was digested with *Mse*I, followed by ligation of an *Mse*I adapter and two rounds of PCR. Selective PCR was performed with an *Mse*+3 primer and one of a set of 74 ^{33}P-labeled, anchored microsatellite primers. Each primer combination revealed complex banding patterns comprising 70 to 150 bands on sequencing gels. As is the case with SAMPL, AMP-PCR, and RAMP markers, a subset of the polymorphisms obtained with MFLP are assumed to be codominant, because they arise from variations in the number of tandem repeats of the amplified microsatellite. Yang et al.[1585] verified this prediction by cloning a number of polymorphic MFLP products from *Lupinus angustifolius*, and showed that cloned MFLP bands can be converted into codominant, locus-specific SSR markers. In a subsequent study, the potential of MFLPs for lupin breeding was demonstrated by the development of MFLP markers that were linked to a gene conferring fungal resistance in *L. angustifolius*.[1586]

The same technology was independently described by Albertini et al.[16] These authors used fluorescence-labeled primers to generate DNA profiles from *Poa pratensis*, and also demonstrated the conversion of cloned fragments into codominant, locus-specific markers.

2.3.8.5   Methylation-Sensitive Amplified Polymorphisms

Still another modification of the AFLP technique was developed for monitoring the state of genomic DNA methylation in fungi[1157] and plants.[223,1580,1583] This type of approach, coined **methylation-sensitive amplified polymorphism (MSAP)** analysis by Xiong et al.,[1580] incorporates the use of methylation-sensitive restriction enzymes. In MSAP, template DNA is usually digested with one rare cutter such as *Eco*RI, and one of a pair of isoschizomers in place of the frequent cutter *Mse*I. Isoschizomers are restriction enzymes that cleave at the same target DNA sequence motif, but display differential sensitivity to the presence of 5-methylcytosine or 6-methyladenine in their recognition sequence.[904] Probably the best known pair of isoschizomers is *Hpa*II and *Msp*I, both of which recognize the tetranucleotide motif 5′-CCGG-3′. Cleavage by *Hpa*II is inhibited by methylation at any of the two cytosine residues, whereas *Msp*I is only inhibited if the external C is methylated.[904] In addition, *Hpa*II (but not *Msp*I) also cleaves if the CCGG motif is hemimethylated (i.e., only one strand methylated). *Eco*RI plus *Hpa*II-*Msp*I were used in most MSAP studies, but *Pst*I-*Mse*I can also be informative.[897]

Pairs of isoschizomers have long been used in conjunction with traditional RFLP analysis and Southern blot hybridization to compare the state of DNA methylation in various tissues or plant species of interest. Reyna-López et al.[1157] were the first to use the DNA methylation assay with isoschizomers in the context of an AFLP approach. Genomic DNA samples from representatives of three major fungal taxa were digested with *Eco*RI in combination with either *Msp*I or *Hpa*II. After ligation of appropriate adapters, preamplification and selective amplification steps were performed with +1/+1 and +1/+2 primers, respectively. Only the *Msp*I-*Hpa*II primer was labeled during selective PCR, leaving *Eco*RI-*Eco*RI fragments undetectable. Methylation at internal cytosines of a CCGG site resulted in polymorphic PCR products obtained with *Eco*RI-*Msp*I vs. *Eco*RI-*Hpa*II. In addition to DNA methylation *per se*, the experiments also revealed differential DNA methylation during different stages of fungal morphogenesis.[1157]

Xiong et al[1580] introduced the MSAP strategy to plants. They monitored patterns of cytosine methylation in rice hybrids and their parental lines as well as in different rice tissues. Increased methylation was detected in the hybrid compared with the parents, and in seedling DNA compared with leaf DNA. In general, methylation patterns could be confirmed by Southern analysis using the differentially amplified fragments as hybridization probes. In another MSAP study in rice, Ashikawa[53] found conserved as well as cultivar-specific methylation at CCGG sites, and demonstrated the stable Mendelian inheritance of methylation patterns over six generations. MSAP has also been applied to the characterization of methylation changes associated with somaclonal variation in oil palm,[897] with micropropagation in apple[1583] and banana,[1067] and with vernalization in winter wheat.[1284] Cervera et al.[223] used MSAP to analyze DNA methylation in different ecotypes of *A. thaliana*. Surprisingly, methylation within ecotypes was almost identical, whereas each ecotype was characterized by a specific methylation pattern. It is tempting to speculate that these patterns reflect specific adaptations to the environmental conditions encountered by each ecotype.

Knox and Ellis[729] devised a somewhat different approach, coined **secondary digest AFLP (SDAFLP)**, to screen for genomic methylation patterns in pea (*Pisum sativum* and related species). For SDAFLP, an unmethylated template DNA is produced by initial *Mse*I digestion of genomic DNA, followed by ligation of *Mse*I adapters and preamplification with *Mse*+0 primers. The amplified products are then digested with *Pst*I, and a *Pst*I adapter is ligated to the restriction fragments. Given that PCR products are generally unmethylated, *Pst*I should be able to cleave all its recognition sites present in this template. This substrate is then used for PCR with *Pst*+2-*Mse*+3 primers, and the products are compared with those obtained from a standard AFLP analysis of *Pst*I-*Mse*I digested genomic DNA. When applied to *Drosophila* and yeast DNA (both of which contain no or little 5-methylcytosine), very similar patterns were observed with AFLP vs. SDAFLP, as expected.[729] However, marked differences were observed between AFLP and SDAFLP patterns from pea. Part of these differences could be attributed to methylated *Pst*I sites in genomic DNA. Contrary to expectations, however, average band sizes were larger, and band numbers were smaller with the SDAFLP variant compared with AFLPs. Obviously, a subset of fragments fails to amplify with SDALFP.[729] A modified SDAFLP protocol using *Eco*RI for the production of a methylcytosine-free template and *Hpa*I-*Msp*II as methylation-sensitive enzymes was applied to banana.[89] Fewer than expected SDAFLP bands were observed in this study, which could have been caused by the inability of *Taq* DNA polymerase to amplify very long *Eco*-*Eco* fragments.

Cytosine methylation has long been known to play an important role for cellular processes in eukaryotes, including the regulation of gene expression, chromatin condensation, cell differentiation, and genomic imprinting (for reviews, see Jost and Saluz,[679] Martienssen and Colot,[889] and Paszkowski and Whitham[1048]). MSAP and

SDAFLP analysis show good reproducibility[89] and both techniques provide convenient PCR-based strategies for monitoring changes in DNA methylation during
development or in response to environmental stimuli. In contrast to most other
methods used to study cytosine methylation, genome sampling specifically targets
methylation-related bands, which can be recovered easily from the gels, reamplified,
and cloned for further study. An unavoidable disadvantage shared with the classical
RFLP-based assays is that only a subset of genomic cytosines is actually targeted
by methylation-sensitive restriction sites.

2.3.8.6 AFLP-Based Expression Profiling

In recent years, molecular marker techniques have increasingly been used for the
analysis of gene expression patterns (so-called transcriptomics). Located somewhere
in between the classical Northern analysis and highly sophisticated methods such
as the serial analysis of gene expression (SAGE),[1460] SuperSAGE,[895] massively
parallel signature sequencing (MSPP),[164] and microarray-based expression profiling,[1233] cDNA fingerprinting methods based on RAPD[816] (see Chapter 2.3.3.5) and
AFLP technologies (see below) have received considerable attention.

The potential of the AFLP technique for generating mRNA fingerprints was first
recognized in the mid-nineties.[64,943] The general strategy to generate an AFLP-based
expression profile of a given tissue involves the isolation of total mRNA, reverse
transcription of mRNA into a population of cDNA molecules; restriction, ligation,
preamplification, and amplification of double-stranded cDNA by a slightly modified
AFLP procedure; and electrophoresis of the final PCR products on sequencing gels.

In the **cDNA-AFLP** variant described by Money et al.[943] for RNA from hexaploid
wheat, the sequence 5'-AGT**CTGCAG**T$_{12}$V-3' was used to prime first-strand cDNA
synthesis. This primer contains a *Pst*I recognition site (typed in bold), so that double-
stranded cDNA can be cleaved with that enzyme. Moreover, the variable 3'-nucleo-
tide (V represents A, C, or G) ensures that synthesis begins exactly at the junction
between the poly(A) tract and the sequence at the 3'-end of the mRNA. After second
strand synthesis, cDNA is digested with *Pst*I and *Mse*I. Ligation to appropriate
adapters, preamplification, and selective amplification with various numbers of selective nucleotides are performed as in the standard AFLP procedure.

Bachem et al.[64] described an alternative cDNA-AFLP procedure that was applied
to expression profiling in developing potato tubers. Their technique differs from the
above method by (1) using standard protocols for double-stranded cDNA synthesis
and (2) digestion of cDNA with *Ase*I-*Taq*I instead of *Pst*I-*Mse*I. Selective amplification with +2/+2 primers resulted in moderately complex banding patterns, with
band sizes from ~100 to 1000 bp. Kinetics of expression of these transcript-derived
PCR fragments during 10 stages of potato tuber development were comparable to
those observed with Northern analysis of the full-sized transcripts. A computer
program, GenEST,[1120] was developed that provides a bidirectional link between the
results obtained by cDNA-AFLP on the one hand, and EST databases on the other.
GenEST not only predicts the sizes of cDNA-AFLP products from EST sequences,
it is also able to find EST sequences that correspond to particular cDNA-AFLP
products of interest.[1120]

Despite the recent development of high-throughput full-genome expression screens (see Chapter 9), cDNA-AFLP still remains a useful technique for several reasons. First, all developmental stages, tissue types, or environmental conditions under investigation can be compared in the same experiment, and the kinetics of the expression of defined mRNAs can be monitored directly. This is particularly advantageous compared with chip-based expression profiling, in which only two tissue states are compared. Second, fragments of interest can easily be recovered from the gel for further analysis. Third, cDNA-AFLP is inexpensive, does not require sophisticated equipment, and can be performed in most laboratories. These three advantages are shared among cDNA-AFLP,[64,943] differential display,[816,817] and arbitrarily-primed RNA PCR.[905,917,1529] In contrast to the latter two strategies that employ random primers and relaxed PCR conditions, cDNA-AFLP favors perfect matches by stringent PCR conditions. This causes an increased confidence in the primary data (smaller expected number of false-positive results), and also allows one to specifically target particular transcripts by increasing the number of selective nucleotides.[64]

An expression profiling technique coined **introduced amplified fragment length polymorphism (iAFLP)** was described by Kawamoto et al.[703] This method aims at high-throughput expression analysis of several tissues in parallel, and is only distantly related to standard AFLP technology. In short, double-stranded cDNAs from each of six different source tissues are digested with *Mbo*I, followed by ligation with *Mbo*I adapters. These adapters are designed to have various sizes due to short insertions into a common sequence. A different adapter size class is applied for each source tissue. PCR amplification of a pool of these cDNAs with one gene-specific primer and one adapter-specific primer results in a cluster of six transcripts that differ by size due to the source-specific adapters. Separation on sequencing gels then allows the comparative quantitation of the source-specific expression of the transcripts. Using three different colors on a fluorescence sequencer, the authors claim that expression profiling of 864 genes across six tissue sources per day could be achieved.[703]

2.3.8.7 Miscellaneous AFLP Variants

Suazo and Hall[1342] suggested a highly simplified AFLP protocol that involves (1) the digestion of DNA and ligation of the adapters in a single reaction; (2) the use of a single, rare-cutting restriction enzyme (*Eco*RI) instead of two enzymes, the use of one adapter and one type of primer; (3) amplification in one step instead of two; (4) separation of fragments on agarose gels; and (5) staining with ethidium bromide. Distinct, reproducible, and polymorphic DNA profiles reminiscent of RAPD patterns were obtained from honey bee (*Apis mellifera*) DNA from Europe and Africa. A similar protocol was also presented by Ranamukhaarachchi et al.[1141]

An AFLP variant employing three restriction endonucleases (i.e., *Xba*I, *Bam*HI, and *Rsa*I), but only two pairs of adapters (i.e., *Xba*I and *Bam*HI), was proposed by Van der Wurff et al.[1447] In this procedure, named **three endonuclease AFLP (TE-AFLP)**, the number of amplified fragments is not only reduced by selective primer extension, but also by selective ligation. This is because no adapters complementary to the ends generated by the frequent cutter *Rsa*I are added. The reduced number of potentially amplifiable fragments eliminates the need for a two-step PCR, which is typically performed in the standard protocol. *Xba*I primers were fluorescence-labeled, and the PCR products were separated on an automated sequencer. Clear-cut DNA fingerprints of relatively low complexity were revealed using template DNA from tomato and the wingless soil invertebrate *Orchesella cincta* (springtails; Collembola). Variation of digestion, ligation, and PCR conditions demonstrated a high level of reproducibility.[1447] In a subsequent study, the TE-AFLP approach, in combination with microsatellite technology, was applied to describe the population genetic structure of *O. cincta* in forest soil.[1448]

2.3.9   Single-Strand Conformation Polymorphism Analysis and Related Techniques

The analysis of single-strand conformation polymorphisms (SSCP) in DNA fragments was first described approximately 15 years ago.[1016] SSCP technology relies on the observation that the mobility of a single-stranded DNA (ssDNA) molecule on a nondenaturing polyacrylamide gel is a function of both its size and its three-dimensional conformation. Because the latter depends on the precise base composition, DNA fragments of identical size but different sequence can be separated from each other.

A standard SSCP experiment involves only a few steps. In the first step, one or more defined or anonymous DNA fragments are amplified by PCR. The amplification products are then denatured by heating, and immediately resolved by polyacrylamide gel electrophoresis. SSCPs are detected as mobility shifts of individual ssDNA fragments relative to each other. The technique is highly sensitive, and DNA fragments differing by mutations in a single nucleotide position can be resolved.[1281] Fragments can be labeled by incorporating radioisotopes[581] or fluorescent dyes during or after the PCR step,[653] and bands are then detected by autoradiography or fluorometry, respectively. Alternatively, unlabeled fragments are visualized by silver staining.[86] Ethidium bromide can also be used, but is less efficient for ssDNA. Gel patterns are easier to interpret when fragments are labeled by PCR with fluorescence-labeled or radiolabeled primers, because only one of the two sister strands is then visible on the gel.

Hayashi[581] gave some technical recommendations to improve SSCP detection. These include the efficient cooling of the gel (an increase in temperature could result in conformational changes and interfere with the reproducibility of results), the inclusion of 5% glycerol in the gel, a relatively low ratio of bisacrylamide to acrylamide, and the use of short fragments (400 bp or less). Under these experimental conditions, more than 90% of mutations are supposed to result in a screenable SSCP.[581,1345] Later studies suggested that fragment size is not too critical.[1084]

Advantages of SSCP analysis include its technical simplicity and high sensitivity for single nucleotide changes. In contrast to other methods that exploit conformational DNA polymorphisms (see below), no specialized equipment is needed. In fact, SSCP is the most rapid and inexpensive method for testing whether or not the sequences of two or more DNA fragments are different. In many phylogenetic and phylogeographic studies, intraspecific variation is undersampled, because many individuals carry the same sequence. In those kinds of studies, SSCP analysis could provide an easy and fast prescreening assay to identify and distinguish organellar haplotypes, and to determine the informative subsets of individuals that are worth sequencing in a population study.[1345] In a population survey of the European wild rabbit, the number and geographical distribution of SSCP haplotypes fully matched the haplotypes obtained from DNA sequencing, but the SSCP approach was an order of magnitude more efficient in terms of time, labor, and resources.[1345]

When applied to cDNAs and/or ESTs, SSCP provides a rapid assay for new mutations in genes, and therefore has a high potential to genetically map genes of known functions.[1084] Running SSCP gels has become a routine prescreening strategy for the identification of single-nucleotide polymorphism (SNP) markers (see Chapter 9.1). A disadvantage is the requirement for careful technical optimization of PCR and electrophoresis parameters. Thus, the precise electrophoresis conditions may determine whether a given PCR fragment runs as a monomorphic or polymorphic band, or even produces a complicated banding pattern on the gel (for an example, see Plomion et al.[1084]). Multiple bands may result from, e.g., the existence of multiple stable ssDNA conformations, partial unfolding, and/or heteroduplex formation.[1345] Because the exact migration behavior of ssDNA fragments is sensitive to a number of parameters, the development of consistent experimental protocols is essential to ensure reproducibility, and the precise running conditions of each SSCP may have to be decided one by one.

Since the first description of the SSCP assay, thousands of research articles have been published on that topic (for a review, see Sunnucks et al.[1345]). SSCP has been most extensively applied in biomedical research or medical diagnosis. Other application areas include linkage analysis and genetic mapping of cDNAs,[1084] increasing the number of polymorphisms detected by arbitrarily primed PCR[903] (see Chapter 4.4), detecting homoplasy in microsatellite analysis,[1345] developing nuclear markers, and physically isolating nuclear haplotypes in phylogeographic analyses.[1345] In plants, the potential of SSCP analysis still seems to be underexploited. An annotated protocol describing the standard PCR-SSCP assay using radioisotopes was presented by Sunnucks et al.,[1345] along with numerous examples of SSCP applications in population biology.

SSCP is only one of several techniques for assaying DNA sequence variation by gel electrophoresis. Other examples include **heteroduplex analysis**,[1539] **denaturing gradient gel electrophoresis (DGGE)**,[932,970] **temperature gradient gel electrophoresis (TGGE)**,[1502] and **temperature sweep gel electrophoresis (TSGE)**[1063] (for a review, see Lessa and Applebaum[802]). All of these methods exploit the conformation dependence of DNA mobility during electrophoresis. For DGGE analysis, polyacrylamide gels are prepared that contain a linear gradient of a denaturant (usually formamide and urea). Double-stranded DNA molecules that migrate through the gradient gel become single stranded at a position that depends on their melting point (which is sequence-dependent). When this position is reached, partial denaturation causes a bubble- or fork-like fragment conformation, which results in a sharp reduction of mobility.[375] The fragment virtually stops. Two fragments of identical size but different sequence will show distinct melting properties, and hence migrate to different points in the gel. Fragments are labeled and detected as in SSCP analysis. DGGE and related methods are technically quite demanding, require highly controlled experimental conditions, and are rarely used for genetic marker generation in plants (for a review, see Dweikat and Mackenzie[375]).

2.3.10 Miscellaneous Techniques

2.3.10.1 Minisatellite Variant Repeat Mapping

A DNA profiling strategy was invented by Jeffreys and colleagues which relies on the internal heterogeneity of human minisatellite repeats rather than length variation. This approach was called **minisatellite variant repeat (MVR) mapping**.[665,667,941] Its principle is based on the observation that certain minisatellite arrays consist of heterogeneous basic units (see Chapter 1.2.1.2). These units are distinguishable from each other by, for example, the presence or absence of a restriction site.

The earliest version of MVR mapping[665] made use of a strategy reminiscent of the chemical DNA sequencing procedure. A DNA fragment comprising the human minisatellite MS32 locus was PCR-amplified from genomic DNA, end-labeled, and then partially digested with *Hin*fI (which cuts once in each 29-bp basic unit) or *Hae*III (which only cuts in the variant units). Electrophoresis and autoradiography produced a continuous *Hin*fI ladder reflecting the basic repeat length on one hand, and a discontinuous *Hae*III ladder on the other, from which the relative positions of the variant units could be deduced. From these data, each MS32 allele could be characterized by a binary code according to the order of variant repeats. Thus, a second source of minisatellite allelic variability became detectable, which allowed the precise designation of alleles. Later, a generalized experimental strategy for MVR mapping based on PCR was created. MVR-PCR was also applied for digital typing of genomic DNA with both alleles superimposed.[667] The allelic variation among different MS32 binary codes was substantial, and even alleles of identical length frequently showed completely different internal maps. Extremely high variability was observed among unrelated human individuals.

Although MVR mapping procedures may provide unprecedented levels of discriminatory capacity (and are therefore valuable, e.g., for forensic analyses in humans), these techniques are not generally useful for DNA profiling in plants. First, establishing MVR (i.e., cloning and sequencing of a minisatellite, design of suitable primers, etc.) is laborious. Second, highly efficient individual-specific discrimination is seldom needed in studies on plants. Third, only a few human minisatellites fulfill the criteria required for MVR mapping, and it will probably not be easy to identify a suitable minisatellite in a particular plant species.

2.3.10.2 Two-Dimensional DNA Typing Methods

A two-dimensional DNA typing approach called **restriction landmark genomic scanning (RLGS)** was introduced by Hatada et al.[577] In this method, genomic DNA is first digested by a rare-cutting restriction enzyme such as *Not*I or *Acc*III. Depending on genome size and complexity, several thousand restriction fragments are generated, which are labeled by filling the ends in the presence of α^{32}P-dNTPs. The size of the labeled fragments is reduced by a second restriction using a six-base cutter such as *Eco*RV. The resulting *Eco*RV fragments are separated on an agarose gel, and a gel strip is excised and subjected to a third restriction step, this time using a four-cutter enzyme. Two-dimension electrophoresis is carried out in polyacrylamide gels, and radioactive spots are visualized by autoradiography. The technique resolves up to a few thousand restriction fragments from all over the genome (i.e., about one fragment per megabase of a typical eukaryotic genome), and allows one to monitor differences between closely related genomes. Although somewhat cumbersome, the technique provides a high level of sensitivity, and the genome is broadly represented. The RLGS technology was adapted to plant genome analysis by Kawase[704] and Matsuyama et al.[896]

2.3.10.3 Single-Nucleotide Polymorphisms

Since the late 1990s, single-nucleotide polymorphisms (SNPs) have become increasingly popular as a molecular marker system (reviewed by Rafalski,[1126,1127] Brumfield et al.,[175] and Morin et al.[953]). This type of sequence variation is characterized by a single base substitution at a particular position, a type of polymorphism that is also recognized by the RFLP technique if the SNP occurs in a restriction enzyme recognition site. Most SNPs are biallelic markers and therefore highly useful for chip-based microarray technology.[1232] To screen for SNPs, genomic DNA from several related test organisms is amplified by PCR with either a specific pair of primers flanking a known sequence or by arbitrary priming. Single base substitutions can be recognized by their impact on the mobility of ssDNA molecules in SSCP gels (see Chapter 2.3.9). PCR fragments that are polymorphic among the test organisms are then sequenced, and the SNP is localized. For many important crops and model species, such as man, mouse, wheat, and *Arabidopsis*, the large amount of DNA sequence data stored in public databases also allows the detection of SNPs *in silico*.[128,1078,1310] SNP and chip technology are treated in more detail in Chapter 9.

3

Laboratory Equipment

This chapter aims at providing researchers with information on necessary equipment used for the various techniques of polymerase chain reaction (PCR) -based and hybridization-based DNA profiling. Before getting started, one should consider to what extent the various methods will be performed in the laboratory, and what resources are already available. If the objective is to perform an occasional DNA fingerprinting experiment, borrowing basic equipment may be more realistic than purchasing high-quality equipment for each step of the process. Conversely, if DNA marker analyses will be among the main activities of the laboratory, the list provided below should give an idea of the equipment required.

Much of this equipment is already available in many biochemical and plant biology laboratories, especially given that PCR technology has become a standard tool in almost any area of biological research. However, some items have relatively specific uses and are not generally found in a non-molecular biology laboratory. To help the reader to set priorities when purchasing equipment for a new laboratory, the importance of each item is classified as follows, with application range as well as available alternatives given within parentheses:

xxx	Essential
xx	Highly recommended
x	Helpful or optional
CLON	Required for tissue culture and/or molecular cloning experiments
HYB	Required for experiments that involve blot hybridization
RAD	Required for experiments that involve the use of radioisotopes

3.1 INCUBATORS

xxx	Water baths (e.g., for growing plant cells, labeling probes, blot hybridization, restriction, ligation, incubation of extracts in some DNA isolation methods)

xxx	Thermal cycler (for automated PCR reactions; heating of probes and samples)
xx	Thermostat-controlled metal heating blocks, which hold test tubes and/or microcentrifuge tubes (for small-scale DNA extraction, incubation, DNA restriction)
CLON, HYB	Dry incubator or rotary shaker (e.g., for growing bacterial cells, blot hybridization, washing Southern blots, staining agarose gels)
HYB	Hybridization oven for rotating tubes (for Southern blot and gel hybridizations; sealed bags in a thermocontrolled rotary shaker or water bath can also be used)

3.2 PLANT AND PLANT CELL GROWTH EQUIPMENT

CLON	Sterile bench, laminar flow hood, and/or biological safety cabinet (e.g., for micropropagation)
CLON	Constant-temperature room available: a rotary shaker with clamps for 100-, 250-, 500-ml and 1- to 2-liter Erlenmeyer flasks
CLON	Constant-temperature room not available: a closed-environment rotary shaker equipped as above

3.3 STERILIZATION

xxx	Autoclave, leak-proof autoclaving bags, sterilization tape (to label sterilized solutions, glassware, or other items)
xxx	Filters and filtration units (0.22-μm pore size; to sterilize solutions that cannot be autoclaved, e.g., vitamins, glucose)
xx	Oven (e.g., to sterilize glassware)
x	Ultraviolet (UV) light source
CLON	Bunsen burner or equivalent (to sterilize inoculation loops)

3.4 WATER PURIFICATION

xxx	Deionization or water purification cartridges and/or glass distillation apparatus
xx	Storage containers

3.5 CENTRIFUGES

xxx	One or more microfuges (optionally refrigerated) that hold standard 1.5-ml and/or 0.5-ml microfuge tubes and/or microtiter plates (for small-scale DNA extraction, PCR, and many other applications). The centrifuges should be equipped with both fixed angle and swinging-bucket rotors

xx	Low-speed centrifuge (10,000 to 20,000 rpm) floor model, or table-top-refrigerated centrifuge for larger tubes (for large-scale DNA extraction)
xx	Vacuum concentrator (e.g., SpeedVac™, for concentrating DNA solutions)
x, HYB	Ultracentrifuge (20,000 to 80,000 rpm) with fixed-angle, vertical, and/or swinging-bucket rotors (for large-scale isolation of highly purified DNA via CsCl gradients)

3.6 REFRIGERATION AND MATERIAL STORAGE

xxx	Refrigerator 4°C (to store, e.g., solutions, DNA, PCR products)
xxx	Freezer –20°C (to store, e.g., solutions, aliquots of PCR reagents, deoxynucleotide triphosphates [dNTPs], DNA, *Taq* DNA polymerase, restriction enzymes, and other DNA-modifying enzymes)
xx	Freezer –70 to –80°C (for long-term storage of frozen tissue samples, reagents, DNA, bacterial cells, X-ray cassettes for autoradiography)
xx	Isothermic storage containers for liquid nitrogen (needed for many DNA isolation procedures) and dry ice
x	A cold room is very useful for performing reactions and centrifugations at 4°C
x	Ice machine
x	Lyophilizer (to freeze-dry plant material, e.g., for long-term storage)

3.7 SAFETY

xxx	Laboratory coats and gloves
xxx	Proper UV shielding and UV-protective glasses
xxx	Fume hood
xxx	Proper storage and containment systems for collection of hazardous waste, such as ethidium bromide, organic chemicals, and radioisotopes
RAD	Plexiglas radiation shield and containers (essential when working with strong β-emitters such as ^{32}P)
RAD	Contamination monitor (Geiger counter)

3.8 PIPETS

xxx	Adjustable micropipets and disposable pipet tips (e.g., 0.5 to 10, 10 to 100, 100 to 1000; 1000 to 5000 µl). If possible, one set of pipets should be designated for each full-time researcher and one set for pre-PCR work exclusively. Pipets or pipet tips can be plugged with cotton on one side to prevent cross-contamination by aerosols. Plugged tips are also commercially available

xxx	Glass or disposable plastic pipets (0.2, 1.0, 5.0, 10.0, and 25.0 ml)
xx	Pasteur pipets
xx	One-channel or multichannel micropipets (e.g., Hamilton pipets) for loading ultrathin polyacrylamide gels
x	Mechanical (Peleus™ balls) and/or electrical pipetting devices for glass pipets

3.9 SUPPLIES (GLASSWARE AND PLASTICWARE)

xxx	Common laboratory glassware and plasticware including beakers, graduated measuring cylinders, Erlenmeyer flasks, reagent bottles of various sizes
xxx	PCR tubes and caps, usually in strips of eight or plates of 96
xxx	Microtiter plates with and without caps (e.g., for PCR, preparing samples for gel loading, DNA cloning)
xxx	Microfuge tubes and caps (0.2, 0.5, 1.5, and 2.0 ml)
xx	Glass or polypropylene centrifuge tubes and adapters (e.g. Corex™ tubes; 15 and 30 ml)
xx	Polypropylene snap-cap and/or sterile screw-cap tubes (15 and 50 ml; for organic extractions)
x	Polyallomer or ultraclear ultracentrifuge tubes (for ultracentrifugation only)
CLON	Culture tubes and caps
CLON	Petri dishes

3.10 DNA DETECTION AND QUANTITATION

xx	Spectrophotometer (UV and visible light)
x	Fluorometer
x	Portable UV source (254 nm)

3.11 ELECTROPHORESIS EQUIPMENT

xxx	Power supply: medium voltage (up to 500 V) for agarose gels
xxx	One or more submarine horizontal gel apparatus (at least 15 × 20 cm; preferably 25 × 30 cm) for agarose gels
xx	Submarine minigel apparatus for agarose gels
xx	UV-transparent trays (for photographing stained gels)
xx	Automated DNA sequencer including the appropriate software for fragment analyses (for DNA sequencing and fluorescent fragment analysis, including amplified fragment length polymorphisms [AFLPs] and microsatellites). Note: If an automated sequencer is available, there is no need to buy the standard polyacrylamide gel

electrophoresis equipment listed below. Such an instrument also circumvents the use of isotopes for fragment analyses and DNA sequencing

xx   Vertical gel apparatus for polyacrylamide gels (e.g., for AFLP and microsatellite analysis)

xx   High-voltage power supply (up to 3000 V) for polyacrylamide gels

xx   Combs, spacers, glass plates, clamps, and/or gel sealing tape for polyacrylamide gels

RAD   Gel dryer and vacuum pump (for drying radioactive polyacrylamide gels)

HYB   Vacuum transfer unit or electrotransfer unit (optional; can be used as a quicker alternative to the classical Southern blot technique)

3.12 DOCUMENTATION OF RESULTS

xxx   Gel documentation system consisting of a UV transilluminator (302- or 366-nm wavelength; to visualize DNA fragments), a digital camera attached to a computer and a (thermo)printer, UV-protective shielding, and appropriate filters for UV photography (red filter for ethidium bromide-stained gels). The individual components can also be purchased separately. Alternatively, the older method for gel documentation by Polaroid photography can also be used.

RAD   X-ray film for radioactive Southern blot hybridizations

RAD   X-ray cassettes, intensifying screens

RAD   X-ray automatic developer or hand developer/fixer baths and plastic trays (an automatic developer is only recommended if large quantities are processed)

RAD   X-ray film hangers

RAD   Phosphorimager including screens, cassettes, and image eraser (as an alternative to autoradiography)

3.13 GENERAL LABORATORY EQUIPMENT

xxx   Microwave oven (e.g., for melting agar and agarose)

xxx   pH meter and pH indicator papers

xxx   Thermometers

xxx   Magnetic heater/stirrer

xxx   Balances (analytical and top loading)

xx   Vortex mixer

xx   Mortars and pestles (for grinding cell material)

xx   Homogenizer (e.g., Ultra-Turrax™)

xx   Bead-mill (e.g., Retsch® MM 300)

xx   Plastic dishes or glass trays (e.g., for staining gels and washing filters)

xx   Alcohol/waterproof markers

x	Desiccators (for storing dried tissue and/or hydrophilic chemicals or reagents)
x	Dishwashing machine
x	Laboratory timers and stopwatches
x	Dialysis clips and tubing (used for cleaning DNA from higher salt concentration, e.g., after CsCl centrifugation)
x	Syringes and needles
x	Plastic wrap (e.g., Saran Wrap™)
x	Aluminum foil (to cover flasks before autoclaving)
CLON	Inoculation loops and needles
HYB	UV cross-linker (for fixation of DNA to nylon membranes; alternatively this can also be done by oven baking or using a standard UV transilluminator)
HYB	Disposable columns (e.g., to separate a labeled probe from the unincorporated nucleotides; alternatively, Pasteur pipets or disposable pipet tips plugged with glass wool can also be used)
HYB	Heat sealer and plastic bags (alternatively used instead of a hybridization oven)
HYB	Nylon hybridization membranes
HYB	Whatman 3MM filter paper or equivalent (for Southern transfer)

4

Methodology

4.1 SAFETY PRECAUTIONS

Working with DNA often involves the use of hazardous chemicals and equipment. The following recommendations are intended to protect the experimenter's health and should be read in advance of starting laboratory work:

1. Chloroform, isoamyl alcohol, phenol, and solutions containing high concentrations (0.2% and more) of β-mercaptoethanol are often used in DNA isolation protocols. Glass plates used for polyacrylamide (PAA) gel electrophoresis are occasionally treated with repel-silane or bind-silane (see Chapter 4.3.5). All of these compounds are toxic upon inhalation and should be handled in a fume hood.
2. Acrylamide and bisacrylamide are neurotoxic. Solutions containing these compounds should be handled with gloves and great care. Contamination of working areas (e.g., during casting of PAA gels) should be avoided. Acrylamide solutions should be polymerized by the addition of N,N,N′,N′-tetramethylenediamine (TEMED) and ammonium persulfate before disposal (see Chapter 4.3.5).
3. Ethidium bromide, generally used for staining of DNA after agarose gel electrophoresis (see Chapter 4.3.6.1) or cesium chloride (CsCl) centrifugation, is a powerful mutagen and carcinogen. Solutions and gels containing ethidium bromide should always be handled with gloves, and contamination of working areas, door handles, computer keyboards, etc., must be avoided. Ethidium bromide-containing waste should be collected and disposed of separately in a legal and environmentally safe way. Sambrook and Russell[1217] describe various ways to decontaminate ethidium bromide solutions.
4. Experiments involving the use of radioactive isotopes (e.g., the powerful β-emitting isotope ^{32}P) should be carried out in a separate laboratory, and behind appropriate shielding (e.g., 1 cm of Plexiglas). Gloves should be worn throughout, and a hand monitor must be routinely used to check for radioactive contamination of the experimenter, the material, and the working area. Radioactive waste must be collected and disposed of in a legal and environmentally safe way.
5. Given that ultraviolet (UV) light is highly mutagenic and destructive, eyes and skin should be protected by wearing glasses and protecting cloth, e.g., when

photographing gels on a transilluminator. UV light should be switched off immediately after taking the photograph. Longer or frequent use of the UV transilluminator makes air circulation a necessity to prevent buildup of toxic ozone. UV light exposure is greatly reduced by documenting the gels with the help of commercial video documentation systems that have widely replaced the former Polaroid cameras.

6. Good standard laboratory practice should be observed when working with bacteria. It is advisable to work with sterile material and solutions and to wear gloves. Genetic manipulation experiments must be carried out according to the regulations in each country.

4.2 ISOLATION, PURIFICATION, AND QUANTITATION OF PLANT DNA

The following section deals with the initial steps of any DNA profiling study, i.e., the collection and appropriate storage of plant material, and the extraction and purification of plant DNA. A number of practical issues should be addressed in the planning stage:

1. Where will the material be collected (e.g., in the field, from botanical gardens, seed banks, herbaria, or grown from seeds in the greenhouse)? If plants need to be sampled in foreign countries, which regulations have to be considered?
2. How many samples are needed? Which sampling strategies should be followed?
3. Which type of tissue is most suitable for collection, storage, and DNA isolation?
4. How should the collected material be documented, preserved, and transported?
5. Which DNA isolation method is most suitable for the species under study?
6. What amount and quality of DNA is needed?
7. Is it possible to obtain DNA from herbarium specimens?

Some ideas pertaining to the above questions (3) to (7) are given in the following sections. For more detailed treatments of strategies how to access, collect, and store plant and algal material, how to deal with regulations, and how to obtain collection permits, the reader should refer to the reviews by Blackwell and Chapman,[137] Dessauer et al.,[332] Loockerman et al.,[841] Milligan,[933] Sytsma et al.,[1355] Taylor and Swann,[1374] and the extensive monograph edited by Guarino et al.[534] Optimization of project design and sampling strategies are also discussed by Baverstock and Moritz.[90]

4.2.1 Collection and Preservation of Plant Tissue in the Field

Quality and yield of plant DNA preparations are to a considerable extent influenced by the condition of the starting material. Whenever possible, fresh, young tissue harvested immediately before DNA isolation should be used. Difficulties may arise when the plant species of interest grow at remote locations, and the collected material has to be stored for days, weeks, or even months before they are returned to the laboratory. Numerous methods have been tested to optimize field collection and preservation of plant material in the absence of laboratory facilities.[7,229,364,454,828,990,1117,1184,1337,1355,1361,1396]

Three useful strategies emerged from such experiments: (1) chemical preservation of plant tissues in solutions containing high concentrations of both cetyltrimethylammonium bromide (CTAB) and NaCl,[990,1184] (2) chemical preservation of plant tissues in ethanol,[454,966] and (3) rapid drying of plant tissues in silica gel or another desiccating agent.[229,828] These strategies are described in more detail below.

4.2.1.1 Starting Material

Most plant DNA isolation protocols make use of young leaves as starting material. However, other tissues may also be considered (or even required) as a source of DNA, depending on the species and problem(s) under study. Etiolated leaves were recommended by Dabo et al.[294] and Michiels et al.[929] because they contain less phenolic compounds than green leaves. Lin and Ritland[822] isolated DNA from petals of *Mimulus guttatus, Lythrum salicaria, Eichhornia paniculata, Aeschynanthus lobbianus,* and *Antirrhinum majus.* Yields were higher, and polymerase chain reaction (PCR) amplification was more reliable as compared with leaf DNA from the same species. Specific DNA isolation methods were designed for various kinds of nonleaf tissue, including dormant buds,[749] pollen grains,[1290] pollen-derived somatic embryos,[616] roots,[178,465,760,825] rhizomes,[1353] tubers,[52,1576] dried corncobs,[1247] and even dry wood.[319,374] For sugar cane, internodal meristem cylinders proved to be a suitable source of DNA.[23] For succulent plant species, Diadema et al.[339] recommended the isolation of DNA from leaf callus. Problems associated with low yields and the presence of PCR contaminants in the highly succulent leaves of *Carpobrotus* species were circumvented by this approach, but tissue culturing prior to DNA extraction was required for each sample.

For a number of reasons, seeds are a particularly attractive source of DNA. First, they often contain less polyphenols and other PCR-inhibitory components.[902] Second, part of the endosperm is usually sufficient for DNA isolation, leaving the viable embryo-containing portion for growing the plant. Marker-assisted selection programs may strongly benefit from such nondestructive sampling.[1479] Third, isolating DNA from seeds instead of leaves also facilitates storage and long-distance transfer of plant material[1479] (see Chapter 4.2.1.2 to 4.2.1.4). Several groups have reported successful DNA isolation from single seeds,[673,1496] half seeds,[254,688,751,902] or even small tissue samples drilled out of seeds[1479] of various plant species. Sangwan et al.[1221] described a method for the simultaneous isolation of DNA and lipids from seeds of opium poppy (*Papaver somniferum*). A comparative analysis of six different small-scale DNA isolation methods applied to nonleaf starting material of six plant species from different genera was presented by Rogers et al.[1182] Most of these methods worked reasonably well with roots, tubers, seeds, and kernels in addition to leaves.

There are a few reports of DNA polymorphism when different tissues are used for DNA extraction. For example, template DNAs isolated from roots vs. leaves consistently yielded primer-dependent random amplified polymorphic DNA (RAPD) polymorphisms in a study on soybean.[238] Genomic alterations during plant development were thought to be responsible for these effects. Donini et al.[356] observed differences in amplified fragment length polymorphism (AFLP) patterns between seed- and leaf-derived DNA templates from wheat. These were interpreted by the authors to be derived from organ-specific variation in DNA methylation of restriction sites. Finally, template-specific variation of AFLP patterns generated from different organs, or from tissues harvested at different times of the year, were reported for strawberry.[44] Microsatellite-primed PCR patterns, conversely, were completely identical for DNA samples derived from young leaves, old leaves, sepals, and rhizomes of strawberry.[44] In the light of these observations, it is advised to start DNA isolation from the same tissue for all individuals that are to be compared.

Given that the amounts of polysaccharides, polyphenols, and other undesirable compounds within plant tissues may vary strongly throughout the course of the year, season of harvesting can also be important.[825] Even the time of the day may have to be considered. For example, Murray and Pitas[966] reported that corn leaf tissue harvested early in the morning consistently yielded the best DNA preparations.

4.2.1.2 Cooling, Freezing, and Lyophilization

Fresh tissue that has to be stored for a short period of time should be kept cool but not frozen (e.g., wrapped in paper or placed in zipped plastic bags, and kept on ice). Cooling may not even be necessary for succulent plant species with a very tight epidermis and cuticula, as long as the storage bags are kept away from light and fluctuating temperatures (see Sytsma et al.[1355] for their field experiences during expeditions to Venezuelan tepuis). Storage for longer periods of time requires either freezing directly in –80°C, shock-freezing with liquid nitrogen followed by storage at –80°C (–20°C is less advisable), shock-freezing with a freeze spray used in histology preparations,[933] freeze-drying (i.e., lyophilization; preferred for tubers),[1576] chemical preservation (see Chapter 4.2.1.3), or drying in silica gel (see Chapter 4.2.1.4). Although fresh or frozen material usually gives higher yields and a somewhat more intact DNA, dried tissue is easier to handle, requires no liquid nitrogen for efficient homogenization, and can be stored for years at room temperature under desiccated conditions.

4.2.1.3 Chemical Preservation

Early work showed that treatments with ethanol, methanol, glycerol, formaldehyde, and other organic solvents are mostly unsuitable for plant tissue preservation, resulting in DNA degradation after a few days.[364,1117] A notable exception to this rule was reported by Rogstad,[1184] who described the successful preservation of small pieces of leaf tissue in saturated, highly viscous NaCl–CTAB solutions. A variant of this procedure (the delayed hot CTAB method) was presented by Nickrent,[990] who prepared a raw extract using a standard CTAB DNA extraction buffer in the field (see Appendix 1A). The homogenized and filtered extract was then stored at ambient temperature until it was returned to the laboratory where the isolation procedure was completed.

Contrary to earlier expectations, two groups reported the successful use of 95% or 100% ethanol to preserve leaves from various plant species. Murray and Pitas[966] were able to isolate nuclei and nuclear DNA from corn, cotton, rapeseed, and sunflower leaves stored in a pickling solution of 1 mM o-phenanthrolene in reagent grade ethanol for years. Flournoy et al[454] showed that leaf tissue of spinach, juniper, and broccoli gave excellent yields of high molecular weight DNA after 11 months of storage in ethanol, provided that a proteinase (Pronase E) was included in the extraction buffer. The authors hypothesized that ethanol treatment causes the formation of insoluble complexes of DNA and denatured proteins (especially histones). During most standard DNA extraction procedures, these complexes will be discarded unintentionally with the protein and cellular debris. This could explain previous failures of the use of ethanol for plant material preservation.[364,1117] Inclusion of proteinase in the extraction buffer presumably disrupts the protein–DNA complexes and liberates the DNA.

A third, possibly useful alternative of chemical preservation is the immersion of tissue in aqueous solutions containing commercial laundry detergent.[68] This method was initially developed for the isolation of DNA from field-collected lizard and snail tissue. Acceptable DNA preparations were obtained after 14 days of tissue storage in the extraction buffer at 37°C. It may be worth testing the performance of laundry detergent solutions as a storage medium for plant materials.

Method

Preservation in NaCl–CTAB solutions is particularly well suited for succulent and other xeromorphic plants, which are difficult to quick-dry with silica gel (see Chapter 4.2.1.4). We successfully used the following variant of the NaCl–CTAB method for collecting succulent, halophytic species of *Suaeda* and other members of the Chenopodiaceae family in Siberia and along the European coasts.[1253] The only equipment needed is a storage bottle filled with saturated NaCl–CTAB solution (we calculate 20 ml per sample), and a set of plastic vials. To prepare the preservative, add solid CTAB powder to a saturated NaCl solution, until the solution becomes highly viscous but not yet gelatinous (about 2.5% CTAB [w/v]). Stir at room temperature until the CTAB is completely dissolved (usually overnight), and fill solution in a leak-proof plastic bottle. Plant material is collected as follows:

1. Fill a plastic vial with 20 ml of preservative (this can be done in advance). Cut plant material of a single individual in 1- to 2-cm-long pieces and submerge the pieces in the medium. The amount of plant tissue should not exceed one fourth of the amount of preservative. Label the plastic vial (use waterproof label or pen).
2. Close the vial tightly. Keep in a dark box, and take care that the plant material remains submerged throughout. Samples can be stored at room temperature for at least 1 month without loss of DNA quality.
3. Process samples as soon as possible after returning to the laboratory. Use a CTAB-based DNA isolation protocol (see Chapter 4.2.3, 4.2.4, and Appendix 1A).

4.2.1.4 Drying

The rapid drying of plant tissues with the help of desiccating agents (usually silica gel) was first suggested by Liston et al.[828] and Chase and Hills.[229] Small pieces of tissue or even whole leaves are placed in sealable plastic tubes or envelopes, together with an excess of desiccant. Inclusion of a moisture indicator will report the occurrence of rehydration, which could lead to DNA degradation.[7] Plastic bags with plant materials in silica gel can be sent easily by mail, and arrive in a suitable condition for DNA extraction. Silica gel treatment is currently the method of choice for preserving field-collected plant tissue for DNA isolation,[183,229] and also works well for red and brown algae.[901,1227] The Missouri Botanical Garden Herbarium established a DNA bank that consists of materials dried in silica gel.[931]

In general, the quality of the isolated DNA will decrease with the duration of the desiccation process. It is therefore advisable to process the samples as soon as possible upon return to the laboratory. Savolainen et al.[1228] stressed that the metabolic and cellular responses to slow drying are similar to those during senescence. Thus, water stress in connection with wounding induces the accumulation of phenolic compounds and browning, which may interfere severely with the isolation of high-quality, PCR-amplifiable DNA (see Chapter 4.2.2.4). Nevertheless, air-drying at room temperature also seems to work for many species (e.g., peach leaves[1396]), and is even preferred by some researchers.[1374] The difficult (but often successful) isolation of DNA from herbarium specimens (see Chapter 4.2.2.8) also demonstrates that slow air-drying preserves useful amounts of high molecular weight DNA.

Other methods of drying involve the crushing of leaf tissue onto a solid support. In the commercial Generation DNA Purification System (Gentra Systems), fresh plant tissue is rubbed onto a collection card, which can be stored for up to 6 months before further processing.[774] In a procedure described by Lin et al.,[821] leaves are crushed against FTA® paper, which is a commercial medium developed for long-term storage of blood stains (available from Life Technologies). FTA paper is impregnated with antimicrobial reagents, and is designed to preserve DNA from biological samples. Disks punched out of the leaf attached to the FTA paper proved to be functional as PCR templates for more than 1 month.[821]

Method

In our hands, the silica gel technique proved to be useful for preservation of leaf material from species of the paleotropical tree genus *Macaranga*, collected under rainforest conditions in East Malaysia.[77,476] The only equipment needed is a batch of telltale silica gel (containing a moisture indicator dye) in a waterproof bag, a set of small paper bags, and a set of zipper-sealable plastic envelopes.

1. Harvest one or a few young leaves, cut them into pieces, and put them into a labeled paper envelope, or a coffee (or tea) filter bag.
2. Insert the paper envelope into a sealable plastic bag, and fill the latter with several grams of dry silica gel (weight ratio of silica gel to plant tissue should exceed 10:1).
3. Seal the plastic bag tightly. Change the silica gel approximately every 6 hours, or when the color changes to pinkish-purple.

After the second or third change, the leaf material is usually completely desiccated. It may be stored in the same paper envelope within the plastic bag, together with a small amount of silica gel. Used silica gel can easily be regenerated (i.e., redried) under field conditions, e.g., using a cooker and a pan. The colored moisture indicator included in the batches of telltale silica gel serves as a warning signal for tissue rehydration, which can lead to DNA degradation.[7,1337] It also helps to avoid raising undue attention from customs authorities and security personnel pertaining to the nature of the white powder.[1355]

4.2.1.5 Preparing Herbarium Vouchers

An important part of any taxonomically oriented sampling procedure is the appropriate labeling of the specimens, and the collection and preparation of herbarium vouchers to document the collected samples (and the DNA isolated therefrom) for further research. The herbarium specimens should be properly dried, labeled, and identified by a competent taxonomist. Duplicate vouchers should be collected because most countries require that the first set of plant specimens be deposited at the national or local herbarium (see Sytsma et al.[1355]).

4.2.1.6 Contamination

Because of the sensitivity of the PCR, extreme caution is necessary to avoid contamination of the plant material with other organisms.[1374] Dirt, fungi, insects, lichens, and other epiphytes should be carefully cleaned from sampled leaf material; if necessary, materials should even be surface-sterilized. Plant material infected by bacterial or fungal parasites may show altered DNA profiles, depending on the type of marker employed. For example, Staub et al.[1326] observed variation in RAPD banding patterns of cucumber plants infected with the fungal parasite *Sphaerotheca fuliginea*, as compared with uninfected plants. Caution is also recommended regarding the reuse of desiccant, given that tiny amounts of residual plant material may cause cross-contamination of samples. Therefore, plant tissues should not be in direct contact with the desiccating agent.

Epiphyllic and endophytic fungi may cause serious contamination problems, especially when PCR products are generated by random, semispecific, or otherwise universal primers.[1206,1252,1613] For example, Zhang et al.[1613] reported the PCR amplification of numerous types of fungal ribosomal RNA gene (rDNA) sequences from (apparently healthy) leaves of eight different bamboo species, irrespective of the primer variant and the PCR conditions. True bamboo rDNA amplicons were only obtained when leaves were successively treated with 95% ethanol and 5% NaOCl, reimmersed in 95% ethanol, and blotted dry prior to DNA extraction. Although epiphyllous mycelia and spores can thus effectively be removed through surface sterilization, contamination by endophytic fungi (which are present in a wide variety of plants without causing obvious symptoms) is almost unavoidable,[1206,1252] and may compromise studies using PCR-generated molecular markers. Saar et al.[1206] recommended that plant genomic DNA should be test amplified with universal primers specific for the internal transcribed spacer 2 (ITS2) region of rDNA, whenever fungal contamination is suspected. The fungal ITS2 sequence is usually shorter than its plant counterpart, leading to a double band on agarose gels.

A major source of contamination in all PCR-based marker methods is the carryover of alien PCR products. This problem may be especially severe when the target

DNA is present in a low concentration and/or poor condition, as is often the case with DNA from herbarium specimens. Numerous measures can be taken to avoid such carryover events, including UV sterilization of materials and solutions, separating pre- and post-PCR laboratories, etc. (see Deguilloux et al[319] and Chapter 4.3.2.6).

In summary, quick-freezing, NaCl–CTAB treatment, or quick-drying using desiccants appear to be the methods of choice if the collected plant material cannot be processed immediately after sampling. However, the efficiency of the various preservation methods described above will certainly vary between species, and it is impossible to predict success in advance. It is therefore mandatory that pilot experiments be performed well before embarking on a collection trip.

4.2.2   Plant DNA Extraction: General Considerations

A variety of problems may be encountered during the isolation and purification of high molecular weight DNA from plant species. These include (1) partial or complete DNA degradation by endogenous nucleases; (2) coisolation of highly viscous polysaccharides, which render the handling of samples difficult and may also inhibit enzymatic reactions; and (3) coisolation of soluble organic acids, polyphenols, latex, and other secondary compounds, which cause damage to DNA and/or inhibit enzymatic reactions. These problems are aggravated when exotic DNA sources such as fossils or old herbarium specimens are used. As a consequence, the quality of DNA preparations obtained by standard procedures is often poor, and yields may range from less than 1 μg to more than 200 μg of DNA per gram of fresh leaf tissue.

Because the biochemical composition of plant tissues and species varies considerably, it is virtually impossible to supply a single isolation protocol that is optimally suited for each plant species. Even closely related species may require quite different isolation procedures. Accordingly, an enormous number of plant DNA isolation protocols (and modifications of existing procedures) have been published (compiled in Appendix 1). The majority of methods aim at isolating total cellular DNA, which is a suitable substrate for almost all PCR-based marker methods. However, there are also numerous protocols that are specifically designed for the isolation of nuclear DNA (Appendix 1D), chloroplast DNA (cpDNA; e.g., Dally and Second,[297] Mariac et al.[880]), and mitochondrial DNA (mtDNA; e.g., Pay and Smith,[1053] Wilson and Choury[1550]), respectively. Maintaining the integrity of nuclei during the first steps of DNA isolation has the advantage that cytoplasmic contaminants can be removed. DNA isolation via nuclei may therefore be an attractive alternative for difficult species, provided that only nuclear DNA profiles are analyzed (see Chapters 4.2.2.7 and 4.2.6). In contrast, methods that yield purified cpDNA or mtDNA are rarely needed in DNA profiling studies (but see Mariac et al.[880]), and will not be treated in detail here.

Plant DNA isolation methods differ in many respects, including the disruption of tissues and cells, the composition of extraction and lysis buffers, and in the way that DNA is purified from other cell ingredients (such as proteins, RNA, membranes, polysaccharides, and polyphenols). Nevertheless, there are a number of common principles, ingredients, and isolation modules that are commonly used in various combinations. These modules are discussed in some detail on the following pages, which are intended to serve as a guideline for tailoring a specific DNA isolation procedure for the species of interest. Two variants of the standard CTAB procedure initially described by Murray and Thompson[965] are described in Chapters 4.2.3 and 4.2.4, respectively. Chapter 4.2.5 gives a standard protocol of the potassium acetate–sodium dodecyl sulfate (SDS) precipitation technique originally described by Dellaporta et al.[323] Finally, Chapter 4.2.6 describes a method involving isolation of nuclei prior to DNA, adapted from Willmitzer and Wagner.[1549] The basic variant of each of these protocols yields DNA preparations that are sufficiently pure for PCR analyses in many plant species. If this is not the case, one or several of the modifications discussed below might help to increase yield and to ensure complete restriction for AFLP analysis and/or reproducibility of PCR.

4.2.2.1 Cell and Tissue Disruption

In most DNA isolation protocols, plant material is shock-frozen in liquid nitrogen and ground to a fine powder, using mortar and pestle. For small-scale isolation (so-called minipreps), grinding is often performed inside microfuge tubes, with either a tight-fitting Teflon pestle or just a yellow pipet tip used to crush the tissue. Teflon pestles may be attached to a powered desktop drill.[586] It is important that the plant material does not thaw before being added to the isolation buffer. Otherwise, cellular enzymes may rapidly degrade the DNA. The use of lyophilized or dried tissue circumvents the problem of thawing and makes grinding more convenient. If large amounts of dry material are to be processed, a coffee grinder could be suitable equipment. For fibrous material, precutting of the leaves with scissors and/or the addition of some sterilized quartz sand may help powdering. Grinding fresh tissue directly in isolation buffer is only recommended if the isolation buffer leaves the organelles intact and does not liberate the DNA. Otherwise, shearing forces may cause reductions of DNA yield and quality. Plant materials stored in concentrated NaCl–CTAB solutions (see Chapter 4.2.1.3) should also be ground at room temperature (e.g., in a sorbitol buffer[1337]), because they are crushed insufficiently in liquid nitrogen.

An alternative to manual grinding is the disruption of tissue in a mechanical tissue homogenizer (e.g., Ultra-Turrax™ Janke and Kunkel) or shaking-mill (e.g., Retsch MM-300®; Mini-BeadBeater® Biospec Product; GenoGrinder 2000®, Spex CertiPrep). Lassner et al.[781] used a sap extractor to homogenize tomato leaves. Automated grinding is especially recommended for tough material such as needles[1283] or herbarium specimens,[367] as well as when large numbers of samples need to be processed. For homogenization in a shaking mill, tissue samples (either fresh plus buffer, dry, or frozen) are placed in a suitable tube (e.g., microcentrifuge tube or 96-well plate) together with small glass or steel grinding beads. Tissues are then homogenized by short periods of shaking at high frequency.[287,866,1283] If frozen tissue is used, shaking periods should be short enough that no thawing occurs. Paris and Carter[1042] added a stainless steel dowel pin to each sample in a 96-well plate and homogenized the tissue in a matrix mill (this instrument induces an electromagnetic field in which the pins move rapidly). Dilworth and Frey[347] placed small leaf disks and three glass beads in the wells of a flat-bottomed microtiter plate and homogenized the tissue in an Eppendorf thermomixer. As a cost-efficient alternative to commercial units, Michaels and Amasino[923] suggested the use of a modified paint shaker carrying two 96-well plates. Mechanical maceration in a microbead mill proved to be the only way to rupture the sporopollenin coat of nongerminated pollen and liberate abundant and nondegraded DNA.[1290]

A simple alternative for tissue grinding in microfuge tubes makes use of ball bearings, liquid nitrogen, a Styrofoam holder, and a vortex mixer.[266] An upscaled variant of this strategy, using ball bearings to grind 16 frozen samples simultaneously in a flask shaker, was described by Karakousis and Langridge.[692] These authors also present a table summarizing the time taken to achieve efficient grinding of leaf material from 45 plant species using the ball bearing method.

Chemical treatment with various compounds has also been reported as a means to disrupt plant cells. For example, Jhingan[669,670] and Williams and Ronald[1545] described the use of potassium or sodium salts of ethyl xanthogenate (an agent used in the textile industry to dissolve cellulose) for the solubilization of cell walls. Tissues can also be lysed (and DNases inactivated) simply by boiling, applying boiling–freezing cycles and/or alkali treatment in some one-step/one-tube procedures[190,636,1397] (see Appendix 1C). If alkali is used, extracts need to be neutralized prior to their use in PCR.[1042] Finally, pregrinding of fresh tissue in ethanol was reported to inhibit endogenous DNAse activities[7] (see below), and was recommended as an alternative to liquid nitrogen.[1279]

4.2.2.2 Lysis of Membranes and Organelles

In the majority of protocols, lysis of membranes and liberation of DNA from nuclei, chloroplasts, and mitochondria is performed during or immediately after tissue homogenization. The extraction buffer then also serves as a lysis buffer, which typically consists of the following ingredients in various combinations:

- **Detergents** are included to destroy membranes, denature proteins, and dissociate proteins from DNA. The most commonly used cationic detergent is CTAB; the most common anionic detergents are sarkosyl and SDS. Standard concentrations of detergents in lysis buffers are 2% for CTAB and 1% for SDS, but optimum values can vary between species (for a discussion, see Mace et al.[866]).
- A suitable **buffer system** (mostly based on Tris-HCl) maintains the pH in a range that avoids the activity optima of degrading enzymes (usually between pH 8 and 9 in plants). Increased buffer concentrations proved to be necessary for the successful isolation of DNA from tree roots.[825]
- **High salt concentrations** (>1 M NaCl) dissociate nuclear proteins (especially histones) from DNA[22] keep polysaccharides in solution during ethanol-precipitation of DNA (see below), and may also salt out PCR inhibitors.[244]
- **Reducing agents** such as β-mercaptoethanol, dithiothreitol, or ascorbic acid inhibit oxidization processes, which either directly or indirectly cause damage to DNA (see Chapter 4.2.2.4).
- **Chelating agents** such as ethylenediaminetetraacetic acid (EDTA), ethyleneglycoltetraacetic acid (EGTA), o-phenanthrolene, or Chelex[1490] are included to capture

bivalent metal ions that stimulate metal-dependent DNases released from the cells. EDTA is routinely included in many DNA isolation buffers and storage solutions.

The lysis buffer may also contain polyphenol oxidase inhibitors and/or polyphenol adsorbents to counteract the detrimental influence of secondary compounds, RNases or high concentrations of lithium chloride to remove RNA, proteinases to digest proteins, and pectinases to digest polysaccharides (for details see Chapters 4.2.2.3 to 4.2.2.7). Published optimization strategies for existing DNA isolation procedures often relate solely to the ingredients (and also the pH) of the extraction and/or lysis buffer.

4.2.2.3 Removal of Proteins and RNA

Proteins are dissociated from DNA by the action of detergents and high salt concentrations in the lysis buffer. Dissolved proteins are then usually removed by one or more rounds of extraction with phenol, phenol–chloroform mixtures, or chloroform–isoamyl alcohol mixtures, followed by centrifugation to separate the phases. Whereas nucleic acids remain in the aqueous phase, most proteins are denatured by this treatment and form an insoluble slurry at the interphase. In the family of protocols introduced by Dellaporta et al.,[323] proteins and polysaccharides are simultaneously precipitated by high concentrations of potassium acetate in the presence of SDS (see Appendix 1B). Some protocols include proteinases in the extraction buffer, usually together with detergents.[22,54,454,688,1237]

RNA often copurifies along with DNA, and may cause problems in the PCR.[1081,1592] RNA is usually removed by an RNase treatment, or by selective precipitation with high concentrations of lithium chloride (e.g., Jobes et al.[672]). The RNA removal step can be inserted early or late in the DNA extraction procedure. A number of techniques have been devised to yield both RNA and DNA from the same isolation procedure.[725,765,875]

4.2.2.4 Removal of Polyphenols and Other Secondary Compounds

Plant species produce a wide range of secondary compounds, including phenols, terpenes, alkaloids, and flavonoids, to name just a few classes. In terms of DNA isolation, phenols and polyphenols are of particular interest because they are easily oxidized by intrinsic enzymatic activities. The resulting quinonic compounds are powerful oxidizing agents that cause the often-observed browning of DNA preparations, damage DNA and proteins,[842] and may also render the DNA inaccessible for some enzymes. The detrimental influence of polyphenols and their oxidation products is most commonly counteracted by one of the four following strategies:

- Inclusion of **polyphenol adsorbents** such as bovine serum albumin (BSA) and soluble polyvinylpyrrolidone (PVP) in the isolation buffer, usually at concentrations from 1 to 6%.[183,278,716,1072,1183,1498] Polyphenol adsorbents are sometimes also added in an insoluble form (polyvinyl polypyrrolidone [PVPP],[178,692] Polyclar AT).
- Inclusion of **phenoloxidase inhibitors** such as diethyldithiocarbamic acid (DIECA) in the isolation buffer.[229,309,619,1050,1072]

- **Inhibition of polyphenol oxidation** by including elevated concentrations of anti-oxidants (such as sodium bisulfite, cysteine, dithiothreitol, ascorbic acid, or β-mercaptoethanol) in the isolation buffer.[17,438,482,684,966] Optimum concentrations of antioxidants can vary considerably among species. For β-mercaptoethanol, reported optimum values range from 0.03% for chickpea[866] up to 5% for the genus *Nelumbo*.[684]
- **High ratios of extraction buffer to plant material** will dilute polyphenols and therefore also reduce their detrimental effects.

4.2.2.5 Removal of Polysaccharides

Pectin-like polysaccharides are often water soluble and tend to coisolate with DNA. They may render DNA preparations highly viscous and inhibit the activity of restriction enzymes[352] and *Taq* DNA polymerase.[324,1037] The following strategies have been developed to remove polysaccharides from DNA preparations:

- Many polysaccharides are more efficiently precipitated by **elevated CTAB concentrations** in the isolation buffer.[365]
- Limiting the heat incubation during CTAB extraction to a maximum of 15 min and **precipitating DNA at room temperature** was reported to leave most polysaccharides in the supernatant.[79]
- Attempts have been made to selectively digest polysaccharides with **pectic enzymes**.[1154,1187] However, such enzyme preparations should be free of DNases, which may not generally be the case.[1187]
- Cheng et al.[242] described the removal of polysaccharides from *Citrus* DNA by extracting the DNA-containing aqueous phase with water-saturated **ether** in the presence of 1.25 M NaCl. Polysaccharides were found to be concentrated in a gel-like interphase between the (top) ether and (bottom) aqueous phases.
- Several studies have exploited the **differential solubility of DNA vs. polysaccharides in salt–ethanol solutions**.[17,428,837,921,924,1087] For example, Fang et al.[428] dissolved the crude DNA pellet in a buffer containing 2 M NaCl, and reprecipitated the DNA with two volumes of ethanol. They observed that, in the presence of high salt, many polysaccharides remained dissolved and were thus discarded with the supernatant. Alcohol precipitation of DNA in the presence of high NaCl concentrations soon became a standard technology for the removal of polysaccharides (e.g., Crowley et al.,[286] Lodhi et al.,[837] Porebski et al.[1087]). Another strategy was suggested by Michaels et al.,[924] who selectively precipitated polysaccharides from *Arabidopsis* DNA preparations by treatment with 0.35 volumes of ethanol under low salt conditions (i.e., 0.25 M NaCl). DNA remained in the supernatant and yield was not affected. This strategy is used in the CTAB protocol described in Chapter 4.2.3.

Given that plant polysaccharides are a highly diverse group of compounds, a subset of these methods may work in some species but not in others. Pilot studies are advisable.

4.2.2.6 Removal of Organic Acids and Endogenous DNase Activities

Huge amounts of oxalic acid are contained in leaves of various plant species, including rhubarb, *Rumex*, *Bryophyllum*, and *Begonia*. Kopperud and Einset[741] found that an initial washing step with a low-salt buffer significantly increased DNA yield from *Begonia* leaves. The authors hypothesized that the low-salt buffer may have washed off the majority of organic acids, which could otherwise render the DNA insoluble.

DNA isolation buffers usually contain EDTA, which chelates magnesium ions and thereby inhibits magnesium-dependent DNases. However, Adams et al.[7] stressed that plant DNases are a very heterogeneous group of enzymes, and that not all DNases are inhibited by EDTA. These authors compared the integrity of DNA from leaves of several plant species that were preground in CTAB buffer, water, or ethanol prior to standard CTAB extraction. Considerable DNA degradation was observed in the water controls, presumably due to the action of endogenous DNases. For some species, especially of the Poaceae family, degradation also occurred in the CTAB buffer, whereas intact high molecular weight DNA was obtained from all samples preground in ethanol. The authors recommended pregrinding of plant tissue in a small quantity of ethanol as a general method for the inactivation of DNases.

Endogenous DNase activities were also observed in nuclei isolated from *Arabidopsis* leaves and embedded into agarose plugs.[832] These activities could be inhibited by treating the plugs with 160 mM L-lysine and 4 mM EGTA, suggesting that Ca^{2+}-dependent DNases are an important source of endogenous nuclease activity. Murray and Pitas[966] reported that endogenous nucleases of corn leaves are inhibited neither by EGTA nor by EDTA, but are efficiently controlled by 10 mM o-phenanthroline in the extraction buffer.

4.2.2.7 General Strategies to Remove Cytoplasmic Contaminants

The majority of cell constituents interfering with DNA isolation reside in the cytoplasm. Some protocols therefore include a **pre-extraction step** in which the tissue is homogenized in a buffer that leaves organelles and membranes intact (i.e., no detergent added; e.g., Jobes et al.,[672] Murray and Pitas[966]; Scott and Playford[1260]). If such a pre-extraction slurry is centrifuged, the DNA is pelleted together with the intact organelles, whereas the vast majority of undesirable cytoplasmic constituents, including RNA, polysaccharides, polyphenols, organic acids, and degrading enzymes, are decanted with the supernatant. In a second extraction step, organelles and nuclei are lysed by the addition of the lysis buffer, including a detergent and (most often) a protease.

An additional advantage of this strategy is the increased physical and chemical protection of DNA during homogenization. As long as the DNA is enclosed in intact organelles, it is largely protected from shearing forces as well as from the action of nucleases. Zhang et al.[1609] showed that 95% of nuclei remain physically intact after grinding in liquid nitrogen and can serve as a source for megabase DNA preparation (see Appendix 1E). Pellets resulting from the first centrifugation in pre-extraction buffer can further be processed to isolate pure nuclei and/or pure cpDNA; e.g., by specifically lysing the organellar membranes with Triton X-100.[880,1072] Numerous plant DNA isolation methods via nuclei have been published, only a small collection of which are compiled in Appendix 1D. A nuclear DNA isolation procedure modified from Willmitzer and Wagner[1549] is described in Chapter 4.2.6.

A second group of strategies to remove cytoplasmic ingredients involves the **selective precipitation of DNA**. Reversible precipitation by at least 0.6 volumes of isopropanol or two volumes of ethanol in the presence of salt is a long-used standard technique of nucleic acid purification. Ethanol and/or isopropanol precipitation steps are therefore routinely included in almost all DNA isolation protocols. However, not all contaminants are removed by this technique, because RNA and many polysaccharides tend to coprecipitate with DNA. Other agents exist that precipitate nucleic acids selectively. The original strategy of CTAB-based DNA isolation procedures exploits the observation that DNA–CTAB complexes are soluble in high salt only.[965] Consequently, these complexes can be precipitated by lowering the NaCl concentration below 0.7%.[80,965,1183] Polyphenols, residual proteins, and many polysaccharides remain in the supernatant. After centrifugation, the pellets are redissolved in buffers containing >1.0 M NaCl, and further purified, e.g., by reprecipitation with ethanol. A variant of this technique involves the use of dodecyl trimethyl ammonium bromide (DTAB) instead of CTAB in the extraction buffer.[1228] DNA can also be freed from contaminants by precipitation with polyethylene glycol (e.g., PEG 8000). A PEG treatment often serves as a final purification step in plasmid or phage isolation,[1217] and has been included by various authors in plant DNA purification protocols.[322,350,1193,1234] Finally, DNA can selectively be precipitated by polyamines such as spermine and spermidine, leaving other macromolecules in solution.[182,613]

A third group of strategies to remove cytoplasmic contaminants involves the reversible **binding of DNA to a solid matrix**. This matrix can be supplied, e.g., by magnetic particles,[1199] an anion exchange resin,[287,1220] or by silica gel particles that selectively bind DNA in chaotropic salt solutions such as guanidine thiocyanate or sodium iodide.[212,398,618,626,793] The silica strategy had originally been designed for the purification of DNA from agarose gels,[1477] but has later been shown to yield genomic DNA from microorganisms, fungi, animals, and plants, as well as from difficult substrates such as dung, soil, and forensic samples (reviewed by Elphinstone et al.[398] and Rogstad[1185]). Höss and Pääbo[618] stressed that the silica matrix should be removed from the sample carefully and completely, because even minute amounts of silica would inhibit the PCR. Such problems were not reported by Rogstad,[1185] who showed that the chaotropic salt solution can be replaced by the standard CTAB isolation buffer. Selective and reversible binding of DNA to a solid matrix is exploited by many commercial kit suppliers (see below). For example, DNA is bound by modified magnetic beads in the DNA isolation kit by Dynal,[1199] by glassmilk in the Fast-DNA™ isolation kit by Qbiogene, and by a silica membrane in the Plant DNeasy® kits by Qiagen.

Finally, we would like to mention two very efficient, but also somewhat tedious ways of selectively purifying DNA from soluble as well as insoluble contaminants. One method is the preparative electrophoresis in agarose[183,825,1184,1227]; the other — and probably the ultimate way to purify DNA from any type of contaminant — is cesium chloride buoyant density gradient centrifugation (e.g., La Claire and Herrin,[765] Murray and Thompson,[965] Stein[1329]). Both techniques are rarely used any more because most DNA profiling techniques are based on PCR and do not require high quantities or extremely high purity of DNA. A detailed protocol of the cesium chloride banding technique can be found in the first edition of this book.

4.2.2.8 Herbarium Specimens and Other Difficult Substrates

The advent of PCR made the minute amounts of DNA present in herbarium specimens accessible to molecular analyses. Herbarium collections became an important source of material for phylogenetic studies, especially for rare, endemic, or even extinct species, which are difficult to collect from their natural environment. DNA analyses of herbarium specimens can assist in clarifying discordances and inconsistencies in taxonomic attribution of historic samples,[307] and offer the possibility to compare the genetic makeup of extant and extinct taxa. Herbarium specimens may also be of interest for population geneticists, because they potentially allow us to compare directly ancient and recent population structures. Pioneering studies in this direction have been performed with DNA from Ice Age brown bears,[797,1022] but not yet on plants. Finally, herbarium specimens infested with pathogens can give valuable information on, e.g., fungal epidemics in the past, and can track pathogen migrations.[1175]

Herbarium vouchers are generally prepared in a way that optimally preserves the morphological and anatomical — but not necessarily the molecular — information content of the living plant. Collectors of earlier days were unaware of the potential of herbarium specimens for molecular systematics, and made no efforts to preserve the integrity of DNA. Hence, DNA preparations from herbarium specimens are often highly degraded, chemically modified, and contaminated by microbial and/or human DNA and by substances inhibitory to PCR.[136,307,367,1183,1228] The pioneering study of Rogers and Bendich[1183] nevertheless demonstrated that a standard CTAB technique can yield DNA from herbarium leaves of various Poaceae and Fabaceae (22 to 118 years old), as well as from mummified seeds and embryos from various angiosperms (up to 40,000 years old). The recovered DNA fragments had average sizes of about 400 bp, but larger molecules were also observed. The extent of degradation appeared to be related to the condition of the leaf material rather than to the age of the specimen. DNA from mummified seeds averaged from 0.2 to 7 kb, and ranged as high as 20 to 30 kb.[1183] It should be noted in this context that the authenticity of subfossil and fossil DNA is controversial, and great care should be taken to verify the ancient nature of the isolated DNA and PCR products resulting therefrom.[272,824]

Savolainen et al.[1228] used four alternative extraction methods to isolate DNA templates from 17 herbarium specimens belonging to various species of the order Celastrales. Only two samples allowed PCR amplification of a 900-bp cpDNA fragment under standard conditions. The failure to amplify the other samples was due to either extremely low DNA concentration or the presence of inhibitors. Removal of inhibitors by adding insoluble PVP (Polyclar AT) or BSA, diluting the extract 100-fold, or using nested primers resulted in successful amplification for some but not all specimens. A subset of PCR products obtained under these conditions proved to be derived from contaminating DNAs. To avoid such contamination, the authors strongly advised that the material and solutions be exposed to UV light prior to PCR. Nine of the specimens were not amplifiable under any of the tested conditions. As above, there was no apparent correlation between the age of the sample and the success of PCR.

Singh and Ahuja[1293] described a variant of the CTAB protocol that was designed for market samples of dry tea leaves, which in a way resemble herbarium specimens. Considerable amounts of only slightly degraded DNA were obtained, but PCR with specific or arbitrary primers was only successful when the dry leaves were prewashed in water and redried before DNA extraction. This procedure removed much of the brown color, and presumably also part of the secondary products interfering with the PCR. Heavily degraded DNA was obtained from tea samples after brewing, but was not a suitable template for PCR.

Drábková et al.[367] compared the performance of seven DNA extraction protocols used on herbarium specimens of *Juncus* and *Luzula* (Juncaceae) collected from 1927 to 1998. The best DNA quality was obtained with tissue ground in a mixer mill and isolated via the DNeasy Plant Mini Kit (Qiagen) or a modified CTAB procedure. Again, there was no obvious correlation between the age of the sample, the quantity of DNA as measured by fluorometry, and the quality of the DNA as determined by PCR of cpDNA target sequences. A modified CTAB procedure performed best (among five methods tested) for extracting DNA from grass samples stored for 25 to 100 years in airtight tins.[136] Yields and quality of DNA samples were low (average size 100 to 600 bp), but PCR amplification of chloroplast microsatellite loci was successful, even though heterologous primers were used.

In our experience, the question of whether a particular herbarium specimen yields DNA and PCR products is a matter of trial and error. We have successfully used the CTAB protocol described in Chapter 4.2.3 for herbarium leaves from various species of the genus *Suaeda* (Chenopodiaceae), which were 2 to 20 years old.[1253] However, the CTAB method failed in the case of *Alexandra,* a genus belonging to the same tribe Suaedeae (Chenopodiaceae). We were also unable to isolate PCR-amplifiable DNA from herbarium specimens of the paleotropical pioneer tree genus *Macaranga* (Euphorbiaceae), which were collected by the so-called Schweinfurth method. In this method, which was routinely used by many collectors traveling in tropical regions, the collected material is treated with 70% alcohol to inhibit molding. Unfortunately, information about the drying method used is often not available, and most herbaria are not able to remedy this situation.[841]

Taken together, the small amounts of template, the mostly degraded state of DNA, the presence of various (and possibly species-specific) PCR inhibitors, and contamination problems render the isolation of DNA from herbarium specimens difficult and unpredictable. To achieve the desired information for the species of interest, the experimenter will have to choose among the various strategies outlined above, and rely on trial and error. Unfortunately, the number of trials is often limited. Some herbarium specimens represent rare and valuable material, and destructive sampling will certainly not comply with the herbarium policies for DNA studies (see Loockerman and Jansen[841] for a survey of these policies, and for guidelines using herbarium specimens in molecular investigations).

Numerous DNA isolation procedures have been developed for other difficult substrates such as fossil animal specimens,[212,618] forensic samples,[1490] animal excrements,[448] and soil (e.g., Ernst et al.,[410] Porteous et al.,[1089] Vázquez-Marrufo et al.,[1457] and references therein). We will not elaborate on these methods because they are only occasionally of interest for botanists. Examples include the assessment of the persistence of transgenic plant material in the soil (e.g., Ernst et al.[410]), and the generation of species- or genus-specific plant root DNA fingerprints from total soil DNA.[178]

4.2.2.9 High-Throughput Procedures

Classical marker techniques such as restriction fragment length polymorphisms (RFLPs) required several micrograms of DNA per assay. DNAs had to be of relatively large size because restriction fragments >10 kb had to be analyzed reliably. Finally, DNA preparations had to be pure because restriction enzymes are highly sensitive to impurities. DNA extraction procedures therefore aimed at obtaining high yields of pure and undegraded DNA. In contrast, most of the presently used DNA profiling techniques are based on PCR and require only nanogram amounts of DNA. Moreover, purity and intactness of DNA preparations are of little concern when specific, short target fragments are amplified, e.g., as in microsatellite analysis. Hence, most of the more recently published DNA isolation protocols are mini- or even micropreps, starting with a few milligrams of plant material.

Minimizing the amount of starting material triggered the development of high-throughput procedures that allow the parallel processing of several hundred samples per day. Some of these methods are standard procedures adapted to a 96-well format[866,1283]; others require only a single tube or plate per sample,[636,1330,1479] and sometimes even consist of a single step.[190,1397] Common purification steps such as ethanol precipitation, or chloroform or phenol extraction are minimized or totally omitted in the latter procedures. Instead, aliquots of crude extracts are used for PCR, either directly or after dilution.

The simplest possible approach is to introduce untreated plant cells or pieces of tissue directly into the PCR assay. Such a strategy worked reasonably well for tobacco leaves[124] and for small amounts of a tomato pollen suspension.[805] This is somewhat surprising, considering the extraordinary chemical and physical stability of the pollen coat.[1290] In other methods, the tissue is treated with alkali[257,723] or squashed on a nylon membrane[775] prior to PCR. Lin et al.[821] described the use of FTA paper to collect and process plant material for PCR (see Chapter 4.2.1.4). Leaves are crushed against the FTA paper, small disks are collected using a punch, and the disks are washed with inorganic reagents and used directly for PCR. DNA preparations obtained by such "quick-and-dirty" approaches often contain PCR inhibitors. In some instances, inhibition may be counteracted just by diluting the extract. If this is not successful, various PCR additives, such as BSA, PVP, etc., will have to be tested for their anti-inhibitory effects (see Chapters 4.2.2.8, 4.3.2.4, and Savolainen et al.[1228]). High-throughput procedures are compiled in Appendix 1C.

Bottleneck problems associated with the time-consuming DNA preparation step are most efficiently solved by using an **automated DNA extraction platform**. Recently, fully automated genomic DNA isolation systems suitable for plasmids, bacterial artificial chromosome (BAC) and yeast artificial chromosome (YAC) clones, and animal and plant tissues have become available from several companies (e.g., AutoGen, Tecan, Boston Biomedica). These instruments rely on various principles. For instance, a pressure cycling technology-based sample preparation system was developed by Boston Biomedica and was introduced to the market in September 2002.[479] Hydrostatic pressure is used to disrupt the plant tissues and the contents are then released into a closed, single-use container. Nevertheless, manual work is still necessary to carry out the later stages of the DNA isolation procedure. HortResearch developed a fully automated system specifically designed for the extraction of DNA from difficult plant tissues such as apple leaves. The system is based on a modified magnetic bead protocol, operates in a 96-well format, and has the capacity to process 1152 leaf samples overnight. It will become commercially available during 2005 (see www.hortresearch.co.nz for more information). Automated DNA extraction systems will be an interesting option for marker-assisted selection and other projects of applied genetics for which large numbers of samples need to be processed at high-throughput rates. However, these instruments are expensive, and not at all suitable for small-scale experiments. We consider it unlikely that automated DNA extractors will replace mortar, pestle, bead mills, and microfuges in the average DNA laboratory in the near future.

4.2.2.10 Commercial Kits

A large number of commercial plant DNA isolation kits are on the market, including:

- Nucleon Phytopure™ Genomic DNA Extraction Kit (Amersham Biosciences[793])
- Fast-DNA and Geneclean® for Ancient DNA (Qbiogene)
- NucPrep™ (Applied Biosystems)
- Puregene® and Generation™ DNA Purification Systems (Gentra Systems; also available from Biozym Diagnostik[774])
- AquaPure™ DNA Tissue Kit (Bio-Rad Laboratories)
- Dynabeads® DNA DIRECT™ (Dynal[1199])
- Plant DNazol® reagent and Easy-DNA™ Kit (Invitrogen Life Technologies[251])
- NucleoSpin® Plant (Macherey-Nagel)
- peqGOLD DNA pure™ PT (peqlab)
- Wizard® Magnetic 96 DNA Plant System (Promega)
- DNeasy Plant kit; DNeasy 96 Plant kit, MagAttract® 96 DNA Plant Kit (Qiagen[1283])
- Extract-N-Amp™ Plant and Seed PCR kits (Sigma-Aldrich)

This list is not necessarily comprehensive. Trying a DNA isolation kit is a convenient, but usually more expensive alternative, especially if only costs of chemicals are taken into account. Using a commercial kit, however, may still be feasible if the alternative is to pay extra money for a laboratory technician to work with a complicated and time-consuming procedure. Commercial kits make use of one or more of the various principles summarized above, and are accompanied by detailed instruction manuals. Some kits are available in a 96-well plate format. They usually work well, and can be successful even for difficult substrates such as herbarium specimens[367] and ancient dry wood.[319,374] A description and discussion of the principles behind the various kits is beyond the scope of this book. We generally recommend that the reader follows the instructions of the suppliers.

4.2.2.11 Megabase DNA Isolation Protocols

The isolation of highly intact, unsheared DNA with average molecular weights in the range of 100 to 10,000 kb (so-called megabase DNA) is essential for physical mapping studies and the generation of large insert genomic libraries. Isolating such DNA from plants is much more difficult as compared with that from animals because plant cells have a rigid cell wall. Most widely used methods involve isolating protoplasts by enzymatic removal of the cell wall (e.g., Honeycutt et al.,[611] Wing et al.[1553]) or the isolation of nuclei (e.g., Hatano et al.,[578] Zhang et al.[1609]). To protect the DNA from physical shearing, isolated protoplasts or nuclei are embedded into low melting point agarose plugs or microbeads, and DNA molecules are isolated and manipulated *in situ*. Restriction digestions are performed by diffusing the enzymes into the agarose. The resulting fragments are then directly subjected to pulsed-field gel electrophoresis or eluted from the gel for further manipulations. Given that megabase DNA is generally not required for DNA profiling studies, we will not elaborate on these techniques here. A selection of relevant reports is compiled in Appendix 1E.

4.2.2.12 Choice of Procedure and Costs

Different kinds of studies demand different speed of processing, and different levels of DNA purity and intactness. Rapid and crude one-tube procedures (Appendix 1C) are the method of choice where robust, sequence-specific markers (e.g., microsatellites) have to be screened in large numbers of samples. This is the case, e.g., in monitoring the results of marker-assisted breeding programs and population studies. Simple high-throughput methods are also the most cost-effective, with costs per sample reported to be as low as Australian $0.60,[692] U.S. $0.25,[866] or even U.S. $0.07 per sample.[1042] Methods based on semispecific or arbitrary PCR such as RAPDs are more sensitive to DNA quality and size. For example, Ikeda et al.[636] found that some of the larger RAPD fragments were missing when a simple DNA isolation method was used, but present when the standard technique of Dellaporta et al.[323] was used. DNA purity is even more important in AFLP studies, which strongly depend on the complete digestion of template DNA by restriction enzymes. The latter techniques therefore require pure and relatively undegraded DNA preparations. In this case, costs will easily increase to U.S. $5 to $15 per sample (see also Csaikl et al.[287]).

High-throughput of samples and DNA preparations of good quality are also usually obtained with automated DNA extractors (Chapter 4.2.2.9) and commercial kits (Chapter 4.2.2.10). The disadvantage of using a kit is the cost, ranging from about U.S. $2 to $5 per sample. Recently, attempts were made to reproduce the efficiency of commercial kits by using do-it-yourself equipment and microtiter plates.[398,939] For example, Elphinstone et al.[398] presented a high-throughput, silica-based DNA isolation procedure for whale skin, processing 96 samples in parallel at one tenth of the cost (i.e., U.S. $0.40 per sample) of commercially available kits. By modifying the extraction step, this procedure may be adaptable for plants as well.

4.2.2.13   Storage of DNA Solutions

If DNA is sufficiently pure and of high molecular weight, it may be stored in 10 mM Tris-HCl, 1 mM EDTA (TE) at 4°C for years. In our laboratory, DNA samples isolated in 1989 by the CTAB procedure described below, and purified by cesium chloride centrifugation, still show no sign of degradation. DNA preparations are much less stable if impurities are left behind, a problem that appears to be most serious with DNA from herbarium specimens[367] and with methods involving Chelex.[1490] DNA samples may also be dried using a SpeedVac centrifuge, and stored under desiccated conditions in microfuge tubes. In general, we recommend long-term storage of dried or dissolved aliquots in desiccated boxes at –80°C.

4.2.3   CTAB Protocol I

CTAB is a cationic detergent that solubilizes membranes and forms a complex with DNA. The CTAB family of protocols dates back to Murray and Thompson,[965] and still represents the most widespread strategy of plant DNA isolation. A large number of variants of the standard procedure have been published (compiled in Appendix 1A). In its simplest version, the method involves the incubation of ground tissue in hot CTAB isolation buffer, followed by chloroform–isoamyl alcohol extraction, and alcohol precipitation of the CTAB–DNA complex from the aqueous phase. The DNA pellet resulting after centrifugation is washed, dried, and redissolved in TE buffer. In the protocol described below, we include an RNase treatment step to remove RNA. Polysaccharides are removed as described by Michaels et al.[924] (see Chapter 4.2.2.5). Additional purification steps may be needed for difficult species.

Solutions (in order of their appearance in the protocol)

Liquid nitrogen

Isolation buffer: 2% w/v CTAB, 1.4 M NaCl, 20 mM EDTA, 100 mM Tris-HCl, pH 8.0, 1% PVP 40,000, 0.2% β-mercaptoethanol (added just before use)

Chloroform–isoamyl alcohol (24 + 1; v/v) (see Safety Precautions, Chapter 4.1)

100% Isopropanol

70% Ethanol

TE buffer: 10 mM Tris-HCl, 1 mM EDTA, pH 8.0

RNase A solution: 10 mg/ml RNase A in 10 mM Tris-HCl, 15 mM NaCl, pH 7.5; boiled for 15 min, cooled to room temperature and stored at –20°C

5 M NaCl

100% Ethanol

Method

1. Grind up to 3 g of fresh or 0.5 g of lyophilized or dried plant material to a fine powder using liquid nitrogen and a mortar and pestle (see Comment 1). Liquid

nitrogen is not essential for lyophilized and dried tissue, but facilitates the grinding procedure considerably. For fibrous material, precutting of the leaves with scissors and/or the addition of some sterilized quartz sand may also help powdering. When using fresh tissue, make sure that the material does not thaw before being added to the isolation buffer. Otherwise, cellular enzymes may rapidly degrade the DNA. Material collected and stored in NaCl–CTAB solutions should be ground in prewarmed (60°C) isolation buffer.

2. Transfer the ground tissue into 15 ml of prewarmed (60°C) isolation buffer in a capped polypropylene tube or a 50-ml Erlenmeyer flask. Suspend clumps with the help of a spatula.

3. Incubate for 30 min at 60°C in a water bath. Mix every 10 min. The optimal incubation temperature and time may vary between different materials and/or species.

4. Add 1 vol of chloroform–isoamyl alcohol, cap the tube, and extract for 10 min on a rotary shaker or by hand. Mixing should be done gently but thoroughly enough to ensure emulsification of the phases.

5. Centrifuge for 10 min ($5000 \times g$, room temperature). Depending on the desired purity of DNA preparations, the (upper) aqueous phase may then be re-extracted one to several times with fresh chloroform–isoamyl alcohol.

6. Transfer the final aqueous phase to a centrifuge tube using a large-bore pipet.

7. Add 0.6 vol of 100% isopropanol, cover with parafilm, and mix gently but thoroughly by inverting the tube several times. At this stage, the DNA–CTAB complex may precipitate as a whitish network. In this case, spool the precipitate out of the solution using a glass hook (e.g., a bent Pasteur pipet), transfer it to 70% ethanol (Step 8), and let dry (Step 9). If the sample appears flocculent, cloudy or water-clear after mixing with isopropanol (which is often the case), collect the precipitate by low-speed centrifugation (10 min, $2000 \times g$, room temperature). If a pellet is visible, continue with Step 8. If not, place the solution at −20°C for 30 min to overnight and centrifuge again, perhaps at higher speed (see Comment 2).

8. Add 20 ml of 70% ethanol, gently agitate the pellet for a few minutes, and collect by centrifugation (10 min, $5000 \times g$, 4°C). Residual CTAB is removed by this step.

9. Invert the tubes and drain on a paper towel for about 1 h. Take care that pellets do not slip down the glass wall. Pellets should neither contain residual ethanol, nor should they be too dry. In both cases, redissolving may be difficult.

10. Add an appropriate volume of TE buffer (e.g., 1 ml; depending on the subsequent purification steps chosen) and let the pellets dissolve at 4°C without agitation. Intactness of DNA and the presence of contaminating polysaccharides are the main determinants of the duration of this step. Solubilization of high molecular weight DNA may take several hours to overnight.

11. Add heat-treated RNase A to a final concentration of 250 µg/ml. Mix, and incubate at 37°C for 2 h.

12. Add 0.05 vol of 5 M NaCl solution (final concentration 0.25 M), mix.

13. Add 0.35 vol of 100% ethanol, mix by inversion, and keep on ice for 10 min. Polysaccharides are precipitated in this step.[924]

14. Centrifuge for 30 min ($5000 \times g$, 4°C). Transfer supernatant to a new tube.

15. Add 1 vol of 100% isopropanol, mix by inversion, and store for 1 h at −20°C.

16. Centrifuge for 10 min ($5000 \times g$; 4°C), wash pellet in 70% ethanol, and centrifuge again.

17. Drain pellet, and dissolve in an appropriate volume of TE buffer.

Comments

1. If only small amounts of material are available, or many samples have to be processed simultaneously, a minipreparation is the method of choice. The protocol given above can easily be scaled down to 300 mg of fresh plant tissue (see also Milligan[933] and Rogers and Bendich[1183]). Homogenization is then best performed in a bead mill in the presence of liquid nitrogen. All incubation, precipitation, and centrifugation steps are performed within microfuge tubes, and all volumes given in the above procedure are reduced to one tenth.
2. It is preferable to perform the isopropanol precipitation (Step 7) at room temperature rather than at 4°C because contaminating compounds such as RNA, polysaccharides, and (if present) latex show less tendency to coprecipitate with DNA at elevated temperatures.[929]

4.2.4   CTAB Protocol II

A shorter miniprep variant of the CTAB method routinely used in our laboratories is based on the protocol of Lassner et al.[781] The plant material may be ground in liquid nitrogen in advance, and either kept frozen or lyophilized. Alternatively, fresh material may be ground directly in the Eppendorf tube with the aid of a small (disposable) grinder, which fits the tube exactly. This can be done with or without liquid nitrogen, quartz sand, or fine glass particles.

Solutions

Extraction buffer: 140 mM sorbitol, 220 mM Tris-HCl pH 8.0, 22 mM EDTA, 800 mM NaCl, 0.8% CTAB, 1% sarkosyl, 0.2% β-mercaptoethanol (added just before use)

Chloroform–isoamyl alcohol (24 + 1; v/v) (see Safety Precautions, Chapter 4.1)

100% Isopropanol

70% Ethanol

TE buffer: 10 mM Tris-HCl, 1 mM EDTA, pH 8.0

Method

1. Combine into a 1.5-ml microfuge tube: 1 ml of extraction buffer, 0.4 ml of chloroform–isoamyl alcohol, and plant tissue (approximately 30 to 50 mg dried tissue or 300 to 500 mg fresh tissue).
2. Incubate with shaking at 55°C for 10 min.
3. Centrifuge in a microfuge (5 min; $12,000 \times g$).
4. Transfer supernatant to a fresh microfuge tube.
5. Precipitate nucleic acids by mixing with 1.2 vol of 100% isopropanol.
6. Centrifuge in a microfuge (10 min; $12,000 \times g$).
7. Decant supernatant, and wash pellet twice with 1 ml 70% ethanol.
8. Drain pellet, and dissolve in 50 µl TE.

This method will usually yield sufficient DNA for numerous PCR reactions.

4.2.5   SDS–Potassium Acetate Protocol

The plant DNA isolation protocol introduced by Dellaporta et al.[323] is a reasonable alternative to the CTAB procedures, and also gave rise to numerous variants (see Appendix 1B). Its key step is the simultaneous precipitation of proteins and polysaccharides by high concentrations of potassium acetate in the presence of SDS. No organic extractions are required, which is a considerable advantage over other methods. Here we give a basic protocol that results in good and PCR-amplifiable DNA preparations for many species. Additional purification steps may be needed for difficult species (see Chapter 4.2.2).

Solutions

    Liquid nitrogen (optional)

    Isolation buffer: 100 mM Tris-HCl pH 8.0, 50 mM EDTA, 500 mM NaCl, 10 mM β-mercaptoethanol (added just before use)

    20% SDS

    5 M Potassium acetate

    100% Isopropanol

    TE buffer: 10 mM Tris-HCl, 1 mM EDTA, pH 8.0

    3 M sodium acetate, pH 8.0

Method

1. Grind up to 3 g of fresh or 0.5 g of lyophilized or dried plant material to a fine powder, using liquid nitrogen, a mortar, and a pestle; see also Step 1 of the CTAB method.
2. Transfer the powder into 15 ml of prewarmed (65°C) isolation buffer in a capped polypropylene tube. Suspend clumps with a spatula.
3. Add 1 ml of 20% SDS and mix thoroughly. Incubate for 20 min at 65°C in a water bath. Mix every 10 min.
4. Add 5 ml of 5 M potassium acetate, mix thoroughly, and incubate on ice for 30 min. In this step, proteins and polysaccharides are precipitated together with the insoluble potassium dodecyl sulfate, leaving the nucleic acids in solution.
5. Centrifuge for 20 min ($20,000 \times g$, 4°C). Filter the supernatant (which might still contain some floating particulate material) through cheesecloth into a new centrifuge tube. Precipitate nucleic acids by the addition of 0.7 vol of ice-cold 100% isopropanol followed by gentle mixing.
6. Incubate for 1 h to overnight at –20°C.
7. Centrifuge for 20 min ($20,000 \times g$, 4°C). Discard supernatant.
8. Invert the tubes and drain on a paper towel for about 1 h. Take care that pellets do not slip down the glass wall.
9. Redissolve pellets in 700 µl of TE buffer (overnight at room temperature, or 1 h at 65°C). Transfer the sample to a microfuge tube, and reprecipitate nucleic acids by adding 0.1 vol of 3 M sodium acetate and 0.7 vol of ice-cold isopropanol followed by mixing.
10. Incubate for 1 h to overnight at –20°C.

11. Pellet nucleic acids in a microfuge (10 min, 12,000 × g). Wash pellet with 70% ethanol, and centrifuge again.
12. Invert the tubes, and drain on a paper towel for about 1 h. Dissolve pellets in an appropriate amount (e.g., 100 μl) of TE buffer.

4.2.6   DNA Preparation via Nuclei

The isolation of plant DNA via nuclei may help circumvent several of the problems mentioned in Chapter 4.2.2. Most importantly, the majority of undesirable constituents such as cytoplasmic polysaccharides, polyphenols, and degrading enzymes are removed by the first centrifugation step, and the DNA is largely protected from shearing forces as well as from the action of nucleases. Isolated nuclei can therefore also serve as a source for megabase DNA.[1609] Here we present a shortened version of a method originally described by Willmitzer and Wagner[1549] for the isolation of nuclei from tissue-cultured tobacco cells. In this method, tissues are disrupted in a buffer containing polyamines, which stabilize nuclear structures. Nuclei are then purified by differential centrifugation, and DNA is prepared by lysis of the nuclei followed by proteinase K digestion. The protocol described below is for 10 g of leaf material, but can be scaled down to about 3 g, or scaled up to more than 50 g. A collection of nuclear DNA isolation methods is compiled in Appendix 1D.

Solutions

Isolation buffer: 250 mM sucrose, 10 mM NaCl, 10 mM morpholinoethane sulfonic acid (MES buffer reagent) pH 6.0, 5 mM EDTA, 0.15 mM spermine, 0.5 mM spermidine, 20 mM β-mercaptoethanol, 0.1% BSA, 0.6% NP-40 (a nonionic detergent). Note: Make this buffer as a 5× concentrated solution (but keeping NP-40 at 0.6%). Sterilize by filtration, and store in a closed bottle in a refrigerator. Dilute with cold distilled water prior to use, and readjust NP-40 concentration to 0.6%.

Floating buffer: Prepare immediately before use. Under constant stirring, add 90 g of cold Percoll® to 12 g of 5× concentrated isolation buffer. Readjust to pH 6.0 by dropwise addition of 1 M HCl. Do this carefully because overtitration may result in irreversible precipitation of Percoll.

Silicone emulsion (e.g., Serva) or 1-octanol (optional)

TE buffer: 10 mM Tris-HCl, 1 mM EDTA, pH 8.0

200 mM EDTA

10% Sarkosyl

2 mg/ml Proteinase K

Method (Note: Steps 1 through 8 are performed at 4°C)

1. Cut 10 g of freshly harvested tissue into 2- to 3-cm pieces, and collect these in a sterilized 200-ml Erlenmeyer flask on ice.
2. Add 50 ml of cold isolation buffer (5 ml/g of tissue) and a drop of silicone emulsion or 1-octanol (to prevent foaming). Homogenize in an Ultra-Turrax® homogenizer with increasing speed. The homogenization step should be brief to prevent the

suspension from warming. For leaf material, a homogenization of 4 times 15 sec is usually sufficient. Homogenization can also be performed using mortar and pestle. In this case, silicone emulsion is not needed, but a small amount of quartz sand should be added to help disrupting the tissue.

3. Filter the slurry through four layers of cheesecloth.
4. Centrifuge the filtrate in a swinging-bucket rotor for 10 min (2000 × g, 4°C). Nuclei and some starch are pelleted by this step.
5. Discard the supernatant. Resuspend the pellet in 20 to 40 ml of isolation buffer by pipetting up and down. Centrifuge as above.
6. Decant and discard the supernatant. Resuspend the pellet in 20 to 40 ml of floating buffer, and centrifuge in a swinging-bucket rotor for 10 min (5000 × g, 4°C). In this step, starch and most residual cell wall materials will sediment, whereas nuclei float on the top of the Percoll solution.
7. Collect the floating nuclei with the help of a Pasteur pipet or a spatula, and resuspend in 20 to 40 ml of isolation buffer. Centrifuge as in Step 4.
8. Resuspend the pellet in TE buffer. Lyse nuclei by adding 0.1 vol of EDTA, sarkosyl, and proteinase K stock solutions to yield final concentrations of 20 mM, 1%, and 200 µg/ml, respectively. At this stage, the DNA is liberated from the nuclei, and the suspension becomes highly viscous.
9. Incubate under gentle agitation for 10 min at 60°C followed by 2 h at 37°C.
10. Centrifuge for 15 min (5000 × g, room temperature) to remove remaining cell wall debris. The supernatant contains high molecular weight nuclear DNA, which may be either directly precipitated by adding 0.1 vol of 3 M sodium acetate and 2 vol of ethanol, or further purified, e.g., by cesium chloride centrifugation (see the previous edition of this book).

4.2.7 Quantitation of DNA

Three procedures are most widely used for estimating DNA concentration. One method is based on the spectrophotometric measurement of UV absorbance at 260 nm. A major disadvantage of this approach is that RNA, oligonucleotides, proteins, residual CTAB,[365,672] and other contaminants interfere with the measurement. Moreover, microgram amounts of DNA are needed to ensure reliable readings, and it is difficult to measure the absorbance in small volumes. The second method is based on the UV-induced fluorescence emitted by ethidium bromide–DNA complexes. Here, a sample of unknown concentration is compared with known standards, usually after electrophoretic separation. This method is less accurate than spectrophotometry, but preferred if only nanogram amounts of DNA are available, or if DNA samples are contaminated, e.g., by proteins or RNA. The third procedure is based on the highly specific binding of DNA to the bis-benzimidazole fluorescent dye Hoechst 33258 (for details, see Labarca and Paigen,[766] Sambrook and Russell[1217]). In this method, different dilutions of the DNA–dye mixture are prepared in microtiter plates alongside with standards, and the measurement is carried out in a fluorescence spectrophotometer. The DNA needs to be larger than ~1 kb, given that the dye binds poorly to small DNA fragments.[1217] Moreover, the DNA preparation must be free of ethidium bromide, which quenches the fluorescence of Hoechst 33258. In the following, brief protocols are presented for quantitating DNA by ethidium bromide staining after separation on agarose minigels and by spectrophotometry.

4.2.7.1 Ethidium Bromide Staining

In this procedure, DNA samples are subjected to agarose gel electrophoresis and subsequently stained with ethidium bromide. The dye intercalates into the DNA double helix, and the intensity of fluorescence induced by UV light is proportional to the amount of DNA in the corresponding lane. Comparison to a dilution series of standards, e.g., λ-DNA, gives an estimate of the amount of DNA in an unknown sample. In contrast to the spectrophotometric method outlined below, this technique allows, at the same time (1) DNA quantitation, (2) estimation of the extent of contamination by RNA, and (3) evaluation of DNA quality and integrity (i.e., the extent of degradation). Several variations of this approach exist, one of which is given below. For alternative procedures, see Sambrook and Russell.[1217] For details on setting up, running, and staining an agarose gel, see Chapters 4.3.4 and 4.3.6.1.

Solutions

> 0.8% Agarose in electrophoresis buffer
>
> Electrophoresis buffer: 1× TAE, 1× TBE, or 1× TPE (see Chapter 4.3.4 for buffer compositions)
>
> Gel-loading buffer: 30% glycerol, 1% SDS; 0.25% bromophenol blue
>
> DNA (e.g., phage λ):   Different concentrations (e.g., from 0.01 to 0.2 µg/µl), diluted in water or TE buffer.
>
> Staining solution: 1 µg/ml ethidium bromide in water (see Safety Precautions, Chapter 4.1)

Method

1. Mix an appropriate amount of the DNA sample (e.g., 5 µl) with 0.2 vol of gel-loading buffer. Mix 5 µl of each of a series of λ-DNA standards (this covers a range of 50 to 1000 ng of DNA if the concentrations given above are used) with 0.2 vol of gel-loading buffer.
2. Load samples along with standards onto a 0.8% agarose gel.
3. Electrophorese until the bromophenol blue dye front has migrated at least 2 cm.
4. Stain the gel for 30 minutes in a tray on a rotary shaker.
5. Evaluate the results on a UV transilluminator at 302 nm, using either a commercial documentation system equipped with a video or digital camera and a computer, or a Polaroid system with an orange filter (see Safety Precautions, Chapter 4.1). Estimate the quantity of the DNA in the sample by comparing the intensity of the fluorescence with the set of standard DNAs.

4.2.7.2 Spectrophotometry

This technique measures the total amount of nucleic acids in a sample (including DNA, RNA, oligonucleotides, and mononucleotides). It is therefore only useful for pure DNA preparations of a reasonably high concentration.

Solutions

> 1× TE: 10 mM Tris-HCl, 1 mM EDTA, pH 8.0, or distilled water

Method

1. Dilute an aliquot of the DNA sample in 1× TE or distilled water (usually in a ratio of 1:100; e.g., 5 μl/500 μl) in a microcuvet.
2. Determine the optical density (OD) at 260, 280, and 320 nm against a blank (1× TE or water).
3. Calculate the DNA concentration in the sample using the formula 1.0 OD_{260} = 50 μg/ml (under standard conditions, i.e., a 1-cm light path). The ratio OD_{260} to OD_{280} provides some information about the purity of the DNA sample. Pure DNA preparations show an OD_{260} to OD_{280} ratio between 1.8 and 2.0. Contamination (e.g., with proteins) results in lower values. The OD_{320} should be close to zero.

4.3 BASIC MOLECULAR TECHNIQUES

A number of basic molecular techniques are required for molecular marker studies, including the isolation of genomic DNA (treated in the previous section), digestion of DNA with restriction enzymes, running and staining of agarose and PAA gels, blot hybridization, radiolabeling of DNA, radioactive signal detection, and the PCR. In this section, we give some simple and general protocols for such standard procedures, together with a few brief comments. For more comprehensive and elaborate surveys and protocols of molecular biological methods, the reader is referred to well-known laboratory manuals such as those by Ausubel et al.,[56] Rapley,[1142] and Sambrook and Russell.[1217]

4.3.1 Restriction of DNA

Digestion of nuclear or organellar DNA with restriction endonucleases still plays a considerable role in many currently used molecular marker techniques, including RFLP (Chapter 2.2.3), cleaved amplified polymorphic sequences (CAPS; Chapters 2.3.2 and 4.9), and especially AFLP and its diverse variants (Chapters 2.3.8 and 4.7). Restriction enzymes are also usually required for the generation of microsatellite libraries.

A number of parameters need to be considered when choosing a particular restriction enzyme for DNA profiling experiments: length of the recognition sequence, frequency of cleavage, cost, and sensitivity of the enzyme to cytosine methylation of the target sequence. The statistical probability that a particular restriction site occurs in DNA is 1:256 and 1:4096 for four-base and six-base sequence motifs, respectively. Restriction enzymes with a four-base recognition sequence, also known as frequent cutters or four-cutters, are mainly used for CAPS, whereas restriction enzymes with six-base specificity (rare cutters or six-cutters) are preferred for classical RFLP analysis. The standard AFLP technique makes use of a pair of restriction enzymes, usually the rare cutter *Eco*RI (recognition site GAATTC) and the frequent cutter *Mse*I (recognition site AATT).

An important parameter for choosing an appropriate enzyme is its methylation sensitivity. A large number of restriction enzymes will not cut if a methylated cytosine is present in the recognition sequence.[904] For certain purposes such as genetic mapping, the use of methylation-sensitive restriction enzymes (e.g., *Pst*I) is recommended because these enzymes preferentially cut in undermethylated, i.e., presumably euchromatic and gene-rich regions (see Young et al.[1601]). Conversely, methylation-sensitive restriction enzymes should be avoided in genetic relatedness studies because the epigenetic nature of DNA methylation may introduce artefactual polymorphisms into the data set.[356] Variants of the RFLP and AFLP technology have been developed that aim at monitoring the extent of DNA methylation by using isoschizomers (i.e., restriction enzymes sharing the same target site, but exhibiting differential sensitivity to methylation)[1157,1580] (see Chapter 2.3.8.5).

In general, restriction digests should be performed under the conditions recommended by the enzyme manufacturers. A 10× concentrated buffer solution is usually supplied with the enzyme. The amount of starting DNA depends on the type of experiment. Whereas 5 to 10 µg of DNA is needed for a single RFLP experiment involving blot hybridization, ~200 ng to 1 µg will be sufficient for many AFLP reactions, and a 10-µl aliquot of the PCR products generated from a pair of specific primers will usually be adequate for CAPS.

To ensure complete digestion, at least 5 to 8 units (U) of enzyme per µg of genomic DNA should be used. For RFLP experiments, control digestions including phage λ-DNA in the sample may be performed. The appearance of λ-derived restriction bands of the expected sizes in the gel, superimposed on a smear of plant DNA fragments, indicates that the enzyme has worked properly. If this was not the case, inclusion of 100 µg/ml BSA or 4 mM spermidine in the restriction buffer may help. Complete digestion of λ-DNA, however, is only a hint but not a definite proof for complete restriction of genomic DNA. For unknown reasons, some target sites of a given restriction enzyme, even in purified genomic DNA, are less accessible than others (also if methylation is not involved), leading to a phenomenon known as hidden partials.[996]

The desired reaction volume is adjusted with distilled sterile water. If the digested DNA is directly applied to a gel, the total volume of the restriction assay must be adjusted to the space limits in the gel slots. In this respect, one has to take into account that restriction enzymes are usually supplied in 50% glycerol. Given that glycerol concentrations exceeding 5% may inhibit the enzyme or cause altered target site specificity, the contribution of the enzyme solution to the total reaction volume should not exceed one tenth. If the DNA sample is too dilute, it may be concentrated in a SpeedVac centrifuge. Digestion may also be performed in a larger volume. DNA is then precipitated after digestion by the addition of 0.1 vol of 3 M sodium acetate and 2 vol ethanol, left at −20°C for 30 min, spun in a microfuge (30 min, 10,000 rpm), washed with 70% ethanol, spun again, drained, and redissolved in an appropriate volume of electrophoresis buffer. Though somewhat more time consuming, digestion of diluted DNA may lead to better results for less pure samples.

The large-scale protocol given below assumes a total reaction volume of 40 µl and 5 µg DNA for each sample. The protocol can easily be scaled up or down, depending on the type of experiment. For example, larger volumes may be required for diluting enzyme-inhibiting impurities in the DNA preparation. Restriction enzymes are sensitive and expensive. They should not be taken out of the freezer unless actually needed, and should always be stored on ice or inside a cooling block. Fresh pipet tips should be used whenever enzymes are dispensed from the original tubes.

Solutions

> 10× concentrated restriction buffer (usually supplied by the manufacturer)
> Restriction enzyme (usually 5 to 50 U/μl)
> Genomic DNA
> Sterile, double-distilled water

Method

1. Pipet in the following order:

Distilled water:	(32 - x μl), where x is volume of DNA used
10× restriction buffer:	4 μl
Genomic DNA (5 μg):	x μl
Restriction enzyme (40 U):	4 μl (see Comment)
Total volume:	40 μl

   Preparing a master mix of water, buffer, and enzyme saves pipet tips and time. A master mix contains all components common to all reactions, while the variable component (here the DNA sample) is aliquoted separately. The mix is distributed into the reaction tubes prior to the addition of DNA.
2. Mix carefully and centrifuge in a microfuge for a few seconds to collect the ingredients at the bottom of the tube.
3. Incubate for at least 3 h to overnight at the incubation temperature recommended by the supplier (37°C for most enzymes).
4. If desired, inactivate the restriction enzyme by incubating the vials at 65°C for 15 min.
5. The sample can either be used directly for further processing, stored at –20°C, or ethanol-precipitated as described above.

Comment

Some experiments (e.g., AFLP analysis) require the simultaneous digestion of the template DNA with two or more enzymes. In this case, a buffer needs to be chosen in which all enzymes are sufficiently active and also retain their specificity (unspecific "star" activity must be avoided!). The various restriction enzyme buffer systems mainly differ in their salt concentration, and lists of percent enzyme activities in different buffers are available from the manufacturer. If no suitable buffer can be found, digestion should be started with conditions optimized for the enzyme requiring the lower salt concentration. After a couple of hours, the buffer concentration is adjusted according to the needs of the second enzyme, and digestion is continued.

4.3.2 Polymerase Chain Reaction

Practically all of the currently used molecular marker systems involve one or more steps of *in vitro* DNA amplification by PCR. The general principle of PCR is outlined in Chapter 2.3.1 and is not repeated here. Detailed PCR protocols are given in the various chapters dealing with the different molecular marker systems, i.e., PCR with arbitrary primers (Chapter 4.4), inter-repeat PCR (Chapter 4.5), AFLP and its variants (Chapter 4.7), locus-specific amplification of microsatellites (Chapter 4.8), and CAPS (Chapter 4.9). Optimizing PCR can be laborious, because numerous parameters influence the outcome. These include the architecture of the primer(s), the activity and amount of the polymerase, the temperature profile, concentrations of primers, template DNA and $MgCl_2$, template quality, the AT content of the template DNA, and the presence of certain additives.[245,631,1192,1274,1341] This section gives some general hints concerning the setup of a PCR and the influence of reaction conditions. For more elaborate treatments of the plethora of PCR protocols see, e.g., Innis et al.,[639] Mullis et al.,[960] Rapley,[1142] Roux,[1192] and Sambrook and Russell.[1217]

4.3.2.1 Primers

Some simple rules help to design efficient PCR primers. Ideally, specific primers should be 15 to 25 bases long, contain 40 to 60% GC, and anneal to the template at about 55°C, slightly below the melting temperature (T_m). Moreover, a primer should not contain sequences that allow hairpin formation and/or base pairing with itself or the complementary primer. Otherwise, so-called primer–dimer artefacts may appear on the gels. Primer sequences should neither contain palindromes nor repetitive motifs. The 3'-end should optimally consist of one or two Cs or Gs. Pairs of primers that are used together in the same reaction should have a similar size, annealing temperature, and GC content. Various computer programs are available that assist in primer design (e.g., Oligo,[1203] Primer_3,[1196] and Prime: GCG, University of Wisconsin), see also Appendix 3.

4.3.2.2 DNA Polymerase

A wide range of brands and types of thermostable polymerases suitable for PCR are commercially available. Most of these enzymes have no 3'- to 5'-exonuclease (proofreading) activity, but instead have a 5'- to 3'-exonuclease activity. Proofreading allows the polymerase to check for correct base pair matching, and, if necessary, to replace a false with the correct nucleotide. Truncated DNA polymerases have been designed that lack a 5'- to 3'-exonuclease activity. For example, the Stoffel fragment of AmpliTaq® DNA polymerase lacks 289 amino acids in the N-terminal region, exhibits a higher thermostability than full-length DNA polymerases, and tolerates higher Mg^{2+} concentrations. Use of the Stoffel fragment in arbitrarily primed PCR was reported to result in the enhanced reproducibility of small fragments.[1303] However, due to a slower processing ability, twice as much of the enzyme is required per reaction. AmpliTaq Gold is a DNA polymerase specifically modified for hot start PCR (see Chapter 4.3.2.3). The enzyme is supplied in an inactive state and requires a preincubation period of 2 to 10 min at 95°C for activation.

4.3.2.3 Thermocycler and Temperature Regimen

Commercially available thermocyclers are usually equipped with 48 or 96 wells (sample positions) in a block, others are adapted for microtiter plates, and some instruments can deal with both kinds of equipment. Thermocyclers can be programmed to various extents. Many applications demand that the transition time between temperatures (also called ramp time or slope) can be programmed (e.g., slow ramp times are recommended for the transition between the annealing and the elongation step in RAPDs; see Chapter 4.4.2.4 and Ellinghaus et al.[392]). In so-called gradient cyclers, different annealing temperatures can be programmed in different wells. This option is particularly useful when annealing temperatures have to be optimized for numerous primers (which is often needed for microsatellite-flanking primers, see Chapter 4.8.5.7). Some thermocycler models use capillary tubes, which allow very short runs (20 sec per cycle) due to the rapid temperature exchange between the reaction mix and the surrounding air stream. Thin-walled tubes also give a better temperature exchange. Modern thermocyclers are equipped with a heated lid, which obviates the need to add mineral oil to the PCR samples.

The temperature regimen should be consistent across all wells. In the past, this has not always been the case.[826,1064] To test for position effects, fragment profiles from an identical template DNA, amplified in several wells, may be compared with each other. Position effects may also be overcome by filling the empty spaces between tubes with dummy vials.[1452] Generally, position effects seem to be negligible in new instruments.

For the temperature profile, the most important point of attention is the annealing temperature. According to Innis et al.,[639] the T_m of primer and annealing site can be calculated from a general rule of thumb (2°C for each AT-pair and 4°C for each GC-pair). However, this rule was originally developed for oligonucleotide hybridization in solutions containing 1 M NaCl (so-called Wallace rule[1391]), and may not be applicable for calculating reliable annealing temperatures in PCR. Computer programs that assist in primer design (see above) also calculate T_m values of primers and primer pairs according to different algorithms.

For specific and semispecific PCR (for definitions, see Chapter 2.3.1), the annealing temperature is usually set to about 5°C below T_m. For PCR with arbitrary primers, low annealing temperatures are typically used that allow a certain extent of primer–template mismatching (see Chapter 4.4.2.4). The efficiency of primer binding depends on various reaction parameters; hence each T_m value calculated by any of the above methods should be considered a first approximation that needs to be tested. Such testing is facilitated if a gradient cycler is available. Highly AT-rich sequences may require reduced extension temperatures (60 or 65°C instead of 72°C[1341]) and longer extension times.

Nonspecific priming and primer–dimer formations can occur when reactions are set up at room temperature. This effect can be counteracted by a so-called **hot start**. The idea behind this approach is to prevent primer, template, and polymerase from interacting with each other before the denaturing temperature is reached. This can be achieved by preamplification heating[293] or by withholding at least one reagent from the reaction mixture until the tube has reached an appropriate temperature.[85,252] In a technique described by Chou et al.,[252] an aliquot of the master mix ($MgCl_2$, Tris-HCl, KCl, gelatin, deoxynucleotide triphosphate [dNTP], primer) is dispensed in each reaction vial, and a piece of wax is added to each vial. The wax is melted by placing the vials in the thermocycler heated to 80°C. When the sample returns to room temperature, a layer of solid wax is formed. The remainder of the reaction mix containing the DNA polymerase is then added together with the template, and the program is started. More recent hot start variants avoid additional handling steps by including a specific anti-*Taq* DNA polymerase antibody in the PCR mixture prior to the addition of the enzyme. This antibody will block the polymerase until it is denatured and deactivated at high temperatures. Finally, enzymes specifically designed for hot start PCR have been designed (see Chapter 4.3.2.2).

Touchdown PCR is a strategy to increase PCR specificity.[354] Cycling is started at annealing temperatures well above the calculated T_m. Sequential cycles are then run at gradually decreasing annealing temperatures. This strategy ensures that the first primer–template hybridization event is as specific as possible, whereas high yields of these initial specific products are obtained at later cycles. Roux[1192] suggested that the annealing temperature range should span about 15°C, starting from a few degrees above the estimated T_m.

4.3.2.4 Template Quality

Template quality may have a considerable influence on the results of a PCR. Various compounds can exert inhibitory effects on PCR; among these are polyphenols, polysaccharides, and RNA. The optimal but also most cumbersome way to avoid such inhibitory effects is the preparation of very clean templates, e.g., by means of CsCl centrifugation[1138] (see Chapter 4.2.2.12 and the previous edition of this book). However, there are also various alternatives that may counteract inhibiting activities:

1. The easiest strategy is a dilution of the template DNA.[1037,1228] This may reduce the concentration of contaminants below a critical threshold, whereas the DNA can often still be amplified.
2. Acidic polysaccharides[1037] and RNA[1081,1592] are all known to inhibit RAPD PCR, whereas neutral polysaccharides are relatively inert.[1037] Choosing a DNA isolation procedure that effectively removes the above compounds may solve problems of PCR inhibition (see Chapter 4.2 and Appendix 1). Inhibition by acidic polysaccharides can also be counteracted to a certain extent by including Tween 20, dimethylsulfoxide (DMSO), or PEG 400 at various concentrations in the PCR.[324]
3. A somewhat unusual remedy is the addition of 1.5 to 5% nonfat milk powder to the PCR mix.[306]
4. Micheli et al.[927] reported that ethanol-precipitable contaminants in genomic DNA from rat tissues are a major source of irreproducibility, and recommend spooling of ethanol-precipitated DNA with a glass rod. Unfortunately, this recommendation will rarely apply to plant DNA preparations, in which insufficient intactness of the DNA molecule often makes winding on a rod impossible.

If the template is heavily degraded (which is often the case for DNA derived from herbarium specimens), the enzyme may jump between templates, thereby forming *in vitro* recombination products.[1023] Such jumping is induced by breaks in the template. PCR products of degraded DNA may therefore reflect chimeric molecules.

4.3.2.5 Yield and Specificity

Unfortunately, not all PCRs result in the production of pure target fragments in high yields. For example, no product at all may be amplified if an inhibitor is present (see above). More often, however, multiple undefined and unwanted products are generated in addition to, or even to the exclusion of the desired fragment. Such undesirable results require optimization of specificity and yield.

Standard approaches to increase PCR specificity include (1) raising the annealing temperature, (2) using a touchdown[354] or a hot start[293] protocol (see above), and/or (3) reducing the concentration of $MgCl_2$. In doing this, the chelating effect of EDTA that may affect Mg^{2+} concentrations in the reaction vial has to be taken into account. To minimize chelating effects, many researchers prefer to store their DNA samples in 10 mM Tris, 0.1 mM EDTA ($TE_{0.1}$) instead of the more commonly used 10 mM Tris, 1 mM EDTA (TE).

Many other strategies to increase PCR specificity have been reported. For example, Sharma et al.[1274] found that digesting genomic or plasmid templates with a restriction enzyme prior to PCR eliminated unspecific amplification. They stressed that any frequent-cutting enzyme can be chosen for this purpose, provided that it does not cleave within the desired amplification product. The molecular basis of the effect remained unclear.

Specificity may also be increased by **nested PCR**. In this strategy, an aliquot of the initial PCR product is reamplified by a pair of internal (nested) primers (one internal and one external primer in semi-nested PCR). Only the legitimate product is expected to be amplified in this second round, because spurious products of the first reaction are unlikely to contain recognition sites for the nested primers used in the second reaction. Nested PCR was used, e.g., to enhance PCR specificity in the transposon display approach described by Frey et al.[461] (see Chapter 2.3.8.2).

Finally, numerous chemicals were reported to enhance PCR specificity and/or yield if added at certain concentrations.[803,1192,1335,1493] Suggested additives include DMSO (2 to 10%), PEG 6000 (5 to 15%), glycerol (5 to 20%), formamide (5%), nonionic detergents such as Tween and Triton X-100 (0.1 to 1%), BSA (0.1%), spermidine (1 mM), and gelatin (0.1%). A frequently used PCR additive is tetramethylammonium chloride (TMAC),[245,631] which is known to equalize the thermal stability of AT and GC base pairs. Hung et al.[631] reported a marked reduction of unspecific priming when TMAC was included in the PCR at a final concentration of 50 µM. Later, Chevet et al.[245] showed that TMAC concentrations between 10 and 100 mM increased both yield and specificity of PCR products, irrespective of the type of thermostable DNA polymerase, source of template, GC content, and annealing temperature of the primers. However, TMAC concentrations above 150 mM completely inhibited the PCR.[245] Finally, Kovárova and Dráber[744] stressed that tetramethylammonium oxalate is an even more efficient specificity and yield enhancer for PCR than TMAC. Kits of PCR specificity enhancers are commercially available from various companies (e.g., Stratagene, Applied Biosystems, Invitrogen).

4.3.2.6 Contamination

The high sensitivity of PCR can also become a major problem, given that artefactual products may result from contamination with unwanted template DNA. Contaminations can originate from numerous sources, including aerosols, skin or hair of the researcher, or product carryover from a previous reaction. Contamination problems can be particularly serious when PCR is used for the detection of small quantities of DNA; when arbitrary, universal, or otherwise unspecific primers are used (as is the case in the RAPD procedure); and when several organisms are living in close association with each other (e.g., in the case of parasites and endophytes[1206,1326,1613]). Several precautions can minimize this problem:

- Pre- and post-PCR steps should be performed in separate rooms, using separate pipet sets.
- High-purity water should be used to prepare the reaction mixtures. Positive-displacement pipets also help to avoid contamination.
- Stock solutions of the reaction components (buffer, dNTPs, MgCl$_2$ solution) should be divided into small aliquots (0.5 to 1 ml) stored at –20 or –80°C. If contamination problems arise, all reaction components should be replaced by new batches, or each batch should be tested individually.
- Clean gloves should be used to handle the pipet tips and to prepare the reaction mixtures.
- The reaction mixture (without DNA polymerase and target DNA) may be irradiated with shortwave UV light prior to PCR to destroy contaminating DNA.
- Negative control reactions (no DNA added) should be included in each run.
- Consider that DNA polymerase preparations may contain traces of DNA from the organism from which they were isolated.

4.3.3   DNA Sequencing

DNA sequencing obviously reveals DNA sequence variation in unparalleled detail, and is therefore the most accurate molecular marker method available. However, variation is only detected in a tiny fraction of the genome. For reconstructing phylogenetic trees and phylogeographic networks, DNA sequence data are nevertheless preferred over fragment-based molecular markers (see Chapter 2.2.2). Sequence data are also needed for the design of microsatellite-flanking primers (Chapter 4.8.5.7), and for the conversion of RAPD or AFLP markers into sequence characterized amplified regions (SCARs; see Chapter 2.3.3.4).

Two basic strategies of DNA sequencing were devised in the mid-1970s. The so-called chemical degradation method of Maxam and Gilbert[899] employs chemicals that cleave behind specific bases in an end-labeled DNA molecule. This treatment generates four nested sets of labeled cleavage products, each terminating at a specific base. After separation on highly resolving, denaturing PAA gels (i.e., sequencing gels; see Chapter 4.3.5.2), these fragment sets are visualized by autoradiography. The resulting sequence ladder can be read directly from the autoradiogram.

The chain termination method described by Sanger et al.[1219] exploits the 5′- to 3′-strand extension activity of a DNA polymerase in the presence of a base-specific chain terminator. A typical Sanger sequencing reaction is set up to contain the denatured template DNA, a target-specific sequencing primer, the DNA polymerase, and the four dNTPs in an appropriate buffer system. Reactions are aliquoted into four microtubes, each containing a specific 2′-3′-dideoxynucleotide (ddATP, ddTTP, ddCTP, or ddGTP). These nucleotide analogs are properly recognized by the polymerase and incorporated into the growing chain. However, because ddNTPs lack a

3′-OH group, they are not extended, and the chain is terminated at the precise position where the ddNTP is added. The concentration ratios of dNTPs and ddNTPs need to be carefully balanced, so that all possible termination points along the DNA chain are represented in the set of reaction products. Labeling is achieved by incorporating either labeled primers or labeled ddNTPs in the reaction. As in the case of the chemical method, a ladder of bands is produced after high-resolution, denaturing PAA gel electrophoresis.

The Sanger method has become the most widely used technique for sequencing DNA, and several variants have been developed. Whereas a genetically engineered form of phage T7 DNA polymerase (Sequenase®) is routinely used in the standard protocol, the so-called cycle sequencing variant[967] employs a thermostable *Taq* DNA polymerase and a thermal profile to generate single-stranded sequence template by asymmetric PCR. Cycle sequencing works equally well with double-stranded and single-stranded templates, and requires less template DNA than the standard methodology.

Although based on the same principle, currently used sequencing protocols differ slightly from each other, depending on the type and conformation of the template (plasmid vs. PCR product; single- vs. double-stranded DNA), type of label employed (radioisotopes vs. fluorescence), target of the labeling reaction (primers vs. terminators), and the type of polymerase used (thermostable vs. thermosensitive; for a survey, see Chapter 12 of Sambrook and Russell[1217]). Given that DNA sequencing is not the primary focus of this book, no attempts are made to provide step-by-step protocols. For those who only occasionally need to generate DNA sequence data, it would be wise to consult a custom sequencing facility, or give the sample to a neighboring laboratory running an automated sequencer. For those who prefer to set up DNA sequencing reactions on their own, we generally recommend the use of a commercial kit. Kits are available from various suppliers (e.g., Amersham Biosciences, Stratagene, Applied Biosystems, Promega), and usually come with a detailed instruction manual. One can choose among kits specifically designed for manual or automated sequencing, labeled primers or labeled terminators, PCR products or plasmids, cycle sequencing or isothermal sequencing, and radiolabeling or fluorescence labeling.

4.3.4 Agarose Gel Electrophoresis

DNA fragments ranging in size from about 100 to 10,000 bp are usually resolved by standard agarose gel electrophoresis. To prepare an agarose gel, powdered agarose is melted in electrophoresis buffer to yield a clear solution, cast into a gel mold, and allowed to solidify. DNA samples are applied to the gel, and a constant electric field is imposed. Under neutral or slightly alkaline conditions, DNA migrates toward the anode. Because the agarose matrix is acting as a molecular sieve, with pore sizes depending on the agarose concentration, DNA fragments up to about 10 to 15 kb can be separated according to size. To resolve larger fragments efficiently, one of the numerous variants of pulsed-field gel electrophoresis is required.[611,1217] After electrophoresis is completed, DNA in the gels is stained with ethidium bromide or another intercalating dye, and the results are documented by photography or by a video camera attached to a computer.

Table 4.1 Optimal Separation Ranges of Agarose
and Nondenaturing PAA Gels

Agarose		PAA	
Percentage	Fragment Size Range (kb)	Percentage	Fragment Size Range (bp)
0.3	5–60[a]	3.5	100–2000
0.6	1–20[a]	5.0	80–500
0.7	0.8–10	8.0	60–400
0.9	0.5–7	12.0	40–200
1.2	0.4–6	15.0	25–150
1.5	0.2–3	20.0	6–100
2.0	0.1–2		

[a] Pulsed-field gel electrophoresis is required for efficient resolution
of DNA fragments > 15 kb.

Source: From Sambrook, J., and Russell, D.W., (2001) *Molecular
Cloning: A Laboratory Manual, 3rd Edition*, Cold Spring Harbor Laboratory Press, Cold Spring Harbor, NY. With permission.

Agarose gels can be prepared in different concentrations, ranging from about
0.3 to 2.5%. Low gel concentrations (~0.4 to 0.8%) are applied for electrophoresing
intact genomic DNA, whereas 1.2 to 2% gels are normally used for RAPDs, microsatellite-primed PCR, CAPS, and RFLPs (Table 4.1). Agarose gels below 0.6% are
soft, slippery, and difficult to handle. In addition to different brands of agarose,
special separation media or mixtures of agarose and other chemicals are commercially available; e.g., Visigel™ (Stratagene), Synergel™ (Diversified Biotech),
NuSieve™ (FMC Biopolymers), and MetaPhor™ (FMC Biopolymers). It is claimed
that these brands give a better resolution than agarose gels and/or a higher strength
than PAA gels, but they are also more expensive than standard agarose.

Three buffer systems are generally used: TAE, TBE, and TPE[1217] (see below for
buffer compositions). The resolving powers of all three buffer systems are almost
identical. TAE has the lowest buffering capacity of the three, and electrophoretic
runs using TAE should not exceed 24 hours without buffer recirculation or exchange.
The higher buffering capacity of TBE allows to use it in a 0.5× concentration in
agarose gels.[1217] Stock solutions of TBE have a tendency to form whitish precipitates
upon storage and should be used up in a few days. According to our own observations,
good RAPD patterns and least so-called smiling of bands are obtained using TPE.

For RFLP analysis, the agarose gel will have to be blotted (or dried) and
hybridized to a suitable probe (see Chapters 4.3.7 to 4.3.10). About 5 to 10 μg of
restriction-digested DNA per lane will be needed for such an experiment. For RAPD,
MP-PCR etc, it is usually sufficient to load about 10 to 15 μl of a 25-μl assay, and
keep the remainder for further analyses. Overloading results in inferior banding
patterns and should be avoided. At least two or three lanes should be loaded with
0.1 to 1 μg (depending on the gel dimensions) of an appropriate molecular weight
marker. A wide variety of markers exhibiting different size ranges are commercially
available. Ladders containing fragments with even size distributions (e.g., 100-bp
ladder, 1-kb ladder) became very popular. Inexpensive alternatives are homemade

size markers produced by cleaving plasmid or phage DNA (e.g., λ-DNA) with a set of suitable restriction enzymes.

Solutions

Electrophoresis buffers (one of the three following):

1× TAE buffer:	40 mM Tris-acetate, 1 mM EDTA, pH 8.0 (adjust pH with glacial acetic acid)
0.5× TBE buffer:	45 mM Tris-borate, 1 mM EDTA, pH 8.0 (adjust pH with boric acid)
1× TPE buffer:	90 mM Tris-phosphate, 2 mM EDTA, pH 8.0 (adjust pH with 85% phosphoric acid)
Note:	Prepare electrophoresis buffers as 5× (TBE), 10× (TPE), or 50× (TAE) concentrated stock solutions and dilute prior to use.
Loading buffer:	0.25% bromophenol blue, 0.25% xylene cyanol, 30% glycerol in electrophoresis buffer or water

Molecular weight marker (see above)

Agarose:	0.3 to 2% in electrophoresis buffer
Staining solution:	1 µg/ml ethidium bromide in electrophoresis buffer or water (see Safety Precautions [Chapter 4.1] and Chapter 4.3.6)

Method

1. Suspend agarose at the desired concentration in an appropriate amount of electrophoresis buffer in a bottle or flask (e.g., 2.8 g of agarose per 400 ml of electrophoresis buffer yields a 0.7% gel). The gel slurry should not occupy more than 50% of the bottle. Flasks should be covered, and bottles should be loosely capped. Metal flasks and aluminium foil must not be used in a microwave oven!
2. Boil the suspension in a microwave oven for 2 to 4×2 min. Swirl the bottle in between. Continue until the agarose is completely dissolved. After complete melting, the solution should be clear and free of particles. Beware of superheating! If heated too long, the agarose solution may suddenly begin to boil violently when swirled. Wear insulating gloves when handling melted agarose.
3. Allow the agarose to cool to 60°C. Stirring helps to prevent uneven cooling. In the meantime, seal the edges of the plastic tray supplied with the electrophoresis apparatus, e.g., using tape. Insert a slot-forming comb. Band resolution is, to some extent, dependent on the shape of the teeth of the comb: sharp teeth yield sharp bands, but also allow less volume to be applied. Check that the teeth are not too close to the bottom of the gel mold. Fine holes in the bottom of a slot might allow your sample to escape in an undesired direction.
4. Make sure that the gel mold is in a horizontal position. For some electrophoresis apparatuses, the gel is best poured with the mold already in place. Slowly pour the agarose into the gel mold, remove small air bubbles with a yellow pipet tip, and allow the agarose to solidify. This usually takes about 1 h at room temperature, or less time in the cold room (4°C).
5. When the agarose is solid, carefully remove the comb and the tape, and insert the gel mold into an electrophoresis apparatus filled with buffer. Electrophoresis runs best if there is not too much buffer on top (about 5 mm). Remove air bubbles

from the slots. Connect the apparatus to a power supply and check whether it is working correctly (before applying the samples).

6. Add 0.2 vol of loading buffer to the DNA samples, mix, and centrifuge for a few seconds in a microfuge to collect the samples at the bottom of the tubes. Mixing of samples and loading buffer can also be done in a microtiter plate, or on a piece of Parafilm. The loading buffer adds color and provides a higher density to the samples, thus allowing them to be applied conveniently to the slots. Moreover, the dyes are moving toward the anode when voltage is applied, and give you an idea how far the electrophoresis has proceeded. Bromophenol blue and xylene cyanol migrate at about the same rate as linear, double-stranded DNA fragments of 300 and 2000 bp, respectively.

7. Slowly load the samples into the submerged slots.

8. Turn on the power supply and start the electrophoresis. For large RFLP gels, running conditions are usually 1 to 2 V/cm (i.e., distance between the electrodes) for 4 to 24 hours, depending on the gel dimensions. Longer runs give better resolution of large fragments. To minimize diffusion, these gels should be preferably run in the cold (4°C) or with water cooling.

9. After the run is completed, remove the gel from the apparatus and stain for 15 to 60 min (depending on the gel thickness) in a tray with staining buffer as described (Chapter 4.3.6.1). If the gel needs to be processed further (e.g., by Southern blotting), place a fluorescent ruler alongside the gel to align marker sizes in the gel with fragment sizes in forthcoming autoradiograms.

4.3.5 PAA Gel Electrophoresis

Polyacrylamide (PAA) gels are prepared from a mixture of monoacrylamide and bisacrylamide. The ratio of these two chemicals influences the specific molecular structure (pore size) of the polymeric net. To start the polymerization, which is based on a free-radical mechanism, TEMED and ammonium persulfate are added to the buffered gel solution. The formation of free radicals is initiated by ammonium persulfate, and catalyzed by TEMED. Because the polymerization process is inhibited by the presence of oxygen, it is advised to remove all oxygen from the gel solution by deaeration with a vacuum pump. Because of the inhibitory action of oxygen, small air bubbles can produce relatively large holes of unpolymerized material in the gel.

PAA gels are generally preferred for achieving high resolution in the low molecular weight range. Gels of different concentrations can be prepared covering a wide range of molecular weights, from 3.5% gels (suitable for 100- to 2000-bp molecules) up to 12% gels (40- to 200-bp molecules; Table 4.1). Diffusion of small molecules is less pronounced than in agarose gels, and size differences of a single base pair may be scored.

4.3.5.1 Nondenaturing PAA Gels

Nondenaturing PAA gels are used for the separation of double-stranded PCR fragments. Fragments are not only separated according to their molecular weight, but base composition and sequence may cause up to a 10% difference in running distance. The protocol given below is for a nondenaturing 5% PAA gel of 15 × 13 × 0.2 cm, but can easily be adapted to other sizes or other percentages.

Solutions and chemicals

Acrylamide stock solution: 38% acrylamide and N',N'-methylene bisacrylamide (19:1 ratio) (monomers of acrylamide and bisacrylamide are neurotoxic; see Safety Precautions, Chapter 4.1!)

1× and 10× TBE buffer   (see Chapter 4.3.5)

TEMED

10% ammonium persulfate solution (freshly prepared; or stored at 4°C for less than 1 week)

Loading buffer: 0.25% bromophenol blue, 60 mM EDTA, 30% glycerol (or loading buffer used for agarose gels; see Chapter 4.3.5)

Method

1. Clamp the two thoroughly cleaned glass plates of the electrophoresis apparatus with spacers in between and seal the spacers with, e.g., 2.5% agarose, PAA sealing gel, tape, or sealing strips (see Comment 1).
2. Mix 5.3 ml of acrylamide stock solution with 4 ml of 10× TBE and 30.42 ml H_2O and deaerate under vacuum for 10 min. A vacuum-proof sidearm flask or a hypodermic needle that is inserted through the rubber top of a small laboratory bottle can be used. The acrylamide stock solution should not be older than 2 to 4 weeks.
3. Add 20 µl of TEMED and 280 µl of 10% ammonium persulfate solution and mix the solution gently, avoiding air bubbles.
4. Pour the solution between the plates (see Comment 1) and put the slot-former in place. A 50-ml syringe may be helpful for casting thin gels. Air bubbles are avoided by tilting the gel mold and starting to pour the gel on one side.
5. Polymerization takes approximately 1 h. The gel can be stored overnight if kept humid (e.g., wrapped in tissues moistened with 1× TBE buffer).
6. Remove the tape from the glass plates and the spacer on the bottom (if there is one). Insert the glass plates with the gel into the electrophoresis apparatus.
7. Fill the electrophoresis tanks with 1× TBE, remove the slot-forming comb, and clean the wells thoroughly using a pipet. Pre-electrophorese the gel for 20 min before loading the samples. Wells have to be cleaned again before sample application.
8. Mix the DNA sample with 0.2 vol of loading buffer. Samples may be concentrated if volumes are too large: (1) by precipitation of the DNA and dissolving in a smaller volume, and (2) by using a vacuum concentrator. Use special narrow tips or a Hamilton syringe to deposit the sample at the bottom of the well. With some experience, it is also possible to use conventional yellow pipet tips. Apply a suitable molecular weight marker to one or more of the lanes.
9. Run the gel at 15 to 100 V (1 to 8 V/cm) at room temperature.
10. After electrophoresis, remove the gel from both glass plates, and stain with ethidium bromide or silver nitrate (see Chapter 4.3.6 and Comment 2).

Comments

1. Different types of PAA gel apparatus are commercially available, and different strategies of how to seal the edges and how to cast the gel are recommended by

the manufacturers. Some models are equipped with a casting tray and others are filled from below using a syringe. We generally recommend that the user follow the instructions of the manufacturer.

2. The gel can also be stained while still attached to one glass plate, which facilitates the handling especially of thin or low percentage gels.[225] In this case, one plate has to be pretreated with bind-silane, and the other with repel-silane following the protocols described in the next section (Chapter 4.3.5.2), or those given by Briard et al.,[166] Sambrook and Russell,[1217] or Tegelström.[1376] The gel will stick to the glass plate treated with bind-silane, and can be dried on this glass plate for a permanent (but very expensive) record.

4.3.5.2 Sequencing Gels

Highly resolving, denaturing PAA gels (also called sequencing gels) are used for separating the sets of single-stranded DNA fragments resulting from DNA sequencing (Chapter 4.3.3). The same kind of gels is also routinely used for AFLP (Chapter 4.7) and microsatellite analysis (Chapter 4.8), as well as for some variants of RAPD (e.g., Welsh and McClelland,[1527] Caetano-Anollés et al.[201,202]; Chapter 4.4.2) and microsatellite-primed PCR (MP-PCR[1621]; see Chapter 4.5.2). Samples are heat-denatured in a buffer containing formamide prior to loading. Including urea at high concentration in the gel and running the gel at high temperatures (50 to 55°C) prevent the DNA fragments from renaturing, and thus DNA mobility is not influenced by base composition.

Solutions and chemicals

Bind-silane

Repel-silane

Gel stock solution: 8 M urea, 6% acrylamide: N',N'-methylene bisacrylamide (19:1 ratio) in 1× TBE (see Safety Precautions, Chapter 4.1). Dissolve 288 g urea, 34.2 g acrylamide, and 1.8 g bisacrylamide in H_2O, add water to 500 ml, dissolve completely by stirring overnight. Add 30 ml of 20× TBE stock solution and fill up to 600 ml. Pass through filter paper and store in a flask wrapped in tin foil. The solution is stable at room temperature for several weeks (see Comment 1).

25% ammonium persulfate solution (freshly prepared; or stored at 4°C for less than 1 week)

TEMED

1× and 20× TBE buffer (same stock solution can be used for gel and running buffer)

0.5 M EDTA, pH 8.0

Loading buffer: 98% formamide, 0.1% bromophenol blue, 0.1% xylene cyanol, 10 mM EDTA. Dissolve 10 mg of each of both dyes in 10 ml formamide (high quality or deionized) and add 200 µl of 0.5 M EDTA stock solution. Mix and store in aliquots at –20°C.

Method

1. After about 10 runs, one or both glass plates need to be siliconized. Treatment with bind-silane reagent ensures that the gel adheres to the glass plate after the end of the run. The other plate is treated with repel-silane. For siliconization, thoroughly clean glass plates with water and detergent, dry with a paper towel, and place the plates in a horizontal position in a fume hood. Apply about 0.5 to 1 ml of bind-silane or repel-silane onto the plate, and evenly disperse the liquid across the glass surface using gloves and a piece of paper towel. After evaporation of the reagent, rinse the plates with deionized water and allow them to dry.

2. Clamp the two thoroughly cleaned glass plates of the electrophoresis apparatus with spacers in between, and seal the spacers with, e.g., 2.5% agarose, PAA sealing gel, tape, or sealing strips (see Comment 2).

3. Fill 60 ml of gel stock solution into a suitable flask, and deaerate under vacuum for 10 min. A vacuum-proof sidearm flask can be used, or a hypodermic needle that is inserted through the rubber top of a small laboratory bottle. The acrylamide stock solution should not be more than 2 to 4 weeks old.

4. Add 60 μl of TEMED and 60 μl of 25% ammonium persulfate solution, and mix the solution gently, avoiding air bubbles.

5. Pour the solution between the plates (see Comment 2). Air bubbles are avoided by tilting the gel mold and starting to pour the gel on one side. Put the slot-former in place. If a shark's-tooth comb is used, it should be inserted with the smooth side facing the gel.

6. Polymerization takes approximately 1 h. The gel can be stored overnight if kept humid (e.g., wrapped in tissues moistened with 1× TBE buffer).

7. Remove the tape from the glass plates and the spacer on the bottom (if there is one). Insert the glass plates with the gel into the electrophoresis apparatus.

8. Fill the electrophoresis tanks with 1× TBE (see Comment 3), remove the slot former and clean the wells thoroughly using a pipet. It is important to remove any unpolymerized acrylamide and urea in the loading area.

9. Pre-electrophorese the gel for 30 to 45 min at 50 to 55 W constant power (corresponding to ~800 to 2200 V and 25 to 35 mA). A stable temperature of 50 to 55°C should be attained (see Comment 4).

10. Denature samples by adding 1 vol of loading buffer, and heating for 3 min at 95°C. After denaturation, immediately transfer samples to ice (see Comment 5).

11. When the samples are ready for loading, switch off the current. If a shark's-tooth comb is used, reinsert the comb with the teeth facing the gel. Teeth should enter the gel to about 1 to 2 mm. Make sure that a sufficiently large gap is left for loading the samples. If appropriate, indicate the sample numbers below the respective wells with a pen. Urea readily diffuses from the gel into the wells, which will prevent the samples from sinking to the bottom. Therefore, the wells have to be cleaned again before sample application.

12. Use special narrow and/or flattened pipet tips or a Hamilton syringe to deposit the sample at the bottom of the well. With some experience, it is also possible to use conventional yellow pipet tips. Normally, the same tip can be used for loading subsequent samples if rinsed in 1× TBE in between (upper buffer chamber). An M13 sequencing ladder can be loaded to one or more separate lanes and serves as a molecular weight marker.

13. Reconnect electrodes and run the gel for 2 to 3 h at 50 to 55 W and 50 to 55°C. Optimal running time and gel resolution depend on the size range of the DNA fragments of interest.

14. After the run is completed, switch power supply off, remove chamber covers, discard running buffer (see Comment 5), disassemble the clamps, and place gel plates horizontally. Cautiously separate the glass plates by lifting the upper plate at one edge, using a spatula or a specific tool supplied by some manufacturers. The gel is expected to adhere to the plate treated with bind-silane (see Comment 6).

15. Dry the gel as outlined in Chapter 4.3.7 and detect radioactive signals by autoradiography as described in Chapter 4.3.11.

Comments

1. Stabilized, ready-to-use, urea-containing PAA solutions (e.g., Sequagel XR, National Diagnostics) avoid the potential dangers of handling poisonous acrylamide and have a longer shelf-life than homemade solutions.

2. Different types of PAA gel apparatus are commercially available, and different strategies of how to seal the edges and how to cast the gel are recommended by the manufacturers. Some models are equipped with a casting tray; others are filled from below using a syringe. We generally recommend that the user follow the instructions of the manufacturer.

3. Qi and Lindhout[1118] reported that higher resolution of AFLP fragments results from buffer gradient electrophoresis. For this, 1× TBE is used as the cathode buffer, and 1× TBE supplemented with 0.5 M sodium acetate is used as the anode buffer.

4. The running temperature must not exceed 60°C because urea may decompose, and glass plates may crack at higher temperatures.

5. If radiolabeled samples are electrophoresed, use appropriate shielding in this and all subsequent steps. Buffers are highly radioactive after the run, especially the one in the lower buffer chamber. Dispose of radioactive waste properly (see Safety Precautions, Chapter 4.1).

6. One of the great advantages of fluorescence-based automated sequencers is that the gels can be discarded at this stage.

4.3.6   Detection of DNA in Gels

A wide variety of techniques are available for the detection of PCR-generated fragments in gels, including staining with intercalating dyes or silver, labeling with radioisotopes, and labeling with fluorescent dyes. DNA profiles generated with radiolabeled probes are usually visualized by autoradiography, whereas fluorescence detection is usually used in conjunction with an automated sequencer (see Chapter 4.3.11). In this section, we describe the two most simple and fast separation and detection systems, i.e., staining of agarose and nondenaturing PAA gels with ethidium bromide, and staining of nondenaturing PAA gels with silver. Other intercalating dyes are commercially available, such as SYBR Green™ (Molecular Probes), SYBR Gold™, and Gelstar® (BMA). These alternatives are claimed to be more sensitive than ethidium bromide, but are also relatively expensive.

Specific requirements may influence the choice of method. For mapping purposes, simple and unambiguously scored patterns are preferred, as obtained with ethidium bromide staining of agarose gels. Conversely, genotype identification may require fingerprints of high complexity. In this case, nondenaturing PAA gel electrophoresis followed by silver staining may be the method of choice. Although both methods are easy to perform and costs are similar, the use of PAA gels and silver staining is more laborious.

4.3.6.1 Ethidium Bromide Staining

DNA fragments can be stained with ethidium bromide whether separated on agarose or PAA gels (including single-strand conformation polymorphism [SSCP] gels[1588]). Ethidium bromide is a powerful mutagen and carcinogen (see Safety Precautions, Chapter 4.1). DNA amounts as small as 10 ng per band can be detected in agarose gels, thus providing sufficient sensitivity for most PCR experiments (except for very small fragments or minor amplification products).

Staining of agarose gels can be achieved in two ways: (1) by adding ethidium bromide to the cooled agarose solution as well as to the running buffer prior to electrophoresis, or (2) by staining the gel after electrophoresis in buffer or water containing ethidium bromide. Ethidium bromide is usually prepared as a stock solution (10 mg/ml) and stored in the refrigerator in the dark. For gels, buffers, or staining solutions, a final concentration of 0.5 to 1 µg/ml is sufficient. Staining takes 5 to 30 min (depending on the gel dimensions) in a tray on a rotary shaker. Rinsing the gel afterwards, or destaining for 15 min in water or 1 mM $MgSO_4$ avoids spillage of ethidium bromide on the UV transilluminator and sometimes produces a higher contrast.[1217] Staining solutions can be reused several times, although the solution is sensitive to light and should be changed regularly. Ethidium bromide waste should be disposed of in a legal and environmentally safe way (according to prescribed regulations).

Ethidium bromide-stained gels are documented on a UV transilluminator at 302 nm using either a commercial documentation system equipped with a video camera and a computer, a Polaroid system with an orange filter and Polaroid type 57 or 667 (3000 ASA) film, or a standard 35-mm camera using the same filter. Because UV light is highly mutagenic and destructive, eyes and skin should be covered by UV safety goggles and protective clothing, respectively (see Safety Precautions, Chapter 4.1).

4.3.6.2 Silver Staining

Electrophoresed DNA fragments can also be detected with silver nitrate staining. With agarose gels, silver does not give better results than ethidium bromide; therefore, it is not generally used.[1376] With PAA gels, conversely, staining with silver enhances sensitivity by about two orders of magnitude, thus enabling the detection of bands containing only 10 to 30 pg of DNA. Consequently, RAPD patterns generally exhibit many more bands when visualized by silver staining of PAA gels compared with ethidium bromide staining of agarose gels (see, e.g., Bassam et al.[86]). Silver staining is also used as one option for AFLP gels[225,249] (see Chapter 4.7.4.3).

Several protocols for silver staining have been described (e.g., Bassam et al.,[86] Blum et al.,[144] Briard et al.,[166] Tegelström,[1376] Ude et al.[1428]), most of which take approximately 2 h. One should be aware that silver not only stains DNA but also RNA and proteins. The presence of restriction enzymes, polymerase, and BSA should therefore be minimized. Commercial kits for silver staining are available from several manufacturers (e.g., Bio-Rad Laboratories), but the technique can easily be performed with homemade solutions. Use high-quality (double-distilled) water and a separate container for each gel. For gels adhering to a glass plate, separate containers should be used for each solution. Glass containers are easier to clean, but plastic containers will also work satisfactorily. Most solutions need to be freshly prepared (the silver solution and the developer during the staining process). Silver solutions should be disposed of in a legal and environmental safe way (according to prescribed regulations).

Solutions and chemicals

> 0.1% CTAB
> 1 M NaOH, freshly prepared
> 25% Liquid ammonia stock solution
> 35% Formaldehyde stock solution (use in fume hood)
> Glycerol
> $AgNO_3$
> Sodium carbonate
> Double-distilled water

Method (according to Tegelström[1376])

1. Rinse the gel in double-distilled water for 3 to 5 min (do not use tap water; traces of chloride may ruin the staining).
2. Soak the gel, under gentle agitation, for 30 min in 0.1% CTAB or in double-distilled water.
3. Incubate the gel in 0.3% ammonia for 15 min under gentle agitation (1.3 ml stock solution per 100 ml solution).
4. Prepare the silver solution in a flask with a magnetic stirrer: dissolve 0.2 g $AgNO_3$ in 125 ml H_2O and add 0.5 ml of 1 M NaOH. The solution will turn cloudy and brownish. Add 0.5 to 0.6 ml of 25% ammonia to the silver solution, drop by drop. When the solution has cleared, add another two drops of 25% ammonia.
5. Pour off the ammonia from the gel and add the freshly prepared silver solution (Step 4). Incubate under gentle agitation for 20 min.
6. Prepare the developing solution (2% sodium carbonate, 0.02% formaldehyde). Sodium carbonate is first dissolved with intense stirring, and then the formaldehyde (60 µl stock solution per 100 ml solution) is added.
7. Rinse the gel briefly in H_2O.
8. Add the developer to the gel. Staining takes 5 to 25 min under gentle agitation.
9. Stop the staining process by a quick rinse with water, and fix the gel in 1.5 to 3% glycerol for 30 min.
10. The gel can now be photographed, and if on a glass plate, dried.

4.3.7 Gel Drying

Denaturing PAA gels containing radiolabeled DNA fragments (e.g., sequencing ladders, AFLP profiles, etc.) are routinely dried on a sheet of filter paper prior to autoradiography. This is usually done with the use of a commercial gel dryer. Although less well known, the same procedure can also be applied to agarose gels, and it has been demonstrated that short oligonucleotide probes[20,937,1391,1421] as well as longer genomic probes[1557] can successfully be hybridized with DNA immobilized in a dried agarose gel matrix. Once dried, the agarose is ultrathin and convenient to store. During subsequent hybridization and washing steps, dried agarose gels rehydrate only partially, and are surprisingly stable and easy to handle.

In this section, we describe the drying of radioactive, denaturing PAA gels prior to autoradiography. For drying of agarose gels, see Ali et al.,[20] Weising and Kahl,[1519] and the first edition of this book.

Solutions

Fixing solution: 10% methanol, 10% acetic acid in H_2O

Method (see Safety Precautions, Chapter 4.1)

1. Transfer the lower glass plate with the gel facing upward to a tray, and cautiously add fixing solution. Make sure that the gel is slightly submerged, but remains attached to the glass plate.
2. After 15 min, gently lift the glass plate out of the fixing solution, pour off excess liquid, and transfer the plate to another tray covered with paper towels. Place one layer of filter paper (Whatman 3MM or equivalent) on top of the gel. Make sure that gel and paper are in close contact with each other.
3. Cautiously lift the filter paper, starting from one end. The gel will remain attached to paper. Cover the gel with plastic (e.g., Saran wrap) and trim it to an appropriate size that fits the gel dryer and the X-ray cassette.
4. For drying, place the gel and filter onto another filter paper in a commercial gel dryer. The Saran wrap should face the rubber gasket. Insert a washing bottle and a cooling device between the gel dryer and the vacuum pump. Location and intensity of radioactive signals on the gel may be monitored at this stage, using a hand-held Geiger counter.
5. Apply vacuum, and turn on the heater to 80°C. Dry for 1 h. Make sure that the gel is completely dry (i.e., flat). Incompletely dried gels may crack and disintegrate when the vacuum pump is switched off.
6. Detect radioactive signals by autoradiography or phosphorimaging as described in Chapter 4.3.11.

4.3.8 Southern Blotting

Southern blot hybridization[1316] has been a key step in traditional marker techniques such as RFLP analysis and oligonucleotide fingerprinting. It is still occasionally used in the context of PCR-based methods. For instance, hybridization of marker-generated PCR fragments with genomic DNA blots gives some indication of the target copy number of the respective fragments in the genome (Chapter 4.6.1). Hybridization of individual RAPD or MP-PCR fragments to a Southern-blotted RAPD or MP-PCR gel can be necessary to verify the homology of comigrating bands (Chapter 4.6.2). Finally, a molecular marker technique, coined random amplified microsatellite polymorphism (RAMPO)[1163] involves hybridization of microsatellite-specific oligonucleotide probes to membrane-bound RAPD or MP-PCR fragments (see Chapters 2.3.7 and 4.6.3).

To generate a Southern blot, electrophoretically separated DNA samples are denatured (i.e., made single stranded) within the gel, and transferred onto a membrane where they are bound at the surface. The original technique described by Southern[1316] makes use of a high-salt buffer, which transfers the DNA to a nitrocellulose filter by capillary forces. During the almost 30 years since its initial description, several variations on the theme have been developed, including alternative transfer buffers (e.g., alkaline buffers), driving forces (e.g., electrophoretic blotting and vacuum blotting), and types of membranes (e.g., nitrocellulose and nylon membranes, charged membranes, hydrophobic membranes). Nylon membranes are now generally used instead of nitrocellulose, mainly because of the high physical resistance of the former. Although the standard Southern blotting variant by capillary transfer in high-salt buffer is somewhat slow, we still find it preferable to other methods in terms of simplicity, reliability, cost, and efficiency of transfer. After blotting (usually overnight), the DNA is irreversibly fixed to the membrane either by heat treatment (2 h at 80°C) or by UV cross-linking (for nylon membranes only) at 0.12 J/cm^2 in a commercial UV cross-linker.

Solutions

Denaturation buffer:	0.5 M NaOH, 1.5 M NaCl
Neutralization buffer:	1.0 M Tris-HCl, 3.0 M NaCl, pH 7.0
Transfer buffer (20× SSC):	3.0 M NaCl, 0.3 M sodium citrate, pH 7.0
6× SSC; 5× SSC; 2× SSC:	Prepare by diluting 20× SSC stock solution with deionized water

Method

1. Document ethidium bromide-stained gel by photography or a video system. Use a UV-reflecting ruler to document marker positions. Transfer the gel to a tray filled with several volumes of denaturation buffer. Incubate for 30 to 45 min at room temperature under agitation.
2. Decant denaturation buffer (which can be reused). Rinse the gel twice with deionized water to remove excess NaOH, and incubate in neutralization buffer for 1 h. Neutralization buffer can also be reused provided that the pH is controlled and readjusted.
3. While the gel is being neutralized, fill another tray with transfer buffer (20× SSC) and cover it with a glass plate so that it forms a bridge. Cut a sheet of filter paper (e.g., Whatman 3MM) to the same width but double the length of the gel. Wet the filter paper extensively with 20× SSC and place it onto the glass plate so that its free ends are immersed in the 20× SSC solution. Cut a nylon membrane to a size 5 mm wider and longer than the gel, and wet the membrane by floating it on distilled water. Do not touch the membrane with your fingers. Use either gloves or forceps.
4. Using two sharp-edged Plexiglass plates, take the gel out of the neutralization buffer, and put it in an inverse orientation (slot openings facing downward) onto the 3MM filter paper on the glass plate. Remove air bubbles between filter and gel by gently rolling a glass rod or a pipet over the gel. Surrounding the gel with Saran wrap, Parafilm, or used X-ray films prevents the direct flow of transfer buffer from the reservoir to the stack of paper towels placed on top of the device.

5. Place the wet nylon membrane on the gel. Remove air bubbles between gel and membrane (see above).
6. Wet two more pieces of filter paper (cut to the same size as the nylon filter) in 20× SSC and put them on top of the membrane. Remove air bubbles.
7. Place a stack of paper towels on top, followed by a glass plate and a weight of about 500 g (e.g., an empty glass bottle). Weights that are too heavy cause early compression of the gel and prevent transfer.
8. Let the transfer proceed overnight.
9. Remove the wet paper towels and the filter papers. Label the positions of slots and gel edges on the membrane as well as the (non-) DNA side using waterproof ink or a pencil. Alternatively, labeling can also be done before blotting.
10. Peel the membrane off the gel, rinse it for a few minutes in 6× SSC to remove residual agarose (which may cause hybridization background), and let it air-dry on a sheet of filter paper.
11. Place the membrane between two sheets of filter paper and bake it for 2 hours at 80°C in an oven or UV cross-link. DNA is irreversibly fixed on the membrane by this step.
12. Store membranes between sheets of filter paper, or in aluminium foil at room temperature under dry conditions.

Variation of Steps 3 to 6 (time saving):

3a. Decant neutralization buffer. Soak five pieces of filter paper (cut to the same size as the gel) with transfer buffer, and put these on the top of the gel, which is resting on a glass plate.
4a. Turn the whole stack upside down and place it on the benchtop.
5a. Place the wet membrane onto the gel. Remove air bubbles.
6a. Place another stack of five filter papers, soaked in transfer buffer, on top of the membrane. Remove air bubbles.

To increase the transfer efficiency of large DNA fragments (> 10 kb), the DNA has to be partially hydrolyzed before transfer. This can either be done by partial depurination in 0.25 M HCl (10 to 20 min, depending on gel thickness) or by short UV treatment (1 to 3 min on a transilluminator). This step needs careful optimization, because treatment that lasts too long may destroy the DNA.

Inverting the gel has two advantages: (1) the DNA side of the blot will then have the same orientation as is shown on the photograph of the gel; and (2) the lower surface of the gel (which is then in contact to the filter) is much smoother than the upper surface, allowing a more intimate contact between both, and avoiding hybridization background (the surface structure of the gel is sometimes visible on autoradiograms).

4.3.9 Generation of Radiolabeled Probes, Primers, and PCR Products

The standard versions of microsatellite analysis, AFLP, and a number of other PCR-based marker techniques involve radiolabeling of the generated DNA fragments, and visualization of the resulting banding patterns by autoradiography or phosphorimaging.[1481,1621] Radiolabeling of PCR fragments with ^{32}P or ^{33}P can be done in various ways. Most often, one of the primers is 5'-end labeled with bacteriophage T4 polynucleotide kinase in the presence of γ^{32}P-ATP or γ^{33}P-ATP (see Chapter 4.3.9.1). Alternatively, radiolabeled α^{32}P- or α^{33}P dNTPs can be included directly in the PCR. End-labeling with polynucleotide kinase is also an appropriate technique for generating radioactive oligonucleotide probes,[20] which can be used, e.g., to generate oligonucleotide fingerprints (see Figure 2.2 and the previous edition of this book), RAMPO patterns (see Figure 2.12 and Chapter 4.6), or for the screening of microsatellite libraries (see Chapter 4.8.5.6).

Other experiments, such as the blot hybridization of RAPD fragments to RAPD blots or genomic DNA blots, require the radiolabeling of larger DNA fragments without PCR. Various techniques based on enzymatic reactions have been developed to label large probes (for an overview, see Ausubel et al.[56] and Sambrook and Russell[1217]). Among these, nick translation[1174] and random priming[433] are still the most useful for the generation of uniformly labeled probes (see Chapters 4.3.9.2. and 4.3.9.3). Kits are commercially available for both methods.

4.3.9.1 End-Labeling of Oligonucleotides

Oligonucleotides are either supplied in a dried or lyophilized form, or dissolved at high concentrations by the manufacturer. For radioactive labeling, we recommend adjusting the concentration to about 2.8 pmol/µl. Oligonucleotides of this concentration can be used directly for the labeling reaction.

Solutions

Oligonucleotide probe:	10 pmol (dried, or dissolved in 3.5 µl distilled water or TE buffer)
10× kinase buffer:	670 mM Tris-HCl, pH 8.0; 100 mM MgCl₂; 100 mM dithiothreitol. This buffer is usually supplied by the manufacturer of the polynucleotide kinase.
γ^{32}P-ATP:	10 µCi/µl stabilized aqueous solution (specific activity ~220 TBq/mmol); e.g., PB10218 (Amersham Biosciences); see Safety Precautions, Chapter 4.1)
T4 polynucleotide kinase	
0.5 M EDTA, pH 8.0	
TE buffer:	10 mM Tris-HCl, 1 mM EDTA, pH 8
10× TE buffer:	100 mM Tris-HCl, 10 mM EDTA, pH 8
0.2 M NaCl in TE buffer	
0.5 M NaCl in TE buffer	
Whatman DE-52 cellulose:	To prepare DE-52 equilibrated in TE buffer, suspend dry DE-52 in 10× TE buffer and let it swell overnight. Let it settle and change the buffer. Repeat several times until the pH of the suspension reaches 8 to 8.5. Then equilibrate the DE-52 material several times with 1× TE buffer and store it in this buffer at 4°C until use.

Method (see Safety Precautions, Chapter 4.1)

1. Dissolve 10 pmol of oligonucleotide in 3.5 µl of distilled water in a microfuge tube. Add 1 µl of 10× kinase buffer, followed by 5 µl (50 µCi) of γ^{32}P-ATP. Draw up radioactive solutions slowly to avoid contamination of the pipet, and discard the tip into the radioactive waste before releasing the pressure.
2. Close the tube, mix cautiously, and spin for a few seconds in a microfuge.
3. Add 2 to 4 units of polynucleotide kinase (usually 0.5 µl, depending on the manufacturer) directly to the mixture, and incubate for 30 min on ice (some suppliers also recommend incubation at 37°C).
4. In the meantime, prepare DE-52 columns. Use either Pasteur pipets or 1-ml pipet tips plugged with glass wool or, more conveniently, disposable plastic columns. Fill the columns with 0.2 to 0.4 ml of DE-52 equilibrated in TE buffer. Wash with several volumes of TE buffer, close the outlet (e.g., with a plastic stopper), and store until use.
5. Stop labeling reaction by adding 1 µl of 500 mM EDTA, and add 90 µl of TE buffer to aid subsequent handling.
6. Remove the plastic stopper from the column, put a small Erlenmeyer flask below the outlet, and apply the labeled oligonucleotide solution to the column. Wash with 4 ml of TE buffer to elute unincorporated γ^{32}P-ATP. When the washing solution has reached the top of the DE-52 cellulose, perform a second wash with 4 ml of 0.2 M NaCl in TE buffer.
7. Replace the Erlenmeyer flask by a 50-ml Falcon tube and discard the flow-through washing solutions into the radioactive waste. Elute the oligonucleotide from the DE-52 cellulose with 2 × 0.5 ml of 0.5 M NaCl in TE buffer. Discard the column into the radioactive waste.
8. Store the labeled probe–primer at –20°C until use.

4.3.9.2 Nick Translation

Labeled deoxynucleotides may be incorporated into double-stranded DNA by nick translation.[1174] Low concentrations of DNase I are used to introduce nicks (i.e., single-strand breaks) within the DNA fragment to be labeled. *Escherichia coli* DNA polymerase I recognizes such nicks. By virtue of its combined 5'-3'-exonuclease and polymerase activity, this enzyme replaces the pre-existing deoxynucleotides in the 3'-direction, resulting in a shift of the nick. Inclusion of radioactive deoxynu-cleotides (e.g., α^{32}P-dCTP) in the reaction results in the generation of efficiently labeled double-stranded DNA molecules. Usually, more than 60% of the labeled deoxynucleotides are incorporated during nick translation, and DNA probes with a high specific activity (10^8 cpm/µg) are generated in this way.

Nick translation kits are commercially available from several companies. It is generally convenient to follow the user's manual included in the kit. Using a kit is easier, but also more expensive per single labeling reaction. The protocol given below can be followed if no kit is available.

Solutions

DNA sample dissolved in 1× TE or water (0.5 to 1.0 µg)

10× nick translation buffer:	0.5 M Tris-HCl pH 7.5; 0.1 M MgSO$_4$; 0.1 M dithio-threitol; 500 µg/ml BSA (fraction V)
DNase I solution:	10 ng/ml in 1× nick translation buffer and 50% glycerol
DNA polymerase I:	5 U/µl
Unlabeled dNTPs:	1 mM dATP; 1 mM dGTP; 1 mM dTTP; (if dCTP is used as a labeled nucleotide)
α^{32}P-dCTP:	10 µCi/µl (specific activity ~110 TBq/mmol; e.g., PB10205 (Amersham Biosciences); any labeled nucleotide can be used if the unlabeled dNTP mix is made up appropriately)
Sterile double-distilled water	
Stop solution:	1% SDS; 10 mM EDTA; 0.25% xylene cyanol; 0.25% bromophenol blue

Method

1. Mix the following components on ice:
    2.5 µl 10× nick translation buffer
    3.0 µl DNA (0.5 to 1.0 µg) dissolved in 1× TE or water
    3.0 µl unlabeled dNTPs
    5.0 µl α^{32}P-dCTP (= 50 µCi)
    2.0 µl DNase I (10 ng/ml)
    2.0 µl *E. coli* DNA polymerase I (5 U/µl)
    7.5 µl sterile water
    The final volume is 25.0 µl.
2. Mix gently and incubate at 14 to 16°C for 60 min.
3. Stop the reaction by adding 1 volume of stop solution.
4. Separate the radiolabeled DNA from the unincorporated dNTPs as described below (Chapter 4.3.9.4).

4.3.9.3 Random Priming

The random priming labeling procedure was developed by Feinberg and Vogel-stein.[433] Oligonucleotides of random sequence (usually a population of synthetic hexamers or octamers) are used as primers for DNA synthesis on single-stranded template DNA. The synthesis of the complementary strand in the presence of labeled dNTPs is catalyzed by the DNA polymerase I Klenow fragment, which lacks the 5'- to 3'-exonuclease activity. Newly incorporated nucleotides are therefore not removed. Probes and primers receive a high specific activity. More than 70% of the labeled deoxyribonucleotide molecules are usually incorporated, and the resulting DNA probes are often longer than those obtained after nick translation.

Random priming kits are commercially available from different companies. It is generally convenient to follow the user's manual included in the kit. Using a kit is easier, but also more expensive per single labeling reaction. The protocol given below may be followed if no kit is available.

Solutions

    DNA sample (10 ng/μl) in 1× TE or water

    10× labeling buffer: 0.25 M Tris-HCl pH 8.0; 30 mM MgCl$_2$; 0.4% β-mercapto-ethanol

    Unlabeled dNTPs: 1 mM dATP; 1 mM dGTP; 1 mM dTTP; (if dCTP is used as a labeled nucleotide)

    α^{32}P-dCTP: 10 μCi/μl (specific activity ~110 TBq/mmol; e.g., PB10205; Amersham Biosciences; any labeled nucleotide can be used if the unlabeled dNTP mix is made up appropriately)

    Primer solution: commercially available random hexanucleotides

    BSA (fraction V): 2 mg/ml

    DNA polymerase I Klenow fragment (0.5 U/μl)

    Deionized water

    Stop solution: 1% SDS; 10 mM EDTA; 0.25% xylene cyanol; 0.25% bromophenol blue

Method

1. Mix on ice:
   3.0 μl DNA sample (30 ng)
   16.0 μl distilled water
   Denature the DNA sample by heating to 100°C for 10 min, and place immediately on ice.
2. Add on ice:
   5.0 μl 10× labeling buffer
   4.0 μl dATP
   4.0 μl dGTP
   4.0 μl dTTP
   5.0 μl primer solution
   2.0 μl BSA
   5.0 μl α^{32}P-dCTP (50 μCi)
   2.0 μl DNA polymerase I Klenow fragment
   The final volume is 50.0 μl.
3. Mix gently and incubate at room temperature for 3 to 4 h.
4. Stop the reaction by adding 1 vol stop solution.
5. Separate the radiolabeled DNA from the unincorporated dNTPs as described below (Chapter 4.3.9.4).

4.3.9.4 *Removal of Unincorporated dNTPs*

To avoid unspecific background hybridization, and to protect the experimenter from unnecessary exposure to radioactivity, unincorporated labeled nucleotides should be separated from probes generated by nick translation or random priming. This is usually done by gel filtration through a Sephadex G-50 (Amersham Biosciences) or

Bio-Gel P-60 (Biorad Laboratories) column. These procedures are suitable for the separation of DNA probes >80 bp from mononucleotides. According to Sambrook and Russell,[1217] the spun-column variant of Sephadex G-50 gel filtration (see below) will also separate oligonucleotides as small as 16 bases from smaller molecules.

Chromatography on Sephadex G-50 is based on gel filtration to separate molecules according to size.[1217] DNA molecules larger than about 80 bp are excluded from the pores of the Sephadex beads, run in the void volume, and pass the column very fast. Small molecules enter the pores and are retained in the column.

Solutions

Sterile double-distilled water
1× TE: 10 mM Tris-HCl; 1 mM EDTA, pH 8.0
1× TEN: 10 mM Tris-HCl; 1 mM EDTA; 100 mM NaCl; pH 8.0
Sephadex G-50: Pretreat as follows:
 1. Add Sephadex G-50 powder (medium or fine) to sterile water, allow to swell overnight at room temperature. Ten grams of G-50 powder result in about 150 ml of swollen resin.
 2. Equilibrate the resin in 1× TE (several changes), autoclave, and store at room temperature or at 4°C.
 3. Prior to use, equilibrate the resin in 1× TEN (several changes).
Dye marker: 0.25% xylene cyanol; 0.25% bromophenol blue in water

Method

 1. Prepare a Sephadex G-50 column in a disposable 1-ml pipet tip, a syringe, or a Pasteur pipet plugged with a small amount of sterile glass wool. Using a Pasteur pipet, fill the column to about 80% of the available volume. Avoid trapping air bubbles.
 2. Wash the column once with 1× TEN.
 3. Apply the labeled DNA sample mixed with the dye marker solution (in a volume of 100 µl or less) to the column.
 4. Add 1× TEN to the column. Follow the separation of the two dyes. The labeled DNA probe runs in front, close to the xylene cyanol.
 5. Collect the probe into a microfuge tube. Discard the column into the radioactive waste.
 6. Store the radiolabeled DNA probe at −20°C until use.

Instead of monitoring the position of the DNA probe with the help of xylene cyanol, the leading (DNA) peak of radioactivity may also be identified by collecting and measuring individual 200-µl fractions into microfuge tubes or by a hand monitor.[1217]

The spin-column technique is also based on gel filtration through Sephadex or Bio-Gel columns. However, packing and running of the column are accomplished by centrifugation rather than by gravity.[1217] In the variant given below, the chromatography column is prepared in a 1.5-ml microfuge tube. It may also be prepared in a disposable 1-ml syringe as described by Sambrook and Russell.[1217] The method is fast and simple, but involves a slightly higher contamination risk than ordinary gel filtration. Therefore, a microfuge should be used that is designated for radioactive experiments only.

Solutions

1× TEN: 10 mM Tris-HCl; 1 mM EDTA; 100 mM NaCl; pH 8.0
Sephadex G-50: Pretreated as described above

Method

1. Using a hot needle, punch a small hole into the bottom of a uncapped 1.5-ml microfuge tube.
2. Plug the bottom of the microfuge tube with sterile glass wool.
3. Place the microfuge tube on the top of another uncapped microfuge tube. Add 800 μl of Sephadex G-50 equilibrated in 1× TEN.
4. Spin for 2 min at 4000 rpm in a centrifuge.
5. Change the lower microfuge tube, add 500 μl of 1× TEN and repeat step 4.
6. Change the lower microfuge tube, and gently apply the labeled DNA sample to the upper tube.
7. Repeat step 4.
8. Remove the labeled DNA probe from the lower tube to a capped reaction tube. Measure the radioactivity of (1) the eluted DNA sample, and (2) the Sephadex material using a hand monitor. The proportion of incorporated radioactivity can be roughly estimated from these values and should be >50%.
9. Store the radiolabeled DNA probe at –20°C until use.

4.3.9.5 Nonradioactive Labeling Procedures

Substantial progress has been made in recent years concerning the development of nonradioactive labeling and detection procedures. For PCR-based marker systems, the exploitation of fluorescent dyes in combination with an automated sequencer is now routine. Kits for labeling DNA with biotin, digoxigenin (DIG), and fluorescent dyes are commercially available. We generally advise that users follow the protocols provided by the manufacturer.

4.3.10 Blot Hybridization

Blot hybridization involves the binding of a labeled, single-stranded DNA probe to complementary, likewise single-stranded DNA sequences attached to a membrane, thereby revealing one or more specific bands. However, single-stranded DNA generally tends to bind to nylon membranes (otherwise Southern blotting would not work), and this unspecific binding of the probe would result in signal generation all over the membrane. To prevent this, membranes are preincubated (i.e., prehybridized) in a buffer containing a variety of high molecular weight blocking agents (e.g., PVP, Ficoll, BSA, nonfat dry milk) and detergents (e.g., high concentrations of SDS[1534]). In most general hybridization protocols, denatured and sonified DNA from unrelated organisms is also included in the hybridization buffer to block the membrane from unspecific binding of the probe.

After prehybridization, the actual hybridization is performed in a buffer of similar composition, but also containing the labeled probe. The results of hybridization are strongly influenced by the applied stringency; i.e., the percentage of base mispairing allowed between probe and target (no mismatch = 100% stringency). Hybridization stringency, in turn, depends on a variety of parameters such as the GC content of the probe–target complex, probe concentration, buffer composition (e.g., salt concentration and the inclusion of formamide), and temperature.[1217] For example, stringency may be increased by lowering the salt concentration, or by increasing the hybridization or washing temperature. Conversely, it may be decreased by including formamide in the hybridization buffer.

In earlier days, hybridization was performed in sealed plastic bags in a (shaking) water bath (or other kind of thermostat). At present, membranes are exposed to the probes in glass cylinders in a roller-bottle oven. Oven hybridization in glass tubes provides effective shielding against radioactivity. Moreover, washing steps after hybridization can be performed within the cylinder, thus avoiding the high contamination risk associated with removing radioactive probes and blots from sealed plastic bags. After hybridization, unbound probe is washed off the membrane. Hybridization stringency is also influenced by the washing steps, i.e., by salt concentration and temperature of washing solutions.

4.3.10.1 Oligonucleotide Probes

Hybridization of microsatellite-specific oligonucleotides with membranes or dried agarose gels carrying restriction-digested genomic DNA was the key step for producing so-called oligonucleotide fingerprints[20,1522] (see Figure 2.2). More recently, oligonucleotide hybridization has been used for the screening of microsatellite libraries (see Chapter 4.8.5.6). A radioactive variant of blot hybridization is described here. For gel hybridization and a nonradioactive (DIG-based) protocol of blot hybridization, see Bierwerth et al.,[133] Weising and Kahl,[1519] and the first edition of this book.

The annealing temperatures (T_m) of the oligonucleotides are calculated according to a rule-of-thumb put forward by Thein and Wallace[1391] (the so-called Wallace rule): 2°C for each AT-pair and 4°C for each GC pair, respectively (assuming a salt concentration of 1 M in the hybridization buffer and an oligonucleotide length of about 16 bp). Hybridization is carried out at T_m −5°C; example; e.g., at 35°C for (GATA)$_4$ and 43°C for (GACA)$_4$. According to Thein and Wallace[1391] and Miyada and Wallace,[937] these conditions result in 100% stringency; i.e., no mismatches are allowed. Although this might not hold true for all oligonucleotides containing simple repeat motifs, we found hybridization results to be reliably reproducible if the conditions were kept constant between experiments.

Solutions

Hybridization buffer:	5× SSPE, 5× Denhardt's solution, 0.1% SDS, 10 µg/ml fragmented and denatured *E. coli*–DNA; sterilize by filtration. Stock solutions that facilitate preparation of this buffer are given below.

Probe:	^{32}P-labeled oligonucleotide (see Chapter 4.3.9.1). Add to an appropriate amount of hybridization buffer at a concentration of 0.5 pmol/ml.
6× SSC (washing solution):	0.9 M NaCl, 0.09 M sodium citrate, pH 7.0

Stock solutions

20× SSPE:	3 M NaCl, 0.2 M sodium phosphate buffer, pH 7.4, 0.02 M EDTA
100× Denhardt's:	2% PVP-40; 2% BSA; 2% Ficoll. Sterilize by filtration and store in aliquots at –20°C.
20% SDS:	Highly concentrated SDS solutions form precipitates at room temperature; heat in a waterbath until the solution is clear.
2.5 mg/ml *E. coli* DNA:	Dissolve in 10 mM Tris-HCl, 1 mM EDTA, pH 8.0; store in aliquots at –20°C and denature by heating (5 min, 100°C) prior to addition to the hybridization buffer.

Method (see Safety Precautions, Chapter 4.1)

1. Wind the membrane onto a 10-ml disposable pipet, transfer it into a hybridization flask filled with 6× SSC, and unroll it to the inner wall of the tube (DNA side facing inward). Discard the 6× SSC and fill the tube with 10 ml of hybridization buffer including the labeled probe.
2. Hybridize for 3 h to overnight at T_m –5°C. Tubes should be closed carefully to avoid contamination and/or loss of probe.
3. After hybridization, decant the probe into a polypropylene tube. The probe may be reused several times. Store at –20°C.
4. Fill the hybridization flask up to one half with 6× SSC, and wash off most of the unbound probe by shaking. Decant the washing solution (radioactive waste). Use gloves to remove the membrane from the tube, transfer it to a tray filled with 6× SSC and wash it in this tray for 3× 30 min in 6× SSC at room temperature.
5. Transfer the membrane to another tray containing 6× SSC prewarmed to hybridization temperature. Wash for 5 min (stringent wash[937]).
6. Transfer the gel to another tray containing fresh 6× SSC at room temperature.
7. Place the membrane on a sheet of Saran wrap, drain excess liquid with filter paper, and wrap it in Saran wrap. Inclusion of a piece of tape between the upper and lower sheets of Saran wrap facilitates future unpacking before reusing the membrane. The blot is now ready for autoradiography (see below, Chapter 4.3.11.1).

4.3.10.2 *Probes Generated by Nick Translation or Random Priming*

Solutions

Radiolabeled DNA probe

6× SSC:	0.9 M NaCl, 0.09 M sodium citrate, pH 7.0
(Pre)hybridization buffer:	7% SDS, 0.263 M Na_2HPO_4 pH 7.2, 1 mM EDTA, 1% BSA (Fraction V)[1534]
Washing solution:	2× SSC, 0.1% SDS (dilute from stock solutions)

Method (see Safety Precautions, Chapter 4.1)

1. Presoak the membrane in 6× SSC.
2. Wind the membrane onto a 10-ml disposable pipet, transfer it into a hybridization flask filled with 6× SSC and unroll it to the inner wall of the tube (DNA side facing inward). Pour off the 6× SSC, and fill the tube with 10 ml of prehybridization buffer. Avoid trapping of air bubbles between the membrane and the tube wall. In this and later steps, hybridization tubes should be closed carefully to avoid contamination and/or loss of probe.
3. Prehybridize for 2 to 3 hours 60°C in a roller-bottle hybridization oven.
4. Add the denatured radiolabeled probe (30 ng per 10 ml of hybridization buffer) to the hybridization tube.
5. Hybridize overnight at 60°C in a roller-bottle hybridization oven.
6. Remove the hybridization solution and rinse twice with washing solution.
7. Transfer the membrane to a tray, and incubate in three changes of washing solution: 15 min at room temperature, 15 min at 60°C, and 15 min at room temperature. Do not allow the membrane to dry at any stage during washing.
8. Transfer the membrane to filter paper, drain off excess liquid, and wrap the damp membrane in Saran wrap. Inclusion of a piece of tape between the upper and lower sheets of Saran wrap facilitates future unpacking before reusing the membrane. The blot is now ready for autoradiography (see Chapter 4.3.11.1).

4.3.11 Signal Detection

4.3.11.1 *Autoradiography*

Radioactive as well as chemiluminescent signals are usually detected by exposing the membrane or dried gel to an X-ray film. For reasons of safety and to avoid artefacts, this is best done in X-ray cassettes. Alternatively, gel–blot and X-ray film can be sandwiched between glass plates, and inserted into light-proof plastic bags. Before applying the film, signal strength should be evaluated using a hand monitor. With some experience, the appropriate exposure time (between several hours and several days) can be roughly deduced from the amount of radioactivity, as indicated by the monitor. If signals are weak, different strategies may be followed to enhance signal intensity. First, different types of X-ray film are available. For example, Kodak XAR is about three times more sensitive than Kodak X-omat S. Second, intensifying screens may be included in the cassettes. At low temperatures, these screens emit photons upon receipt of radioactive β-particles, thereby increasing signal strength several-fold.[1217] As with X-ray films, screens with different degrees of intensification are available. Disadvantages of using intensifying screens are the need for a –80° freezer for exposure, and bands on the autoradiogram appear less sharp.

Method (see Safety Precautions, Chapter 4.1)

1. Insert the dried sequencing gel or membrane into an X-ray cassette (with or without intensifying screens, depending on signal strength). Use appropriate shielding to protect yourself against β-radiation.
2. Evaluate signal strength using a hand monitor.

3. In the darkroom, place a sheet of X-ray film between gel or membrane and intensifying screen. Autorad markers (e.g., Glogos™, Stratagene) facilitate the alignment of the autoradiogram with the gel.
4. If screens were used, store the cassette at −80°C.
5. After an exposure of several hours to several days (depending on signal strength), remove the cassette from the freezer, let it warm to room temperature, and develop the film as recommended by the supplier. Handle X-ray films carefully because they are sensitive to scratching, especially when wet.

4.3.11.2 Phosphorimaging

Phosphors are chemical substances that emit visible light after induction by short-wave radiation. In contrast to fluorescence, phosphorescence persists after the induction ceases. When a membrane is exposed to a phosphorimaging screen, the pattern of radioactive signals is stored in the screen. Upon excitation by light of a certain wavelength in a phosphorimaging apparatus, the stored pattern is released and immediately transferred to a computer.

The phosphorimaging technology introduced by Molecular Dynamics in 1989 offers several advantages over the traditional method using X-ray films and intensifying screens: (1) Storage phosphors are 10 to 250 times more sensitive to incident radiation than X-ray film, resulting in greatly reduced exposure times. For example, samples that require overnight exposure to X-ray film can be imaged accurately after only 1 h of exposure to a storage phosphor screen. Because maximum sensitivity is obtained at room temperature, −80°C facilities are not required. (2) Multiple samples can be exposed simultaneously. (3) Storage phosphor screens are quantitatively accurate over five orders of magnitude, compared with only two orders of magnitude for X-ray film. Multiple exposures to compensate for the limited dynamic range of X-ray films are therefore not required. (4) Special treatments, chemicals, or a darkroom are not needed because the reading of the image from a storage phosphor screen is carried out in a phosphorimager. A storage screen is reusable up to about 1000 times. Phosphorimaging systems and accessories are commercially available from several suppliers (e.g., Amersham Biosciences, Packard, Bio-Rad, Fuji, Kodak), but the widespread use of the technique is still limited by its relatively high cost (US $26,000–$40,000 for the complete package, and between US $1,000 and $4,000 for each storage screen).

4.3.11.3 Automated DNA Sequencers

Fluorescence-labeled PCR products are best analyzed on an automated DNA sequencer. The availability of such an instrument will replace the standard PAA gel electrophoresis equipment for most applications, including DNA sequencing, AFLP, and microsatellite analysis. DNA fragments are resolved by high-voltage electrophoresis in either PAA gels or capillaries. Automated DNA sequencers are commercially available from various companies, including Amersham Biosciences, Applied Biosystems, Beckman Instruments, and LI-COR. The cost ranges between about U.S. $50,000 and $250,000, depending on the model and the software included in the package.

4.4 PCR WITH ARBITRARY PRIMERS

This section describes the methodology of PCR with arbitrary primers. A protocol is presented that is based on the original RAPD article by Williams et al.,[1546] with some modifications. The protocol is followed by a detailed discussion of the impact of reaction conditions, with special attention paid to the reproducibility and robustness of the technique. The discussion is focused on RAPD analysis, which is the most widely used variant of arbitrarily primed PCR, but many remarks are also valid for AP-PCR,[1527] DNA amplification fingerprinting (DAF),[201,202] and other modifications of the basic technology. See Chapter 2.3.3 for a more detailed treatment of the general principles of PCR with arbitrary primers and the properties of the anonymous DNA markers generated by this procedure.

4.4.1 Standard RAPD Protocol

For general precautions in PCR experiments, see Chapter 4.3.2.

Solutions (see Comment 1)

Taq DNA polymerase:	5 U/µl
10× buffer:	200 mM Tris-HCl, pH 8.3, 500 mM KCl, 20 mM MgCl$_2$, 0.01% gelatin. Ten-fold concentrated PCR buffer is usually supplied by the manufacturer of the enzyme. It may or may not contain magnesium chloride and additional ingredients, depending on the brand of the enzyme.
dNTP stock:	2 mM each of dATP, dCTP, dGTP, and dTTP. Ready-made solutions containing all four dNTPs are commercially available from several suppliers.
PCR primer:	5 µM (5 pmol/µl) random 10-mer oligonucleotides (see Comment 2)
Template DNA:	5 to 20 ng/µl

Method

1. Use thin-walled PCR tubes to set up a reaction with 25-µl volume containing 20 mM Tris-HCl pH 8.3, 50 mM KCl, 2 mM MgCl$_2$, 0.001% gelatin, 200 µM of each dATP, dCTP, dGTP, and dTTP, 0.8 µM primer (20 pmol per reaction), 1 unit *Taq* DNA polymerase, and 15 to 100 ng of template DNA. Pipetting errors are minimized by preparing master mixes for all samples (see Comment 3). Set up a negative control, in which water replaces the DNA. Use a specially designated pre-PCR pipet set. Microtiter plates are a useful alternative to PCR tubes (see Comment 4).
2. Mix the contents, and centrifuge the vials briefly (see Comment 5). Microtiter plates can be centrifuged in specially equipped centrifuges.
3. Insert the tubes or the microtiter plate into a thermocycler and start the desired program. We use the following program (but see Chapter 4.4.2.3):
   3 min 94°C (initial denaturing step)

45 cycles each consisting of 15 sec at 94°C (denaturing), 30 sec at 35°C (annealing), and 90 sec at 72°C (elongation)

3 min 72°C (final elongation step)

Ramp times are set to "minimum" except for the transition between annealing and elongation step (90 sec).

4. After amplification, the tubes can be stored at 4°C for a couple of days (or for a longer period, if necessary, at −20°C). An aliquot of each sample is mixed with loading buffer and electrophoresed on a 1.4% agarose gel along with a suitable size standard (see Chapter 4.3.4). Gels are stained with ethidium bromide or another intercalating dye (e.g., SYBR Green, Molecular Probes). Bands are visualized under UV light and documented by photography, or using an electronic imaging set-up equipped with a video camera (see Chapter 4.3.6).

Comments

1. Frozen aliquots of concentrated PCR buffer, dNTPs, arbitrary primer, and (if necessary) magnesium chloride, should be thawed immediately before use, mixed by vortexing, and kept on ice. Thorough mixing is essential; RAPD failures often result from inadequate mixing of freshly thawed stock solutions.

2. Sets of arbitrary 10-mer primers can be purchased as sets from various companies such as Operon, Roth, University of British Columbia (UBC primer sets 1 to 8; at http://www.biotech.ubc.ca/frameset.html).

3. A master mix includes the enzyme, 10× PCR buffer, magnesium chloride, and dNTPs. Master mixes are briefly vortexed, centrifuged, and aliquoted into each tube. Depending on the setup of the experiment, primer and template are either included in the master mix or added separately for each tube. If n samples are to be analyzed, sufficient master mix should be prepared for n + 1 samples.

4. Most thermocyclers are compatible with 96- or 384-well plates, which increases sample throughput considerably.

5. Older thermocyclers may not be equipped with a heated lid. In this case, the reaction solution needs to be overlaid with two or three drops of mineral oil to prevent evaporation.

4.4.2 Influence of Reaction Conditions and Components

Optimization of RAPD protocols can be laborious and problematic, given that many reaction components as well as any part of the PCR program may affect the results. Numerous articles describe how optimization can be achieved (e.g., Aldrich and Cullis,[17] Caetano-Anollés,[194] Micheli et al.,[927] Munthali et al.,[961] Williams et al.,[1547] Wolff et al.,[1561] and Yu and Pauls[1603]). In general, it is advisable to determine optimal RAPD conditions empirically by performing a set of pilot experiments. Possible influences of the most important reaction parameters on RAPD banding patterns are discussed in the following paragraphs.

4.4.2.1 Primers

The standard RAPD approach uses single 10-mer primers to amplify genomic template DNA. Despite the word *random* used in the acronym RAPD, these primers should not be fully random. For example, RAPD primers should not be self-complementary and should have a GC content >40%.[1546] Sets of standard 10-mer RAPD primers are commercially available from several manufacturers (see above). User-defined primers can also be designed by a spreadsheet computer program called Random Oligonucleotide Construction Kit (ROCK).[1338] This program is running under Microsoft Excel® and can be downloaded from http://www.sru.edu/depts/artsci/bio/ROCK.htm.

Much shorter primers (down to 5 bases)[201,202] and much longer primers (20 bases and more)[314,496,867,1527,1590] also generated complex banding patterns under the appropriate experimental conditions. The efficiency of primer binding decreases with decreasing primer length, probably imposing the minimum length of five nucleotides, as demonstrated in the DAF variant.[198] Ye et al.[1590] compared the performance of short (i.e., 10 bases) and long (i.e., 17 to 24 bases) RAPD primers under identical conditions, using pear or grapevine DNA as a template. In general, they found that long primers generated more fragments, a wider range of fragment sizes, and a larger number of polymorphic fragments per primer. Gillings and Holley[496] found that pairs of long PCR primers, directed toward the enterobacterial repetitive intergenic consensus (ERIC) sequence, amplified polymorphic PCR products from a wide range of species, including bacteria, plants, fungi, and vertebrate and invertebrate animals. There was no obvious correlation between genome complexity and the number of bands generated, which is typical for RAPDs. Similar results were obtained with other long primers, using the same conditions as in ERIC-PCR. From our experience, we also conclude that primers of any size can be used to produce RAPD fragments from any organism. The only prerequisites are (1) the absence of a specific target sequence in the template DNA, and (2) the use of relatively nonstringent PCR conditions.

Primer concentrations have a profound influence on fragment yield and the quality of banding patterns,[397,867,962,1547] and should be maintained constant throughout a given set of experiments. For the standard RAPD protocol, primer concentrations are generally good between 0.2 and 2 µM. However, about 5 to 10 times higher primer concentrations are used for the DAF variant (i.e., 2 to 15 µM).[198] Caetano-Anollés[197] stressed that high primer-to-template ratios, as used in DAF, result in more complex DNA profiles and a more stable amplification reaction.

Most combinations of RAPD primers and genomic DNA produce banding patterns of moderate complexity on agarose gels, but there will always be a subset of primers that generates poor patterns, or even fail to amplify a given template DNA. It is therefore common practice to prescreen a large number of RAPD primers with a small set of template DNAs to identify useful primers for a particular study. Fritsch et al.[464] assessed the so-called amplification strength of 480 10-mer primers in three plant species and found that a high GC content (especially in the four bases closest to the 3′-end) was positively correlated with primer performance. Kubelik and Szabo[754] also found twice as many amplification products with primers of 80 to 100% GC compared with the standard RAPD primer set. However, differences among species are considerable, and we generally recommend that the performance of RAPD primers be tested in pilot experiments.

In the so-called **mini-hairpin primers** designed by Caetano-Anollés and Gresshoff,[198,199] a short and highly stable hairpin structure at the 5'-terminus is connected to an arbitrary core sequence at the 3'-terminus. These primers were used to amplify a wide variety of templates, ranging in size from a few hundred to a few thousand base pairs (e.g., plasmids or PCR products) to billions of base pairs (genomic DNA from soybean and bermudagrass), using the DAF variant of arbitrarily primed PCR. After separation on denaturing, electrophoresis with 20% PAA/7 M urea gels, and silver staining, complex fingerprints were obtained with core sequences as short as three nucleotides, using high concentrations of magnesium sulfate (3 to 6 mM) and primer (up to 30 μM). Simulation studies using small plasmid DNA templates showed that a perfect match of the first three nucleotides at the 3'-end is required, and demonstrated an extraordinary match of expected and observed amplicons. The complexity of DNA profiles could be tailored by the inclusion of degenerate bases in the primer sequence. In soybean and centipede grass, the use of mini-hairpin primers instead of unstructured arbitrary primers greatly increased the number of polymorphic DNA fragments detected.[198,199]

The use of **primer combinations** instead of a single RAPD primer was suggested by several groups, and was often (but not always) shown to increase the discriminatory potential.[118,313,614,622,926,1527,1547] RAPD reactions using two different primers usually result in patterns that are quite different from the patterns generated by each primer alone. Theoretically, the higher number of priming sites targeted in the primer combinations compared with single primer reactions should lead to more complex (in part additive) banding patterns and smaller average fragment sizes. In practice, however, numerous additional fragments appear, and banding patterns are not additive. This may be explained by a considerable degree of primer–template mismatch, paired with the competitive nature of primer–target site selection[557] (see Chapter 2.3.3.2). Fragments with annealing sites for two different primers at the ends do not form hairpin structures, and thus primers will not be outcompeted by internal hairpin formation, as was suggested to be the case with fragments having identical primers at both ends.[204,205]

In a segregating population of *Brassica napus*, Hu et al.[622] showed that two-primer RAPD products and the respective single-primer products were unlinked, suggesting that their origins are from independent genomic regions. Thus, using primers pairwise increases the number of independent polymorphisms that may be generated from a limited set of primers. Combinations of long RAPD primers (15- to 20-mers) seem to be particularly promising in this respect.[313]

Caetano-Anollés and Gresshoff[199] described the generation of "fingerprints from fingerprints" by reamplification of DAF products with mini-hairpin primers (see above) or 5'-anchored microsatellite primers (see Chapter 4.5). The usefulness of this approach, which the authors coined **arbitrary signatures from amplification profiles (ASAP)**, was illustrated by a bulked segregant analysis[928] of the *nts-1* supernodulation locus in soybean. Bulks of wildtype and mutant plants, respectively, were first amplified with an arbitrary octamer primer, and then reamplified with a mini-hairpin primer. Whereas no differences between the two bulks were visible after the first round of PCR, several polymorphisms were identified by the second round.[199] The ASAP technique was also used to discriminate cultivars of *Pelargonium*,[1325] to identify somatic mutants and radiation-induced sports in chrysanthemum,[1419] and to characterize genetic instabilities in bermudagrass off-types.[195,196] In all of these studies, genotypes were difficult to distinguish by DAF, but readily discriminated by ASAP. Reproducibility depended on appropriate optimization of PCR components, and especially of primer concentrations.[199]

4.4.2.2 Polymerase

Different thermostable DNA polymerases often give rise to different RAPD products.[1235,1303,1547] Schierwater and Ender[1235] compared the amplification patterns of *Daphnia* DNA, obtained from a total of 13 commercially available thermostable DNA polymerases. Although patterns obtained by the various enzymes looked quite similar at first glance, there were qualitative and quantitative differences, and some fragments were only amplified by certain polymerases. Some scientists prefer the Stoffel fragment, which lacks 5′- to 3′-exonuclease activity (see Chapter 4.3.2.2). Fragment patterns obtained with the Stoffel fragment were reported to be more polymorphic and more reproducible than those obtained with full-length polymerases.[361,1303] However, a comparison of Ampli*Taq* and the Stoffel fragment using identical primers and reaction conditions resulted in completely different RAPD fragments, which also mapped, with one exception, to different genomic positions.[1303] These results emphasize that the choice of polymerase is important, and switching to another type of enzyme in the course of a project is not recommended.

4.4.2.3 Thermocycler and Temperature Regimen

Running the same RAPD program on different thermocyclers was reported to result in different amplification patterns.[587,867,1064] This phenomenon is likely caused by different temperature profiles in the reaction tubes.[587] However, we performed identical RAPD reactions using three different thermocyclers (Perkin Elmer 480, MJ Research PTC100, and Pharmacia) side by side, and obtained the same fragment profiles with all three instruments (Wolff, unpublished results).

Originally, a temperature profile of 1 min at 94°C, 1 min at 36°C, and 2 min at 72°C was suggested for RAPD analysis.[1546] Our protocol given above uses shorter time intervals for each step, and generally works very well for thermocyclers with an optimal temperature transfer between block and reaction tubes. Even more condensed programs may work in some instruments. Yu and Pauls[1603] optimized reaction times using an MJ Research PTC100 thermocycler and found that 5 sec at 94°C, 30 sec at 36°C, and 60 sec at 72°C gave better results than did programs that required more time. Shorter periods at 94°C prolong the lifetime of the polymerase. Conversely, longer elongation times will favor the production of larger fragments.[472]

Transition times between the different steps of a cycle should not be too short.[472,1256] This is especially important for the ramping time between the annealing (36°C) and extension temperature (72°C). For example, we found that 55 sec on a Perkin Elmer 480 thermocycler gives unreliable results, whereas a ramp time of

90 sec results in reproducible fragment patterns. If the temperature is increased too fast, a primer–template complex with 10 to 20% mismatches may denature again before the polymerase has elongated the DNA strand to a length that is sufficiently resistant to 72°C. Ellinghaus et al.[392] also reported a profound increase in number and yield of RAPD fragments from mouse DNA if transition times between the annealing and elongation step were increased to 7 min. A similar effect was observed when a transition interval was introduced between melting and annealing steps.[1256]

A major point of attention is the annealing temperature (see Chapter 4.3.2.3). Usually, the annealing temperature is set to 5°C below T_m, which can be estimated with the help of specific computer programs, see Appendix 3.[1196,1203] For RAPD primers, an annealing temperature of 36°C may be chosen as a preliminary value.[1547] However, amplification using arbitrary 10-mer primers is also possible at much higher annealing temperatures, provided that the reaction conditions are properly optimized.[54,397,472] In chrysanthemum, even primers with as little as 50% GC content resulted in useful patterns at an annealing temperature of 40°C, and primers with 80 or 90% GC still amplified DNA at 44°C. We also observed that primers with a high GC content resulted in too many bands (i.e., a smear) with an annealing temperature of 36°C, whereas fewer and more distinct bands were produced at 40 or 42°C (Wolff, unpublished results). Higher reproducibility by increasing the annealing temperature was also reported by Levi et al.[803]

Touchdown PCR has been proposed to avoid spurious priming in specific PCR protocols.[354] In touchdown PCR, initial annealing temperatures are quite high, but are decreased by 1°C or less in every cycle until the proper T_m is reached (see Chapter 4.3.2.3). A touchdown protocol involving a range of annealing temperatures from 55 to 45°C resulted in an improved reliability of RAPD patterns obtained from species of various taxa, including mouse, rat, fish, oak, and yeast, obtained with three different brands of thermocyclers in two laboratories.[472] However, it was not clear whether the touchdown program or the use of relatively high annealing temperatures was responsible for the increased reproducibility.

RAPD experiments are usually performed with a cycle number of 40 to 45, but 35 cycles have also been recommended by some authors.[961] According to our own experience with chrysanthemum, significantly higher yields of RAPD products are obtained after 45 cycles than after 40 or fewer cycles (Wolff, unpublished results).

4.4.2.4 Template Concentration and Quality

Optimization of template concentration is extremely important for obtaining good RAPD patterns (see, e.g., Doulis et al.[361]). Initially, from 500 pg to 500 ng of template DNA may be tried; changes in fragment patterns and background levels should be checked. Negative controls (no template) should also be included. In theory, RAPDs can be generated from only minute amounts of DNA and/or tissue. For example, Benito et al.[108] showed that a small piece of cereal endosperm is sufficient for 60 RAPD reactions, and Brown et al.[174] demonstrated that RAPDs can even be performed with single tobacco protoplasts. However, reproducibility of RAPD patterns is often poor in the very low concentration range.[397,1547] In some cases, deviating patterns, or even complete lack of bands, were also encountered in the higher DNA concentration range,[961] perhaps due to residual impurities in the template (such as polyphenols and polysaccharides), or residues of CsCl, silica gel, or other compounds used for purifying the DNA.[618]

There is usually a rather wide concentration window in which patterns are stable. Thus, Koller et al.[735] reported that patterns changed only quantitatively within a 200-fold change of DNA concentration, and Schlegel et al.[1236] observed identical AP-PCR patterns of mammalian cell lines, when template DNA concentration varied between 50 ng and 1 μg. Using optimized reaction conditions and a touchdown PCR protocol, Gallego and Martínez[472] found no differences in amplification patterns between 10 and 400 ng of template DNA from various organisms. However, narrower windows have also been reported.[333,867] A template concentration of 20 to 50 ng per 50-μl reaction is often considered optimal. The Stoffel fragment appears to be relatively insensitive to any variation in DNA concentration, producing identical RAPD patterns for template DNA amounts ranging from 5 to 200 ng.[361]

Template quality is an important issue in all sorts of PCR (see Chapter 4.3.2.4). There are numerous examples in which low template quality negatively affected the generation and reproducibility of RAPD patterns. In *Dioscorea bulbifera*, for instance, crude DNA preparations yielded only unreliable results, whereas highly complex, reproducible RAPD fingerprints were obtained with CsCl-purified DNA.[1138] At present, most studies require high throughput, and CsCl centrifugation is no longer the method of choice. However, various alternatives for removing and/or inactivating inhibiting activities from PCR are available (see also Chapters 4.2.2 and 4.3.2). These include diluting the template DNA extract[1037,1228] and removal of polyphenols, RNA, and acidic polysaccharides which are all known to inhibit RAPD-PCR.[1037,1081,1592] In a study on fungal DNA, RAPD patterns were greatly improved if the template DNA was pretreated with RNAse.[1592] This was explained by the detrimental effect of RNA priming on the RAPD reaction.

4.4.2.5   Magnesium Concentration

Changes in magnesium concentration are well known to affect RAPD fragment patterns.[397,1045,1326,1547] In some studies, a relationship was found between primer and magnesium concentration on the one hand, and the average size of RAPD products on the other. Thus, smaller fragments became stronger, and larger fragments because weaker with increasing primer and magnesium concentrations.[397] Similar observations were made in microsatellite-primed PCR with unanchored primers[1524] (see Chapter 4.5.2). Magnesium concentrations may be tested from about 1.5 mM up to 10 mM in pilot experiments. A concentration of 2 mM generally seems to be a good starting point. The chelating effect of EDTA may affect Mg^{2+} concentrations if the DNA is dissolved in TE.

4.4.2.6   PCR Additives

Numerous additives have been reported to improve specificity and/or yield of PCR reactions, and to counteract inhibitory effects (see Chapter 4.3.2.5). Levi et al.[803] reported that the inclusion of 1% Triton X-100 and 0.1% gelatin resulted in more reliable RAPD patterns from woody plants, as compared with the standard PCR ingredients and reaction conditions. Annealing temperatures could be increased to 48°C under these conditions, which probably stabilizes DNA–DNA hybrids. Stommel et al[1335] found that different types of gelatin had opposite effects on the quality of RAPD patterns from different plant species. Certain gelatins even inhibited DNA amplification, at least at the 0.1% concentration suggested by Levi et al.[803] Conversely, 0.1% BSA in the RAPD assay increased both yield and specificity of the RAPD reaction, as evidenced by sharper banding patterns obtained from tomato, potato, and blueberry template DNA. Stommel et al.[1335] therefore recommended that BSA be substituted for gelatin in the protocol of Levi et al.[803]

The phage T4 gene protein 32 (Gp32) is a single-stranded DNA-binding protein that prevents nonspecific primer annealing at low temperatures.[1255] Inclusion of Gp32 in the master mix increased the quality of RAPD patterns in vertebrates.[132]

4.4.2.7 Reproducibility and Quality of Banding Patterns

The influence of the various parameters outlined above on the reproducibility of RAPD patterns has been addressed in numerous studies (e.g., Benter et al.,[118] Caetano-Anollés,[194] Chen et al.,[238] Ellsworth et al.,[397] Gallego and Martínez,[472] Muralidharan and Wakeland,[962] Penner et al.,[1064] Staub et al.,[1326] Wolff et al.[1561]). Because RAPDs use nonstringent PCR conditions, the technique is notoriously sensitive to changes in experimental conditions. Artefactual polymorphisms may result, which in turn may lead to an overestimation of the levels of variation.

Lack of reproducibility may, for instance, be a consequence of poor template quality,[927,1138] presence of too much RNA in the sample,[1592] inconsistent interpretation of mixed-intensity banding patterns,[132] competition between primer target sites[557] (see Chapter 2.3.3.2), generation of heteroduplex molecules[61,995] (see Chapter 2.3.3.2), and generation of primer-derived, nonspecific amplification products[1035,1124] (see below). Penner et al.[1064] studied the reproducibility of RAPD analysis among six different laboratories and found considerable variation. Using the same protocol, but different instrumentation, scientists often were unable to amplify DNA with many of the selected primers. Conversely, allowing each scientist to use his or her own optimized protocol increased the mean reproducibility to 77% over all five primers (in a range of 36 to 100% for the different primers). The thermocyclers appeared to be the most important source of variation.

An obvious (though labor-intensive) measure to enhance reproducibility is to carry out replicate experiments, and exclude inconsistent bands from the analysis. A second measure is to keep reaction conditions perfectly constant within each set of experiments, and to process all samples to be compared simultaneously. Slow transition from the annealing to the extension steps seems to increase reproducibility[118,1256] (see above), as do high initial annealing temperatures and a touchdown PCR protocol.[472]

Given that the outcome of an RAPD experiment is influenced by many interacting variables, complete optimization can only be achieved if each component is tested independently and across a wide concentration range. Caetano-Anollés[194] applied the so-called **Taguchi methods**, which are widely used in industrial process design,[1360] to optimize the DAF protocol. The Taguchi strategy successfully identified variables with major effects on product yield and the number of bands. Reproducibility windows were defined for the concentrations of primer, template, magnesium chloride, dNTPs, and DNA polymerase, as well as for the annealing temperature. The optimized DAF protocol proved to be transportable across laboratories, and yielded identical amplification patterns of soybean template DNA using three different thermocyclers. The author stressed that other marker technologies that depend on several variable factors could also benefit from the application of Taguchi methods. However, there seems to be little response in the literature, probably because most researchers are not familiar with Taguchi's system of experimental design.

A final point of concern is the frequent appearance of bands in negative control reactions without template DNA[776,1035] (see Figure 2.9). These nonspecific amplification products (sometimes called **ghost bands**) appear to be primer-derived; they are not observed when primers and template are omitted from the RAPD-PCR mixture.[1035] The origin of these fragments is unknown, although the condition is probably aggravated by the high number of PCR cycles commonly used in the RAPD method. Contamination of solutions by foreign DNA is probably not responsible, given that repeated amplifications of the negative control using the same solutions never result in identical banding patterns. However, it has been suggested that extremely low amounts of DNA from *E. coli* or the plasmid used to multiply and produce the *Taq* DNA polymerase may be involved.[145] In general, the presence of adequate quantities of template DNA seems to prevent the occurrence of ghost bands,[551,776] but exceptions to this rule have also been reported.[1035] We believe that there is no reason for alarm as long as the (weak) patterns in the negative control are different each time, and contain no bands similar to the major patterns obtained with template DNA.

4.4.3   Modifications

Numerous modifications of the basic RAPD methodology have been described. One group of modifications relates to the type of primers used (see Chapter 4.4.2.1); another group pertains to strategies of **electrophoretic separation and fragment detection**. In the standard RAPD approach of Williams et al.,[1546] fragments are separated on agarose and stained with ethidium bromide. Conversely, both DAF[201,202] and AP-PCR[1527] use PAA gel electrophoresis. PCR fragments are visualized by autoradiography in AP-PCR, and by silver staining in DAF. Other combinations were also described, including (1) silver staining of AP-PCR gels[1236]; (2) separation of RAPD fragments on denaturing gradient gels (DGGE)[375,376]; (3) separation of RAPD fragments on temperature sweep gels[1063]; (4) separation of AP-PCR fragments on nondenaturing PAA gels to detect SSCPs[903]; and (5) fluorescent detection of RAPD fragments separated in automated DNA sequencing machines, either based on PAA gels[275] or capillary electrophoresis.[788]

The latter variant certainly provides the highest resolution. Thus, bands that appeared as single RAPD products on agarose gels were resolved into up to seven products when they were run on a PAA gel.[275] Furthermore, automated sequencers often allow the use of in-lane markers, which considerably improve the precision of fragment sizing. Leamon et al.[788] presented a high-throughput protocol for the multiplex analysis of fluorescently labeled RAPD fragments by capillary electrophoresis in an ABI PRISM™ 310 DNA sequencer. Sophisticated software tools were used to analyze the series of peaks, and to distinguish true signals from background noise. The output file for each sample consisted of a binary string, where each data point indicated the presence or absence of a band of a particular size class.[788] These strings could be imported directly into computer programs commonly used for the analysis of population genetics, such as POPGENE (see Chapter 5.6 and Appendix 3).

The availability of fast methods is essential for routine screening of large numbers of plants; e.g., in population surveys or for marker-assisted selection in plant breeding programs. For such purposes, Penner et al.[1065] suggested the replacement of gel electrophoretic separation of RAPDs by a **dot blot** assay. Total RAPD products obtained from wheat and barley DNA were dotted onto a nylon membrane, and hybridized with a nonradioactively labeled diagnostic DNA fragment generated from the same RAPD primer. Clear presence–absence signals were observed with each of 12 fragments tested. Some of these fragments were diagnostic for a certain chromosome region and others were linked to a trait of interest. Dot blot hybridization clearly reduces time and cost of large screening programs. However, not all RAPD fragments may be suitable for such studies (e.g., artificial cross-hybridization may result from the presence of repetitive DNA on a fragment), and proper controls have to be set up to avoid false-positives and false-negatives.

Another group of RAPD modifications involves the **digestion of DNA with restriction endonucleases**. This could be done either before or after the amplification step (pre- or postdigestion), and may either result in less complex patterns that are easier to evaluate,[1166,1516] or reveal increased levels of polymorphism.[205,731] Postdigestion can also be used to convert RAPD fragments into codominant markers.[463]

The above-described methods and modifications certainly have the capacity to increase the number of observed polymorphisms. However, it should also be considered that any additional experimental steps and sophisticated procedures of data evaluation detract from the main advantages of RAPDs: speed, low cost, and user-friendliness.

4.5 MICROSATELLITE-PRIMED PCR

A tremendous variety of DNA profiling methods have been developed that make use of PCR primer pairs specific for certain kinds of repetitive DNA (see Chapter 2.3.5 for a survey). However, only a minority of these techniques became well established members of the molecular marker family. To date, PCR with anchored or unanchored microsatellite-complementary primers is probably the most widely applied variant of inter-repeat PCR. The popularity of this approach, variously referred to as **inter-simple sequence repeat PCR (ISSR-PCR)**, [1621] **single primer amplification reactions (SPAR)**,[542] or **MP-PCR**,[1276,1475,1524] is due largely to the ubiquitous occurrence of microsatellites in eukaryotic nuclear genomes. Thus, sets of generic, microsatellite-complementary primers can be used in any species, without prior sequence information. In contrast, most other variants of inter-repeat PCR do require such information (e.g., inter-retrotransposon amplified polymorphism [IRAP][681] and other techniques employing transposon-specific primers). The use of 5'-anchored microsatellite primers has the additional advantage (at least in theory; see Chapter 2.3.5.3) that flanking (and potentially hypervariable) microsatellites are incorporated into the PCR products.

In this section, we describe a simple protocol that works for PCR with both anchored and unanchored microsatellite primers.[1139,1524] PCR products are separated on agarose gels and stained with ethidium bromide. The protocol is followed by a discussion of the influence of reaction conditions on the reproducibility and robustness of the technique, and of the various modifications regarding primer choice, fragment detection, and visualization. For a more comprehensive description of the different variants of inter-repeat PCR, see Chapter 2.3.5.

4.5.1   Standard Protocol of Microsatellite-Primed PCR

Solutions (see Comment 1)

Taq DNA polymerase: 5 U/µl

10× PCR buffer:	200 mM Tris-HCl, pH 8.3, 500 mM KCl, 20 mM MgCl₂, 0.01% gelatin. Ten-fold concentrated PCR buffer is usually supplied by the manufacturer of the enzyme. It may or may not contain magnesium chloride and additional ingredients, depending on the brand of the enzyme
dNTP stock:	2 mM each of dATP, dCTP, dGTP, and dTTP; ready-made solutions containing all four dNTPs are commercially available from several suppliers
PCR primer:	5 µM (5 pmol/µl) of a single microsatellite-complementary oligonucleotide (see Comment 2)
Template DNA:	5 to 20 ng/µl

Method

1. Use thin-walled PCR tubes to set up a PCR with 25-µl volume containing 20 mM Tris-HCl pH 8.3; 50 mM KCl; 2 mM MgCl₂; 0.001% gelatin; 200 µM each of dATP, dCTP, dGTP, and dTTP; 0.8 µM primer (20 pmol per reaction); 1 unit *Taq* DNA polymerase; and 15 to 100 ng of template DNA. Pipetting errors are minimized by preparing master mixes for all samples (see Comment 3). Set up a negative control to which water is added instead of DNA. Use a specially designated pre-PCR pipet set. Microtiter plates are a useful alternative to PCR tubes (see Comment 4).

2. Mix the contents and centrifuge the vials briefly (see Comment 5). Microtiter plates can be centrifuged in specially equipped centrifuges.

3. Insert the tubes or the microtiter plate into a thermocycler and start the desired program. We use a modified touch-down program, with 58 → 50°C for 5'-anchored primers and GC-rich primers, and 48 → 40°C for AT-rich primers[1139] (see Comment 2). Ramp times are set to "minimum" except for the transition between annealing and elongation step (90 sec):

1 min	94°C (initial denaturing step)
1 cycle with	94°C for 20 sec
	58°C for 30 sec
	72°C for 40 sec
1 cycle with	94°C for 20 sec
	54°C for 30 sec
	72°C for 40 sec
1 cycle with	94°C for 20 sec
	52°C for 30 sec
	72°C for 40 sec
37 cycles each with	94°C for 20 sec
	50°C for 30 sec
	72°C for 40 sec
Final extension	72°C for 150 sec
3 min at	72°C (final elongation step)

4. After amplification, mix an aliquot of each sample (usually about half the volume) with loading buffer and electrophorese on a 1.4% agarose gel along with a suitable size standard (see Chapter 4.3.4). Stain gels with ethidium bromide or another intercalating dye (see Chapter 4.3.6). Bands are visualized under UV-light and documented by photography, or using an electronic imaging set-up equipped with a video camera.

Comments

1. Frozen aliquots of concentrated PCR buffer, dNTPs, primer, and (if necessary) magnesium chloride should be thawed immediately before use, mixed by vortexing, and kept on ice.

2. We use three different types of primers: (1) unanchored primers between 25% and 50% GC (e.g., [GATA]$_4$); (2) unanchored primers with 50% GC (e.g., [GACA]$_4$, [GGAA]$_4$); (3) 5′-anchored, degenerate primers (e.g., BDB[CA]$_7$C, DVD[TC]$_8$, B[CAA]$_5$). In the latter, B = C, G, or T; D = A, G, or T; H = A, C, or T; and V = A, C, or G. Sets of anchored and unanchored microsatellite-specific primers can be purchased from various companies, e.g., University of British Columbia (UBC primer set no. 9 is available at www.michaelsmith.ubc.ca/services/NAPS/Primer_Sets/).

3. A master mix includes the enzyme, 10× PCR buffer, magnesium chloride, and dNTPs. Master mixes are briefly vortexed, centrifuged, and aliquoted into each tube. Depending on the set-up of the experiment, primer and template are either included in the master mix or added separately for each tube. If n samples are to be analyzed, sufficient master mix should be prepared for n + 1 samples.

4. Most thermocyclers are compatible with 96- or 384-well plates, which increase sample throughput considerably.

5. Older thermocyclers may not be equipped with a heated lid. In this case, the reaction solution needs to be overlaid with two or three drops of mineral oil to prevent evaporation.

4.5.2 Influence of Reaction Conditions and Components

It is well known that reaction conditions can have a considerable impact on the quality and reproducibility of RAPD results (see Chapter 4.4.2). In their seminal

paper describing microsatellite-primed PCR for the first time, Meyer et al.[922] emphasized that MP-PCR would be more reproducible than RAPD analysis because of the
higher stringency of annealing. However, a detailed examination of the influence of
primer, template, Mg^{2+} concentration, and annealing temperature on the quality of
banding patterns suggested that the reproducibility of both techniques is actually
quite similar, and that mismatch priming may also play a prominent role in MP-
PCR.[1524] Whereas changes in template concentration had a relatively small influence,
increasing Mg^{2+} or primer concentrations resulted in the gradual disappearance of
large bands, accompanied by an increase in the intensity of smaller bands, as
commonly encountered in RAPDs[397] (see Chapter 4.4.2).

Increasing the annealing temperatures also produced qualitative changes in banding patterns.[1524] A number of bands amplified from the *E. coli* genome by the
$(GACA)_4$ primer even survived at 64°C annealing temperature. This was the highest
temperature at which a 100% matching positive control plasmid was still amplified.
Given that the *E. coli* genome contains only a single tetranucleotide repeat with n
≤4,[546] mismatch-primed target sites must have been successfully amplified, even at
64°C. These findings were supported by Gillings and Holley,[496] who generated
distinct PCR products from various bacterial template DNAs, using the 3'-anchored
microsatellite primer $(GT)_7GG$ at an annealing temperature of 55°C. This is 5°C
above T_m, as calculated according to the Wallace rule (i.e., 2°C for each AT pair,
and 4°C for each CG pair[639,1391]). Mismatch priming is presumably also responsible
for the generation of organelle-specific PCR products from sugar beet, using
$(GATA)_4$ and $(GACA)_4$ primers.[843,844]

High annealing temperatures considerably exceeding those calculated by the
Wallace rule (see above) were reported to improve the quality of banding patterns
generated by the nonanchored primers $(GATA)_4$, $(GACA)_4$, $(CAA)_5$, and $(CAG)_5$ in
Brassica oleracea and other plant species.[150] In our experience, using the touchdown
PCR protocol presented in the previous section increases the specificity and expands
the reproducibility windows where identical products are obtained.[1139]

4.5.3   Modifications

4.5.3.1   Anchored vs. Unanchored Primers

As with RAPDs and AFLPs, primer performance is a matter of trial and error and
needs to be determined empirically. The different variants of microsatellite-primed
PCR are mainly distinguished by the type and length of the anchor attached to the
primer. Whereas most tri- and tetranucleotide repeat-specific primers also work well
without any anchor,[150,542,922,1524] dinucleotide repeat-specific primers need to be
equipped with a degenerate or nondegenerate 5'- or 3'-anchor.[1621] Moreover, anchored
and unanchored microsatellite primers can be used singly or in combination with
arbitrary primers (RAMP).[1574] In RAMP, special PCR programs need to be designed
that cope with the different annealing temperatures of both types of primers.[894,1574]

PCR with unanchored dinucleotide repeat primers and AT-rich trinucleotide
repeat primers usually resulted in a smear on agarose gels, or even failed
completely[1524] (see also Figure 2.10). AT-rich primers also did not amplify soybean

DNA, even if supplied with a 5′-anchor.[1495] Bornet and Blanchard[150] claimed to have observed clear patterns with unanchored dinucleotide primers, but results were not shown. In an extensive study on Douglas fir (*Pseudotsuga menziesii*) and sugi (*Cryptomeria japonica*), Tsumura et al.[1422] compared the efficiency of 96 different microsatellite primers of various lengths and sequences, including dinucleotide repeats anchored at their 5′- and 3′-ends as well as unanchored tri-, tetra-, and pentanucleotide repeats. More than 60% of the primers yielded interpretable banding patterns after separation on agarose gels and staining with ethidium bromide. Anchored dinucleotide repeats based on GA and GT motifs produced the most useful banding profiles. Inheritance analysis showed that all bands behaved as dominant markers, and that 96% of all bands were inherited according to Mendelian expectations. In a study on soybean, Wang et al.[1495] scored 91% of AMP-PCR markers as dominant and 9% as codominant.

Anchor sequences are often designed to contain degenerate base positions (i.e., BDB[CA]$_8$, with B = C, G, or T; and D = A, G, or T), which are expected to increase the total number of amplified fragments.[1621] This expectation was confirmed by comparative AMP-PCR analyses in soybean.[1495] In this study, the use of degenerate instead of nondegenerate anchor sequences not only resulted in an increase of PCR products, but also in an increase of the proportion of polymorphic fragments.

According to theoretical expectations, the presence of a 5′-anchor ensures that the targeted microsatellite is part of the product. However, it was rarely tested whether this is indeed the case. Wang et al.[1495] sequenced two RAMP products and found the exact sequence of the 5′-anchored microsatellite primer at one end of the fragment, without any additional repeats. They also found that a single base change in the 5′-anchor sequence did not significantly modify the banding pattern. Dávila et al.[302] cloned and sequenced a number of RAMP and AMP-PCR products from barley. In about 50% of the fragments, the same numbers of microsatellite repeats occurred in the PCR products as in the primer sequences. However, a slightly higher number of tandem repeats were found in the PCR products of the other 50%, suggesting that the 5′-anchor was functional in these cases.

Difficulties may arise if the anchor is too short.[446] Instead of binding to the 5′-end of the repeat, the primer may then slip to the 3′-end and exclude the microsatellite from the PCR product. Primers with comparatively long 5′-anchors may prevent such a slippage, because they allow the use of high hybridization stringencies. Fisher et al.[446] designed a degenerate primer with the sequence KKVRVRV(CT)$_6$, where K = G or T; V = G, C, or A; and R = G or A. This primer contains a 7-base anchor including five blocking bases adjacent to the repeat. With this primer, stringency proved to be an important parameter. No amplicons were produced at excessive stringency, whereas too low stringency resulted in slippage of the primer to the 3′-end. Reliable anchoring was only observed within a relatively small window of ±1 to 4°C. The strong 5′-anchors described by Fisher et al.[446] were also successfully used by Brachet et al.[158] to amplify polymorphic AMP-PCR products from *Fraxinus excelsior*.

In a study on sweet potato, Huang and Sun[625] compared the PCR product patterns obtained from a combination of two microsatellite-specific primers (i.e., HVH[TG]$_7$ + [AG]$_8$T, and BDB[CA]$_7$ + [AG]$_8$T) with those obtained with each single primer. The number of fragments amplified by mixed primers was much higher, and numerous additional bands of low molecular weight (<250 bp) were observed. Presumably, these novel fragments are derived from intervening sequences between the two different microsatellite motifs targeted. However, the MP-PCR profile generated with mixed primers was not simply a summary of the products obtained by single primers, similar to the results obtained with RAPD primer combinations (see Chapter 4.4.2.1). Liu and Wendel[831] used a set of three 3'-anchored primers [i.e., (GA)$_9$C, (GA)$_9$T, and (GA)$_9$A] either singly, or in all possible combinations, to generate DNA profiles from a number of cotton cultivars. When agarose gels were used to separate the fragments, the use of mixed primers did not result in a significant increase of the number of bands. However, when fluorescently labeled primers were combined and PCR products were resolved on an automated sequencer, the same increase in fragment number, particularly in the low molecular weight region, was observed, as described by Huang and Sun.[625] Finally, 5'-anchored microsatellite primers were also used to generate fingerprints of fingerprints in the ASAP approach described by Caetano-Anollés and Gresshoff[199] (see Chapter 4.4.2.1).

4.5.3.2 Fragment Separation and Detection

Various gel electrophoresis and detection systems have been used for the visualization of inter-repeat PCR polymorphisms. These include ethidium bromide staining of 1.5 to 2% agarose gels,[150,542,922,1524] PAA gel electrophoresis and autoradiography of radiolabeled fragments,[502,542,1495,1621] silver staining of PAA gels,[228,502,1218] SYBR Green I staining of PAA gels,[1542] and separation of fluorescence-labeled fragments on an automated sequencer.[625,831]

Using 3'-anchored microsatellite primers to amplify banana DNA, Godwin et al.[502] found the detection to be least sensitive on agarose with ethidium bromide, intermediate on 10% PAA minigels with silver staining, and most sensitive on sequencing gels with radiolabeling. Liu and Wendel[831] compared the results obtained by agarose–ethidium bromide staining vs. automated sequencer–fluorescence labeling and found ~3.3 times as many clear-cut bands with the latter approach. Finally, Wiesner and Wiesnerová[1542] compared the MP-PCR patterns generated from flax plants, as visualized by agarose–ethidium bromide, PAA–silver staining, and PAA–SYBR Green I, respectively. In this study, the SYBR Green I method was most efficient in terms of the number of polymorphic bands observed. These results illustrate that the levels of polymorphism observed in a particular system are not only determined by the type of marker, but also by the methodologies of fragment separation and detection. As in the case of AFLPs (Chapter 4.7.4.3), microsatellites (Chapter 4.8.3.1), and other molecular marker techniques, fluorescence technology in combination with an automated sequencer is likely to provide the highest resolution of MP-PCR fragments, but also requires the most expensive equipment.

4.6 PCR AND HYBRIDIZATION: COMBINATORY TECHNIQUES

There are a few occasions for which gel-separated PCR products need to be transferred to a membrane, followed by blot hybridization with a specific probe. These include screening of microsatellite-enriched genomic libraries by colony PCR and oligonucleotide hybridization (see Chapter 4.8.5.6), assessing the genomic copy number of a PCR amplicon of interest (see below), and homology testing of comigrating RAPD and MP-PCR bands (see Chapters 4.6.2 and 5.1.2). Furthermore, a molecular marker strategy coined RAMPO,[1163] RAHM (random amplified hybridization microsatellites),[255] or RAMS (randomly amplified microsatellites)[401] has been developed that is based on a combination of PCR and blot hybridization (see Chapters 4.6.3 and 2.3.7). Because the underlying methodology is based on simple combinations of techniques described in Chapters 4.3, 4.4, and 4.5, no specific protocols are provided here.

4.6.1 Assessing the Genomic Copy Number of PCR Amplicons

Southern blot analysis can be used to establish whether any particular PCR fragment of interest represents single-copy, mid-repetitive, or highly repetitive DNA sequences. In a typical experiment of this kind,[313] an agarose gel containing PCR fragments is blotted onto a nylon membrane (see Chapter 4.3.8) and hybridized with radiolabeled total genomic DNA (see Chapters 4.3.9 and 4.3.10). Fragments representing single- or low-copy sequences will be visible only after prolonged exposure to X-ray film, whereas strong signals are produced by fragments harboring repetitive DNA sequences. In general, RAPD fragments were shown to represent low- as well as high-copy DNA, with no obvious preference.[890,1151,1547] Interestingly, the use of longer primers (i.e., 15- and 20-mers) led to a decrease in the number of RAPD fragments containing repetitive DNA in roses.[313] Amplification products corresponding to single copy sequences may be converted into codominant RFLP markers, thereby circumventing the traditional cloning process.[890,1151] Copy number determinations can also be done the other way round. Individual PCR fragments are then isolated from a gel, and hybridized with a Southern blot containing genomic DNA (e.g., Rabouam et al.[1124]).

4.6.2 Testing the Homology of Comigrating Bands

Homology of comigrating bands is an important issue, especially for systematic studies[1173] (see also Chapter 5.1.2). With all multilocus DNA profiling techniques, the evaluation of genetic relationships among organisms is solely based on the shared presence of similar-sized amplification products. However, it is entirely possible that completely unrelated DNA fragments have the same mobility on gels by chance, and are therefore misinterpreted as being homologous. For RAPD, MP-PCR, and related techniques, the extent of homology between comigrating fragments can exemplarily be tested by hybridizing (radio)labeled, purified PCR fragments of interest with a Southern blot containing RAPD or MP-PCR products generated by the same primer(s).[777,1123,1173,1398,1543] The same strategy can also be used to check for the presence of artefactual extra bands resulting from internal priming.[1123,1124]

To isolate a particular PCR fragment, the band of interest is marked under UV light, and excised from or punched out of the agarose or PAA gel, e.g., with a sterile Pasteur pipet. The fragment is then either reamplified from a small part of the plug, using the original primer(s),[1543] or purified by a commercial kit (PCR product purification kits are available from numerous suppliers). Purified fragments are (non)radioactively labeled, and hybridized with a Southern blot of the RAPD or MP-PCR pattern of origin, as described in Chapters 4.3.9 and 4.3.10.

4.6.3   Random Amplified Microsatellite Polymorphism (RAMPO)

RAMPO,[1163] RAHM,[255] and RAMS[401] are closely related techniques that combine arbitrarily primed or microsatellite-primed PCR with microsatellite hybridization (see Chapter 2.3.7). In the first step of the procedure, an RAPD or MP-PCR experiment is performed as described in Chapters 4.4.1 and 4.5.1. PCR fragments are separated on agarose gels, stained with ethidium bromide, and documented by photography or a video-based imaging system as described in Chapters 4.3.4 and 4.3.6. In conventional RAPD or MP-PCR, gels are discarded at this stage. In RAMPO, a secondary level of information is detected by blotting the gel on a nylon membrane, and hybridizing the blot with a (non)radioactively labeled, microsatellite-specific oligonucleotide as described in Chapters 4.3.8 to 4.3.10. Oligonucleotide hybridization can also be performed within dried agarose gels (see Chapter 4.3.7). The same blots and gels can be reused several times after stripping off the probe by treatment with 5 mM EDTA at 60°C. Several independent RAMPO patterns can therefore be generated from a single RAPD gel.

  In general, RAMPO patterns obtained after autoradiography are completely different from RAPD and MP-PCR patterns observed after staining, and are presumably composed of low-copy, microsatellite-harboring PCR fragments (see Figure 2.12 in Chapter 2.3.7). Oligonucleotides consisting of mono-, di-, tri-, and tetranucleotide repeats were successfully used as probes to discriminate different species and cultivars of yams (genus *Dioscorea*).[1139,1140] RAMPO patterns show good reproducibility and are usually less complex than the underlying ethidium bromide staining patterns, especially when based on RAPD gels.[1140]

4.7   AMPLIFIED FRAGMENT LENGTH POLYMORPHISM

In this section, we describe the basic methodology of generating AFLP markers. Two protocols are provided. One protocol is based on the original method described by Vos et al.[1481] and involves the use of radioactively labeled primers. AFLP products are separated on denaturing PAA gels and detected by autoradiography. The second protocol is a modification of the fluorescence-based AFLP protocols described by Debener and Mattiesch[314] and Myburg et al.,[969] which is routinely used in one of our laboratories. One of the two primers used in the selective PCR step is labeled with an infrared dye, and amplification products are separated and detected on an automated LI-COR sequencer.

  The protocols are followed by a discussion of technical aspects regarding the key steps of the AFLP procedure (i.e., restriction of genomic DNA, preamplification and amplification with selective primers, and fragment separation and detection). We give a brief survey of options to modify and tailor these steps according to the needs and requirements of a particular experiment. Finally, we review a number of studies dealing with the important issues of AFLP robustness and reproducibility.

For a description of the general principles of the AFLP technique and its variants, see Chapter 2.3.8 and Figures 2.13 and 2.14 therein.

4.7.1 Standard AFLP Protocol Using Radioisotopes

The protocols given in this and the following section are each divided into four steps. The first step involves the preparation of restriction site-specific adapters, digestion of genomic DNA, and ligation of adapters to both ends of the genomic restriction fragments. In the second step, the restricted and ligated DNA template is preamplified by PCR with a pair of preselective primers. Selective PCR amplification is performed in the third step, whereas the fourth step comprises fragment separation by denaturing PAA gel electrophoresis, and fragment detection by autoradiography or fluorescence technology. Commercial AFLP kits are available from, e.g., Applied Biosystems and Invitrogen. These kits are accompanied by detailed manuals. See Chapter 4.3.2 for general comments on PCR, and for precautions associated with contamination in PCR.

4.7.1.1 Step 1: Restriction of Template DNA and Ligation to Specific Adapters

This protocol employs *Hin*dIII as a rare cutter, and *Mse*I as a frequent cutter. Restriction and ligation are performed in a single step.

Solutions

Genomic template DNA: 250 ng (concentration should be ~20 ng/µl) (see Comment 1)

10× RL buffer: 100 mM Tris-acetate (pH 7.5), 100 mM magnesium acetate, 500 mM potassium acetate, 50 mM dithiothreitol (sterilize by filtration) (see Comment 2)

*Hin*dIII adapter and *Mse*I adapter (see below for preparation of adapters)

*Hin*dIII restriction enzyme (10 U/µl)

*Mse*I restriction enzyme (10 U/µl)

10 mM ATP

T4 DNA ligase (2 U/µl)

Double-distilled sterile H_2O

Method

1. Prepare a double-stranded adapter specific for *Hin*dIII sites by mixing the following complementary oligonucleotides:

   1 nmol (5.1 µg) *Hin*dIII oligo 1 (top strand): 5′-CTCGTAGACTGCGTACC-3′
   1 nmol (5.5 µg) *Hin*dIII oligo 2 (bottom strand): 5′-AGCTGGTACGCAGTC-TAC-3′ in 200 µl sterile H_2O (the final *Hin*dIII adapter concentration is 5 pmol/µl [5 µM]). Heat at 98°C for 5 min, let the solution slowly cool down to room temperature. The annealed *Hin*dIII adapter has the following conformation (see Comment 3):

   5′-CTCGTAGACTGCGTACC-3′
   3′-CATCTGACGCATGGTCGA-5′

2. Prepare a double-stranded adapter specific for *Mse*I sites by mixing the following, complementary oligonucleotides:

   10 nmol (49.3 µg) *Mse*I oligo 1 (top strand): 5'-GACGATGAGTCCTGAG-3'
   10 nmol (42.1 µg) *Mse*I oligo 2 (bottom strand): 5'-TACTCAGGACTCAT-3' in 200 µl sterile H_2O (the final *Mse*I adapter concentration is 50 pmol/µl [50 µM]). Heat at 98°C for 5 min, let the solution slowly cool down to room temperature. The annealed *Mse*I adapter has the following conformation (see Comment 3):

   5'-GACGATGAGTCCTGAG-3'
   3'-TACTCAGGACTCAT-5'

3. For restriction and ligation, mix the following components in a microfuge tube. To avoid pipetting of volumes < 1 µl, prepare a master mix including all components except the DNA, and add aliquots of this mix to the DNA samples. Incubate 4 h to overnight at 37°C, and then heat the reaction for 10 min at 65°C to inactivate the enzymes. Digested and ligated DNA can be stored at −20°C for extended periods of time.

Ingredient	Stock Concentration	Final Concentration or Amount	Volume Needed for One Reaction (µl)
DNA	250 ng total	10 ng/µl	x
RL buffer	10×	1×	3.00
*Hind*III	10 U/µl	2.5 units	0.25
*Mse*I	10 U/µl	2.5 units	0.25
*Hind*III adapter	5 µM	2.5 pmol	0.50
*Mse*I adapter	50 µM	25 pmol	0.50
ATP	10 mM	0.2 mM	0.60
T4 DNA ligase	2 U/µl	0.1 unit	0.05
Sterile H_2O			24.85 − x
Total volume (µl)			30

Comments

1. DNA preparations need to be of reasonable quality to ensure complete digestion by the restriction enzymes. See Chapters 4.2.2 and 4.3.2.4 for DNA purification methods and methods to deal with enzyme-inhibiting substances, respectively.
2. In our laboratory, 10× RL buffer can be replaced with 10× T4 ligase buffer (usually supplied with the enzyme) with similar results.
3. The adapters described here are identical to those of Vos et al.[1481] except that *Eco*RI-restriction site-specific bases at the 5'-end of the bottom strand oligo were replaced by *Hind*III restriction site-specific bases.

4.7.1.2 Step 2: Preamplification

In this protocol, preamplification is performed with one selective base at each primer (referred to as N below). We routinely use *Hind*III+A primers and either *Mse*I+A or *Mse*I+C primers for preamplification. Depending on genome size, preamplification may also be performed with +0/+0, +0/+1, +2/+1 or +2/+2 primers (see Chapter 4.7.4.2).

Solutions

Digested and ligated template DNA

10× PCR buffer: 100 mM Tris-HCl (pH 8.3), 500 mM KCl, 0.01% gelatin.

Ten-fold concentrated PCR buffer is usually supplied by the manufacturer of the enzyme. It may or may not contain magnesium chloride and additional ingredients, depending on the brand of the enzyme.

50 mM MgCl$_2$

dNTP stock: 2 mM each of dATP, dCTP, dGTP, dTTP. Ready-made solutions containing all four dNTPs are commercially available from several suppliers.

Taq DNA polymerase (5 U/µl)

PCR primer HindIII+1: 50 ng/µl (5'-GACTGCGTACCAGCTTN-3')

PCR primer MseI+1: 50 ng/µl (5'-GATGAGTCCTGAGTAAN-3')

Double-distilled sterile H$_2$O

Method

1. Dilute an aliquot of the digested and ligated template DNA 1:5 with double-distilled, sterile water.
2. Mix the following components in a microfuge tube. To avoid pipetting volumes < 1 µl, prepare a master mix including all components except the DNA, and add aliquots of this mix to the DNA samples.

Ingredient	Stock Concentration	Final Concentration or Amount	Volume Needed for One Reaction (µl)
Digested and ligated DNA aliquot			5.0
10× PCR buffer	10 x	1 x	5.0
MgCl$_2$	50 mM	1.5 mM	1.5
Primer HindIII+1	50 ng/µl	1.5 ng/µl	1.5
Primer MseI+1	50 ng/µl	2.5 ng/µl	2.5
dNTPs	2.5 mM	0.2 mM	4.0
Taq DNA polymerase	5 U/µl	1 unit	0.2
Sterile H$_2$O			30.3
Total volume			50 µl

3. Perform preamplification PCR with the following program (see Comment):
   Set ramp rate to 0.8°C/sec
   20 cycles each with 94°C for 30 sec
                       60°C for 30 sec
                       72°C for 60 sec
   Final extension at 72°C for 7 min
4. Run 3 µl of the resulting preamplification PCR products in 1% agarose gels. A homogenous, light smear should appear. The presence of discrete bands at this stage indicates disproportionate amplification.
5. Dilute the preamplification products 1:20 in double-distilled sterile H$_2$O or in TE. Store at –20°C or use immediately for selective amplification.

Comment

Twenty cycles of preamplification may not be sufficient for species with very large genomes.

4.7.1.3 Step 3: Selective Amplification

Selective amplification is performed with two primers, each with +3 selective nucleotides at their 3'-end, using a touchdown PCR protocol.[354,1481] The *Hind*III primer is radiolabeled by phosphorylating its 5'-end with T4 polynucleotide kinase in the presence of γ^{32}P-ATP or γ^{33}P-ATP. The *Mse*I primer remains unlabeled.

Solutions

PCR primer *Hind*III+3:	50 ng/µl (5'-GACTGCGTACCAGCTTNNN-3') (see Comment 1)
PCR primer *Mse*I+3:	50 ng/µl (5'-GATGAGTCCTGAGTAANNN-3') (see Comment 1)

T4 polynucleotide kinase (10 U/µl)

10× kinase buffer:	670 mM Tris-HCl (pH 8.0), 100 mM MgCl$_2$, 100 mM dithiothreitol; this buffer is usually supplied by the manufacturer of the enzyme

γ^{32}P-ATP (~220 TBq/mmol) (e.g., PB10218; Amersham Biosciences) or
γ^{33}P-ATP (~110 TBq/mmol) (e.g., BF1000; Amersham Biosciences) (see Comment 2)

10× PCR buffer:	100 mM Tris-HCl (pH 8.3), 500 mM KCl, 0.01% gelatin

50 mM MgCl$_2$

dNTP stock:	2.5 mM each of dATP, dCTP, dGTP, dTTP

Taq DNA polymerase (5 U/µl)
1:20 diluted aliquot of preamplified DNA
Double-distilled sterile H$_2$O

Method

1. For radiolabeling the selective *Hind*III+3 primer, mix the following reagents in a microfuge tube (total volume 25 µl). Draw up radioactive solutions slowly to avoid contamination of the pipet, and discard the tip into the radioactive waste before releasing the pressure:
   *Hind*III primer (50 ng/µl), 5 µl
   γ^{32}P-ATP or γ^{33}P-ATP, 5 µl
   T4 polynucleotide kinase, 2.5 µl
   10× kinase buffer, 2.5 µl
   Double-distilled sterile H$_2$O, 10 µl
2. Add 75 µl of sterile H$_2$O, resulting in 100 µl of labeled, diluted (2.5 ng/µL) *Hind*III primer
3. Incubate at 37°C for 30 min and stop the reaction by heating at 70°C for 5 min (inactivation of the kinase). The labeled primer may be used immediately for selective PCR, or stored at –20°C until use.
4. Mix the following components in a microfuge tube. To avoid pipetting volumes < 1 µl, prepare a master mix including all components except the DNA, and add aliquots of this mix to the DNA samples.

Ingredient	Stock Concentration	Final Concentration or Amount	Volume Needed for One Reaction (µl)
1:20 diluted, preamp-lified genomic DNA			2.5
Triton X-100	100%	0.3%	0.03
10× PCR buffer	10×	1×	1.0
MgCl₂	50 mM	2 mM	0.4
Radiolabeled *Hind*III+3 primer	2.5 ng/µl	0.25 ng/µl	1
Unlabeled *Mse*I+3 primer	50 ng/µl	1.5 ng/µl	0.3
dNTPs	2.5 mM	0.2 mM	0.8
Taq DNA polymerase	5 U/µl	1 unit	0.2
Sterile H₂O			3.77
Total volume			10 µl

5. Perform selective PCR with the following program:
   Set ramp rate to 0.8°C/sec
1 cycle with	94°C for 30 sec
	65°C for 30 sec
	72°C for 60 sec
11 cycles each with	94°C for 30 sec
	65°C for 30 sec (touchdown of 0.7°C per cycle)
	72°C for 60 sec
24 cycles each with	94°C for 30 sec
	56°C for 30 sec
	72°C for 60 sec
Final extension at	72°C for 7 min.

Comments

1. The innermost of the three selective bases must be the same as used in the preamplification reaction.
2. Resolution of AFLP patterns on X-ray films is better with [33]P. However, ^{33}P is more expensive than ^{32}P, and requires longer exposure times.

4.7.1.4 Step 4: AFLP Fragment Separation and Autoradiography

After PCR, radiolabeled AFLP products are mixed with formamide-containing loading buffer, and separated on 4.5 to 6% denaturing PAA gels as described in Chapter 4.3.5.2. Gel loading is facilitated by the use of multichannel pipettors (e.g., multichannel gel loading syringe; Hamilton). After the run, the gel apparatus is disassembled, and gels are dried and exposed to X-ray films as described in Chapters 4.3.7 and 4.3.11. In general, overnight exposure will be sufficient for ^{32}P-gels, whereas ^{33}P-gels may take several days.

4.7.2 AFLP Protocol Using Fluorescence-Labeled Primers

The first two steps of the following protocol (i.e., preparation of adapters, restriction, ligation, and preamplification) are identical to the procedure described above

(Chapter 4.7.1). For selective multiplex PCR, infrared dye-labeled primers are used instead of radiolabeled primers, and PCR products are separated and detected using a two-dye 4200 LI-COR automated sequencer.

4.7.2.1 Steps 1 and 2

These steps are the same as those listed in Chapter 4.7.1.

4.7.2.2 Step 3: Selective Amplification

Solutions

> 1:20 diluted aliquot of preamplified template DNA
>
> 10× PCR buffer: 100 mM Tris-HCl, pH 8.3, 500 mM KCl, 0.01% gelatin
>
> MgCl₂ solution: 50 mM
>
> dNTP stock: 2 mM each of dATP, dCTP, dGTP, dTTP
>
> *Taq* DNA polymerase: 5 U/µl (e.g., Invitrogen)
>
> PCR primer *Hin*dIII+3, labeled with IRDye700: 2 ng/µl (see Comments 1 and 2) (5'-GACTGCGTACCAGCTTNNN-3')
>
> PCR primer *Hin*dIII+3, labeled with IRDye800: 2 ng/µl (see Comments 1 and 2) (5'-GACTGCGTACCAATTNNN-3')
>
> PCR primer *Mse*I+3: 50 ng/µl (see Comment 2): (5'-GATGAGTCCTGAGTAACNNN-3')
>
> Double-distilled sterile H₂O

Method

1. Mix the following components in a microfuge tube (see Comment 3). To avoid pipetting volumes < 1 µl, prepare a master mix including all components except the DNA, and add aliquots of this mix to the DNA samples

Ingredient	Stock Concentration	Final Concentration or Amount	Volume Needed for One Reaction (µl)
Preamplified DNA			2.5
Triton X-100	100%	0.3%	0.03
10× PCR buffer	10×	1×	1.0
MgCl₂	50 mM	2 mM	0.4
Primer *Hin*dIII-IRDye700+3	2 ng/µl	0.3 ng/µl	1.5
Primer *Hin*dIII-IRDye800+3	2 ng/µl	0.36 ng/µl	1.9
Primer *Mse*I+3	50 ng/µl	2.5 ng/µl	0.6
dNTPs	2.5 mM	0.2 mM	0.8
Taq DNA polymerase	5 U/µl	0.25 units	0.05
Sterile H₂O			1.22
Total volume			10

2. Perform selective PCR using the following program:
   Set ramp rate to 1°C/sec
   12 cycles each with 94°C for 30 sec
                        65°C for 30 sec (touch down of 0.7°C per cycle)
                        72°C for 60 sec
   24 cycles each with 94°C for 30 sec
                        56°C for 30 sec
                        72°C for 60 sec
   Final extension   72°C for 7 min.

Comments

1. IRDye-labeled primers are commercially available from MWG Biotech. Infrared dye-labeled primers are sensitive to light. Store in the dark, and avoid exposure to any kind of illumination.
2. The innermost of the three selective bases must be the same as that used in the preamplification reaction.
3. The differential labeling of the two *Hin*dIII primers (IRDye700 vs. IRDye800), which are combined with a single unlabeled *Mse*I primer, allows multiplexing of two sets of AFLP products in the same reaction. Some primer pair combinations do not produce scorable patterns when multiplexed in a single PCR[969] (our own observations). This is perhaps due to differences in the T_m, which result in competition effects during the touchdown phase of the PCR program. Pilot experiments need to be done to avoid such nonfunctional combinations.

4.7.2.3 Step 4: AFLP Fragment Separation and Fluorescence Detection

After PCR, AFLP fragments are mixed with formamide-containing loading buffer and resolved on an automated two-dye sequencer (LI-COR 4200 IR^2). A detailed description of the apparatus is beyond the scope of this book. We generally recommend that users follow the instructions of the manufacturer for casting, loading, and running gels on the LI-COR (see also Myburg et al.[969]).

Solutions

Loading buffer: 98% (v/v) formamide, 20 mM EDTA pH 8.0, 0.025% basic fuchsin

Gel: 6% Sequagel XR (ready-to-use gel mix, containing acrylamide, bisacrylamide, TBE buffer, urea and TMED); polymerization is started by the addition of 1/100 volume of ammonium persulfate stock solution (10 mg/ml)

Running buffer: 1× TBE[1217] (see Chapters 4.3.5 and 4.3.5)

Method

1. Mix products of the selective PCR with 1 vol of loading buffer, denature at 85°C for 5 min, and quickly cool on ice before loading.
2. Apply 1-μL samples of each reaction on a denaturing 41 cm × 0.2 mm PAA gel, attached to an automated LI-COR sequencer. The gel should be prerun for 30 min

just before loading the samples. IRDye700- and IRDye800-labeled fragments are visualized on different channels. Throughput can be increased by loading the same gel twice or even several times.

3. Store gel images electronically for further analysis.

4.7.3 Selective Amplification of Microsatellite Polymorphic Loci and Microsatellite AFLP Protocols

Selective amplification of microsatellite polymorphic loci (SAMPL) and microsatellite AFLP (MFLP) combine the advantages of AFLP and microsatellite marker technologies in a single assay[16,950,1556,1585] (see Chapters 2.3.8.3 and 2.3.8.4). In SAMPL, an AFLP adapter-specific primer and a compound microsatellite primer are combined in the selective PCR step. SAMPL produces multiplex patterns of amplified fragments in the range between 40 and 60 bands per assay, of which about 15 to 25% may be polymorphic. Variation arises from length polymorphisms within the microsatellite repeats and mutations in the recognition site of the restriction endonuclease used to cut the template DNA. The exact number of polymorphic bands depends upon the general sequence variability in the target genome. MFLP is closely related to SAMPL, but a 5′-anchored microsatellite primer is used instead of a compound microsatellite primer (see Figure 2.14C and 2.14D).

The standard AFLP protocols given in Chapters 4.7.1 and 4.7.2 can easily be converted into SAMPL and MFLP assays. We routinely use the same preamplification products for AFLP, SAMPL, and MFLP analysis by combining a labeled HindIII+1, +2, or +3 primer with either an unlabeled MseI+3 primer (for AFLP), an unlabeled compound microsatellite-specific primer (for SAMPL), or an unlabeled, 5′-anchored microsatellite primer (for MFLP). Some useful SAMPL primers are compiled in Table 4.2. Figure 4.1 shows a comparative analysis of AFLP-, SAMPL- and MFLP-banding patterns, obtained from the same preamplification products of a set of *Suaeda maritima* specimens (Chenopodiaceae).

Table 4.2 Examples of SAMPL Primers

Primers	Ref.
$A[CA]_7[TA]_2T$	950
$T[GT]_7[AT]_2T$	950
$A[GA]_4[GT]_4G$	950
$G[TG]_4[AG]_4A$	950
$[CT]_8ATA$	950
$[GAAA]_3[GACA]_2$	852
$[TC]_2[TG]_7$	852
$[GAA]_3[GT]_2G$	1031
$[TA]_5[CA]_2C$	1031

4.7.4 Technical Aspects and Modifications

4.7.4.1 Restriction and Ligation of Template DNA

All AFLP experiments start with the digestion of genomic DNA with (normally two) restriction enzymes that produce sticky ends, followed by ligation of suitable adapters that do not restore the restriction sites (see Figure 2.13). In the original procedure, restriction of template DNA and ligation of adapters are performed in two successive steps,[1481] but both reactions can also be performed in a single step[37,314,1342] (see protocols given above).

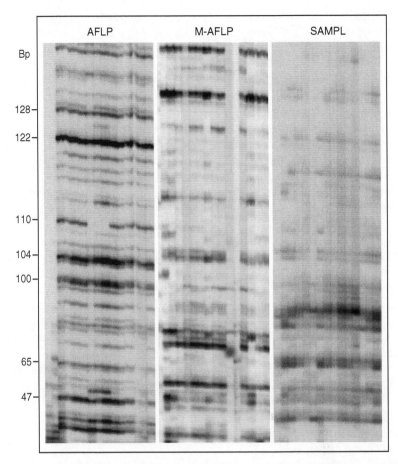

Figure 4.1 Comparative analysis of AFLP,[1481,1605] MFLP,[1585] and SAMPL[950] banding patterns, obtained from the same preamplification products of a set of *Suaeda maritima* specimens collected along the coasts of Central Europe (Chenopodiaceae). This species exhibits very low intraspecific variation with most marker systems (Weising, Schütze, and Prinz, unpublished data). Genomic DNA aliquots were digested with *Hind*III and *Mse*I, ligated to the corresponding adapters, and subjected to two rounds of PCR. Preamplification was carried out with *Mse*+C and *Hind*+A primers. Selective primer combinations were *Mse*+CAC plus *Hind*+AG for AFLP, DBD(AC)$_7$A plus *Hind*+AG for MFLP, and G(TG)$_4$(AG)$_4$A plus *Hind*+AG for SAMPL (D = A, G, or T; B = C, G, or T).The *Hind*+AG primer was labeled with the fluorochrome IRDye800 in all experiments. PCR products were separated on a 6% denaturing PAA gel in an automated LI-COR sequencer. Positions of size markers (lane M) are indicated (Bp, base pairs).

Restriction enzymes are quite sensitive, and their activity can be partially or even totally inhibited by impurities in the DNA preparation (see Chapter 4.3.1). Partial digestion is difficult to detect, but will often result in artefactual variation of AFLP banding patterns. The problem of hidden partials is well known from RFLP fingerprinting based on blot hybridization.[996] The best precaution to avoid partial digestion is the use of good quality DNA for AFLP studies. According to our experience with various genera of Bromeliaceae, Chenopodiaceae, and Euphorbiaceae, DNAs prepared from fresh or dried leaves using the Qiagen Plant DNeasy kit or the CTAB protocol I described in Chapter 4.2.3 yielded highly reproducible banding patterns without further purification. However, problems were commonly encountered when DNAs derived from herbarium specimens were used as templates. With such templates, we often observed an under-representation of small AFLP fragments, compared with AFLP patterns derived from related taxa. A lack of small fragments could be indicative of partial digestion. An additional problem with herbarium specimens is the frequently degraded state of the template DNA, which could lead to an under-representation of large AFLP fragments. We therefore assume that the applicability of AFLPs for the analysis of herbarium samples is limited.

In general, two different restriction enzymes (a rare cutter and a frequent cutter) are used in combination to digest genomic target DNA. The frequent cutter generates small fragments that amplify well, whereas the rare cutter limits the total number of fragments to be amplified. Digestion results in three different types of restriction fragments, of which those with two different ends are amplified preferentially.[1481] The use of two enzymes instead of one also increases the number of possible primer combinations in subsequent PCRs. Most commonly, the six-cutter *Eco*RI is used together with the four-cutter *Mse*I, but *Hin*dIII-*Mse*I[314] (see protocols above), *Pst*I-*Mse*I,[1028,1480,1601] *Pst*I-*Taq*I,[249] and *Sac*I-*Mse*I[165] have also been applied. Reineke and Karlovsky[1149] replaced *Mse*I by the less expensive *Tru*II, which cleaves the same target sequence (i.e., AATT). Suazo and Hall[1342] described a simplified AFLP variant in which only one rare-cutting restriction enzyme (*Eco*RI) was used instead of a pair of enzymes (see Chapter 2.3.8.7). The relatively small number and large average size of *Eco*RI-*Eco*RI fragments allowed the authors to separate the AFLP fragments on agarose gels. Conversely, Van der Wurff et al.[1447] proposed an AFLP variant in which three enzymes (two rare cutters and one frequent cutter), but only two adapters are used (see Chapter 2.3.8.7).

The choice of enzyme combination can have a profound influence on the outcome of the experiment because different enzymes may preferentially cut in different genomic regions. For example, Breyne et al[165] found that *Sac*I-*Mse*I fragments were much less polymorphic than *Eco*RI-*Mse*I fragments among the same *Arabidopsis thaliana* ecotypes. This could partly be explained by the fact that *Sac*I restriction sites are less frequent in coding than in noncoding regions of the *Arabidopsis* genome.

Many restriction enzymes are sensitive to DNA methylation, and will not cut if 5-methylcytosine or 6-methyladenine is present in their recognition sequence. Important examples include *Pst*I, *Hpa*II, and *Msp*I. Because of the pleiotropic nature of DNA methylation, such enzymes are certainly not the primary choice in studies of genetic relatedness. Donini et al.[356] observed reproducible differences between AFLP fingerprints obtained from *Sse*83871-*Mse*I-digested leaf vs. seed DNA of bread wheat (*Triticum aestivum*). The same phenomenon was obtained with root vs. shoot DNA of seedlings of *Aegilops speltoides* and *Aegilops mutica*.[356] *Sse*83871 is an eight-cutter with a recognition sequence of 5′-CCTGCAGG-3′, and will not cut if any of the cytosines or adenines are methylated.[356] The authors hypothesized that the tissue-dependent differences in banding patterns could be due to the differential methylation state of the various organs used for DNA extraction. If methyl-sensitive restriction enzymes are used for AFLP-based relatedness studies, template DNAs should be obtained from physiologically uniform tissues of the same developmental stage. Otherwise, epigenetic variation may be misinterpreted as true sequence polymorphisms.

Conversely, the use of methylation-sensitive enzymes can be advantageous for genetic mapping. In soybean, *Eco*RI-*Mse*I-derived AFLP markers were unevenly distributed across a genetic map, forming dense clusters in regions of reduced recombination around the centromeres.[706] Such clusters were not observed when AFLPs were instead generated with *Pst*I and *Mse*I.[1601] Because centromeric regions are assumed to be heavily methylated, the difference in the behavior of both types of markers is likely due to the sensitivity of *Pst*I to cytosine methylation. Markers generated with methylation-sensitive enzymes such as *Pst*I are likely to be underrepresented in centromeric heterochromatin, but enriched in nonmethylated, single-copy, gene-rich regions. A related effect was reported from Norway spruce (*Picea abies*), in which highly complex AFLP banding patterns with a low signal-to-noise ratio were produced by the standard *Eco*RI-*Mse*I pair.[1028] The pattern complexity was strongly reduced when *Eco*RI was replaced by *Pst*I. The authors interpreted these observations by the presence of long tracts of repetitive and highly methylated DNA in the large spruce genome. *Pst*I, unlike *Eco*RI, does not cut within these sequences, and the number of AFLP bands in the low molecular weight range is hence reduced. Paglia and Morgante[1028] generally recommended methylation-sensitive restriction enzymes for AFLP studies in species with large genomes.

The sensitivity of certain restriction enzymes to DNA methylation has prompted the development of AFLP variants designed to monitor the extent and dynamics of cytosine methylation in plant and fungal genomes (i.e., MSAP[1157,1580] and SDAFLP[729]; for details, see Chapter 2.3.8.5).

4.7.4.2 Preamplification and Selective Amplification

AFLP analysis of complex genomes usually involves two consecutive PCR steps, which are commonly referred to as preamplification and selective amplification, respectively. Performing a preamplification step not only guarantees optimal primer selectivity, it also provides an inexhaustible supply of DNA template. Preamplification is usually done with +2, +1, or +0 primers or combinations thereof, depending on the size and complexity of the genome (numbers refer to selective bases at the 3'-end of primers; see Chapter 2.3.8.1 and Figure 2.13). Most studies use +1/+1, but +0/+2 also worked well in papaya.[1449]

An alternative strategy for preselection of restriction fragments was described in the original AFLP procedure of Zabeau and Vos.[1605] In this variant, restricted template DNA was ligated with a biotinylated *Pst*I adapter and an unmodified *Mse*I adapter. Fragments carrying one or two *Pst*I sites at their ends were then captured on streptavidin-coated magnetic beads and used for selective amplification. This method selects against the presence of the (presumably small) *MseI-MseI* fragments and is still used by some authors (e.g., Barker et al.,[79] Beismann et al.[100]). Lan and Reeves[770] designed unique adapters, which resulted in complete suppression of *MseI-MseI* fragments, whereas *Eco*RI-*MseI* fragments were exponentially amplified, and *Eco*RI-*Eco*RI fragments were only linearly amplified. However, such selection does not seem to be necessary, because *Eco*RI-*MseI* fragments are also preferentially enriched during preamplification using standard adapters.[1481]

PCR products resulting from preamplification are usually diluted five- to 50-fold, and small aliquots are subjected to a second, selective PCR. The number of AFLP products generated during the selective PCR depends both on the number of selective bases at the 3'-ends of the primers and on the genome size of the organism investigated.[1481] For large genomes, more selective bases will be needed to produce a banding pattern of appropriate complexity than for small genomes. According to the guidelines provided by Vos and Kuiper,[1480] +0/+1 combinations are adequate for bacterial artificial chromosomes, +1/+2 or +2/+2 combinations are a good starting point for fungi and bacteria (but see Lin et al.[819]), whereas +3/+3 combinations will be appropriate for most plant species. For example, Ren and Timko[1152] found that +2/+3 selective bases produced too dense patterns in species of the genus *Nicotiana*, whereas +3/+4 generated too few amplification products. Two systems are commercially available (Invitrogen): one with +2/+3 selective bases for organisms with small genomes, one with +3/+3 selective bases for organisms with larger genomes.

The use of +3/+3 combinations may still be inadequate for species with very large genomes, such as many monocots and conifers.[565,566,1409] For example, the total number of bands obtained with +3/+3 combinations in *Hemerocallis* species (Liliaceae; genome size ~4500 Mb) ranged from 82 to 136.[1409] Several primer combinations had to be omitted from the analysis because banding patterns were too complex and/or too fuzzy. In the genus *Alstroemeria* with haploid genome sizes equaling ~25,000 Mb, about 12,000,000 different *Eco*RI-*MseI* fragments can be expected, assuming random distribution of restriction sites, and ignoring the effects of repetitive DNA. Theoretically, the addition of a total of six, seven, and eight selective nucleotides to the core primers should then result in ~3000, 720, and 180 bands, respectively. In practice, the average number of AFLP products obtained with six, seven, and eight selective nucleotides was 109, 87, and 91.[565] Thus, expected and observed values were in close correspondence to each other only in the case of eight selective nucleotides (+4/+4), whereas much fewer bands than expected were observed for +3/+3 and +4/+3.[565] Mixing experiments showed that comigration of unrelated bands was in part responsible for the lower than expected complexity in the latter case.[565]

Early experiments had shown that mismatches are tolerated when more than three selective bases are added to the core primer.[1481] These mismatches occur at the fourth (and higher) nucleotide position(s) from the 3'-end of the selective primers, and lead to a loss of specificity and accuracy of fingerprints. If, for example, eight selective nucleotides are used to reduce the AFLP banding pattern complexity obtained from large genomes to an adequate level, such mispairing should be circumvented by performing preamplification with +2/+2, followed by selective amplification with +4/+4 selective primers.[565,566,1453] The same conditions should be applied to all samples to be compared: Han et al.[565] showed that various distributions of the same eight selective nucleotides (+AGCC/+CAGG) across two, three, or four amplification steps result in qualitative differences of banding patterns, with a two-step protocol (+2/+2 followed by +4/+4) producing the clearest pattern.

As in the case of RAPDs, some AFLP primers and primer combinations perform better than others.[569,1118] For example, when analyzing 256 selective primer combinations in the genus *Beta*, Hansen et al.[569] found that a major proportion of the total polymorphism was revealed by a small number of combinations. Prescreening primer combinations for polymorphisms and distinctness of banding patterns may thus considerably reduce subsequent labor (see also Qi and Lindhout[1118]). Hansen et al.[569] also found a strong correlation between (1) the number of bands, (2) the number of polymorphisms, and (3) the AT content of the selective bases, indicating that AT-rich regions of the genome are more common and variable than GC-rich regions. A reduced number of bands with GC-rich selective nucleotides was also observed in soybean.[706] In *Alstroemeria* species, the same correlation was only seen in the case of +3/+3 and +3/+4 fingerprints, whereas the GC contents of +4/+4 fingerprints had no influence on the average band number. To conclude, it is generally advisable to screen for primer combinations with optimal numbers and sequences of selective bases before embarking on a large-scale study.

4.7.4.3 Fragment Separation and Detection

In the standard AFLP procedure of Vos et al.,[1481] one of the primers used for selective amplification is labeled by a radioisotope (usually ^{32}P or ^{33}P), products are separated on sequencing gels, and banding patterns are visualized by autoradiography or phosphorimaging. Sharper bands are achieved with ^{33}P, but exposure times are significantly longer. Higher resolution was also observed when AFLP fragments from barley were separated by buffer gradient electrophoresis, using 1× TBE as a cathode buffer, and 1× TBE supplemented with 0.5 M sodium acetate in the anode buffer.[1118] Primer labeling can be replaced by incorporation of α^{32}P- or α^{33}P-dNTPs in the selective PCR.[1149] In general, disadvantages of using radioisotopes include problems planning experimental schedules, constraints caused by the decay period of radioactivity, the requirement of specific isotope laboratory facilities and protective shielding, the waste problem, and the potential health hazards.

Another option for the detection of AFLP patterns is silver staining of denaturing PAA gels.[166,225,249,838,839,1411,1428,1441] Cho et al.[249] showed that silver staining and ^{32}P-labeling essentially yielded identical AFLP banding patterns, but resolution of silver-stained gels was better. Chalhoub et al.[225] used 30 ng of selective *Eco*RI+3 primer instead of 5 ng for the radioactive assay, and doubled the loading volume to 4 µl. One of the glass plates was treated with bind-silane, and silver staining was performed with the gel attached to the glass plate using the protocol of Bassam et al.[86] An important advantage of the silver staining technique is that individual bands can be recovered from the gel simply by adding a few microliters of water on top of the stained gel (which rehydrates the underlying DNA), transferring the drop to a PCR tube, and reamplifying the desired band.[225] In this way, bands of interest can be recovered weeks after the experiment, and the dried gel provides a permanent record. Conversely, silver-staining also has several disadvantages. First, all fragments become visible instead of only a subset of fragments defined by a labeled primer. Therefore, the pattern may become too complex. Second, both strands of the DNA are stained, resulting in doublet bands or fuzzy, broad bands on denaturing gels, which may complicate the interpretation of results. Third, there is a gradient in staining intensity: low molecular weight bands are less intensely stained than high molecular weight bands. Fourth, the staining process is sometimes difficult to control and may give inconsistent results. Fifth, there is only one record per experiment, whereas several autoradiograms with different exposure times can be produced with the radiolabeling procedure.

The combination of DIG-labeled AFLP primers with a commercial AFLP™ kit was suggested by Vrieling et al.[1484] DIG-labeled AFLP products from three plant species were resolved on a sequencing gel, blotted onto a nylon membrane by capillary transfer, and cross-linked to the membrane by UV light. Signals were visualized with the use of a DIG detection kit and a chemiluminescent substrate (CSPD™, Tropix). A comparison with radiolabeling showed identical AFLP profiles, but a higher background obtained with the digoxigenin procedure. Only fuzzy patterns were obtained when DIG-labeled dUTPs were used instead of DIG-labeled primers. A similar strategy was proposed by Lin et al.,[820] who separated the AFLP fragments on sequencing gels, followed by capillary blotting. However, the fragments remained unlabeled, and the membrane was hybridized with a universal AFLP probe. This probe is complementary to the *Eco*RI primer sequences and carries an alkaline phosphatase molecule bound to its 5′-end. Signals were eventually detected by the alkaline phosphatase-induced chemiluminescence of CSPD. Disadvantages of these two techniques are the lengthy blotting, hybridization, and detection procedures.

Scott et al.[1259] reported the separation of AFLP fragments from the oomycete *Phytophthora infestans* on ready-to-use, nondenaturing PAA minigels, followed by ethidium bromide staining. Banding patterns were less complex and less intense than those obtained with the standard procedure using radioisotopes, but the authors stressed that recovery and reamplification of fragments of interest were much easier. Suazo and Hall[1342] suggested an even more simplified AFLP protocol (see Chapter 2.3.8.7), that involves separation of fragments on a mixture of agarose and Synergel™ (Diversified Biotech) instead of PAA, again followed by staining with ethidium bromide. Agarose gel electrophoresis and ethidium bromide staining were also used by Ranamukhaarachchi et al.,[1141] who also suggested the inclusion of 2% formamide in the PCR mix to generate more intense and more uniform bands. In general, separating AFLPs on agarose gels combines some advantages of RAPDs (such as speed, simplicity, and easy reamplification, and cloning of polymorphic bands of interest) with those of standard AFLPs (i.e., high reproducibility due to stringent PCR conditions). Although it provides less resolution (and hence detects less variation) compared with the standard protocol, this strategy may be an interesting option when radioactive facilities or automated sequencers (see below) are unavailable.

As is the case with other DNA markers, the most elaborate alternative to the use of radioisotopes is the use of primers that are 5′-labeled with fluorescent dyes. Banding patterns are then detected with a phosphorimager[1188] or an automated sequencer.[624,969,1254] The use of fluorescence-labeled primers in combination with an automated sequencer appears to be the method of choice for AFLP (and also microsatellite) genotyping for several reasons:

1. Products labeled with different fluorochromes can be multiplexed in the same lane, allowing a much higher throughput compared with the conventional assay[263,969,1254,1275] (see protocols given in Chapters 4.7.2 and 4.8.2). Multiplexing can be done either by mixing the differentially labeled PCR products prior to electrophoresis or by simultaneous PCR amplification.
2. Throughput can be further enhanced by multiple reloading of gels.[969]
3. Internal standards allow for the precise determination of band sizes. Fluorescence-labeled size standards are commercially available (e.g., SequaMark DNA size marker Research Genetics, GeneScan® TAMRA500 Applied Biosystems), but can also be made cost-effectively in the laboratory.[1431]
4. All bands are detected at the bottom of the gels where resolution is maximal. As a result, a higher number of discrete bands are usually detected as compared with ^{32}P or ^{33}P labeling using the same primer combinations.[1254]

On the negative side, fluorescence-based detection requires elaborate equipment; fluorescence-labeled primers are relatively expensive; and fragments of interest are not easily accessible for cloning, reamplification, and sequencing. Lazzaro et al.[783] devised a method of postlabeling RFLP and AFLP fragments with fluorochromes using a commercial sequencing kit. This method circumvents the purchase of expensive primers, but also involves additional reaction steps.

4.7.5 Robustness and Reproducibility

AFLPs are produced under relatively stringent PCR conditions. From a theoretical standpoint, they may therefore be considered more reproducible and more robust than most other multilocus marker systems currently available. Results obtained so far generally met these expectations, provided that high-quality DNA was used to ensure complete restriction.[37,69,569,969,1141,1405] AFLP patterns also appear to be quite insensitive to variations in template concentration.[1029,1030,1481,1618]

A few case studies illustrate the extent of AFLP reproducibility. Tohme et al.[1405] extracted DNA from a duplicate set of 20 *Phaseolus* genotypes, and subjected it to identical AFLP conditions. More than 97% of bands were reproducibly obtained in both assays. Arens et al.[37] compared AFLP patterns from DNA of the same individuals of *Populus nigra*, isolated by three different DNA extraction procedures and typed with the same primer pair combinations. Percentages of irreproducible bands (i.e., bands appearing with one isolation procedure, but not with the two others) ranged between 0.1 and 3.9%. Similar values were reported by Winfield et al.[1552] based on duplicate leaf samples of *Populus nigra* isolated by the same DNA extraction procedure. Hansen et al.[569] investigated the reproducibility of AFLPs in sugar beet and related species of the genus *Beta*. Ten sugar beet breeding lines and five wild beets were screened with 256 primer combinations. Of the >11,000 bands obtained, 96.4% were polymorphic. Repeated analysis of all steps from DNA isolation to data scoring yielded an overall reproducibility of 97.6% in 5088 comparisons. The errors fell into three categories:

1. Typing mistakes caused 0.3% of the erratic bands (human error).
2. In one experiment, 1.5% of the bands were clearly resolved as double bands but ran together in the replication (gel resolution error).
3. In one experiment, 0.5% of bands were present, but they were absent in the other experiment (error rate intrinsic to the AFLP technique itself).

Hansen et al.[569] also tested the influence of competition by designing a mixing experiment. An artificial F_1 was generated by mixing the DNA of two parents prior to amplification and comparing the patterns of the mixed sample and the pure samples. In 99.8% of the cases the expected band was present in the artificial F_1. In this study, competition errors were approximately 0.2%, which is considerably less than the 14% competition errors reported for RAPDs in *Brassica*[557] (see Chapter 2.3.3.2). High reproducibilities are also obtained when fluorescence-labeled AFLP fragments are resolved on automated sequencers. For example, Myburg et al.[969] analyzed two independent DNA samples from the same *Eucalyptus* tree by multiplex PCR with fluorescent primers on a two-dye LI-COR automated sequencer. Of a total of 1465 AFLP bands, 1452 (99.1%) were identical between the two samples. Finally, Baurens et al.[89] tested the reproducibility of AFLPs with methylation-sensitive restriction enzymes (methylation-sensitive amplified polymorphism [MSAP] and secondary digest AFLP [SDAFLP]; see Chapter 2.3.8.5) and found an error rate of 0.2% per analyzed band. In summary, AFLPs appear to be significantly more reproducible than RAPDs, provided that good quality DNA is used as starting material.

4.8 GENERATION AND ANALYSIS OF MICROSATELLITE MARKERS

Microsatellites combine several properties that are considered desirable for molecular markers in Chapter 2.1:

- Microsatellites are highly polymorphic single-locus markers.
- Nuclear microsatellites are inherited in a codominant manner.
- Locus-specific primer pairs allow the unambiguous assignment of alleles and ensure a high degree of reproducibility.
- Data on primer pairs and PCR product sizes are easily exchanged among laboratories.
- The marker technique is universally applicable because microsatellites are ubiquitous and abundant components of all eukaryotic and some prokaryotic genomes.
- Given that most microsatellites reside in noncoding DNA, they may be assumed to be selectively neutral.

Because of all of these advantages, locus-specific PCR amplification of micro-satellites became the marker technique of choice for many types of studies. In fact, this technique suffers from only two major disadvantages. First, microsatellite assays require sequencing gels and are technically somewhat more demanding than, e.g., RAPD analyses. Second, unless sufficient database information is available, micro-satellites have to be cloned and their flanking regions sequenced for every species (or at least every genus; see Chapter 4.8.4.3) under study. These requirements made the establishment of microsatellite markers for a new species quite cumbersome, and have been a main obstacle to using the marker system on a large scale. Never-theless, information on microsatellite-flanking sequences has accumulated rapidly during the last decade, and elaborate enrichment cloning techniques have been developed that facilitate the generation of markers.

This section starts with the description of standard protocols for analyzing nuclear and organellar microsatellites, assuming that suitable primers are already available (Chapter 4.8.1 and 4.8.2). These protocols are followed by a discussion of technical aspects and problems regarding the key steps of microsatellite analysis (Chapter 4.8.3). We then discuss various strategies to identify microsatellite-flanking regions and design locus-specific primer pairs. Before embarking on a microsatellite study in a particular species, the following options should be considered:

1. Microsatellite primers might already be available for the species under investigation.
2. Microsatellites of the species under investigation might be found in DNA sequence entries submitted to Genbank, EMBL, and DDBJ databases.
3. Microsatellite markers developed for related species may be transferable to the species under investigation.
4. Microsatellites might be cloned from a standard small insert library.
5. Microsatellites might be cloned from a library enriched for simple repeats.
6. Microsatellite marker and library development might be ordered from a commer-cial company.

The options (1) to (3) listed above do not require cloning; they are described in Chapter 4.8.4. Microsatellite cloning strategies are the subject of Chapter 4.8.5, in which a protocol is given for the establishment of small-insert genomic libraries enriched for microsatellites. The biology of microsatellites is discussed in Chapter 1.2.2. The general principles of locus-specific microsatellite analysis and general properties of microsatellite markers are described in Chapter 2.3.4.

4.8.1 Microsatellite Analysis Using Radioisotopes

The following protocol may be used to amplify specific microsatellite loci from the nuclear, chloroplast, or mitochondrial genome, provided that primer pairs specific for microsatellite-flanking regions are available. PCR fragments are labeled by the inclusion of α^{33}P- or α^{32}P-dCTP in the PCR. Microsatellite fragments are resolved on sequencing gels, and detected by autoradiography. See Safety Precautions for work with radiolabeled DNA (Chapter 4.1). For general comments on PCR, and for precautions associated with contamination in PCR, see Chapter 4.3.

Solutions

10× PCR buffer: 100 mM Tris-HCl (pH 8.3), 500 mM KCl, 0.01% gelatin

50 mM MgCl$_2$

PCR primers: 5 µM stock solution of each forward and reverse primer (see Comment 1)

dNTP stock: 2.5 mM dATP, 2.5 mM dGTP, 2.5 mM dTTP, 0.25 mM dCTP (see Comment 2)

α^{32}P-dCTP   ~110 TBq/mmol; e.g., PB10205; Amersham Biosciences or

α^{33}P-dCTP   ~92.5 TBq/mmol; e.g., BF1005; Amersham Biosciences; see Comment 3 and Safety Precautions (Chapter 4.1)

Taq DNA polymerase (5 U/µl)

Template DNA: ~25 ng/µl

Double-distilled sterile H$_2$O

Method

1. Mix the following components in a microfuge tube. To avoid pipetting volumes <1 µl, prepare a master mix including all components except the template DNA, and add aliquots of this mix to the DNA samples. Draw up radioactive solutions slowly to avoid contamination of the pipet, and discard the tip into the radioactive waste before releasing the pressure.

Ingredient	Stock Concentration	Final Concentration or Amount	Volume Needed for One Reaction (µl)
Template DNA			1.0
10× PCR buffer	10×	1×	1.0
MgCl$_2$	50 mM	2.5 mM	0.5
Forward primer	5 µM	0.5 µM	1.0
Reverse primer	5 µM	0.5 µM	1.0
dNTPs	2.5 mM	0.2 mM	0.8
α^{32}P-dCTP	370 kBq/µl	1.5 kBq/µl	0.04
Taq DNA polymerase	5 U/µl	0.1 U/µl	0.2
Sterile H$_2$O			4.46
Total volume			10 µl

2. Start a PCR using the following program:
   Initial denaturation at 94°C for 3 min
   30 cycles each with   94°C for 30 sec
                         50 to 65°C for 30 sec (see Comment 4)
                         72°C for 1 min
   Final extension   72°C for 7 min
3. After PCR, mix samples with formamide-containing loading buffer, and separate on a sequencing gel as described in Chapter 4.3.5.2. Gel loading is facilitated by the use of multichannel pipettors (e.g., multichannel gel loading syringe; Hamilton).
4. After the run, disassemble the gel apparatus, dry the gel, and expose to X-ray film as described in Chapters 4.3.7 and 4.3.11. In general, overnight exposure will be

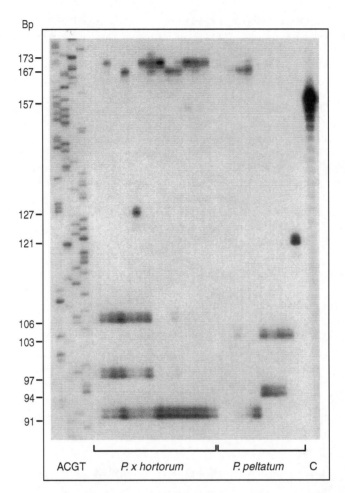

Figure 4.2 Microsatellite analysis of *Pelargonium* cultivars. Genomic DNA aliquots from 13 cultivars of *Pelargonium X hortorum* and nine cultivars of *P. peltatum* were amplified using a primer pair specific for the microsatellite locus PhSSR45.[94] Radiolabeled PCR products were resolved on a sequencing gel and visualized by autoradiography. Complex patterns comprising one to four bands are resolved, depending on the ploidy level of the plant material and the state of heterozygosity of the targeted locus. Several allele size classes are revealed. The largest size class is also the most variable one. A typical stutter pattern (see Chapter 4.8.3.2) is observed in all lanes, especially in the positive control of lane C (PCR product amplified from the bacterial clone). Positions of size markers (M13 sequencing ladder) are indicated (Bp, base pairs).

sufficient for ^{32}P gels, whereas ^{33}P gels may take several days. A typical microsatellite gel pattern obtained with ^{32}P-labeling and the above protocol is shown in Figure 4.2.

Comments

1. As an alternative to the inclusion of α^{33}P- or α^{32}P-dCTP in the PCR, the 5′-end of one of the primers may be radiolabeled with polynucleotide kinase in the

presence of γ^{32}P-ATP prior to the PCR (see Chapters 4.3.9.1 and 4.7.1 for labeling protocols). Using labeled primers instead of labeled nucleotides sometimes reduces background and results in a more even distribution of signal intensities on autoradiograms.[630] On the negative side, this variant requires an additional experimental step.

2. When 5′-labeled primers are used instead of α^{33}P- or α^{32}P-dCTP, all cold dNTPs must be provided in equimolar concentrations.[630]

3. Resolution is better with ^{33}P. However, ^{33}P is more expensive than ^{32}P, and requires longer exposure times.

4. The annealing temperature will depend on the primer sequence and the highest possible temperature (~60 to 65°C) should be chosen to ensure locus-specific amplification. Annealing temperatures may have to be optimized for individual loci. Specificity may also be increased by using a touchdown protocol as described in Chapter 4.7.1. Lower temperatures (~50°C) may be required when heterologous or degenerate primers are used (see Chapter 4.8.4.3). Longer annealing times and higher Mg^{2+} concentrations may also help. The availability of a gradient thermocycler will facilitate the optimization process.

4.8.2   Microsatellite Analysis Using Fluorochromes

The protocol given in this section may be used to amplify specific microsatellite loci from the nuclear, chloroplast or mitochondrial genome, provided that primer pairs specific for microsatellite-flanking regions are available. Infrared dye-labeled primers are used instead of radiolabeled primers or nucleotides, and PCR products are separated and detected using a two-dye, model 4200 LI-COR automated sequencer. The differential labeling with two fluorochromes allows the combination of two primer pairs in a single reaction (multiplexing). See Chapter 4.3.2 for general comments on PCR, and for precautions associated with contamination in PCR.

Solutions

10× PCR buffer: 100 mM Tris-HCl (pH 8.3), 500 mM KCl, 0.01% gelatin

50 mM $MgCl_2$

Forward primer I (labeled with IRDye700): 5 μM stock solution (see Comments 1 and 2)

Reverse primer I (unlabeled): 5 μM stock solution

Forward primer II (labeled with IRDye800): 5 μM stock solution (see Comments 1 and 2)

Reverse primer II (unlabeled): 5 μM stock solution

dNTP stock: 2.5 mM each of dATP, dGTP, dTTP and dCTP

Taq DNA polymerase (5 U/μl)

Template DNA: ~25 ng/μl

Double-distilled sterile H_2O

Method

1. Mix the following components in a microfuge tube. To avoid pipetting volumes <1 μl, prepare a master mix including all components except the template DNA, and add aliquots of this mix to the DNA samples.

Ingredient	Stock Concentration	Final Concentration	Volume Needed for One Reaction (μl)
Template DNA			1.0
10× PCR buffer	10×	1×	1.0
MgCl₂	50 mM	2.5 mM	0.5
Forward primer I (labeled with IRDye700)	5 μM	0.5 μM	1.0
Reverse primer I (unlabeled)	5 μM	0.5 μM	1.0
Forward primer II (labeled with IRDye800)	5 μM	0.5 μM	1.0
Reverse primer II (unlabeled)	5 μM	0.5 μM	1.0
dNTPs	2.5 mM	0.2 mM	0.8
Taq DNA polymerase	5 U/μl	0.1 U/μl	0.2
Sterile H₂O			2.5
Total volume			10 μl

2. Start a PCR using the following program:
   Initial denaturation at 94°C for 3 min
   30 cycles each with   94°C for 30 sec
                    50 to 65°C for 30 sec (see Comment 3)
                    72°C for 1 min
   Final extension at   72°C for 7 min

3. After PCR, mix samples with formamide-containing loading buffer, and separate on an automated two-dye LI-COR sequencer as described in Chapter 4.7.2. A detailed description of the apparatus is beyond the scope of this book. We generally recommend that the user follow the instructions of the manufacturer for casting, loading, and running gels on the LI-COR (see also Myburg et al.[969]).

4. IRDye700- and IRDye800-labeled fragments are visualized on different channels. Throughput can be further increased by loading the same gel twice or several times.

5. Store gel images electronically for further analysis.

Comments

1. IRDye-labeled primers are commercially available from MWG Biotech. Infrared dye-labeled primers are sensitive to light. Store in the dark, and avoid exposure to any kind of illumination.

2. Primer pairs labeled with two different fluorochromes are multiplexed in the present protocol. Higher multiplexing rates can be obtained when larger numbers of fluorochromes are available (e.g., on automated sequencers constructed by Applied Biosystems). They may also be achieved by designing primers in a way that the resulting PCR products occupy nonoverlapping size ranges. With increasing multiplex rate, there is also an increasing chance of unwanted pairwise interactions between primers. The possibility of such interactions needs to be checked using suitable computer programs (e.g., Oligo,[1203] Primer3,[1196] Prime: GCG, University of Wisconsin). If primer interactions are a problem, PCRs may be performed separately, and the reaction products mixed before electrophoresis.

3. The annealing temperature will depend on the primer sequence and the highest possible temperature (~60 to 65°C) should be chosen to ensure locus-specific amplification. Annealing temperatures may have to be optimized for individual loci. Specificity may also be increased by using a touchdown protocol as described in Chapter 4.7.1. Lower temperatures (~50°C) may be required when heterologous or degenerate primers are used (see Chapter 4.8.4.3). Longer annealing times and higher Mg^{2+} concentrations may also help. The availability of a gradient thermocycler will facilitate and speed up the optimization process.

4.8.3   Technical Aspects and Modifications

4.8.3.1   Fragment Separation and Visualization

Separation and detection of microsatellite markers can be achieved in various ways, depending on the availability of equipment and the requirements of the experiment. Standard protocols of microsatellite analysis use radioisotopes, sequencing gels, and autoradiography to detect the PCR products (Chapter 4.8.1). This method allows the discrimination of fragments that differ by a single base pair. Agarose gels in combination with ethidium bromide staining are easier to handle, but will not allow accurate size determination. Nevertheless, agarose gels are worth considering when allele size differences are sufficiently large (e.g., in mapping projects), and/or when markers are based on tri- or tetranucleotide repeats. High-percentage agarose gels or commercial agarose preparations such as MetaPhor™ (FMC Bioproducts) were used for microsatellite studies by, e.g., Akkaya et al.,[15] Bell and Ecker,[104] Mellersh and Sampson,[913] Routman and Cheverud,[1191] and White and Kusukawa.[1538] Kristensen and Børresen-Dale[752] reported an improved electrophoretic separation of microsatellite markers in agarose gels containing bis-benzimide. An accurate size standard (10 or 25 bp ladder) is needed to score fragment size, and an allele-ladder (PCR products of known alleles) aids accurate genotyping.

Radioactive labeling can also be avoided by resolving microsatellite markers on nondenaturing or denaturing PAA gels and staining with ethidium bromide,[1265] SYBR Green,[952] SYBR Gold,[1129,1535] or silver.[193,285,724,906,984] Separation of microsatellite markers from *Thuja plicata* in discontinuous PAA gels followed by staining with SYBR Gold was reported by White et al.[1535] SYBR Gold staining is simpler and less expensive than silver staining, but the image quality may not be optimal.[1129] A more sophisticated microsatellite typing procedure involves primer extension combined with matrix-assisted laser desorption–ionization time-of-flight (MALDI-ToF) mass spectrometry.[160,1043]

As in the case of AFLPs and other marker technologies, the most efficient technique to resolve microsatellite markers involves the use of fluorescence-labeled primers or dNTPs in combination with an automated DNA sequencer and the appropriate software.[161,348,714,718,750,936,1258,1435,1620] Fluorescent PCR products are detected by real-time laser scanning during electrophoresis. The allelic information is immediately stored in a computer as the PCR fragments pass by the detection window. No gel handling is required after electrophoresis.

In automated sequencers, fragments are either resolved in sequencing gels or in capillaries. Capillary DNA analyzers allow both high-precision microsatellite genotyping and high throughput.[876,1531] An additional advantage of capillary electrophoresis is that every sample is run separately, and therefore a spillover between wells, as in gel-based electrophoresis, is impossible. Cross-contamination between lanes can become a real problem when banding patterns are complicated (e.g., in polyploids; see Figure 4.2), and stuttering is extensive.

In sequencers that can detect two or more dyes (e.g., Applied Biosystems, LI-COR), markers labeled with different fluorochromes can be multiplexed in a single lane. One dye may be attached to an internal size marker that allows the computer to generate a calibration curve for automated allele sizing and quantitation. This obviates problems of lane-to-lane and gel-to-gel variation (e.g., band shifts and smiling effects), but can still yield inaccurate size estimates.[550] By exploiting the availability of different colors and different product size ranges, Schwengel et al.[1258] reported the multiplex analysis of up to 24 different human microsatellite loci per lane. A fluorescence-based heptaplex system including a sex marker was reported as a highly sensitive human identification system with random matching probabilities of 10^{-8} to 10^{-10} (Urquhart et al.[1435]). Multiplex PCR is not a trivial task, and PCR parameters such as relative primer concentrations need to be balanced carefully. Critical parameters of the simultaneous amplification of several target loci have been reviewed by Henegariu et al.[592]

Multiplex PCR and fluorescence-based allele sizing of microsatellite markers were first applied to plants by Mitchell et al.,[936] who reported the successful coamplification of 11 SSR loci from *Brassica napus* in a single reaction. More recently, fluorescent-labeled microsatellite panels for multiplex PCR have been developed for several major crop species, including soybean,[974] barley,[865] rice,[139] and sunflower.[1363] One obvious disadvantage of the fluorescence technology is that expensive equipment, primers, and software are required. It is advisable to first order unlabeled primers and test the functionality of the markers on agarose or Metaphor gels.

4.8.3.2 Stutter Bands and Other PCR Artefacts

Instead of yielding well-defined, distinct fragments, the enzymatic amplification of mono- and dinucleotide repeats often results in clusters of bands that are separated from each other by one- or two-base pair intervals. These so-called stutter bands (or shadow bands) are the most common artefact of microsatellite analysis (see Figure 4.2). The additional bands most probably result from slippage of the *Taq* DNA polymerase during the PCR.[580,830,968,1298] Sequence analyses of shadow bands generated by PCR of dinucleotide[968] and tetranucleotide repeats,[1491] respectively, showed that the sequence becomes scrambled only in the repeat region. In general, the largest and most intense fragment (i.e., the highest peak on a DNA sequencer) is the real fragment.

The problem of stutter bands increases with the numbers of repeats. Therefore, microsatellites with very high numbers of repeats (>20) are generally hard to score. Compound microsatellites (e.g., a combination of mono- and dinucleotide repeats) may result in particularly confusing patterns of a one-base stutter superimposed on a stutter of two bases. Stutter bands can interfere with the correct interpretation of allele sizes, especially when two alleles differ by a single repeat unit, and when homo- and heterozygotic states have to be discriminated (see, e.g., Smulders et al.[1302]). Differences between samples are often recognized quite easily because the whole clusters of bands are shifted relative to each other (see Figures 2.7 and 2.8 in Chapter 2.3.4.2, and Figure 4.2). However, the determination of absolute fragment sizes is sometimes difficult.

Spencer et al.[1317] stressed that stuttering could be reduced by varying the primer and dNTP concentrations. However, Smulders et al.[1302] concluded that the ladder artefacts were not easily removed by changing the PCR conditions. Ordering of alleles according to size may help in assigning allelic states correctly (e.g., Hüttel et al.,[633] Saghai-Maroof et al.[1210]). According to Neilan et al.,[984] stuttering is almost absent when PCR products are separated on nondenaturing 10% PAA gels followed by silver staining. Microsatellites consisting of tri-, tetra-, or pentanucleotide repeats are generally easier to analyze. Slippage is less severe, and amplicons are more clearly resolved (see Figure 2.7, lower panel), even on high-percentage agarose gels.[382,714,718]

In general, *Taq* DNA polymerases will add an extra A at the end of a PCR product. However, this does not necessarily happen to all fragments, and two fragments differing by one base pair are sometimes seen on the autoradiogram (or double peaks on a sequencer). The extent of this problem depends on the type and brand of DNA polymerase, PCR conditions, and primers. If A-addition is a real problem (which is rarely the case), the inclusion of a very long final extension step in the PCR (such as 30 min at 72°C) will ensure that all PCR fragments get the additional adenosine residue, and single bands are obtained.

4.8.4 Generating Microsatellite Markers Without Cloning

4.8.4.1 Literature Screening

Microsatellite markers are being developed continuously for a wide range of species, and even for exotic species, solely due to an interest in the species' ecological or evolutionary characteristics or its rarity. In the last decade, hundreds of reports describing novel plant microsatellites appeared in the literature. It is entirely possible that markers already have been developed for the species of interest, or for a closely related species (see Chapter 4.8.4.3), and an intense screening of the literature should be the very first step taken in the frame of any microsatellite project. Novel primers flanking plant microsatellites are published in a wide range of journals dealing with genetics, ecology, evolutionary biology, and plant breeding, most notably the *American Journal of Botany*, *Annals of Botany*, *Conservation Genetics*, *Crop Science*, *Euphytica*, *Genetic Resources and Crop Evolution*, *Genetics*, *Genome*, *Heredity*, *Journal of Heredity*, *Molecular Breeding*, *Molecular Ecology*, *Molecular Ecology Notes*, *Molecular Genetics and Genomics*, *Plant Breeding*, *Plant Molecular Biology*, and *Theoretical and Applied Genetics*. Some journals, such as *Molecular Ecology*

Notes and *Conservation Genetics*, have a special section for short notes describing novel microsatellites. The home pages of these journals should be checked, as well as traditional literature databases such as Medline, Web of Science, and NCBI/Genbank. Blackwell Science Publishers have created a database containing all details for reported loci (i.e., primer sequences, PCR conditions, polymorphism levels, cross-species amplification, and literature citations) in a searchable format (available from http://tomato.bio.trinity.edu/home.html). Currently, this database contains all primer submissions to *Molecular Ecology* and *Molecular Ecology Notes*. For the future, it is planned to include relevant submissions from other journals as well.

4.8.4.2 Database Mining

Myriad publicly available DNA sequences are stored in large databases such as EMBL, GenBank, and DNA Data Bank of Japan (DDBJ) (for a recent survey, see Galperin[473] and the 2004 database issue of *Nucleic Acids Research*). It is worth screening these databases for the presence of microsatellites in the nuclear genome of the species of interest (and also of closely related species); e.g., by using the BLAST search algorithm at the National Center for Biotechnology Information (NCBI)/GenBank home page at http://www.ncbi.nlm.nih.gov/. Database mining already resulted in hundreds of microsatellite markers for a considerable number of plant species, including potato,[930] tomato,[1302] and rice.[250]

Microsatellites are surprisingly common in the vicinity of genes, and trinucleotide repeats preferably occur in exons.[951] Recently, the feasibility of generating microsatellite markers from expressed sequence tag (EST) databases has been demonstrated in rice,[250] wheat,[417,475,609] barley,[609,1392] various other members of the grass family,[274,474,690] grapevine,[45,317,1261] apricot,[317] kiwifruit,[460] watermelon,[535] and the model species, *Medicago truncatula*.[418] EST- and cDNA-derived microsatellites have two important advantages over anonymous markers. First, they are physically linked to a gene, which may encode a trait of interest. Second, primer target sequences that reside in transcribed regions are expected to be relatively conserved, thus enhancing the chance of marker transferability across taxa (see below). On the negative side, the association with coding regions may limit the polymorphism of EST-derived microsatellite markers,[417] and microsatellites in 5'- or 3'-untranslated regions are often located close to the cloning site, which makes primer design difficult.[39]

Databases are also commonly exploited to generate microsatellite markers for organellar DNA. Fully sequenced chloroplast and mitochondrial genomes are available from an increasing number of plant species, and a large amount of sequence data has accumulated for noncoding regions of plant cpDNA and mtDNA. These regions often contain polymorphic mononucleotide repeats.[1092,1093,1114] If no cpDNA or mtDNA sequence data are available for the species of interest, one may also refer to consensus primers that flank cpDNA microsatellites in unrelated species (see Chapter 4.8.4.4). Sets of such primers have been developed for both gymnosperms[1462] and angiosperms,[253,1520] as well as for particular angiosperm families (e.g., Solanaceae[180] and Poaceae[1115]).

4.8.4.3   Marker Transferability between Species: Nuclear Microsatellites

As mentioned above, microsatellite markers are often available in the public domain for a species that is closely related to the species of interest. However, these markers may or may not be functional in heterologous species. Two important prerequisites need to be fulfilled for the successful transfer of a microsatellite marker from a focal species (i.e., where the marker originates) to a nonfocal species:

1. The primer binding sites have to be sufficiently conserved for amplification to occur. Intraspecific sequence variation, which is commonly found in microsatellite-flanking regions,[938] may interfere with primer annealing. Nonamplifying markers result in so-called null alleles that may disturb the results of parentage and population studies in heterozygous situations[208,1061] (see Chapter 2.3.4.1). Problems associated with null alleles are enhanced if primers are derived from heterologous sources.[1026,1061]
2. A polymorphic microsatellite should be present between the primer binding sites also in the nonfocal species. If there is no microsatellite or only a very small one, the amplified fragments are likely to be monomorphic or exhibit a low level of variation.

The presence of a microsatellite in a heterologous PCR product can be detected by either Southern or dot blot analysis using a microsatellite-specific probe[1533,1541] or by direct sequence analysis.[694,1054,1454] One may also simply test whether there is intraspecific variation in the nonfocal species.

Studies in this direction often revealed a tendency of microsatellites to be shorter and less polymorphic in species other than that from which they were first isolated. The reasons for this behavior are controversial,[27,390,391,1198] but a sampling bias (also called ascertainment bias) for exceptionally long microsatellites during the cloning procedure generally is considered a more likely explanation than directional evolution. Mechanistically, the decrease of polymorphism observed in nonfocal species is often caused by the interruption of long, contiguous microsatellite arrays by base substitutions, and/or by shortening of perfect arrays through slippage events[81,141,753,1454] (see also Chapter 1.2.2).

In animals, there has been a fair amount of success with transferring microsatellite markers among taxa. Thus, cross-species amplification has been demonstrated among genera of the same family (e.g., between goat, cattle, and related Bovidae[1066]), among families of the same order (e.g., in pinnipeds[483]), and even among species belonging to different orders (e.g., in marine turtles[447]). Conservation appears to be especially pronounced in marine animals, where microsatellite loci were shown to be conserved within whales,[1239] marine turtles,[447] and fish[1165] during 300, 400, and 470 million years of evolution, respectively. By combining published cross-priming data from birds[1100] and mammals,[945,1066] Primmer et al.[1100] found a significant inverse relationship between microsatellite performance, the proportion of polymorphic loci among those that amplify, and the evolutionary distance between the pair of species compared.

In plants, transferability of microsatellites between congeneric species has been demonstrated in numerous taxa, including *Actinidia*,[627,1525] *Brassica*,[750,769,847] *Camellia*,[702] *Citrus*,[714,859] *Clusia*,[554] *Olea*,[1134] *Pinus*,[694,761,1282] *Prunus*,[317] *Quercus*,[1332] and *Vitis*.[45,317,336,1261,1393] Usually, a low percentage of markers also amplifies fragments from species belonging to other genera from the same family, as has been shown, e.g., for various Asteraceae,[1541] Brassicaceae,[1454,1533] Cucurbitaceae,[700] Fabaceae,[1038,1054] Fagaceae,[1332] Limnanthaceae,[721] Mimosaceae,[304] Oleaceae,[311] Poaceae,[274,609,758,1179,1392] and Vitaceae.[45,317,337]

For example, Van Treuren et al.[1454] examined the performance of 30 microsatellites from *Arabidopsis thaliana* in two species of *Arabis*. PCR products were generated by about 50% of the primers, but alleles were generally shorter, repeat numbers were smaller, and the level of intraspecific variation was lower in the nonfocal compared with the focal species. These results concur with the ascertainment bias phenomenon suggested by Ellegren et al.[390] (see above). Peakall et al.[1054] surveyed the transferability of 31 microsatellite markers from soybean to three wild relatives of the genus *Glycine* and to several other Fabaceae. Up to 65% of primer pairs amplified products from *Glycine* species, but only 3 to 13% were successful in other genera. With the exception of *G. clandestina*, size variation was low in nonfocal species. In the genus *Clusia*, eleven loci amplified across all 17 *Clusia* species tested, whereas none amplified in a sister species of the genus *Chrysochlamys*.[554] Fragment sizes were polymorphic in the nonfocal species wherever tested. Rossetto[1189] summarized the data from a large number of studies and found that 58% of microsatellites were polymorphic within the same family and 78% within the same subgenus.

It is yet unclear why microsatellites and their flanking DNA are relatively conserved in some taxa, but not in others. For example, Decroocq et al.[317] showed that transferability is higher between *Vitis* species than between Rosaceae species. Karhu et al.[694] and Kutil and Williams[761] demonstrated an unusually high conservation of primer binding sites among a number of pine species over a period of more than 140 million years. Kutil and Williams[761] claimed that markers based on imperfect repeats are likely to be less conserved than those harboring perfect repeats. This argument is based on the observation that the death of a microsatellite is accompanied by the accumulation of interruptions and/or by large deletions,[753] and compound microsatellites could therefore represent a late stage in microsatellite evolution. Such a relationship, however, was not supported in a study on *Clusia*, where transferability among species seemed to be uncorrelated to the perfectness of the repeat.[554]

One important determinant of the extent of marker transferability across species is the source and characteristics of the library. Thus, primer binding sites are expected to be more conserved when the microsatellite flanking sequences are maintained under selective constraints. This is most obviously the case for transcribed regions. Consequently, microsatellites within genes provide good chances to design primer pairs that are more broadly applicable, e.g., within plant families. Trinucleotide repeats are the predominant type of microsatellites in exons, and frequently have been exploited as markers.[12,761,1264,1311] More recently, microsatellite markers have been generated from EST) sequences available from public databases[250,317,474,554,609,1261,1392] (see Chapter 4.8.4.2). EST-derived microsatellite markers generally are less polymorphic than genomic microsatellites, but often show an increased level of conservation among taxa.[45,250,609,1392] For example, Holton et al.[609] designed primer pairs that flank microsatellites found in barley ESTs. Many of these markers were transferable to wheat. In addition to their frequent location in coding regions, markers based on trinucleotide repeats have the additional advantage of exhibiting less stuttering (see Chapter 4.8.3.2).

Taken together, cross-species transferability of microsatellite markers appears to be less successful in plants compared with animals, which makes exploitation of heterologous microsatellites a plausible option for congeneric species, but not for higher taxa. With the possible exception of microsatellites residing in coding regions, the evolutionary divergence time across which microsatellites are conserved in the nuclear plant genome seems to be restricted to a maximum of 15 to 30 million years.[694,1541] In *Plantago*, we observed an extremely low conservation of dinucleotide microsatellites, namely 1 to 2 million years (Wolff, unpublished data), whereas in *Clusia*, we calculated a maximum divergence time of 11.9 to 19.5 million years over which the loci were conserved (Hale and Wolff, unpublished data). Microsatellite markers derived from EST, cDNA-, low-copy, or undermethylated DNA libraries exhibit a better transferability than those derived from noncoding regions.[45,609,1180,1282] However, there is also a trade-off between variability and conservation: high conservation of primer binding sites is frequently accompanied by low levels of polymorphism for the enclosed microsatellite.[1282]

4.8.4.4 Marker Transferability between Species: Chloroplast Microsatellites

Studies employing chloroplast microsatellites (cpSSRs) are also limited by the necessity of sequence data for primer design. In earlier studies, primer sequences flanking cpSSRs were usually inferred from fully or partially sequenced chloroplast genomes of a handful of plant species, including rice,[646,648,1106,1107] maize,[1108] barley,[1110] wheat,[647] potato,[1109] tobacco,[180] soybean,[1093,1096] and *Arabidopsis thaliana*.[1105] In general, primer pairs used in these studies produced polymorphic PCR fragments in the species of origin and their close relatives, but transferability to more distant taxa was limited. For example, tobacco-derived cpSSR primer pairs worked well in potato[1109] and other *Solanum* species.[1442] Conversely, cpSSRs that are perfect and polymorphic in rice frequently proved to be degenerate, interrupted, and monomorphic in other Poaceae,[646] and primer pairs flanking conifer cpSSRs were nonfunctional or produced monomorphic bands in angiosperms.[218,1520]

Weising and Gardner[1520] developed a set of consensus chloroplast microsatellite primers (ccmp1 to ccmp10) with the aim of amplifying cpSSR regions in the chloroplast genome of dicotyledonous angiosperms in general. These primers produced single and distinct bands from species of many different taxa, and revealed interspecific polymorphism within the genera *Nicotiana*, *Lycopersicon*, and *Actinidia*. Sequence analysis of PCR products showed that cpSSR variation was the major cause of polymorphism in the target species. The universal applicability of ccmp primer pairs among dicots was verified by numerous subsequent studies. High levels of cross-priming were observed in 20 species belonging to 13 angiosperm families,[529] and intraspecific variation at ccmp loci was reported from *Medicago sativa* (Fabaceae),[915] *Silene paradoxa* (Caryophyllaceae),[916] *Olea europaea* (Oleaceae),[125,126] *Hedera helix* (Araliaceae),[527] *Corylus avellana* and *Carpinus betulus* (Corylaceae),[528,1032] *Camellia sinensis* (Theaceae),[702] *Vitis vinifera* (Vitaceae),[51,524] *Caryocar brasiliense* (Caryocaraceae),[263] various species of *Macaranga* (Euphorbiaceae),[1476] *Prunus domestica* (Rosaceae),[318] and *Anacamptis palustris* (Orchidaceae),[280] a monocot. In *Caryocar brasiliense,* intraspecific variation was detected with 10 of 10 ccmp primer pairs tested,[263] whereas only monomorphic ccmp products were found in *Calluna vulgaris.*[1153] In many cases, sequence analysis indicated the presence of microsatellites in the amplification products, with up to 22 A residues at the ccmp10 locus of *Carpinus betulus.*[528]

An expanded set of 23 cpSSR-flanking consensus primers for angiosperms was designed by Chung and Staub,[253] using a similar strategy as Weising and Gardner[1520] but accepting a lower threshold number of consecutive As or Ts in a repeat (i.e., $n \geq 7$). Other panels of universal chloroplast microsatellite primers have been developed at the family level, e.g., for Solanaceae,[180] Fagaceae,[320,1266] and Poaceae.[1115]

4.8.5 Microsatellite Cloning

For quite some time, the requirement for tedious molecular cloning procedures has been a major bottleneck that restricted a widespread application of microsatellite markers to exotic germplasm. Recently, several families of strategies have been developed to establish genomic libraries enriched for microsatellites (reviewed by Zane et al.[1606]). These techniques considerably enhanced the yield and efficiency of marker generation, and at the same time reduced the time and effort spent with the cloning procedure itself. Enrichment is especially worthwhile (1) for isolating markers based on the less frequent but relatively easy-to-type tri- and tetranucleotide repeats, (2) for organisms in which microsatellites are less abundant, such as birds (see survey in Zane et al.[1606]), and (3) when large numbers of markers are needed, e.g., as in mapping projects.[1136,1362,1378]

4.8.5.1 Conventional Libraries

Conventional strategies to develop microsatellites from small-insert partial genomic libraries encompass the following steps (see, e.g., Rassmann et al.,[1143] Röder et al.,[1179] Weber and May,[1513] Weising et al.[1525]): (1) creating a partial genomic library in a phage or plasmid vector, (2) screening several thousands of positive plaques or colonies by blot hybridization with repeat-specific probes, (3) identifying positive clones on duplicate filters, followed by isolating plasmid and sequencing DNA, (4) designing primers and, (5) analyzing and identifying PCR polymorphisms. According to a literature survey made by Zane et al.,[1606] the number of positive clones (i.e., clones containing one or more microsatellites) obtained by these procedures ranges from less than 0.04 to 12%, with particular low yields in birds. For plants, reported cloning efficiencies range from 0.059 to 5.8%.

For species with very large genomes that mainly consist of repetitive DNA, the output of markers can be increased significantly by cloning microsatellites from the hypomethylated fraction of the genome. This can either be done by (1) predigestion with a methylation-sensitive restriction enzyme such as *Pst*I, and constructing the library from the — presumably undermethylated — low-molecular weight fraction,[1180] or by (2) denaturing and self-annealing the restriction fragments to be cloned, and subsequent removal of the fast-annealing, repetitive DNA fraction by hydroxyapatite chromatography.[399,400] For example, Röder et al.[1180] reported a more than two-fold increase in the proportion of functional markers obtained from bread wheat when the source DNA was enriched for hypomethylated single-copy sequences by *Pst*I digestion and isolation of *Pst*I fragments in the size range from 2 to 5 kb. Nonetheless, still higher cloning efficiencies are usually achieved when libraries are enriched for the presence of microsatellites by one of the procedures outlined below.

4.8.5.2 Microsatellite Enrichment Based on Primer Extension

One strategy of microsatellite enrichment relies on the enzymatic extension of repeat-specific oligonucleotide primers, using a primary library of single-stranded genomic DNA as template.[1018,1024] In the method described by Ostrander et al.,[1018] a small-insert phagemid library is superinfected into an *E. coli* host strain deficient in UTPase (*dut*) and uracil-*N*-glycosylase (*ung*) genes. In such strains, dUTP efficiently competes with dTTP for the incorporation into DNA. Closed circular single-stranded phagemid DNA is isolated and second-strand synthesis primed with $(CA)_n$ and $(TG)_n$ oligonucleotides, using a thermostable DNA polymerase. The resulting double-stranded plasmids are repaired by ligase, and the products are used to transform a wildtype *E. coli* strain. Because of a strong selection against single-stranded, uracil-containing DNA molecules in the new host, the resulting library primarily consist of double-stranded products primed by the microsatellite. Colony hybridization demonstrated an approximately 50-fold enrichment for CA-repeats compared with a conventional library, with ~50% of the sequenced clones containing a repeat. Using the same strategy, Bell and Ecker[104] observed an about 10-fold enrichment of CA- and GA-repeats from *Arabidopsis thaliana*.

A similar principle was exploited in a protocol described by Paetkau.[1024] In the first step of this procedure, single-stranded template DNA is isolated from a small-insert M13 phage library. Microsatellite-containing clones are then labeled with biotin by annealing and extending a 5′-biotinylated, repeat-specific primer with the help of the Klenow fragment of DNA polymerase I. Biotinylated clones are captured by streptavidin bound to magnetic beads, eluted, made double-stranded by a second round of primer extension, and transformed into *E. coli* to produce a secondary library. Sequencing of randomly selected clones indicated 100% enrichment (24 positive clones of 24 tested), when the dinucleotide repeat $(CA)_{18}$ was used for extension, whereas only one of 12 positive clones were obtained with the tetranucleotide repeat motif $(ATGG)_n$.

The primer extension procedures described by Ostrander et al.[1018] and Paetkau[1024] mainly differ from other protocols described below in that microsatellites are enriched from established libraries rather than from PCR-amplified genomic DNA.

Problems associated with duplicates (i.e., the same clones isolated and sequenced several times, see Koblizkova et al.[730] and Chapter 4.8.5.8) are therefore less pronounced. On the negative side, these methods involve a relatively large number of steps, and their efficiency will also depend on the size of the primary library.

4.8.5.3 Microsatellite Enrichment Based on Selective Hybridization

The most popular family of enrichment strategies is based on hybridization selection of microsatellites prior to cloning (e.g., Armour et al.,[42] Connell et al.,[271] Cordeiro et al.,[273] Edwards et al.,[386] Fischer and Bachmann,[445] Hamilton et al.,[559] Inoue et al.,[640] Jakše and Javornik,[655] Kandpal et al.,[687] Karagyozov et al.,[691] Kijas et al.,[713] Koblizkova et al.,[730] Lyall et al.,[860] Prochazka,[1104] Reusch et al.,[1155] Waldbieser,[1489] and White and Powell[1536]). Enrichment cloning by hybridization selection generally involves the following steps:

1. Genomic DNA is fragmented, either by sonication,[271,687,691] digestion with a single restriction enzyme (most authors), or digestion with a cocktail of different restriction enzymes.[559,655]
2. The resulting DNA fragments are ligated to specific adapters. Depending on the fragmentation procedure chosen, the ends of the fragments may need to be polished before ligation, e.g., by removing single-stranded overhangs with mung bean nuclease.[655] Special adapters carrying recognition sites for restriction enzymes were constructed to facilitate later cloning steps,[445] or to allow the specific cleavage of unwanted adapter dimers generated during the ligation process.[559]
3. Ligation products are amplified by PCR with adapter-specific primers; denatured and hybridized with single-stranded, microsatellite-specific oligonucleotides bound to a nylon membrane[42,386,655,691]; or attached to magnetic beads via biotin and streptavidin (Kijas et al.[713] and most other authors).
4. After washing off unbound DNA, hybridizing fragments (which are expectedly enriched for microsatellites) are eluted from the membrane (or beads) and reamplified using adapter-complementary primers (optional). For protocols using nylon membranes, the washing temperature may be critical for the efficiency of enrichment.[273]
5. An optional step is to repeat the enrichment cycle (i.e., steps 3 and 4).
6. The enriched, PCR-amplified DNA fraction is then digested with a suitable restriction enzyme to produce vector-compatible ends, ligated into a vector, and transformed into *E. coli*. Restriction can be omitted when a PCR fragment cloning vector with a T overhang is used (e.g., TOPO-TA cloning kit, Invitrogen).
7. Transformants are plated, then insert-containing clones are selected by blue-white screening,[1217] amplified by colony PCR, and analyzed for the presence of a microsatellite by Southern hybridization. Positive clones are selected for sequencing. If the enrichment efficiency turns out to be very high, the screening step may be omitted and clones selected randomly for sequencing.

Optionally, a size-selection step may be included (either after fragmentation or after ligation) to collect fragments in an optimal size range for DNA cloning and sequencing (between ~400 and 800 bp). Size selection involves agarose gel electrophoresis, excision, and purification of the desired size fraction.

Hybridization selection was generally reported to result in highly enriched libraries, with ~5 to >80% of clones containing a microsatellite. Different types of repeat motifs served as targets. Whereas pure $(GT)_{15}$ oligonucleotides were used by Karagy-ozov et al.,[691] Edwards et al.[386] spotted a mixture of many different di-, tri- and tetranucleotide-repeat-complementary oligonucleotides onto the hybridization membrane. Armour et al.[42] and Jakše and Javornik[655] preferred long repeat arrays (>200 bp) as targets, to favor the isolation of relatively long microsatellites.

Kijas et al.[713] first used biotinylated, microsatellite-specific oligonucleotides attached to streptavidin-coated magnetic particles as targets for hybridization selection. In the first step of the procedure, size-selected (300 to 1500 bp), MboI-digested genomic citrus DNA was ligated into the BamHI site of a pGEM vector. From this primary ligation library, single-stranded DNA was produced by asymmetric PCR with the vector-derived forward- and reverse-sequencing primers in a 10:1 ratio. The PCR products were mixed with streptavidin-coated magnetic beads carrying the bound microsatellite motif, and hybridized at 30°C for 20 min. After several washing steps, bound DNA fragments were released by alkali treatment, neutralized, desalted, reamplified with vector primers, digested with MboI, religated into fresh vector, and transformed into E. coli. About 20% of the clones contained TAA-repeats.

Various modifications of the magnetic bead technique have since been published:

- Kandpal et al.[687] used sonicated human DNA fragments ligated to EcoRI linkers, or MboI-digested human DNA ligated to MboI adapters as starting material. Fragments were hybridized with biotinylated $(CA)_{15}$, $(CAG)_{15}$, or $(GATA)_{11}$ oligonucleotides in solution, captured with a Vectrex–avidin matrix, and eluted after several washing steps. The eluate was amplified by PCR, and subjected to a second round of hybridization, capture, elution, and PCR. The final PCR product was digested with MboI or EcoRI, and cloned. Enrichment efficiencies were very high (about 90% for CA repeats), but there was also some incidence of duplicates. Duplicates were also reported in other enrichment studies,[94,730,986,1136,1362] and most likely are a consequence of the PCR steps applied in these procedures (see Chapter 4.8.5.8).
- Waldbieser[1489] described a combination of affinity capture enrichment for $(ATA)_n$ and $(GATA)_n$ microsatellites from channel catfish and a PCR-based library screening procedure for positive clones. Screening involved the creation of ordered libraries, separate pooling of columns and rows, and the amplification of colony pools by a combination of vector and microsatellite primers.
- Prochazka[1104] described a microsatellite hybrid capture technique to enrich simultaneously for various microsatellite repeat motifs in human DNA in a single experiment. PCR-amplified products were directly cloned using a TA cloning kit.
- Fischer and Bachmann[445] combined some elements of the methods described by Edwards et al.,[386] Kandpal et al.,[687] Prochazka,[1104] and Waldbieser[1489] into a new technique that proved to be suitable for plant species with large genomes, such as Allium. This technique was also effectively applied to the banana fungal pathogen Mycosphaerella fijiensis[986] and the ornamental plant genus Pelargonium.[94]

Enrichment efficiencies provided by selective hybridization protocols are often high enough to allow the picking and direct sequencing of random recombinant clones (e.g., Tang et al.[1362]). Alternatively, inserts form randomly isolated clones are amplified by colony PCR, blotted, and hybridized with repeat-specific probes to confirm the presence of a specific microsatellite before sequencing.[94] A protocol for microsatellite enrichment by selective hybridization is given in Chapter 4.8.5.6.

4.8.5.4 Microsatellites from Cloned PCR Products

In another group of protocols, the generation of a genomic library is circumvented by cloning PCR products generated from RAPD primers,[255,401,856,1430] (see also Chapters 2.3.3 and 2.3.7), microsatellite primers[158,446,815] (see also Chapter 2.3.5.3), or AFLP primers[552,583,1606] (see also Chapter 2.3.8). Attempts in this direction were met with varying success. For example, more than 50% of AMP-PCR fragments cloned from *Pinus radiata* proved to contain internal microsatellites,[446] whereas only few internal repeats were found in AMP-PCR products from *Fraxinus excelsior.*[158] Ueno et al.[1430] used the RAMPO approach to isolate CA- and GA-containing RAPD fragments from *Camellia japonica*, with acceptable yields. From a total of 339 RAPD primer amplifications, 73 fragments yielded strong hybridization signals, 30 were cloned and sequenced, and 21 were found to contain a microsatellite repeat. PCR primers were designed for 12 clones, and four pairs yielded single-locus polymorphic products.

In a technique coined **PCR isolation of microsatellite arrays (PIMA)**, Lunt et al.[856] used a set of three arbitrary 10-mer primers to generate large numbers of RAPD fragments from the fish *Gadus morhua*. PCR fragments were purified and cloned in a T-vector (pCR2.1, Invitrogen), recombinant clones were transferred to replica plates, and arrayed clones were screened by colony PCR using microsatellite-specific and vector-specific primers. Only those reactions with extra-amplification products indicating the presence of a repeat were sequenced. Clearly, the efficiency of such strategies will depend on the extent of microsatellite accumulation in certain PCR products.

Lian et al.[815] developed a method for generating codominant microsatellite markers from ISSR products. In their procedure, which was coined **dual-suppression PCR**, genomic DNA is first amplified with a single, unanchored microsatellite primer such as $(AC)_{10}$. The result PCR fragments are cloned and sequenced, and a set of two nested primers (IP1 and IP2) defining one microsatellite-flanking region is designed from this sequence. The so-called walking method of Siebert et al.[1289] is then used to determine the sequence of the unknown other flank. This involves the construction of a set of genomic libraries of restricted, adapter-ligated genomic DNA from the same organism. The adapters consist of two strands of different lengths (48 vs. 8 bp), and extension of the short adapter strand is blocked by an amino residue at its 3'-end. A second set of nested primers (AP1 and AP2) is designed from the sequence of the long adapter strand. A single fragment containing a specific microsatellite is then amplified from one of the libraries by two rounds of PCR with pairs of external (AP1+IP1) and nested primers (AP2+IP2), respectively. The sequence of this final PCR product is used to construct the second microsatellite-flanking primer (see Siebert et al.[1289] for details of the procedure). The feasibility of the technique was demonstrated with *Salix reinii*,[815] *Pinus densiflora*,[815] *Robinia pseudoacacia*,[814] and the mangrove *Rhizophora stylosa*.[649] Dual-suppression PCR does not require enrichment and screening procedures, but is nonetheless quite complicated. A simplification introduced by Shibata et al.[1286] uses inverse PCR[1007,1420] to determine the unknown microsatellite-flanking sequence, thus again circumventing the need to create genomic libraries.

Hakki and Akkaya[552] described a procedure based on AFLP markers. AFLP products were generated from wheat (using *Eco*RI+3 and *Mse*I+3 selective primer pairs), and enriched for microsatellites using a modification of the biotin-streptavidin method of Fischer and Bachmann.[445] The method requires no cloning, but involves the recovery of individual AFLP bands from PAA gels and the use of radioactivity. In a method called **selectively amplified microsatellite (SAM) analysis**, Hayden and Sharp[583] designed single primers from sequenced microsatellite AFLP bands, and used these primers in combination with a 5'-anchored microsatellite primer[446] to generate locus-specific markers.

In addition to reviewing existing microsatellite cloning techniques, Zane et al.[1606] also contributed an enrichment protocol based on the AFLP procedure. In their so-called **fast isolation by AFLP of sequences containing repeats (FIASCO)** procedure, genomic DNA is digested with *Mse*I and ligated to *Mse*I adapters as described in Chapter 4.7.1. The resulting fragments are amplified using four adapter-specific primers, each carrying a different nucleotide (i.e., A, C, G, or T) at its 3'-end (*Mse*I+N). Several parallel PCRs are performed, each with a different number of cycles. PCR conditions producing a visible smear (but no distinct bands) on an agarose gel are considered optimal, and are repeated to collect sufficient amounts of PCR product. About 250 to 500 ng of amplified DNA is then hybridized in solution to a biotinylated, repeat-specific oligonucleotide [$(CA)_{17}$ in the original protocol], and products are captured by streptavidin-coated magnetic beads. DNA recovered from washing and denaturation steps is precipitated, redissolved, and reamplified with *Mse*I+N primers. Final PCR amplicons are cloned in a T-vector, and clones are sequenced directly. The percentage of positive clones ranged from 50 to 95%, depending on the organism. One of the advantages of the FIASCO protocol is that the primer mixture can be changed if the amplified DNA shows one or more strong bands superimposed on the smear, which is indicative of highly repetitive DNA and may lead to the over-representation of one or a few fragments in the final library.

4.8.5.5 *Miscellaneous Enrichment Procedures*

In the triplex affinity capture technique proposed by Ito et al.,[651] a small-insert restriction fragment library was established prior to enrichment. When total super-coiled plasmid preparations from such a library were mixed with a biotinylated oligonucleotide such as $(CT)_{13}$, target DNA molecules [i.e., double-stranded DNA regions consisting of $(GA)_n:(CT)_n$ repeats] interacted with the biotinylated probe via intermolecular triple helix formation. Positive clones were captured by streptavidin-coated magnetic beads, released by alkali, and plasmids were purified and used for secondary transformation. The use of this somewhat exotic technique is limited to sequence motifs that are capable of triple helix formation *in vitro* (such as GA- and GAA-repeats; see also Milbourne et al.[930]).

A technique that combines the principles of 5'-anchored microsatellite-primed PCR and serial analysis of gene expression (SAGE)[1460] was proposed by Hayden and Sharp[583] and Hayden et al.[585] In a relatively complicated procedure coined **sequence-tagged microsatellite profiling (STMP)**, a library of concatenated, short

(16 bp) sequence tags representing microsatellite-flanking regions is generated with the help of the type IIs restriction endonuclease *Bsg*I. A large number of tags is sequenced in the same reaction. Together with a flanking *Pst*I site, each tag contained sufficient sequence information for the design of a specific PCR primer. This primer was then used together with a 5′-anchored microsatellite primer to generate a codominant STMP marker. Additional PCR steps are required to convert the STMP markers into conventional microsatellite markers defined by two flanking primer pairs.[585]

4.8.5.6 Protocol: Microsatellite Enrichment Cloning Using Magnetic Beads

The following protocol is a modification of the biotin–streptavidin procedure of Fischer and Bachmann.[445] The main steps are summarized in Figure 4.3A and B. In short, about 6 μg of genomic DNA are digested with a frequent-cutting restriction enzyme producing blunt ends (such as *Rsa*I or *Alu*I) and ligated with a pair of adapters (Step 1). The products of the restriction-ligation step are purified with a Geneclean kit (Qbiogene), and hybridized with a mixture of biotinylated, di- and trinucleotide-specific oligonucleotides. Hybridized fragments are captured with streptavidin-coated magnetic beads (Dynal, Step 2), concentrated with the help of Microcon® YM-30 spin filters (Millipore), and amplified with adapter-specific primers (Step 3). PCR products are size-selected, ligated into a T-vector (Step 4), and transformed by a heat-shock treatment into competent *E. coli* cells (Step 5). Positive clones are identified by blue-white screening (for details on bacterial transformation and screening of recombinants, see Sambrook and Russell[1217]). Random clones are amplified by PCR with vector-specific primers, and the presence of a microsatellite is checked by Southern blot analysis (Step 6). Positive clones are sequenced.

Step 1: Preparation of Restricted and Ligated DNA

See Chapter 4.3.1 for precautions associated with the use of restriction enzymes.

Solutions

Genomic template DNA:	6 μg (concentration should be ~0.5 μg/μl) (see Comment 1)
10× RL buffer:	100 mM Tris-acetate (pH 7.5), 100 mM magnesium acetate, 500 mM potassium acetate, 50 mM dithiothreitol (sterilize by filtration) (see Comment 2)
Adapter:	See below for preparation of the adapter

*Rsa*I restriction enzyme (6 U/μl)

10 mM ATP

T4 DNA ligase (1 U/μl)

Double-distilled sterile H$_2$O

Geneclean II DNA Purification Kit

TE buffer:	10 mM Tris-HCl, 1 mM EDTA, pH 8.0

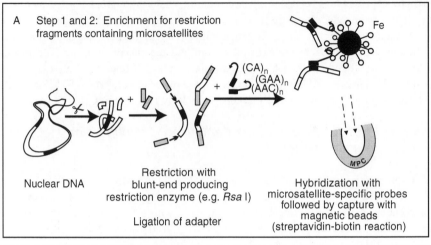

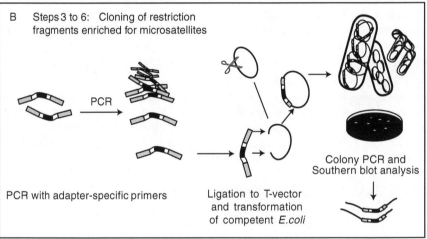

Figure 4.3   Summary of Steps 1 to 6 of the microsatellite enrichment cloning procedure[445] described in Chapter 4.8.5.6. See text for details.

Method

1. Prepare a double-stranded adapter by mixing equimolar amounts of the following complementary oligonucleotides in 100 μl sterile H_2O (resulting in an adapter concentration of 8.3 μM):

    0.83 nmol oligo 1 (21-mer): 5′-CTCTTGCTTACGCGTGGACTA-3′
    0.83 nmol oligo 2 (25-mer): 5′-pTAGTCCACGCGTAAGCAAGAGCACA-3′
    The oligo2 needs to be phosphorylated at its 5′-end.

2. Heat at 95°C for 3 min, let the solution slowly cool to room temperature. The annealed adapter has the following conformation (see Comment 3):

    5′-CTCTTGCTT**ACGCGT**GGACTA-3′
    3′-ACACGAGAACGAA**TGCGCA**CCTGAT-p-5′

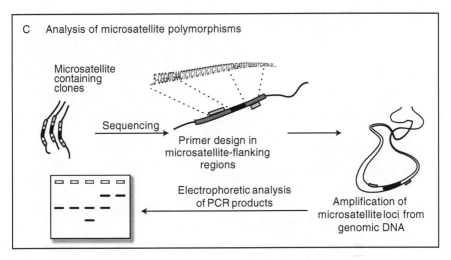

C   Analysis of microsatellite polymorphisms

Microsatellite containing clones

...5'-CGGATGAACTCTCTCTCTCTCTCTCTAGATGTGGGTCATA-3'...

Sequencing

Primer design in microsatellite-flanking regions

Electrophoretic analysis of PCR products

Amplification of microsatellite loci from genomic DNA

Figure 4.3   (continued).

3. For restriction and ligation, mix the following components in a microfuge tube:

Ingredient	Stock Concentration	Final Concentration or Amount	Volume Needed for One Reaction (µl)
DNA	0.5 µg/µl	6 µg total	12
RL buffer	10×	1×	3.5
*Rsa*I	6 U/µl	18 units	3.0
Adapter	8.3 µM	~0.71 µM	3.0
ATP	10 mM	1 mM	3.5
T4 DNA ligase	1 U/µl	6 units	6.0
Sterile H$_2$O			4.0
Total volume			35 µl

4. Incubate 4 h to overnight at 37°C, then heat the reaction for 10 min at 95°C to inactivate the enzymes. Immediately proceed with the next step.
5. Purify restriction fragments with a Geneclean II DNA Purification Kit or equivalent, following the instructions of the kit manufacturer.
6. Add 10 µl of TE buffer to the final glassmilk pellet and gently resuspend with a pipet. Spin for a few seconds, and transfer the eluate to another microfuge tube. Repeat the elution process with the same volume of TE buffer to obtain ~20 µl of purified DNA. This solution will be hybridized to microsatellite-specific oligonucleotides in the next step.

Comments

1. DNA preparations need to be of reasonable quality to ensure complete digestion by the restriction enzyme. See Chapters 4.2.2 and 4.4.2.5 for DNA purification methods, and methods to deal with enzyme-inhibiting substances, respectively.
2. Select a buffer that allows good activities for both T4 DNA ligase and the corresponding restriction enzyme. In our laboratory, 10× RL buffer can be replaced with 10× T4 ligase buffer (usually supplied with the enzyme) with similar results.

3. The adapter contains a recognition site for the restriction enzyme *Mlu*I (written in bold type). In the original article by Fischer and Bachmann,[445] the *Mlu*I site was cut prior to cloning, and the resulting restriction fragments were ligated in the presence of *Mlu*I into the *Bss*HII site of a dephosphorylated plasmid vector. The *Mlu*I site is not required for cloning in a T-vector, as described in the present protocol.

Step 2: Affinity Capture of Microsatellite-Containing DNA Fragments

Solutions and Materials

Purified restricted and ligated DNA fragments
3'-biotinylated oligonucleotides: 10 μM each of $(CA)_{10}$, $(CAA)_8$, and $(GAA)_8$
20× SSC: 3.0 M NaCl, 0.3 M sodium citrate, pH 7.0
10% Sodium dodecyl sulfate (SDS)
5 M NaCl
1 N NaOH
1 N HCl
6× SSC and 1× SSC: Prepare by diluting 20× SSC stock
2× SSC, 0.1% SDS: Prepare by diluting 20× SSC and 10% SDS stocks
TE buffer: 10 mM Tris-HCl, 1 mM EDTA, pH 8.0
Elution buffer: 100 mM NaOH, 100 mM NaCl (freshly prepared
                                  from 5 M NaCl and 1 N NaOH stocks)
Neutralization solution: 0.2 N HCl (freshly prepared from 1 N stock solution)
100 mM Tris-HCl pH 7.5
M-280 Streptavidin Dynabeads (Dynal): 10 mg/ml (1 mg Dynabeads are supposed
    to bind 200 pmol single-stranded DNA)
Magnetic particle collector (MPC) (Nunc)
Heating block set to 95°C
Water bath set to 74°C (see Comment 1)
Microcon YM-30 spin filters (Millipore)

Method

1. Dispense 35 μl of Dynabeads per capture reaction in a 2.0-ml microfuge tube, wash twice with 700 μl of TE buffer, and equilibrate twice in 700 μl of 6× SSC. Between each wash, collect beads with the MPC and discard the supernatant.
2. After the last wash, resuspend Dynabeads in 35 μl of 6× SSC (see Comment 2).
3. Prepare a pool of biotinylated microsatellite-specific oligonucleotides (oligo mix) by mixing 1 μl each of $(CA)_{10}$, $(CAA)_8$, and $(GAA)_8$ stock solution (1 pmol of each oligo) with 6.5 μl of distilled water.
4. Mix in a microfuge tube:
    9.5 μl oligo mix
    19.5 μl 20× SSC
    20.0 μl restricted and ligated genomic DNA
    16.0 μl distilled water
    Mix well, denature in a heating block at 95°C for 5 min, and transfer to a water bath set at 74°C (annealing temperature, see Comment 1). Let the hybridization mixture cool slowly to approach the annealing temperature (74°C).

5. Add 35 μl of equilibrated and prewarmed Dynabeads suspension (from Step 2) to each hybridization mixture, mix gently (total volume 100 μl), and incubate for 20 min
6. Capture beads with the MPC and discard the supernatant.
7. Wash the beads twice with 200 μl of 2× SSC, 0.1% SDS (5 min each; room temperature), followed by two washes with 200 μl of 1× SSC (5 min each; room temperature), and one wash with 200 μl of 1× SSC (3 min at 74°C; so-called hot wash). After each wash, capture the beads with the MPC and discard the supernatant.
8. Immediately after the hot wash, add 20 μl of elution buffer (100 mM NaOH, 100 mM NaCl). Mix well, spin briefly in a microcentrifuge to pellet the beads, and transfer the supernatant to a fresh microfuge tube. **Caution:** Do not discard the supernatant at this stage, it contains your DNA!
9. Add 10 μl of 0.2 N HCl to the eluted DNA to neutralize the samples.
10. Add 2.2 μl of 100 mM Tris-HCl (pH 7.5) to adjust and stabilize the pH.
11. Mix gently, and transfer the solution to the wider part of a Microcon YM-30 spin filter attached to its corresponding tube. Add 450 μl TE buffer to the DNA on the filter, and spin for 6 min at $12,000 \times g$ at 4°C. Discard the flow-through.
12. Add another 450 μl TE buffer to the DNA on the filter, and spin for 6 min at $12,000 \times g$ at 4°C. Discard the flow-through.
13. Invert the filter and place it onto a new microfuge tube with the wider part facing down to the interior of the tube. Spin at $1000 \times g$ (~3700 rpm) for 2 min to collect the concentrated DNA solution (~20 μl) in the microfuge tube.

Comments

1. Annealing temperatures depend on the length and sequences of the oligonucleotides used for hybridization. Pooling is only advised for oligonucleotides with roughly similar melting temperatures (T_m). Annealing temperatures (T_a) should be set about 5°C below T_m. For the oligonucleotides used in the present protocol, T_m values as calculated with the Oligo program[1203] ranged from 78.0°C [for $(GAA)_8$] to 82.2°C [for $(CAA)_8$], and T_a was set to 74°C.
2. Dynabeads should not run dry for more than a few seconds. Immediately add the next washing solution after discarding the supernatant.

Step 3: PCR Amplification of Captured DNA Fragments Using an Adapter-Specific Primer

Method

1. Set up a PCR with the following ingredients:

Ingredient	Stock	Final Concentration	Volume Needed (μl)
Captured DNA fragments			2.0
10× PCR buffer	10×	1×	2.5
MgCl$_2$	25 mM	1.5 mM	1.5
Adapter-specific primer (21-mer), see Step 1	5 μM	1 μM	5.0
dNTPs	2.5 mM	0.2 mM	2.0
Taq DNA polymerase	5 U/μl	1 unit	0.2
Sterile H$_2$O			11.8
Total volume			25

2. Perform PCR using the following program:

$$94°C \text{ for } 5 \text{ min}$$

28 cycles each with 94°C for 48 sec

56°C for 1 min

72°C for 2 min

Final extension at   72°C for 7 min

3. Run 4 µl of the resulting preamplification PCR products in 1.5% agarose (see Chapter 4.3.4). A homogenous, light smear should appear. The presence of discrete bands at this stage indicates disproportionate amplification.

4. Purify the remainder of the PCR products using a Geneclean II DNA Purification Kit as described above.

Comment

Optionally, the purified DNA can now be subjected to a second round of hybridization, capture, and PCR amplification. In this case, use 10 µl of the purified PCR fragments as a template for the oligonucleotide hybridization (Step 2). A second round of enrichment increases the yield of positive clones (which may be required for rare microsatellite motifs), but also bears an increased risk of cloning the same microsatellite more than once (See Chapter 4.8.5.8 for problems associated with duplicates, and Table 4.3 for an example of the efficiencies of one vs. two rounds of microsatellite enrichment in *Pelargonium*).

Step 4: Ligation of Purified DNA Fragments to a pGEM-T Vector (Promega)

Solutions and Materials

Purified amplified DNA fragments

Microcon YM-100 spin filters (Millipore)

TE buffer: 10 mM Tris-HCl, 1 mM EDTA (pH 8.0)

The following components are contained in the pGEM-T vector kit of Promega:

2× T4 DNA ligase buffer

T4 DNA ligase

pGEM-T vector

Double-distilled sterile H_2O

Method

1. Apply the solution to the wider part of a Microcon YM-100 spin filter attached to its corresponding tube. Add 450 µl of TE buffer to the DNA on the filter, and spin for 15 min at $500 \times g$ at 4°C. Discard the flow-through (see Comment).

2. Add another 450 µl of TE buffer to the DNA on the filter, and spin for 15 min at $500 \times g$ at 4°C. Discard the flow-through.

3. Invert the filter and place it onto a new microfuge tube with the wider part facing down to the interior of the tube. Spin at $1000 \times g$ (~3700 rpm) for 2 min to collect the concentrated DNA solution (~20 µl) in the microfuge tube.

Table 4.3 Efficiency of Microsatellite Enrichment Cloning and Primer Design in a Study on *Pelargonium* (see also Becher et al.[94])

Motif for Selection	Enrichment Efficiency: Positive Clones of Total Screened		Efficiency of Primer Construction						
	After One Round of Selection	After Two Rounds of Selection	Total Positives	Total Sequenced	Presence of SSR Verified	Duplicates	Primers Designed	Primers Functional on Agarose	Primers Functional on PAA
CA	2 (43) = 5%	11 (46) = 24%	13	13	10	1	9	8	8
GA	13 (90) = 14%	18 (46) = 39%	31	28	25	3	16	7	5
CAA/GAA	5 (90) = 5%	37 (116) = 32%	42	30	29	1	19	14	9
Total	20 (223) = 9%	66 (208) = 32%	86	71	64	5	44	29	22

Note: A more than threefold increase in the relative numbers of positive clones was observed after two vs. one round of hybridization selection. Primer functionality was first tested by PCR with three to six genomic templates and a positive control on agarose gels. Primer pairs yielding single (or two) bands within the expected size range were then used to amplify the same template set in a radioactive PCR, and amplification products were separated on denaturing PAA gels. Primers were considered functional if one to four (some plants were tetraploid) polymorphic bands in the expected size range were detected.

4. Mix the following components in a microfuge tube, incubate for 1 h at room temperature, and then overnight at 4°C:

Ingredient	Volume Needed for One Reaction (µl)
PCR products	1
2× T4 DNA ligase buffer	5
pGEM-T vector	1
T4 DNA ligase	1
Sterile H₂O	2
Total volume	10 µl

Comment

This is an optional size-selection step. Microcon YM-100 spin filters have a cut-off point of 100 kDa. Smaller molecules such as oligonucleotides and very short PCR fragments (< 300 bp) pass through the filter and are discarded.

Step 5: Transformation of Competent E. coli Cells

Solutions and Materials

100 mg/ml ampicillin (store at –20°C)

100 mM isopropylthio-β-D-galactoside (IPTG) (store at 4°C)

50 mg/ml X-gal (5-bromo-4-chloro-3-indolyl-β-D-galactopyranoside) in dimethylformamide

LB medium:	Mix 10 g tryptone, 5 g yeast extract, 5 g NaCl, and 20 g agar per liter H₂O, dissolve by autoclaving
LB agar plates:	Allow the LB medium to cool to 50°C after autoclaving; add 1 ml ampicillin, 0.5 ml IPTG and 1.6 ml X-gal stock solutions per liter LB medium, resulting in final concentrations of 100 µg/ml ampicillin, 0.5 mM IPTG and 80 µg/ml X-gal; pour the medium onto the plates under a sterile flow bench and allow it to solidify
Competent E. coli cells:	Heat-shocked DH5-α competent cells (Invitrogen) (store at –70°C)

DNA fragments ligated in pGEM-T vector (see above)

TE buffer:	10 mM Tris-HCl, 1 mM EDTA (pH 8.0)
SOC medium:	Prepare according to Sambrook and Russell[1217] (page A2.3)

Glycerol (autoclaved)

Water bath set to 42°C

Rotary shaker for microfuge tubes

Incubator set to 37°C

Tray with wet ice

Method

1. Remove the competent cells from the freezer and thaw on ice.
2. Gently mix the cells and dispense 50-µl aliquots into prechilled microfuge tubes.
3. Dilute 1 µl of ligated DNA fragments 1:5 in TE buffer. Add 1 µl of the diluted fragments to a 50-µl aliquot of competent cells. Mix 1 µl of undiluted DNA fragments with a second aliquot. Set up a positive control by mixing control DNA (e.g., pUC18 vector) with a third aliquot of cells. Mix the DNA fragments with the cells by gently pipetting up and down.
4. Incubate cells on ice for 30 min.
5. Heat-shock cells for 45 sec in the 42°C water bath. Do not shake.
6. Transfer cells to ice and incubate for 2 min.
7. Add 0.9 ml of SOC medium prewarmed to room temperature.
8. Shake at 225 rpm for 1 h at 37°C.
9. Plate 100 to 400 µl of the cell suspension on LB agar plates supplemented with 100 µg/ml ampicillin, 0.5 mM IPTG, and 80 µg/ml X-gal. To the remaining cell suspension, 1/10 vol of autoclaved glycerol is added (final concentration 10%). After mixing, the cell suspension is transferred and kept at −70°C for future plating.
10. Incubate plates overnight at 37°C, and then transfer to 4°C to visualize blue vs. white colonies. Only white colonies are expected to carry a DNA insert in the pGEM-T vector (for details on X-gal/IPTG screening of recombinant clones, see Sambrook and Russell[1217]).

Step 6: Identification of Clones Containing a Microsatellite (see Comment)

Solutions and Materials

Plated colonies on LB agar

Fresh LB agar plates containing 100 µg/ml ampicillin, 0.5 mM IPTG and 80 µg/ml X-gal

Adapter primer (21-mer oligo; see above) 5 µM

10× PCR buffer

25 mM MgCl$_2$

2.5 mM dNTPs: 2.5 mM each of dATP, dCTP, dGTP, and dTTP

Taq DNA polymerase (5 U/µl)

Double-distilled, sterile H$_2$O

Method

1. Dispense 50-µl aliquots of sterile water into PCR tubes. Prepare as many aliquots as bacterial colonies are to be screened.
2. Touch a single white colony with a sterile pipet tip and wash it into the 50-µl water aliquot. Transfer another aliquot of the colony to a new LB plate to create an ordered library for later reference.
3. After having picked all colonies of interest, close the PCR tubes and lyse the cells for 10 min at 98°C in a thermocycler. Use 5 µl of the lysate for the PCR below.

4. Grow the ordered library on the fresh LB plates overnight at 37°C and store at 4°C.
5. Set up a PCR with the following ingredients:

Ingredient	Stock	Final Concentration	Volume Needed (µl)
Template DNA (lysed bacterial colony)			5.0
10× PCR buffer	10×	1×	5.0
MgCl₂	25 mM	1.5 mM	3.0
Adapter-specific primer (21-mer), see Step 1	5 µM	0.5 µM	5.0
dNTPs	2.5 mM	0.2 mM	4.0
Taq DNA polymerase	5 U/µl	1 unit	0.2
Sterile H₂O			27.8
Total volume			50 µl

6. Perform PCR using the following program:

94°C for 2 min
30 cycles each with 94°C for 48 sec
60°C for 1 min
72°C for 1 min
Final extension   72°C for 5 min

7. Resolve PCR products on 1.5% agarose gels and stain with ethidium bromide as described in Chapters 4.3.4 and 4.3.6.1. An insert-specific PCR product should be visible in each lane. An example is shown in Figure 4.4 (upper panel).
8. Blot the gel to a nylon membrane as described in Chapter 4.3.8.
9. Generate 5′-end-labeled oligonucleotide probes specific for the enriched microsatellite motifs as described in Chapter 4.3.9.1.
10. Hybridize the membrane carrying the colony PCR products with the labeled oligonucleotide probes as described in Chapter 4.3.10.1.

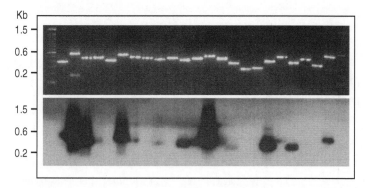

Figure 4.4 Example of a test for positive clones in a *Pelargonium* genomic library enriched for microsatellites.[445] Individual bacterial clones were picked, aliquots lysed by heating, and plasmid inserts amplified by colony PCR with adapter-specific primers. PCR products were separated on a 1% agarose gel, blotted onto a nylon membrane, hybridized with a mixture of ^{32}P-labeled (GAA)₅ and (CAA)₅ probes, and autoradiographed. Clones exhibiting a strong hybridization signal were sequenced; more than 90% contained the expected microsatellite (see also Table 4.3). Positions of size markers are indicated (kb, kilobase pairs).

11. Detect positive clones carrying the respective microsatellite motif by autoradio-
    graphy as described in Chapter 4.3.11.1. (A typical result is shown in Figure 4.4,
    lower panel.)
12. Repeat steps 1, 2, 3, 5, and 6 with the positive clones, using the colonies of the
    ordered library produced in step 4 as a template.
13. Purify PCR products using a commercial kit (e.g., Qiaquick® columns, Qiagen).
14. Sequence PCR products using a commercial kit, or have them sequenced com-
    mercially.

Comment

In a study on *Pelargonium* using the protocol given here, 32% of clones contained
microsatellites after two cycles of enrichment (see Table 4.3). Even higher yields
were reported from other species and/or with other microsatellite motifs. Very high
yields allow the direct sequencing of randomly selected clones. We nevertheless
advise screening of clones by colony PCR and agarose gel electrophoresis prior to
sequencing to select clones with an optimal insert size and avoid chimeric clones
that carry more than one insert.

4.8.5.7 Primer Design

If the sequenced clone contains a microsatellite, the next important step is the design
of flanking PCR primers. Various computer programs are available that help design
primer pairs for microsatellite amplification (e.g., Oligo,[1203] Primer3, see Appendix
3[1196]). In general, it is advised that users follow the rules outlined in Chapter 4.3.2.1.
For microsatellite analysis, primers are preferentially designed to amplify fragments
between 100 and 250 bp, to ensure unambiguous allele designation on sequencing gels.

Unfortunately, not all positive clones are actually useful for primer design. Thus,
some clones may not contain a microsatellite at all, whereas others contain a microsatellite so close to one of the insert–plasmid boundaries that no flanking primer
can be designed (see Chapter 4.8.5.7 below). This is often the case for EST-derived
markers.[39] To eliminate such useless positives and minimize unnecessary sequencing,
a PCR-based prescreening procedure for the presence and position of a microsatellite
was developed by Taramino and Tingey.[1366] In this procedure, five PCR reactions
are set up, using the cloned DNA as a template:

1. Vector-derived forward and reverse primer
2. Vector forward and microsatellite primer A (carrying a 5′-degenerate anchor)
3. Vector forward and microsatellite primer B (having the complementary sequence
   of primer A, also carrying a 5′-degenerate anchor)
4. Vector reverse and microsatellite primer A
5. Vector reverse and microsatellite primer B

The first reaction informs about the insert size, the second to fifth reactions
inform about the presence, location, and orientation of the microsatellite. This allows
one to determine which end should be sequenced with priority. The efficiency of
the screen was illustrated by the study of Huang et al.,[627] who were able to discard

25% of the positive clones of a microsatellite-enriched *Actinidia chinensis* library prior to the sequencing step.

The number of useful microsatellite loci is further restricted by nonfunctional primer pairs. In general, primer performance is tested in a small number of individuals, often in conjunction with electrophoresis on high-percentage (1.6 to 2%) agarose gels. Template DNA from the original clone should be used as a positive control. If a PCR product is obtained from the clone, but not from genomic DNA, the clone may represent a chimera (see below). Primer pairs that produce distinct single-locus patterns on agarose are eventually tested on sequencing gels. Application of a rating system may be helpful to select those markers that are most useful for genetic analyses. Smulders et al.[1302] rated the quality of microsatellite markers developed for tomato on a scale from 1 to 5. Quality scores 1 and 2 were assigned to well-scorable fragments, quality 3 to fragments forming stutter ladders with equal intensities, quality 4 to multilocus patterns, and quality 5 to fuzzy bands or no amplification at all. The same rating system was also applied, e.g., to barley[1136] and potato.[930] Ramsay et al.[1136] recommended the use of only the top two classes for genotyping, whereas minor quality levels may be sufficient for mapping purposes.

4.8.5.8 Factors Affecting the Efficiency of Microsatellite Cloning

Developing microsatellites basically is not difficult, although the process often consists of a large number of steps, each of which can be a hurdle. Generating and screening the library can take any time between 1 week and 2 months. Sequencing may take several weeks, depending on the available equipment and funds. A few weeks also will be required for designing and ordering primers, and testing primer performance. Depending on the rate of success associated with the above steps, it may take any time between 2 months and 2 years to obtain a set of polymorphic microsatellite markers for a new species by molecular cloning.

Unfortunately, many potential markers are lost along the long and winding road leading from a sequenced positive clone to a well-performing microsatellite marker. Squirrell et al.[1322] evaluated this process based on a large number (71) of studies from various sources, but predominantly *Molecular Ecology* and *Molecular Ecology Notes*. The first problem encountered is that not all clones contain a microsatellite, despite high enrichment efficiencies and/or positive hybridization signals of colony PCR products. Second, sequencing sometimes reveals the presence of duplicate clones. This problem usually pertains to less than 10% of the clones, but can be considerable under certain circumstances. Third, chimeric clones may have been generated by, for example, concatenation of two inserts. Clone redundancy and chimerism are treated in more detail below. On average, the three problems mentioned so far together cause a loss of 36% of clones, notwithstanding the origin of the library (AFLPs, RAPDs, restriction fragments) and the enrichment technique used.[1322]

In the next stage, the experimenter attempts to develop primers that flank unique microsatellite-containing sequences. Because the cloning site is randomly positioned with regard to the microsatellite, a subset of the sequenced clones will be unsuitable for primer design because the microsatellite is too close to the cloning site on one or both sides. In general, this problem is less serious with libraries carrying, on average, longer inserts. It is also possible that there is sufficient flanking DNA available in the clone, but the underlying sequence is unsuitable for primer design. This can be due to an extremely high A/T content, repetitiveness, and/or cryptic simplicity of these sequences. The number of candidates is further reduced by the unsuitability of the microsatellite itself. For example, mononucleotide repeats often show a lot of stutter fragments and may become uninterpretable in the assay (see Chapter 4.8.3.2). Very long microsatellites may also turn out to be problematic because it is difficult for the *Taq* DNA polymerase to run through long stretches of repeats.[38] Squirrell et al[1322] conclude that, on average, only 35 candidates remain from 100 sequenced clones at the primer design stage.

The final stage of microsatellite marker development involves testing the performance of the ordered primer pairs. The main problems at this stage are nonamplification of some or all genomic template DNAs, the amplification of multiple fragments, and the absence of polymorphism. Nonamplification could result from chimerism (see below) and/or from mutations within one of the primer binding sites in certain alleles (so-called null alleles; see Pemberton et al.[1061] and Chapters 2.3.4.1 and 4.8.4.3). The problem of null alleles potentially can be solved by redesigning primers in other stretches of the flanking DNA, avoiding the mutated primer binding site.[208,413,645,676,1026] Multiple fragments may occur when the locus is duplicated or part of a larger repeat (this problem is often observed in species with large genomes such as conifers), or when primers are not sufficiently specific.

Problems associated with a lack of polymorphism can potentially be minimized by selecting perfect microsatellites above a certain threshold length (e.g., >10 uninterrupted repeat units) for primer design. Squirrell et al.[1322] indicate that, on average, another 50% of primer sets are lost at this stage, leaving 17 successful microsatellite primer sets. Even primer pairs that seem fine and polymorphic with a small number of test samples may produce problems when applied to other samples or subjected to different PCR conditions (e.g., multiplexing). All in all, it is fair to say that one should sequence about 100 clones to end up with 10 sets of well-performing primer sets. See Lowe et al.,[851] Ramsay et al.,[1136] Röder et al.,[1181] and Table 4.3 for examples of attrition rates encountered during the development of microsatellite markers.

Redundancy of clones is sometimes a problem, especially when large numbers of microsatellite markers are isolated. For example, Chen et al.[239] found that 25 of 122 microsatellites (i.e., ~20%) cloned from a size-selected, unenriched rice library were isolated more than once. Given the total estimated number of 5000 to 10,000 microsatellite loci in the rice genome,[906] such a high level of redundancy would not have been expected. One possible explanation is a bias introduced by the restriction enzyme, which may preferentially cut certain genomic regions into fragments of the selected size range (i.e., 300 to 800 bp).[239]

Two kinds of strategies may be followed to increase the percentage of nonredundant clones. One involves the generation of several libraries in parallel, using different DNA preparations as a source, such as (1) DNA digested by a single, frequent-cutting enzyme producing sticky ends; (2) DNA digested by a combined set of different enzymes producing blunt ends (e.g., Hamilton et al.,[559] Jakše and Javornik[655]); (3) sheared DNA (e.g., Connell et al.,[271] Karagyozov et al.[691]); and (4) cDNA libraries, preferably from different tissues (e.g., Saha et al.,[1211] Scott et al.,[1261] Woodhead et al.[1568]). Alternatively, the number of (mainly useless) micro-satellites being part of larger repeats may be reduced by selectively cloning under-methylated low-copy DNA[399,400,1181,1616] (see Chapter 4.8.5.1).

The problem of duplicate clones can become especially pronounced with mic-rosatellite-enriched libraries (see, e.g., Kandpal et al.,[687] Kishore et al.[721] Lowe et al.,[851] Tang et al.[1362]). In one extreme example, Kishore et al.[721] found that 541 of 1237 sequenced clones from a microsatellite-enriched library of meadow foam (*Limnanthus alba*) were redundant (43.7%). Most enrichment methods include one or more PCR steps, often simply to get more DNA for the next step of the procedure. Each PCR will generate artificial duplicates; i.e., clones containing exactly the same fragment. To minimize this problem, the number of cycles in the PCR steps should be kept at a minimum.[271,655]

Still another problem is the generation of **chimeric clones**; i.e., clones in which different genomic regions have been joined together to form a single insert. Chimeras can sometimes be identified by computerized sequence analysis because the recog-nition site of the enzyme used for cloning is re-formed at the joining site. The problem appears to be more pronounced when the library is established from sheared DNA or from DNA cut by several different enzymes.[239,1036] Koblizkova et al.[730] described a mechanism of chimera formation, which appears to be specific for the PCR step during microsatellite enrichment procedures. According to this model, few residual microsatellite-specific oligonucleotides remaining in the sample from the hybridization selection step may pair to a microsatellite-containing genomic frag-ment, and act as a primer in the subsequent PCR. One portion of the target is then amplified together with the adapter primer. In the next round of amplification, the resulting PCR product may again pair to a microsatellite present in another fragment, and is then extended toward the other side. The resulting hybrid molecule then contains parts of both original restriction fragments involved, with a common mic-rosatellite motif in the middle (for illustration, see Figure 1 in Koblizkova et al.[730]). Such chimeras will normally remain undetected. Chimerism should always be suspected if primer pairs are only functional with plasmid DNA, but not with genomic DNA.

4.8.5.9 Commercial Development of Microsatellite Libraries

As a completely different alternative, it is worth considering the purchase of a library of microsatellite-flanking primers from a commercial company. An increasing number of suppliers (including academic institutions) offer the establishment of enriched microsatellite libraries from any species of interest for about US $10,000 (see list in Appendix 2B). This is a good (but expensive) option if the skills, knowledge, time, and equipment are not available, and/or if microsatellites are used only occasionally.

4.9 CAPS ANALYSIS OF CHLOROPLAST AND MITOCHONDRIAL DNA

CAPS markers are generated by the digestion of PCR products with restriction enzymes[368,738,1415,1516,1548] (see also Chapter 2.3.2). The experimental approach, also known as PCR-RFLP, is fairly simple. In the first step of a standard CAPS experiment, defined PCR products are amplified from nuclear or organellar DNA, using primers complementary to known sequences. In the second step, PCR products are digested with one or more restriction enzymes, and restriction site polymorphisms are displayed by agarose gel electrophoresis and ethidium bromide staining. Nondenaturing PAA gels and SSCP gels have also been used[660] (see Chapter 2.3.9). To identify suitable combinations of amplicons and restriction enzymes, a wide range of PCR primer pairs (see Chapter 4.9.2) and restriction enzymes need to be screened during the initial phase of a CAPS project, using a small set of templates. Combinations that reveal informative polymorphisms are then applied to the full set of organisms under investigation.

4.9.1   Standard CAPS Protocol

We have used the following protocol to generate CAPS markers from noncoding cpDNA regions of various plant species, using primer pairs described by Demesure et al.,[326] Dumolin-Lapègue et al.,[372] and Grivet et al.[529] The protocol can easily be adjusted to other organisms and genomes. For general precautions associated with PCR experiments, see Chapter 4.3.2.

Solutions

Taq DNA polymerase:	5 U/µl
10× buffer:	200 mM Tris-HCl, pH 8.3, 500 mM KCl, 20 mM MgCl$_2$, 0.01% gelatin. Ten-fold concentrated PCR buffer is usually supplied by the manufacturer of the enzyme. It may or may not contain magnesium chloride and additional ingredients, depending on the brand of the enzyme
dNTP stock:	2 mM each of dATP, dCTP, dGTP, and dTTP. Ready-made solutions containing all four dNTPs are commercially available from several suppliers
PCR primer:	5 µM of each forward and reverse primer (see Chapter 4.9.2)
Template DNA:	5 to 20 ng/µl

10× concentrated restriction buffer (usually supplied by the manufacturer)

Restriction enzyme (5 to 20 U/µl; see Comment 1)

Method

1. Use thin-walled PCR tubes to set up a PCR with 50-µl volumes containing 20 mM Tris-HCl pH 8.3, 50 mM KCl, 2 mM MgCl$_2$, 0.001% gelatin, 200 µM of dNTPs, 0.4 µM of each primer, 2 units *Taq* DNA polymerase, and 15 to 50 ng of template DNA. Pipetting errors are minimized by preparing master mixes for all samples (see Comment 2).
2. Mix the contents, and centrifuge the vials briefly (see Comment 3).
3. Insert the tubes into a thermocycler and start the desired program. We use the following program:

3 min at 94°C (initial denaturing step)
30 cycles consisting of:
    30 sec at 94°C (denaturing)
    45 sec at 50 to 65°C (annealing; see Comment 4)
    90 sec at 72°C (elongation)
    3 min 72°C (final elongation step)

4. Set up separate restriction assays for each product–enzyme combination, containing 9 µl of distilled water, 2 µl of 10× restriction buffer, 8 µl of PCR product (see Comments 5 and 6), and 1 µl of restriction enzyme (5 to 20 U). Pipetting errors are minimized by preparing master mixes containing the restriction enzyme, 10× restriction buffer and water.
5. Mix carefully and centrifuge for a few seconds to collect the ingredients at the bottom of the tube.
6. Incubate for at least 3 h at the incubation temperature recommended by the supplier (37°C for most enzymes).
7. Mix with 0.2 vol of loading buffer and resolve fragments on a 1.0 to 2.0% agarose gel, along with a suitable size standard (see Comment 7 and Chapter 4.3.4).
8. Stain with ethidium bromide, and document results as described in Chapter 4.3.6.1.

Comments

1. Frequent-cutting enzymes with four-base specificity are generally used to increase the chance of detecting cleavage sites within a fragment of unknown sequence. Other points to consider are the cost of the enzyme, and its activity in the PCR mix[140] (see Comment 5).
2. A PCR master mix includes the enzyme, 10× PCR buffer, magnesium chloride, and dNTPs. Master mixes are briefly vortexed, centrifuged, and aliquots are dispensed into each tube. Depending on the set-up of the experiment, primer and template DNA are either included in the master mix or added separately. If n samples are to be analyzed, sufficient master mix should be prepared for n + 1 samples.
3. Older thermocyclers may not be equipped with a heated lid. In this case, the reaction solution needs to be overlaid with two or three drops of mineral oil to prevent evaporation.
4. Annealing temperatures depend on the length and GC content of the primers. A gradient cycler facilitates pilot experiments to optimize annealing temperatures. If unexpected bands indicate specificity problems, try the touchdown protocol described in Chapter 4.5.1.
5. Many restriction enzymes are sufficiently active in standard PCR buffers.[140] PCR and restriction can then be performed sequentially in the same tube. However, it is advisable to perform pilot experiments to avoid problems resulting from unspecific cleavage (star activity) under suboptimal conditions.[140] If star activity is suspected, or if the restriction enzyme of choice proves to be inactive in the PCR mix, PCR products need to be purified prior to digestion. This is most conveniently done with a commercial kit (e.g., QiaQuick spin columns).
6. A 50-µl PCR assay is sufficient for digestion with five different restriction enzymes, each combined with 8 µl of PCR product. If desired, the remaining 10 µl can be used to check for the correct size of the intact fragment.
7. The optimal agarose concentration depends on the size distribution of the CAPS markers. In the pilot phase of a CAPS project, electrophoretic conditions may be

optimized by splitting the digested samples into two to several aliquots, each of which is separated on agarose gels of different percentage (e.g., 1 and 2%). High resolution of fragments <400 bp is achieved, e.g., by using mixtures of 1% conventional agarose and 2% NuSieve GTG agarose (or another commercial agarose designed for use at high concentrations).

4.9.2 Choice of CAPS Primers

CAPS primer pairs are usually targeted at specific genomic regions with known sequences. For example, Konieczny and Ausubel[738] designed 18 sets of primers based on mapped and sequenced nuclear DNA fragments of *Arabidopsis thaliana*, and applied these markers for genetic mapping. Semispecific or arbitrary primers may also be used (e.g., in a modification of the RAPD technique[1516]; see Chapter 4.4.3). CAPS markers derived from organellar genomes have become particularly popular for phylogeographic and phylogenetic studies (reviewed by Wolfe and Liston,[1558] Schaal et al.,[1229] and Newton et al.[988]). In plants, these studies have been facilitated by the availability of numerous sets of universal primer pairs targeted at conserved coding regions of both the mitochondrial genome (e.g., Demesure et al.,[326] Dumolin-Lapègue et al.,[372] Duminil et al.,[371] Jeandroz et al.[660]) and the chloroplast genome (e.g., Demesure et al.,[326] Dumolin-Lapègue et al.,[372] Grivet et al.,[529] Saltonstall[1216]). Chen and Sun[237] showed that an enhanced throughput of CAPS analysis of *Spiranthes* species (Orchidaceae) can be achieved by multiplexing three pairs of mtDNA-specific primer pairs, amplifying intronic regions of the nicotinamide adenine dinucleotide (NADH) dehydrogenase gene.

5

Evaluation of Molecular Marker Data

In Chapter 4, we presented a variety of techniques to generate DNA fragment patterns. The present chapter deals with the evaluation of these patterns, i.e., their translation into biological meaning. Several commonly used approaches are described, and some inevitable problems and pitfalls are discussed. Once the positions and matches of DNA bands have been scored, the data are ready to be evaluated quantitatively with the help of various statistical methods. Data can be analyzed in many ways, and a large number of computer programs have been developed, some of which are freely available on the Internet (see review by Labate[767] and Appendices 3 and 4). Examples for the application of these techniques are given in Chapters 6 and 7. The three main application areas are (1) the identification of genotypes; (2) the assessment of genetic diversity and/or relatedness (including phenetic, phylogenetic, population genetic, and phylogeographic analyses); and (3) segregation and linkage analysis for genetic mapping.

5.1 ROBUSTNESS AND REPRODUCIBILITY

Some of the commonly employed molecular marker methods, especially random amplified polymorphic DNAs (RAPDs), have been severely criticized for lack of robustness and reproducibility[397,962] (see Chapter 4.4.2.7). For this reason, some scientific journals have become somewhat reluctant about accepting RAPD-based articles. Most of the initially encountered reproducibility problems, however, have been reduced considerably through improved laboratory procedures. For example, changing the thermocycler or brands (or even just batches) of chemicals during a set of connected analyses should be avoided.[238,1296] Moreover, recently reported problems with robustness and reproducibility in other marker systems[44,516,1506] show that none of them are perfect, and that a critical evaluation of the results for possible artefacts is always necessary.

5.1.1   Reliability

In an effort to improve the reliability of results obtained from molecular marker studies, some authors advocate the repetition of various steps of the laboratory protocol (usually amplification and electrophoresis) twice or even three times, to allow the selection of only fully reproducible bands. In many cases, this is a suboptimal allocation of resources because data variances decrease much faster by doubling the number of analyzed bands than by running each sample twice with the same primer.[1296] In addition, band scoring appears to be considerably less reproducible than the amplification reaction.[177] Hiring a second person to score the bands would therefore be more efficient than repeating all the runs. Reliability can be improved further by scoring only comparatively strong bands, given that the scoring error rate was shown to decrease with an increase in band intensity.[1295,1296] Moreover, the smallest and the largest bands in multilocus fragment patterns are often less reliable[1333] and should therefore be excluded from the analysis.

Competitive priming remains a serious problem for RAPDs[557] (see Chapter 2.3.3.1), as is the occurrence of artefactual bands representing rearranged fragments produced by nested primer annealing and interactions within and between DNA strands during polymerase chain reaction (PCR).[1124] These problems have not been widely addressed, and no general solutions have been presented.

Amplified fragment length polymorphism (AFLP) and intersimple sequence repeat (ISSR) markers are often more reliable than RAPD markers,[569,677,1030,1481,1621] presumably because the former methods employ more stringent PCR conditions. However, insufficient DNA quality can interfere with all kinds of analyses, and is most critical when restriction enzyme digestion is involved, such as in AFLP analysis.[516] Band differences that occur when, e.g., DNA from roots is compared with DNA from leaves of the same genotype, have thus been explained by overall differences in DNA quality as well as organ-dependent methylation patterns.[44,238,356]

The reproducibility of multilocus methods such as AFLP, RAPD, and ISSR across laboratories is sometimes limited.[1064] This is rarely a critical issue, because none of these techniques are used regularly for between-laboratory comparisons or for creating data banks to be shared among several users. Instead, microsatellite DNA analysis currently is considered the method of choice for these purposes. Microsatellite markers are generally claimed to be quite reproducible. Moreover, due to the locus specificity of these markers and the few, small-sized bands exhibited in a single amplification, problems caused by, e.g., insufficient DNA quality, usually can be identified early on. However, evaluation of microsatellite bands is sometimes obscured by severe stuttering, which is especially common for loci with short repeats (see Chapter 4.8.3.2). Another problem is the occasional occurrence of nonamplifying alleles — so-called null alleles[1061] (see Chapter 4.8.4.3).

5.1.2   Band Homology

Multilocus DNA profiles have been used for the evaluation of genetic relatedness and diversity in a tremendous number of studies (see Chapters 6.3 and 6.4). Such experiments rely on the assumption that comigrating bands on a gel actually represent identical states of a homologous character. However, this is not necessarily the case. Size homoplasy arises when two or more bands of identical length and obtained with the same primer or primer pair are not identical by descent. Homoplasy may occur with any type of DNA data — both multilocus fragment patterns[1004,1459] and microsatellites.[366,415,416,480,1017]

In RAPD and ISSR analysis, one main reason for false matches is the limited resolving power of the commonly used agarose gels. Another reason is that unrelated fragments often comigrate incidentally (see Chapter 4.6.2). This issue is less problematic with high-resolution gels, which are routinely used for AFLPs but can be adapted to any other multilocus marker technique (see Chapters 4.4.3 and 4.5.3). However, size homoplasy was also observed in AFLP analysis,[1004] especially for small fragment sizes.[1459] The difference in variability of the different size classes causes additional problems, given that many equations rely on the assumption of uniform and low frequencies of alleles and mutation rates.

Numerous studies have verified that most comigrating RAPD and AFLP fragments are identical by descent, at least at the intraspecific level.[778,1169,1573] However, the risk of homoplasy increases with both the mutation rate of the DNA sequence underlying a certain polymorphism and the taxonomic distance between the taxa involved. Thus, quantitative estimates of pairwise similarity between samples from **different** species showed larger variance than corresponding estimates between samples from the **same** species, indicating that homoplasy is more common in interspecific comparisons.[1319] The effects of mistakenly inflated levels of band homology depend on the type of data and analysis. For example, it has been shown that up to 20% of artefactually comigrating, nonhomologous RAPD bands do not interfere appreciably with calculations of genetic relationships.[6]

Problems with size homoplasy are encountered frequently with microsatellites,[415,416,480,1017] mainly because of their often high mutation rates (see Chapter 1.2.2.3). These problems may become particularly pronounced when microsatellite allele sizes are compared across species boundaries. Thus, comparative sequencing in *Arabis* and *Arabidopsis* species showed that the structure and/or average length of the actual microsatellite locus can differ considerably, despite being amplified with the same primer pair.[1454] Size homoplasy of microsatellite markers may go unnoticed if, e.g., there is a nucleotide substitution in a repeat, or multiple backward and forward mutations in a microsatellite stretch exist, or if a change in the repeat number is compensated by an indel in the flanking DNA or other repeat in the fragment. Some forms of homoplasy, also termed molecularly accessible size homoplasy (MASH),[416] may be detected by sequencing alleles, and some may become apparent by superimposing the repeat lengths on a phylogenetic tree independently obtained with other markers.[1017] Fortunately, homoplasy at microsatellite loci is rarely a problem within a species (but see Ortí et al.[1017]), and the large amount of structured variation at multiple microsatellite loci usually compensates for the random noise created by undetected homology.[416]

5.1.3 Band Linkage and Neutrality

A problem with all kinds of DNA markers stems from the fact that bands treated as unrelated may in reality be linked with each other. Consequently, their respective alleles are not transmitted independently but instead as a haplotype. Linkage dise-quilibrium is a deviation from the random association of alleles in a population, caused by population substructuring or high levels of inbreeding. The occurrence of linkage disequilibrium is especially problematic for population studies, the estima-tion of relatedness of individuals, and paternal exclusion.

Theoretically, problems associated with allelism and linkage may be approached by analyzing the offspring of a biparental cross for transmission frequencies, seg-regation, and linkage of fingerprint fragments. Another step frequently taken in multilocus studies capitalizes on a set of linear correlation tests among all pairs of bands. Closely linked bands can then be detected and removed from further analysis, so that they do not inflate the final results. Various other approaches have been suggested for detection of codominant bands in multilocus data, such as Fisher's exact test[78,148,1039] or χ^2 analysis.[971,1403,1591] Fortunately, the presence of some unidenti-fied codominant bands in a data set does not seriously affect, e.g., cluster analyses.[1500] At present, many computer programs (e.g., Arlequin) can be used for detecting linkage disequilibrium.

Putative problems with linkage caused by a nonrandom distribution of markers across the genome have been reported for several marker systems. Thus, RAPD loci are sometimes clustered in specific chromosomal regions[493,709] and may occasionally produce large aggregates of coinherited bands.[1403] Similarly, microsatellites prefer-entially mapped to chromosomes 1 and 5 in *Arabidopsis thaliana*,[258] and about half of the mapped AFLP-markers in rye (*Secale cereale*), belong to one of four different clusters, located on three chromosomes in proximity to the centromere.[1205]

Marker bands may also be linked to non-neutral traits and therefore be subjected to selection. It has been argued that ISSR and microsatellite markers may be more neutral than RAPDs because they are mainly derived from repeated DNA in which genes are less likely to reside.[414] However, it has also been suggested that microsat-ellite regions instead are a major source of eukaryotic evolution and an important substrate for evolutionary changes.[697] The origin of the investigated microsatellite clones may play an important role in this respect. For example, most microsatellite loci developed from genomic libraries of tomato proved to reside in heterochromatic, centromeric regions, whereas marker loci developed from expressed sequence tags (ESTs) were preferentially associated with euchromatic, presumably gene-rich regions[39] (see also Chapter 1.2.2.2). Evidence for non-neutrality of microsatellite loci has been reported for a number of grass species, presumably as a consequence of an interaction between natural selection and restricted recombination due to predominant selfing.[425,652,810,812,1425]

5.2 FRAGMENT SIZING AND MATCHING

The evaluation and comparison of marker data from different samples requires that individual bands within a lane are assigned to particular positions (which is often done by molecular weight marker-assisted sizing), and different lanes are screened for comigrating (i.e., matching) bands. This section deals with problems associated with fragment sizing and matching.

5.2.1 General Precautions

The precision and accuracy of band scoring strongly depends on several methodological parameters, including DNA quality, completeness of restriction, electrophoretic conditions (especially gel electrophoresis versus capillary electrophoresis), and means of signal detection. Gel electrophoresis deserves particular attention in this respect, given that the mobility of DNA fragments is often somewhat uneven across a gel. One example is the so-called smiling effect resulting from lower electrophoretic mobility in the outermost lanes. Such mobility artefacts may be caused by irregularities in the electric field, residual impurities in the DNA preparation (e.g., proteins), or differences in heat dissipation from the gel. The resulting band shifts can lead to severe misinterpretations of band matching. Variation caused by band shifts can be detected by scoring replicate pairs of individual samples, and the inclusion of in-lane molecular weight standards (see Chapter 5.2.2).

Whenever possible, samples to be compared should be run in adjacent lanes. This is particularly important for complex multilocus band patterns. Large sample numbers, however, will require band matching to be determined between lanes widely separated on the gel, and even between lanes derived from different gels. In these cases, appropriate standards should be included at several positions on each gel. Whereas molecular weight markers typically are used for this purpose, one or several of the investigated samples may also serve as standards, especially if they contain invariable bands present in all individuals.

Even if highly efficient standardization procedures are used, the decision to regard two closely spaced bands as different or identical, respectively, is always somewhat subjective and prone to error. Additional precautions should therefore be taken to minimize the number of misinterpretations. First, only unambiguously scorable bands should be considered for the analysis. Second, bands that cannot be scored accurately throughout all (or most) lanes to be compared should be excluded from the analysis. Third, comigrating fragments of different intensity should not be treated as identical if the intense band is much stronger than the faint band (a factor of 2, however, might only reflect the homo- vs. heterozygous state of the same band).

An automated DNA sequencer will allow much more accurate fragment length determinations than agarose gels, and usually provides single base-pair resolution. However, electrophoresis artefacts may still occur. For example, overloading was reported to cause error in microsatellite analysis on an ABI 377 sequencer.[439] When gel-based DNA sequencers are used, samples may also spill over from one well to the next. Because the detection of fluorescently labeled fragments is very sensitive, even a small spillover may be confused with a poorly amplified AFLP or microsatellite fragment. This is not a problem with capillary-based sequencers, in which different samples are run in separate capillaries. One must be aware that any change in running conditions or the particular fluorescent label of a primer may have a slight influence on the fragment mobility. Therefore, a set of standard samples should be included in each analysis.

5.2.2 Equipment

Fingerprint patterns may be evaluated either by the eyes and hands of the investigator or by automated methods (image analysis, see below). Fragment sizing and matching by eye is most often done by scoring the autoradiograms or photographs directly, usually with the help of a transparent ruler. Alternatively, bands may be copied onto an acetate transparency. This is especially helpful if in-lane molecular standards are visualized by a separate hybridization reaction. Usually only the presence vs. absence of a fragment at a particular position is documented, although occasionally additional information may be obtained by noting fragment intensities.[325]

In recent years, a variety of computerized image analysis systems have appeared on the market, which, among other specifications, were also designed for the evaluation of DNA fragment patterns. Image analysis is usually based on a high-quality video camera, digital camera, or a scanning device for visible and ultraviolet (UV) light, radioactivity, or fluorescence, respectively. Suitable software systems allow the interactive editing of the primary image on the computer screen, including background reduction, band sizing with the help of in-lane or external molecular weight standards, band matching, and comparison across gels. Primary images as well as processed data can be stored in the computer and used for later comparisons. Most importantly, image analysis allows the investigator to set intensity thresholds for the bands to be scored, and mobility thresholds for recognizing a match between two bands.[495] Alternative image analysis techniques have also been employed. For example, phosphorimaging is not only an efficient way of detecting signals (Chapter 4.3.11.2), but may also be used for sizing of fragments and storage of data in a computer.

An increasingly common technique to visualize PCR-generated fragments uses fluorescence-labeled PCR primers and real-time laser scanning with an automated DNA sequencing device,[139,161,349,624,936,1620] combined with specific fragment analysis software. Several distinct fluorescent dyes are available for fragment detection, whereas an in-lane molecular marker may be labeled with a separate dye. This technique is used mainly for AFLP and microsatellite analyses. Variations in fragment sizes can be assigned to corresponding categories using the fragment analysis software or spreadsheet programs such as Microsoft Excel®. Minimum values for acceptable peak heights and marker frequencies can be defined. It should be mentioned that inaccurate allele size differences are sometimes also observed with fluorescent in-lane markers, especially at microsatellite loci.[550]

Taken together, accurate fragment sizing and matching is a difficult step of DNA marker technology, and precautions must be taken to avoid misinterpretations. To determine which level of accuracy is needed, the aim of the study must be considered. For example, deciding whether a suspect should be prosecuted for murder is very different from deciding whether a specific plant is the putative "father" of a set of seedlings. A decision also has to be made concerning the methodology of fragment evaluation. Although it is costly, image analysis and direct acquisition of data into a computer are very helpful if large numbers of data have to be handled. For a comparison of, e.g., small sets of offspring plants with their putative parents, pattern evaluation by eye is sufficient and more cost effective.

5.3 MULTILOCUS VS. SINGLE-LOCUS APPROACHES

The method used for generating fragment patterns (e.g., type of PCR primers) largely determines to what extent the obtained bands can be assigned to particular genomic loci. The difference among data generated by multilocus and single-locus techniques, respectively, is mirrored by a difference in the type of applicable data analysis.

5.3.1 Multilocus Markers

Markers generated by multilocus methods such as AFLP, RAPD, and ISSR are generally treated as dominant, although the occurrence of up to 5%, or in extreme cases even 15%, of codominantly inherited fragments has been reported.[1501] Attempts were made to score AFLP bands as codominant markers when collecting data for genome mapping.[1080] However, these investigations were based on material with known allelic distribution, obtained from controlled crossings between parents with known genotypes. Codominant scoring of bands in multilocus patterns will probably not become very useful when information on the precise level of genetic relatedness among samples is lacking.

Various approaches have been taken for the interpretation of multilocus marker data, such as cluster analysis and principal coordinate analysis. The resulting studies have been quite informative. For example, correlations between geographic and genetic distances are often higher when the latter are derived from RAPDs compared with allozymes[340] or morphological traits.[177] An increasing number of studies employ data evaluation methods designed for defined single loci when analyzing multilocus patterns, especially in the fields of ecology and evolution. In highly inbred species, most loci are in a homozygous state, and the effect of dominance is minimal.[437] In outcrossing species, Hardy-Weinberg equilibria must be assumed for calculating population parameters from dominant marker data. Bayesian methods do not have such requirements.[607]

5.3.2 Single-Locus Markers and Polyploids

In general, variation at allozyme, RFLP, and microsatellite DNA loci is codominantly inherited. Thus, all single-locus types of analyses are applicable, although problems with determining the correct mutational model may affect the choice of appropriate statistical methods for microsatellite data.[469,508,1297]

Many plant species are polyploid. Whereas this generally does not cause any problems with multilocus data (see Chapter 6.3.3.5), the analysis of polyploids can become complicated in microsatellite-based studies. Polysomic inheritance frequently results in the simultaneous occurrence of several alleles of a single microsatellite locus (see Figure 4.2). Estimating the exact number of copies of individual alleles in such loci is usually not attempted, and data are often entered as a binary matrix and treated as representing dominant markers.[914] For more in-depth genetic analyses, the microsatellite allele counting–peak ratios (MAC-PR) approach allows one to analyze allelic configuration also in polysomically inherited loci, provided that the experimental data hold sufficiently high quality.[413,1003] Several computer programs are able to analyze marker data for polyploid species. For example, SPAGeDi[572] allows the inclusion of both diploid and tetraploid individuals in the same analysis. AUTOTET is a program designed to analyze the simplest situation, namely that of an autotetraploid.[1399]

5.4 BAND SHARING AND GENETIC DISTANCES

Many statistical analyses are based on genetic distances and rely on estimates of phenotypic diversity, thereby obviating the need for locus-specific data. As a first step in these analyses, multilocus band patterns are subjected to one of various strategies to quantify pairwise similarity of the genotypes represented in the different lanes. Most commonly, a similarity index is calculated from band sharing data of each pair of fingerprints. These indices can be used to quantify the amount of variation between pairs of samples directly. They can also be grouped according to the origin of the compared samples, and depicted in a frequency table.[918] More often, matrices of pairwise similarity (or distance) are used as an input file for various subsequent multivariate analyses.

5.4.1 Coefficients of Similarity

One of the most commonly used similarity indices is Dice's coefficient, which is also known as Nei and Li's coefficient[981]:

$$S = \frac{2n_{ab}}{n_a + n_b} \qquad (5.1)$$

Here, n_a and n_b represent the numbers of bands present in lanes a and b, respectively, and n_{ab} represents the number of bands shared by both lanes. S can acquire any value between 0 and 1, where 0 means no bands in common, and 1 means patterns are identical. Another, perhaps even more commonly used index is Jaccard's coefficient:

$$S_J = \frac{n_{ab}}{n_a + n_b - n_{ab}} \qquad (5.2)$$

These two indices take only positive matches (both bands are present) into account, and often yield closely correlated results. However, Dice's index places a weight of 2 on shared bands, which purportedly permits a better differentiation of individuals with low levels of similarity. Given that the absence of, e.g., an RAPD band, may have several different causes, it has been argued that using the mutual absence of bands is improper for calculating similarity. Nevertheless, the simple matching coefficient, which includes double zeros, has also been used in some cases:

$$S_S = \frac{n_{ab} + n_{AB}}{N} \qquad (5.3)$$

Here, n_{ab} represents the number of bands shared by both lanes, n_{AB} represents the total number of bands that are absent in both lanes a and b (but present in some other lanes), and N is the total number of bands.[312] Various other indices have also been used, but Dice's and Jaccard's indices have frequently been reported to produce the most logical results.[868]

5.4.2 Dissimilarity Coefficients and Genetic Distances

Genetic distances between samples are often calculated as the complement of the previously mentioned similarity indices. Thus, the complement of the simple matching coefficient is known as Gower's index. One of the most commonly applied distance estimators, especially for data matrices in, e.g., cluster analyses, is the Euclidean distance, in which squared distance estimators are summed over all bands scored in the two samples involved:

$$D_E = \sum \left(x_a - x_b \right)^2 \qquad (5.4)$$

Here, $x = 0$ when a band is absent in lanes a or b, and $x = 1$ if the band is present. This parameter is actually identical to Gower's index multiplied by the total number of scored bands.

A matrix containing pairwise genetic distances between samples can be used for comparisons with other matrices based on, e.g., geographic distances or habitat variability. Thus, the spatial or habitat-dependent patterning of individuals within populations as well as larger-scale patterns can be analyzed for possible associations with the patterning of genetic variation. The most commonly taken approach is the Mantel test,[877] which is frequently applied for the analysis of isolation-by-distance (IBD) scenarios. The Mantel statistic (r_M) is interpreted as a Pearson correlation coefficient and, unfortunately, is sensitive only to linear relationships of spatial autocorrelation. Relationships that may alter along, e.g., a geographic gradient, are better described by some kind of spatial autocorrelation analysis (e.g., correlograms and variograms; see also Chapter 5.6.7).

5.4.3 Identity and Uniqueness

A similarity index derived from multilocus data is usually not zero even for completely unrelated individuals. The level of this so-called background band sharing strongly depends on the combination of marker origin and species, and has to be taken into account if relatedness is deduced from similarity indices. If x is the proportion of bands shared by unrelated individuals, then the probability that the offspring inherited a specific band from one of the parents is[666]:

$$P = \frac{2x - 1 + (1 - x)^{3/2}}{x} \tag{5.5}$$

If all bands represent statistically independent characters, the probability that n bands would all match in two randomly taken individuals of a population, assuming a background band sharing level of $\bar{S}$, is:

$$P = \bar{S}^n \tag{5.6}$$

However, this equation is an overestimate, because it ignores the fact that two individuals do not always have the same number of bands. Alternatively, the probability of a random match can be estimated with the following equation:

$$P = \frac{\left[\bar{S}^2 + (1 - \bar{S})^2 \right] \times \bar{n}}{\bar{S}} \tag{5.7}$$

This value is an important estimate of how accurately cultivars or clones are identified. In other words, it represents the likelihood of falsely identifying a cultivar or clone. It should be noted, however, that such estimates largely depend on the experimental setup. When using multilocus approaches, comparisons are generally carried out only within the same study, using identical experimental conditions.

To compare S and n among species or populations, sampling variances can be calculated.[861,862] If a sufficient number of bands per individual are scored, n and S behave like normally distributed variables. Therefore, the sampling variance of the number of bands per individual is:

$$\mathrm{Var}(n) = \frac{k \left(\overline{n^2} - \bar{n}^2 \right)}{k - 1} \tag{5.8}$$

where k is the sample size. The sampling variance of the average number of bands, $\bar{n}$, is $\mathrm{var}(\bar{n})/k$. The sampling variance for S is calculated as:

$$\mathrm{Var}(\bar{S}) = \frac{\sum (S_{xy} - S)^2}{a - 1} \tag{5.9}$$

In Equation (5.9), a is the number of pairwise comparisons and S_{xy} are the similarity indices of all pairs x and y. This calculation is only correct if all individuals are used only once in the pairwise comparisons, e.g., when compared only with an adjacent lane. If instead all lanes are compared with all other lanes (which is usually done), the calculations become far more complicated because of dependence of S_{xy}.

The number of markers needed for the detection of at least one difference between a pair of suspected duplicates in, e.g., a gene bank, can be calculated easily according to Virk et al.[1471]:

$$\left(x/y\right)^z = 1 - a\left(x\right) \tag{5.10}$$

Here, x represents the number of monomorphic markers, y is the total number of markers, z is the number of markers required, and a is the confidence level. Formulas defining the minimum number of bands and individuals required to detect a difference between two populations (consisting, e.g., of allogamous cultivars) are also available.[490]

A large number of statistical methods have been developed to determine identity with microsatellites, and the most sophisticated ones are related to forensic science.[420] For forensic as well as biological questions, it is important to ensure that the correct reference population is used to determine allele frequencies. For biological questions, the most frequently applied equation to calculate the probability of identity P_{ID} is[1025]:

$$P_{ID} = \sum P_i^4 + \sum \sum \left(2 \times p_i \times p_j\right)^2 \tag{5.11}$$

In Equation (5.11) p_i and p_j are the frequencies of the alleles i and j, and $i \neq j$.[1025]

Waits et al.[1488] developed these equations further for natural populations, also taking population structure into account. Recently, Ayres and Overall[63] released a program, API-CALC, which not only considers population structure, but also inbreeding and the presence of close relatives.

5.4.4 Clonal Structure

A special problem for genotype identification relates to the fact that many plants reproduce vegetatively and thus consist of several genets, each of which is made up of a number of ramets sharing the same genotype. Brookfield[169] showed that DNA fingerprinting may be effective to distinguish whether individuals are members of the same genet or not, provided that the frequency of sexual reproduction is considerably higher than the mutation rate of the fingerprint loci. This is also an important prerequisite for differentiating among vegetatively propagated cultivars such as berries, fruits, potatoes, and many woody ornamentals. The probability that the identified genotypes represent unique genets can be estimated for populations in Hardy-Weinberg equilibrium.[1354] The probability that multilocus genotypes are unique, P_{gen}, can be calculated as[191,1354]:

$$P_{gen} = \prod p_i \tag{5.12}$$

Here, p_i is, for each locus in the multilocus genotype, the frequency of band presence or absence among all plants sampled.[191,1354] For codominant loci, the corresponding probability P_{gen} is instead calculated as[1046,1300]:

$$P_{gen} = \left(\Pi\, p_i \right) 2^h \tag{5.13}$$

where p_i is the frequency of each allele observed across all investigated loci, and h is the number of heterozygous loci.[1046,1300] Both Equations (5.12) and (5.13) calculate the probability with which a second copy of a particular genotype is present in the investigated material. Consequently, the probability of drawing $(n - 1)$ copies is $(P_{gen})^{n-1}$.

Based on the number and frequency of individual genets detected with DNA markers, various estimators of genotypic diversity are calculated for clonal species. The proportion of distinguishable genets P_D is calculated as

$$P_D = G/N \tag{5.14}$$

where G is the number of genotypes detected and N is the sample size.[396] The frequency of unique genotypes (which is zero in a population composed of a single genotype, and 1 in a population where every plant has a unique genotype) can be estimated with the complement of Simpson's index corrected for finite samples:

$$D = 1 - \sum \frac{n_i \left(n_i - 1 \right)}{N \left(N - 1 \right)} \tag{5.15}$$

where n_i is the number of plants with genotype i and N is the total sample size.[1079] Finally, an evenness measure (E is 0 in a population where all plants represent different genotypes or where one genotype dominates and the rest are represented by one plant each, and E is 1 in a population where all genotypes are represented by the same number of plants) can be calculated as described by Fager[423]:

$$E = \frac{D_{obs} - D_{min}}{D_{max} - D_{min}} \tag{5.16}$$

where D_{obs} is calculated according to Equation (5.15), and D_{min} is obtained as:

$$D_{min} = \frac{\left(G - 1 \right)\left(2N - G \right)}{N \left(N - 1 \right)} \tag{5.17}$$

and D_{max} is obtained as:

$$D_{max} = \frac{N(G-1)}{G(N-1)} \qquad (5.18)$$

Two computer programs for detecting multilocus identity, MLGsim and Geno-Type/Genodive, use a simulation approach to determine the chance of finding identical multilocus genotypes in a population solely due to chance and therefore calculate the likelihood that multilocus genotypes are the result of vegetative propagation (clones) or sexual reproduction.

5.5 ORDINATION, CLUSTERING, AND DENDROGRAMS

In other sections of this chapter, we describe how DNA fingerprint data can be used to calculate genetic distance; e.g., between individual plants, between accessions of crops, and between populations of wild species. When large numbers of samples are involved, it is difficult to interpret these genetic distances. In this section we describe how such data can be analyzed to infer the phenetic and phylogenetic relationships between samples.

Ordination, clustering, and dendrograms have long been used in research fields where multiple character data sets are generated. The main purpose of multivariate statistics is to condense the differences between the entries for many characters into fewer characters and to visualize these entries in a multidimensional space. Thereby, the complexity of data is reduced, and at the same time as many of the characters as possible are used to differentiate among entries. These techniques can also be employed to identify the most discriminating characters.

Apart from the more obvious applications (i.e., estimating relatedness among operational taxonomical units [OTUs] such as cultivars, species, and populations), phenetic analyses also serve for more specialized purposes, such as selecting a smaller core collection with preserved genetic diversity from within a larger gene bank. This can be done by, e.g., ordination techniques[1280] and cluster analysis,[623] but the maximum diversity algorithm developed by Marita et al.[884] may be even more useful for this purpose.[478] Several different statistical techniques are available for the analysis of phenetic and, to a lesser extent, phylogenetic patterns based on DNA fingerprint data. Specialized books should be consulted for a more comprehensive treatment of this subject (e.g., Manly[873]). See also Appendix 3.

5.5.1 Ordination Techniques

For AFLP, RAPD, and ISSR data, a presence–absence matrix of bands from the analyzed samples can serve as a starting point for the ordination techniques. This type of matrix must be treated as nonparametric data. If the data set instead consists of allele frequencies in, e.g., populations or collections of accessions, it can be analyzed in a parametric fashion, provided that the data are normally distributed and variances are equal. However, we can also start from a distance matrix, calculated by, e.g., Gower's dissimilarity index (see above). Nonparametric data (0/1 matrix and genetic distances) can be subjected to **principal coordinates analysis (PCO)** and discriminant functions. For parametric data (allele frequencies), the more powerful **principal components analysis (PCA)** and **canonical variates analysis (CVA)** are appropriate.

Among the ordination methods, PCA is probably the most well-known technique. The starting point is quantitative data (allele frequencies) for multiple markers from two or more taxa or populations. This method makes use of a multidimensional solution of the observed relationships. PCA finds hypothetical variables (components), that account for as much of the original variance in the multidimensional data set as possible. If the first two or three components explain a large proportion of the variance, the analysis has been successful. The successive principal axes, representing the first major axis, the second major axis, etc., account for the greatest, the second greatest, etc., amount of variation. The original data (taxa or populations) can now be plotted into the (two or three dimensional) coordinate system, revealing the degree of similarity among entries by the way they group together.

A related technique is the PCO or metric multidimensional scaling. Here, the data matrix is not derived from the presence or absence of all bands in all OTUs, but rather from the distances (or similarities) between the OTUs. Given that a distance matrix is the starting point, PCO does not use as much information as is used by PCA. The data points describe a hyperellipsoid in a multidimensional space. To simplify the description of these clouds of points, the (principal) axes through the hyperellipsoid are calculated. Again, the first dimension of the solution gives the best fit to the full set of data. A scatter plot of the data depicts the varying levels of similarity among the entries. One can also calculate how well samples cluster in groups with a biological meaning, such as region of origin.

A third ordination method is **multidimensional scaling (MDS)**, which aims at preserving the rank order of magnitude of distances between the points, so that, for example, larger distances in the original model are also larger in the simplification. Compared with PCA, MDS generally reveals a larger proportion of the variability already present in the first two dimensions.[1319]

Various other kinds of analyses have been used that integrate the correlated variations into a few vectors. Among these are factorial correspondence analysis[120,710,786] and Hayashi's Quantification Theory III analysis, which is designed for discontinuous variables and has been applied to microsatellite data.[3] CVA is a parametric technique that aims at distinguishing samples in the data set that fall into different groupings. For example, if accessions of a crop can be divided into two subgroups based on their origin, we can use CVA to find the character (marker) that best distinguishes these two predefined groups. Subsequently, discriminant function analysis can be used to place additional samples into the predefined groups.

Numerous computer programs are available that perform PCA, PCO, MDS, and CVA. Among these are standard statistical packages, such as SPSS and SAS. NTSYS-pc, Genetix, ADE-4, GenAlEx, and PCAgen are more specialized programs that not only calculate multivariate statistics but also display the outcome of the analyses in a graphical format (see also Appendix 3).

5.5.2   Construction of Dendrograms

Several classical books describe molecular phenetic and phylogenetic analyses in great detail (e.g., Hillis et al.[597] and Soltis et al.[1308]). More recently, several others appeared, each with its own emphasis. Thus, Page and Holmes[1027] and Nei and Kumar[982] describe tree reconstruction, starting from models of evolution, and Hall[556] is very helpful on practical aspects of tree reconstruction, especially for beginners in the field. The most recent addition to this collection is the long-awaited book, *Inferring Phylogenies*, by Felsenstein,[435] who also devotes a chapter to RAPD, AFLP, and microsatellite data. Together, these books review the methodology of molecular phylogenetics in great detail.

The aim of producing a dendrogram is to visualize the best representation of the phenetic (overall similarity) or phylogenetic (evolutionary history) relationships among a group of so-called operational taxonomic units (OTUs). These can be individuals, cultivars, populations, or species. Three major strategies for constructing dendrograms can be distinguished.

The first strategy comprises the **distance methods**, also referred to as **cluster analyses** or **phenetic methods** (see, e.g., Avise[57] and Swofford et al.[1352]). The starting point is a data matrix of pairwise distances, which is calculated from the primary data by, e.g., the Dice or Jaccard algorithm (see Chapter 5.4). In the first step of the tree-building procedure, the two OTUs with the minimal distance to each other are grouped together, and the distance matrix is reduced by one row and one column. After this, a second round of clustering is performed, again grouping the two closest OTUs together. This process is reiterated until a single OTU remains. The resulting dendrograms express phenetic similarities among the OTUs and are therefore called **phenograms**. They do not necessarily reflect phylogenetic relationships.

The most frequently used distance matrix algorithm is the so-called **unweighted pair group method using arithmetic average (UPGMA)**. UPGMA assumes a rigid molecular clock, which means that the evolutionary rates along all branches of the tree need to be identical. Saitou and Nei[1213] developed a more sophisticated clustering method called **neighbor-joining (NJ)**. This algorithm produces additive trees and does not assume identical evolutionary rates along all branches.

All clustering methods are based on the overall similarity between pairs of OTUs, and therefore information about particular characters (markers) is lost. No fragments are considered to be more informative than others. Only one optimal phenogram is produced from a given data set, and there is no way to compare or rank suboptimal phenograms. The clustering methods are computationally simple and can handle large data sets. A variety of general and specialized software packages is available, e.g., SPSS, SAS, NTSYS, and Phylip (see Appendix 3). The resulting tree can be a good starting point for more elaborate maximum-likelihood (ML) approaches, see below. Recently, the cluster structure exhibited by multilocus-based NJ trees has also been used to make inferences about population processes.[606]

The second strategy for constructing dendrograms involves **parsimony methods** and aims at reconstructing phylogenetic patterns (for review, see Hall,[556] Page and Holmes,[1027] and Swofford et al.[1352]). Parsimony trees are reconstructed according to the principles of cladistics (see Page and Holmes[1027] and Felsenstein[435]) and are also called **cladograms**. They are based on the character matrix itself — no distance calculations are involved. The maximum parsimony (MP) criterion is applied to compare alternative tree topologies; i.e., only those trees that explain the data set with the smallest number of mutational changes (the shortest or most parsimonious trees) are selected for further comparison. Specific algorithms are used to compute trees that are in maximum concordance with this criterion. In contrast to distance matrix-based methods, some characters (such as fragments or base positions) may be more important for inferring a tree than others, and the extent of homoplasy present in the data set can be inferred. With large data sets, heuristic rather than exhaustive search strategy methods need to be used to find the best (i.e., shortest) tree.

Unfortunately, parsimony trees can be biased by multiple substitutions and variations in the evolutionary rate among the individual characters.[259] Also certain problems arise with analyzing DNA fingerprint data via parsimony, including dominance, oversimplified evolutionary models, and interdependence.[435] Nevertheless, this topic is controversially discussed, and parsimony methods have successfully been applied to reconstruct phylograms from AFLP fingerprint data by various authors.[77,693,740] Several software packages can perform parsimony analyses. Most frequently used are Hennig86 (Farris), Phylip (Felsenstein), and PAUP (Phylogenetic Analysis Using Parsimony; Swofford); see also Appendix 3.

The third strategy for tree construction relies on the **maximum likelihood (ML)** method, using standard statistical methods for a probabilistic model of evolution (reviewed by Huelsenbeck and Crandall[628]). It aims at finding the tree with the largest probability to reflect the actual data set on the background of an appropriate evolutionary model. ML is only suitable for data for which such a model can be described. This is certainly the case for DNA sequence data, but only rarely for DNA fingerprint data. ML is also computationally very demanding, and large data sets are difficult to analyze. More recently, **Bayesian inference** of phylogeny has become very popular (reviewed by Archibald et al.,[33] Hall,[556] and Huelsenbeck et al.[629]). The Bayesian approach attempts to find a tree with maximum *a posteriori* probability (rather than *a priori* probability as in ML), is much faster than ML, and handles larger data sets. Recently, the Bayesian probability approach has also been applied to multilocus fingerprint data derived from an AFLP analysis.[170] The resulting tree successfully discriminated the four subspecies of *Leonardoxa africana*, an ant-plant species native to Central Africa.

The statistical support for individual branches of phylogenetic trees can be estimated with various methods, such as resampling with replacement (bootstrapping),[434] resampling without replacement (jackknifing and parsimony jackknifing),[429] and the decay index.[163,954] Bootstrap and jackknife can also be applied to phenograms and ML trees. A more detailed description of these methods is beyond the scope of this book. For more information, the reader should consult the books and reviews cited above.

To summarize, although several methods are at hand to construct a dendrogram from DNA fingerprint data, distance methods are used in the vast majority of studies. Among these, NJ appears to be the most appropriate algorithm, because it does not assume equal evolutionary rates of the different characters (i.e., fingerprint bands) analyzed. A drawback of distance data is that we do not know what the character state was in the ancestral OTU. Most dendrograms are therefore drawn unrooted. Depending on the type of data collected, it may well be worth also running a parsimony analysis, and comparing the resulting cladogram with the phenogram obtained by NJ. Robust branches are expected to be resolved with both tree-building strategies[740] and can therefore be recognized as such.

5.6 POPULATION GENETIC ANALYSIS

Population genetic methods are employed to analyze genetic variation observed in, e.g., a wild species or in a collection of accessions of a crop (see the review in Lowe et al.[850]). Genetic variation is generally distributed in a hierarchical way; i.e., within an individual, between individuals within a population or a collection of accessions, between populations within a region of origin, and between all populations and all regions inhabited by the respective species. The extent of genetic variation in a species and its distribution among populations (or other entities of subdivision) is determined by a large number of factors, including the mating system, the demographic history, the effective population size, and the extent of gene flow by, e.g., migration or seed dispersal between populations. At each level, the genetic variants are united through interbreeding at some time point in the past. By analyzing the amount of variation and its partitioning over these hierarchical levels, we can draw important conclusions about the biology of a species or the domestication history of a crop.

Population genetic analyses are preferably performed on codominant data sets derived from, e.g., microsatellites, RFLPs, or allozymes. In fact, many techniques to measure variation originate from early studies of allozyme variation. It is generally recommended that one first test whether natural populations are in Hardy Weinberg (HW) equilibrium and whether there is linkage of markers, before more detailed and complex analyses are undertaken.[576] Deviations from HW can result from, e.g., inadvertent sampling across more than one random mating population (population structure), self-fertilization, or selection. A locus that is not in HW equilibrium because of, e.g., selection at this locus or selection for a gene linked to it, or due to the occurrence of null alleles, should be omitted from the analysis.

Valuable and informative estimates of variation and relatedness can also be obtained from dominant data, given that the large number of loci typically analyzed provides good genome coverage and a large number of data points (reviews in Nybom[1000] and Nybom and Bartish[1001]).

5.6.1 Measures of Variation

When banding patterns or genotypes have been scored, the observed variation in a group of cultivars or in populations of a species can be quantified. The simplest quantitative measure is the number of polymorphic markers, P, often expressed as the percentage of all markers scored in a set of samples. Another simple measure is allelic diversity, A, which is sometimes reported from studies on multiallelic loci such as microsatellites. A is the number of alleles per locus, averaged over all loci tested. However, this measure is sensitive to sample size: the larger the sample size is for a population, the higher is the chance that new (rare) alleles are detected. Clearly, a population harboring three alleles with similar frequency is more polymorphic than a population with one very frequent allele and two infrequent alleles. Therefore, a better measure than A is A_e, the effective number of alleles that corrects the absolute number of alleles by taking into account the allele frequency:

$$A_e = \frac{1}{\sum p_i^{\,2}} \qquad (5.19)$$

In Equation (5.19), p_i is the frequency of each allele observed across all investigated loci.

Wild plant populations as well as crops may undergo so-called bottlenecks, which often result in loss of genetic variation, proportional to the severity and length of the bottleneck.[91,1109] Allelic richness measures actually provide a more sensitive tool for detecting recent genetic bottlenecks with microsatellite markers than the more commonly used allelic evenness measures such as heterozygosity.[1133,1537] Tests have been designed to detect bottlenecks and their severity using microsatellite data; e.g., the computer program BOTTLENECK.

Another measure of variation is the observed frequency of heterozygous samples averaged over loci, H_o. In plants, this may not be a very useful measure, especially if the investigated plants reproduce at least partially by selfing. Thus, inbred lines of, e.g., maize, are homozygous ($H_o = 0$), although a collection of different lines may contain ample variation. An increase in H_o may also be a consequence of the occurrence of null alleles.[282,1536]

One of the most frequently used measures of genetic variation is Nei's gene diversity,[978] which also takes into account the allele frequencies. The gene diversity h is the probability that two randomly chosen alleles (or haplotypes) are different in the sample. This is calculated per locus as:

$$h = 1 - \sum p_i^{\,2} \qquad (5.20)$$

Gene diversity is summed from i = 1 to k, where p_i is the frequency of the ith allele and k is the number of alleles. Gene diversity h is then averaged over all loci and is often denoted as H. In an ideal and randomly mating population, H is identical to the expected heterozygosity H_e. According to a compilation of 79 microsatellite-based studies, grand means for H_o and H_e are almost identical (0.58 and 0.61, respectively), but H_o was nevertheless lower than H_e in 64 of these studies.[1000]

Nei's gene diversity can also be calculated for dominant, biallelic data:

$$h = 1 - p_i^2 - q_i^2 \qquad (5.21)$$

Here p is the frequency of the visible allele and q is the frequency of the null allele for the marker i. There are two situations in which allele and genotype frequencies can be estimated correctly with dominant markers. The first pertains to species that are highly selfing: all individuals are then homozygous and allele frequencies become identical to band frequencies.[863] In the second (ideal) situation, a separate codominant data set has already provided evidence for random mating and Hardy Weinberg (HW) equilibrium:

$$h = 2q_i\left(1 - q_i\right)$$ (5.22)

In this case, q is generally derived from $x^{1/2}$ (the so-called square-root method), where x is the frequency of individuals that lack the marker.[863] Lynch and Milligan[863] also proposed a more unbiased gene diversity estimate for dominant markers, i.e.,

$$h = 2q_i\left(1 - q_i\right) + 2\mathrm{Var}\left(q_i\right),$$ (5.23)

with

$$\mathrm{Var}\left(q_i\right) = \frac{1-x}{4N}$$ (5.24)

The null allele frequency, q, is then estimated for N sampled individuals from:

$$q = x^{1/2}\left[\frac{1 - \mathrm{Var}\left(x\right)}{8x^2}\right]^{-1}$$ (5.25)

Finally, Lynch and Milligan[863] also showed that removing (so-called pruning) loci with low levels of polymorphism improved the estimates, and recommended that analyses be restricted to bands, whose observed frequencies are less than 1 − (3/N). The computer program AFLP-SURV is designed for dominant data. It calculates diversity measures and lets the user decide whether to use the Lynch and Milligan[863] method.

Some multilocus studies report H_e values that have been calculated across all scored bands (i.e., including monomorphic bands), and others report values that are calculated on the basis of polymorphic loci only. When comparing different estimates of H_e, it should be realized that dominant markers such as RAPD and AFLP can only produce two alleles at each locus, and therefore the maximum value is 0.5, whereas multiallelic markers such as microsatellites can produce values up to 1.

Still another measure of diversity, which is often used for dominant data, is Shannon's index of diversity. It is generally calculated as:

$$H' = -\sum p_i \log_2\left(p_i\right)$$ (5.26)

where p_i is the frequency of the ith allele in the population, group of accessions or species, and both alleles (presence and absence) of each locus must be taken into account. Alternatively, the natural logarithm is used instead of $\log_2$. There are, however, a number of different ways in which this statistic has been calculated,[192] and care must be exerted when making comparisons across studies.

Various other routes for calculations of within-population diversity have been taken. For example, Zhivotovsky[1615] advocated the use of a Bayesian approach for the analysis of RAPD data, and Krauss[748] demonstrated that the standard square-root method, as well as the Lynch and Milligan approach,[863] are just as efficient as the Bayesian approach in producing within-population diversity estimates, that are very similar to H_o values.

Whenever diversity estimates are obtained for more than one group or population, appropriate computer programs (see Appendix 3) can be used to determine whether the investigated groups or populations differ in their level of variation. However, great care should be exerted when comparing levels of variation that have been revealed with different sets of markers. Only when large numbers of studies are reviewed (such as in the work of Hamrick and Godt[563] and Nybom[1000]) are differences between studies solely due to marker choice evened out, and the general patterns become visible.

5.6.2   Genetic Differentiation between Populations

Separate populations of a wild species or regional groups of accessions of a crop usually differ in the relative allele frequencies of genes and markers. Random drift, selection, founder effects, and bottlenecks can cause populations to differentiate, whereas high migration rates between populations will prevent or slow down differentiation. To learn more about these factors and their effects on our plant material, we can calculate the partitioning of genetic variation, within and between populations or collections of crop accessions. Many traditional measures for population differentiation were originally developed for allozymes. A good introduction into the topic of population structure is given by Hartl and Clark,[576] whereas Balloux and Lugon-Moulin[76] have reviewed the methodology of estimating population differentiation with microsatellite markers.

Several publicly available computer programs can be used to calculate population divergence estimates, such as Arlequin, FStat, R_{ST}, Popgene, Genepop, GDA, and Microsat (see Appendix 3).

5.6.2.1   F Statistics and Related Measures

Differentiation of populations causes a reduction in the proportion of observed heterozygotes compared with the number of expected heterozygotes (Hartl and Clark[576]). The extent of this reduction can be used to obtain a measure of population structure employing Wright's F statistics.[1572] The fixation index F can be calculated for each hierarchical level; e.g., between populations within regions and between regions on continents. For every hierarchical level, the expected heterozygosity is calculated from the allele frequency at that level using Equation (5.20).

A common notation is to have subpopulations (S) within a total population (T), and then calculate the fixation index F as:

$$F_{ST} = \frac{H_T - H_S}{H_T} \tag{5.27}$$

in which H_T is the expected heterozygosity in the total population (species) and H_S is the mean expected heterozygosity in the subpopulations (populations). F_{ST} ranges from zero to 1. For more information about F_{ST} and several definitions of this parameter, see the review by Balloux and Lugon-Moulin.[76]

Nei[979] extended this analysis to multiple loci and more than two alleles at a locus using the average heterozygosities (H_T and H_S) in a population over different loci. As an estimate equivalent to F_{ST}, Nei[979] defined the coefficient of gene differentiation as

$$G_{ST} = \frac{H_T - H_S}{H_T} \tag{5.28}$$

in which H_T and H_S are averaged over loci for any number of alleles. However, in many publications, G_{ST} (or F_{ST}) is calculated in a slightly different way, by first estimating G_{ST} for each locus, and then averaging the obtained values over all loci.[562] Both methods were used for calculating population differentiation in allozyme studies, with those following Nei's formula producing slightly higher values.[291]

It must be noted that G_{ST} values depend on the variability within subpopulations as well as in the total population. Microsatellite loci usually have such large numbers of alleles, that both H_S and H_T may get close to 1 and therefore G_{ST} becomes very small, even when the subpopulations have different sets of alleles.[590]

The classical work on F_{ST} and G_{ST} was developed for allozymes, assuming an infinite alleles model (IAM; see Chapter 6.3.2.1). However, microsatellite DNA evolves much faster than allozymes, and also by different mechanisms (see Chapter 1.2.2.3). The mode of evolution must therefore be taken into account when population structuring is estimated with microsatellite data.[76,511] If we assume that microsatellites evolve by stepwise mutations (stepwise mutation model [SMM]; see Chapter 6.3.2.1), then allele size differences are relevant for the calculation of distances between loci and genotypes. Slatkin[1297] and Goldstein et al.[508] developed similar SMM-based methods, called R_{ST} and delta mu squared distances ($\delta\mu^2$), respectively. Michalalakis and Excoffier[925] developed Rho$_{ST}$, which is analogous to R_{ST}, but is based on molecular variance components. A collection of computer programs for calculating R_{ST} and other estimates of genetic differentiation from microsatellite data was presented by Goodman.[511] Delta mu squared distances ($\delta\mu^2$) can be calculated with the program MICROSAT.

Simulation studies demonstrated that either the IAM or SMM approach may be preferable depending on, e.g., population size, sample size, number of analyzed loci, and number of alleles per locus.[469] It was suggested that genetically close populations should be analyzed with F_{ST}, and more distant populations should be analyzed with R_{ST}, to dissect the effects of IBD (best detected with R_{ST}) from those of mutation (best detected with F_{ST}).[1012,1144,1155] It was also proposed that loci with few alleles are better suited for analysis with R_{ST}, whereas loci with many alleles are better suited for an analysis with F_{ST}.[652] However, simulation studies showed that R_{ST} estimates are inaccurate, and often misleading, with small numbers of loci (fewer than five) and small sample sizes (fewer than 10).[416] F_{ST} and R_{ST} values often differ considerably when calculated on the same data set, usually with lower values obtained for F_{ST}.[262,1155]

Modifications of the standard G_{ST} estimates allow one to analyze data obtained from organellar genomes; e.g., the chloroplast genome.[403] If no assumptions about the type of mutation can be made, a feasible alternative is to apply chord distances (see Chapter 5.6.3).

5.6.2.2 Analysis of Variance

At present, the genetic structure of populations is often investigated by analysis of variance instead of using the F_{ST} and G_{ST} estimators. Weir and Cockerham[1517] describe an estimate, which is an analog of F_{ST}. A hierarchical analysis of variance is performed on allele frequencies, whereby covariance components can be distinguished because of interindividual differences, interpopulation differences, and higher levels of partitioning.

The **analysis of molecular variance (AMOVA)**[421] approach (available in the computer program Arlequin) is often used with dominant marker data. It works with distances between individuals and calculates the variance between and within predefined groups, and allows one to test various particular genetic structures. Originally, squared Euclidean distances were preferred,[421] but very similar results have also been obtained with, e.g., Jaccard's coefficient of similarity. Variance components can be calculated for each hierarchical level. AMOVA proved to be very useful for partitioning of variation in wild species and among groups of cultivars originating from different regions. Although the principles behind F_{ST} and G_{ST} on the one hand and the AMOVA on the other hand are rather different, they usually produce very similar estimates when applied to the same set of marker data (reviewed by Nybom[1000]).

5.6.2.3 Shannon's Index

Shannon's index of diversity calculates population differentiation in a way that is quite similar to Nei's G_{ST} and Wright's F_{ST}. First, Shannon's index of diversity is calculated for the populations, H_{pop}, and for the species, H_{sp}.[192] The component of within population diversity is H_{pop}/H_{sp}, and the component of between population diversity is $(H_{sp} - H_{pop})/H_{sp}$.

Shannon's index of diversity has been calculated in different ways; e.g., over individual loci or over primers.[192] In addition, monomorphic bands are sometimes taken into account by giving them zero diversity, whereas other studies have not been clear about this. Therefore, these estimates cannot easily be compared across studies, or with other diversity measures. Nybom et al.[1002] compared Shannon's values of population differentiation with G_{ST} and F_{ST} values obtained with AMOVA in the same plant material, and found that values obtained with Shannon's diversity index ranged wider and were often higher than those obtained with F_{ST} or G_{ST}.

5.6.3   Genetic Distances between Populations

In many cases we are interested in determining the genetic distance between two populations or two collections of cultivated plants, which can be visualized subsequently by constructing a dendrogram or by using an ordination method. To calculate this distance, we need a measure that is related to the collective evolutionary distance of our samples. Pairwise F_{ST} values (or an analog) are most often used, and are calculated as described above. A generalized approach, which is independent of markers, is the use of chord distances.[219] Populations are positioned in a hypersphere as determined by their allele frequencies, and distances between populations are then distances between multidimensional points in a hypersphere.[128] GENDIST is a population genetics computer program in which this method is incorporated (see Appendix 3).

Probably the best known distance estimator is Nei's unbiased genetic distance,[980] D, which is based on the probability that a randomly chosen allele from each of two populations will be different, relative to the probability that two randomly chosen alleles from the same population will be different.[980] This distance is calculated as:

$$D = -\ln S \tag{5.29}$$

For calculation of S, see Equation (5.1). For comparative studies using different distance and identity measures see Kalinowski,[683] Tomiuk and Loeschke,[1407] and Tomiuk et al.[1408] Pooling or bulking DNA from several individual plants has been used occasionally for estimations of between-population distances with dominant, multilocus markers.[733] The resulting band profiles are somewhat simpler than the combined profiles of individual plants. Replicate bulk samples had almost identical band patterns, indicating that this method is quite robust.

5.6.4   Inbreeding Coefficient and Mating Systems

The mating system of a species can be determined from codominant DNA marker data by an indirect method based on the deviation of genotype ratio from HW equilibrium. To calculate the selfing rate s or the outcrossing rate t from genotype ratios, we must assume that the population is in equilibrium and that there is no substructure in the population (Wahlund effect). The outcrossing rate t can then be calculated as[1518]:

$$t = \frac{1-f}{1+f} \tag{5.30}$$

in which f is the coefficient of inbreeding, which can be calculated from expected heterozygosity, H_e, and observed heterozygosity, H_o, in a population:

$$f = 1 - \frac{H_o}{H_e} \tag{5.31}$$

The selfing rate s is related to t: $s = 1 - t$. This formula provides only a rough estimate because deviation from HW ratios can be caused by other factors, such as population structure or selection. It is also rather generalized because effects over (several) generations and a whole population are averaged, which may be an advantage or disadvantage, depending on the question asked. In addition, when only adult plants are genotyped, selection at the seedling stage (e.g., due to heterozygote advantage) cannot be excluded.

To obtain more specific and detailed estimates of outcrossing rates, a direct method may be preferred, based on collecting seeds from single mothers and determining genotypes of mothers and offspring. The standard program for estimating outcrossing rates is Kermit Ritland's MLTR, which accommodates codominant data, and MLDT, which is designed for dominant data. Milligan and McMurry[934] reported that dominant markers are useful for these types of analyses, but that 1.5 to 2 times more data are needed as compared with codominant markers.

Under some circumstances, direct inbreeding values can be estimated from the loss of heterozygosity over time, based on generational data:

$$F_e = 1 - \frac{H_j}{H_m} \tag{5.32}$$

In Equation (5.32), H_j is the heterozygosity of juvenile plants and H_m is the heterozygosity of mature individuals.[458]

5.6.5   Estimation of Relatedness and Paternity Testing

Related individuals are expected to share a higher number of fragments than unrelated individuals. Estimating the true relatedness from observational data is, however, problematic, because these can only present information about identity by state as opposed to the more desirable identity by descent. For multilocus data, Lynch and Milligan[863] nevertheless provided a set of relatively unbiased equations that allow to calculate a coefficient of relatedness. For microsatellite DNA data, Moran's I has been used for estimating relatedness.[571] Analysis of pairwise relatedness and parentage can also be achieved by various computer programs based on ML methods.[854]

Paternity analysis is obviously easier if, for example, the mother genotype is known and if only a limited number of possible fathers exist. Highly variable polymorphic markers such as microsatellites are most informative for this purpose. If only a few putative fathers have to be taken into account, paternity can be determined by exclusion. Otherwise, indirect estimation procedures are needed. For recent reviews on kinship analysis with molecular markers, see Blouin,[143] Jones and Ardren,[675] Neff,[977] and Van de Casteele.[1440] Several computer programs have been developed, often with animal subjects in mind, such as Identity, IDENTIX, Delirious, MER, Relatedness and Kinship, CERVUS, PARENTE, and FaMoz (see Appendix 3). The latter program allows the use of either codominant or dominant (and also plastid) markers, whereas the other programs mainly capitalize on codominant data.

5.6.6 Migration and Hybridization

Measures of population differentiation allow one to estimate the migration rate between populations, assuming that these populations are in equilibrium (e.g., no selection, identical mutation rates, and generation time). In a diploid organism, the relationship between population differentiation F_{ST} and migration rate is given by the following equation[1571]:

$$F_{ST} = \frac{1}{1 + 4Nm} \tag{5.33}$$

Here, N is the population size, m is the migration rate, and Nm is the number of individuals migrating per generation.[1571] This equation can also be expressed as:

$$Nm = \frac{1 - F_{ST}}{4F_{ST}} \tag{5.34}$$

Real populations, however, are unlikely to fulfill all of the assumptions listed above, and Nm values are therefore only rough estimates.[1540]

In plants, migration rates actually correspond to gene flow through seeds and pollen. Both can be monitored separately, if F_{ST} values are obtained from biparentally inherited (nuclear) as well as uniparentally inherited (plastid) markers (see also Chapter 6.3.2.4). Given that chloroplast DNA is maternally inherited in angiosperms and paternally inherited in conifers (with some notable exceptions), markers in the chloroplast genome are generally recommended for this purpose.[404] Chloroplast microsatellites are particularly useful in this respect, because they tend to be polymorphic within species (reviewed by Provan et al.[1112,1114]; see also Chapters 2.3.4.2, 6.3.2.4, and 6.5.1). With caution, migration rates estimated from a haploid organism (or a uniparentally inherited marker) can be calculated from the F_{ST} for the haploid marker:

$$F_{ST} = \frac{1}{1 + 2Nm} \tag{5.35}$$

The computer program TWOGENER is designed to calculate gene flow through pollen, taking spatial distribution of plants into account, but without determining paternity of offspring.[55]

For the identification of recently migrated individuals and for pinpointing their most likely origin, an assignment test can be undertaken.[1504] Such a test can also determine the number of markers needed to discriminate between populations and to provide estimates of how diverged are the populations. This is achieved by calculating how often an accession or sample can be linked to its population or region of origin. Assignment tests can also be applied if hybridization or introgression between two taxa is suspected. Individual genotypes can then be tested and placed in the correct taxon. Several computer programs can assign samples to the most likely population of origin; e.g., Arlequin, Structure, NewHybrid as Genetix, GENECLASS, and Assignment calculator Doh (see Appendix 3).

5.6.7   Gene Flow, Isolation by Distance, and Spatial Structure

In the absence of barriers, populations that are close together, in general, will experience more gene flow than more distant populations, thus leading to a positive correlation between genetic and geographic distances. This effect is known as **isolation-by-distance (IBD)**. The scale at which IBD occurs is strongly dependent on the characteristics of the respective species.

A Mantel test[877] is commonly used to detect a possible correlation between geographic and genetic distance. The original Mantel test involves two distance matrices that are multiplied element by element. The sum of these products is tested against a value expected under the assumption of no association between the two matrices. The testing is usually performed by either assuming a normal distribution (as suggested by Mantel[877]) or by constructing a null distribution with a Monte Carlo procedure (rows and columns of one of the matrices are permutated randomly). A multiple correlations extension has been described by Smouse et al.,[1301] in which Mantel's cross-product statistics are normalized into a Pearson's product-moment correlation coefficient, r_M. A major problem with the Mantel test is that only linear relationships are tested. Mantel tests can be calculated with many computer programs such as Arlequin, NTSYS, and Genetix (see Appendix 3).

When populations are sampled on a small scale and the relative position of every individual in space is known, spatial autocorrelation analysis can be performed.[1304] For multilocus data, a matrix of genetic distances is first calculated using, e.g., Jaccard's similarity index. Then Mantel's Z statistics are computed between the genetic distance matrix and a series of different geographic distance matrices. Genetic distances in one geographic distance class at a time are then compared with those among all investigated observations. Positive autocorrelations are obtained for observations that are less distant than the mean genetic distance, whereas negative autocorrelations hold for pairs of observations that are instead more distant than the mean genetic distance. When using codominant, multiallelic microsatellite markers, Moran's I can be calculated.[1340]

For a review of different ways to measure spatial structure in plants, see Escudero et al.[412] The computer program SPAGeDi[572] calculates the most commonly used autocorrelation measure (Moran's I), whereas PSAWinD combines three spatial indices, i.e., Moran's I, SND,[1304] and NAC.[122]

5.7 PHYLOGEOGRAPHY AND NESTED CLADE ANALYSIS

Phylogeographic studies aim at unraveling the historical processes that led to the current distribution of a species and therefore often deal with intraspecific phylogenetic information[58,59,850] (for examples of such studies, see Chapter 6.5). In general, non-recombining markers such as mitochondrial and chloroplast markers are most appropriate, although in plants a problem with insufficient variation of the chloroplast genome exists[1229] and nuclear phylogeographies have been attempted successfully[1010,1011] (see Chapter 6.5.2). In general, networks are considered more appropriate for phylogeographic analyses than trees.[1090] Such networks can be generated from binary sequence or marker data, e.g., by the computer program TCS.[260]

Templeton et al.[1380–1382] developed a method called **nested clade analysis (NCA)**, which simultaneously analyzes genetic structure in space and time by mapping geographic information onto phylogeographic networks. An NCA analysis can be performed with the computer program GeoDis.[1091] Although NCA has become the standard procedure for analyzing phylogeographic data, it has been criticized because it may result in false inferences and does not allow rigorous statistical testing.[728] An improved method integrating intra- and interspecific phylogeographic inference is now available.[1381]

The analytical tools of phylogeography are expanding as rapidly as their applications, and a detailed treatment of the topic goes beyond the scope of this book. For further information, the reader should consult the book by Avise,[58] and two special issues of *Molecular Ecology* [i.e., volumes 7(2), 1998, and 13(4), 2004].

5.8 STATISTICAL TESTING OF HYPOTHESES: ANALYTICAL AND COMPUTATIONAL METHODS

Population genetic data are used to calculate estimates of genetic variation and differentiation, as described in the previous sections. Most of the computer programs employed for these purposes will also test a certain hypothesis; e.g., whether an estimate is significantly different from zero. There are many ways to accomplish this, and the statistical, analytical, and computational methods to perform such tests are becoming more and more sophisticated and complex. A description of these methods is beyond the scope of this book. For more information, see the reviews of Beaumont and Rannala,[92] Beebee and Rowe,[99] Holsinger et al.,[607] Rousset and Raymond,[1190] Shoemaker,[1288] and the special issue of *Molecular Ecology* [13(4), 2004].

In most computer programs, hypothesis testing can be carried out with traditional statistical tests, such as the t-test and analysis of variance (ANOVA). In cases for which the distribution of variation is unknown, resampling methods such as bootstrapping may be used to estimate confidence intervals. Other methods include randomization (permutation) tests, in which a large number (often >1000) of randomly reordered data are generated, and the test statistics of this new set are compared with the value obtained for the original data to set confidence intervals.

Maximum likelihood methods for testing the probability of obtaining a particular data set, given specific assumptions, are now more accessible because computing power is less limiting. The usefulness of Bayesian statistics for genetic analyses is still under discussion, but it is now becoming an alternative way of testing multiple hypotheses using prior information.[92,1288] Markov Chain Monte Carlo (MCMC) simulations efficiently explore probabilities of a particular state expected under a particular hypothesis.

6

Applications of DNA Fingerprinting in Plant Sciences

In this chapter we present a survey of the various applications of polymerase chain reaction (PCR) -based DNA fingerprinting, especially in the fields of genotype identification, population genetics, plant systematics, and phylogeography. Because of space limitations, only a few representative examples can be portrayed from each field of applications. For a closer description of the laboratory procedures, see Chapter 4. For statistical evaluation of DNA fingerprints and calculations, see Chapter 5. Linkage analysis and genetic mapping are treated in Chapter 7, and the most commonly used DNA profiling methods are compared in Chapter 8.

6.1 A BRIEF HISTORY OF DNA FINGERPRINTING

6.1.1 Minisatellite and Oligonucleotide DNA Probes Detect Genetic Variation

Based on Southern blot analysis and the restriction fragment length polymorphism (RFLP) technique, the so called DNA fingerprinting methodology was first introduced to plant genome analysis in 1988. The initial experiments used either the M13 repeat probe discovered by Vassart et al.[1456] or the human minisatellite probes 33.6 and 33.15 developed by Jeffreys et al.[662,663] (see Chapters 1.2.1 and 2.2.3.1). In the very first DNA fingerprint report dealing with plants, Ryskov et al.[1204] demonstrated DNA fragment pattern differences between two varieties of barley (*Hordeum vulgare*), following Southern blot hybridization of *Hae*III-digested DNA samples with the M13 probe. The same probe was also used by Rogstad et al.[1186] to generate DNA fingerprints from a panel of gymnosperms and angiosperms. In a third article appearing in the same year, Dallas[295] applied the human 33.6 minisatellite probe to distinguish rice cultivars.

In 1989, synthetic oligonucleotides that recognize simple repetitive DNA sequences (i.e., microsatellites; see Chapter 1.2.2) were introduced to plant DNA fingerprinting.

Weising et al.[1521] showed that polymorphic DNA fragment patterns were produced when restriction-digested barley or chickpea DNA was separated by agarose gel electrophoresis, and the dried gels were hybridized with a radiolabeled (GACA)$_4$ or (GATA)$_4$ probe. Numerous articles were published subsequently on this so-called oligonucleotide fingerprinting approach in plants, showing that the level of detected variation was highly dependent on the chosen probe.[129,329,1277,1522,1523] For reviews on plant DNA fingerprinting with mini- and microsatellite-complementary hybridization probes see Nybom,[997,998] Weising et al.,[1522] and Weising and Kahl.[1519]

6.1.2   PCR-Based Methods Enter the Stage

The introduction of PCR-based methods constituted a new milestone in the field of DNA fingerprinting. Two methods using primers with arbitrary sequence were published in 1990,[1527,1546] and a third one was published in 1991.[201,202] Arbitrary primers were shown to generate anonymous PCR amplicons from genomic DNA, resulting in polymorphic banding patterns after gel electrophoresis and staining. The random amplified polymorphic DNA (RAPD) approach developed by Williams et al.[1546] has become the best known variant of this prototype of PCR-based DNA profiling (see Chapter 2.3.3).

Only a few years later, a promising new method, coined amplified fragment length polymorphism (AFLP) analysis, was presented by Zabeau et al.[1605] and Vos et al.[1481] (see Chapter 2.3.8). This method incorporated elements of both RFLP and RAPD and, although technically more demanding than RAPD, produced very high numbers of polymorphic bands. A third group of PCR-based DNA profiling techniques guides the PCR amplification to certain types of (mostly repetitive) DNA, without the need to develop species-specific primers. This type of approach is best exemplified by the inter-simple sequence repeat (ISSR) -PCR technique presented by Zietkiewicz et al.[1621] (see Chapter 2.3.5).

The rapidity with which large numbers of samples can be processed made PCR-based methods increasingly popular. AFLP, RAPD, and ISSR are still broadly used, although RAPDs in particular have sometimes been criticized for problems with reproducibility and competitive priming (see Chapters 2.3.3.2 and 4.4.2). These problems are less pronounced for AFLP, which is currently regarded as the method of choice when high numbers of bands are desired. All three methods usually arrive at very similar estimates of genetic diversity and genetic distances, when applied to the same plant material. However, the detected loci are mostly biallelic (a band is present or absent), and initial attempts to distinguish hetero- from homozygotes by band intensity have largely been abandoned. Consequently, the bands generated by these multilocus techniques must be treated as dominant markers, which reduces their potential for use in population genetics and in-depth genetic analyses.

For reviews of the early work on PCR-based methods with arbitrary primers, see Newbury and Ford-Lloyd,[987] Rafalski and Tingey,[1128] Tingey and del Tufo,[1402] Waugh and Powell,[1509] and Williams et al.[1547] More recent reviews were provided by Doré et al.,[359] Nybom,[1000] Nybom and Bartish,[1001] and Wolfe and Liston.[1558]

6.1.3 Microsatellite DNA Analyses Yield Codominant Markers

PCR with primers complementary to the DNA sequences flanking hypervariable microsatellites were introduced in 1989,[829,1298,1367,1513] and became increasingly attractive in animal genetics in the early 1990s. The availability of locus-specific, codominantly inherited bands with high levels of polymorphism soon prompted botanists to explore the potential of this approach for the genetic analysis of plants (see Chapter 2.3.4). The feasibility of PCR amplification of microsatellites in plants was first demonstrated in soybean in 1992.[14] Microsatellites soon proved to be excellent tools for discriminating between plant genotypes, for population studies, gene tagging, and linkage mapping. The major drawbacks of microsatellite markers are the time and cost involved with developing species-specific primer pairs[1322] (see Chapters 4.8.4 and 4.8.5). Fortunately, several studies have demonstrated the possibility of microsatellite marker transfer to congeneric species, or occasionally, even to other genera (e.g., Arnold et al.,[45] Peakall et al.[1054]; see Chapter 4.8.4.3).

6.1.4 Universal Organellar DNA Primers Produce Uniparental Markers

A series of PCR primers were developed beginning in the early 1990s that allow the amplification of cpDNA and mtDNA introns and intergenic spacers in a wide array of plant species.[326,371,372,1358] Because of the conservation of binding sites within coding regions, many of these primer pairs are universal; i.e., they are transferable across species, genera, and even families. PCR products amplified with organelle-specific primer pairs are either sequenced directly (see Chapter 2.2.2) or digested with restriction enzymes in the so-called cleaved amplified polymorphic sequences (CAPS) approach[738] (see Chapter 2.3.2). The lack of genetic recombination of the chloroplast genome allows the combination of polymorphisms observed at several loci to form a so-called haplotype. Haplotypic variation can then be exploited, e.g., for phylogeographic analyses (see Chapter 6.5).

Although the first universal organellar DNA primers targeted previously identified genes, a more recent approach aimed at amplifying chloroplast microsatellites.[1092,1093,1520] These appear to be less variable in repeat number than their nuclear counterparts,[1113] but still reveal considerable intraspecific variation (see Chapter 2.3.4.2 and Provan et al.[1114] for a review).

6.2 GENOTYPE IDENTIFICATION

The ability to identify individual plants is at the core of many applications, of which only a selection can be highlighted below. In many cases, it is essential to find a method that can discriminate all of the sampled genotypes from one another. If we assume that a genotype arises from the fusion of two generative cells (e.g., an egg and a pollen cell), then each genotype will be different, with the possible exception of individuals belonging to the same, highly inbred line.

6.2.1   Individual-Specific DNA Fingerprints

Most of us are aware of the importance of individual-specific human DNA finger-prints in forensics, which, for instance, help to identify individuals present at a crime scene or involved in an immigration dispute. There are also examples where an individual-specific DNA fingerprint is required for tracing the origin of a particular plant sample. In **forensic botany**, samples of plant material are used to solve criminal cases, and plant DNA fingerprints have been used as evidence to link the individual, on whom the plant material was found, to a crime scene.[279,742,932a]

When a decision needs to be made about whether two particular samples are identical to each other, then the overall variation in the group of origin (e.g., taxon) must first be established. In principle, many approaches from allozyme electrophoresis to various DNA-based methods can be used in forensic botany. RAPDs and AFLPs are most suitable in this respect, given that they are relatively inexpensive and easily applied to any unknown organism without prior knowledge of DNA sequences. For example, Congiu et al.[270] was able to detect illegal growing of a patented strawberry variety based on a data set derived from only six RAPD primers. Also using RAPDs, Korpelainen and Virtanen[742] found out that two samples of the vegetatively propagating moss species, *Brachythecium albicans* and *Ceratodon purpureus*, most likely originated from a crime scene.

If, however, the DNA is degraded, neither RAPDs nor AFLPs will perform well. In such cases, microsatellite DNA loci, with their small PCR fragment size, are better suited. Microsatellite markers have the additional advantages of being species-specific (i.e., insensitive to contamination by foreign DNA) and highly reproducible. Moreover, results from microsatellite analyses are easily managed and compared in databases. In a study on *Cannabis sativa*, Gilmore et al.[497] demonstrated the potential of microsatellite analysis for forensic investigations. According to an analysis of molecular variance (AMOVA), 25% of the total genetic variation existed between accessions, and 6% existed between the two major *C. sativa* groups used for fiber and drug production, respectively. These results showed that microsatellite DNA fingerprinting might aid in determining agronomic type, geographic origin, and production locality of these clonally propagated drug crops.

Exploiting the high sensitivity of modern PCR analysis, it is even possible to determine which tree is the cause of subsidence of a building. Roots from the foundations of the subsiding building and the above-ground parts of neighboring trees are collected and analyzed. If the trees in question belong to different species, it is sufficient to sequence a variable part of the chloroplast genome (such as the *trn*L intron) to unequivocally assign a root to the correct species. If the trees that purportedly caused the damage belong to the same species, microsatellite or RAPD analysis are methods of choice. Figure 6.1A shows the results of an RAPD analysis of three ash trees (*Fraxinus excelsior*) and one root. Triplicate extractions of the root were used to check for artefacts in the RAPD pattern. Figure 6.1B shows an electropherogram of two microsatellite loci amplified from genomic DNA of three oak trees and two roots, clearly identifying which tree belongs to which root (Wolff, unpublished data).

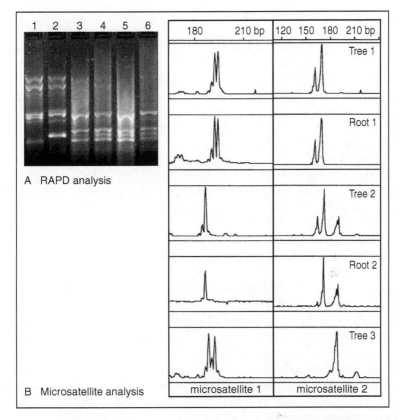

Figure 6.1 (A) RAPD analysis of a building subsidence case. DNA aliquots from three candidate ash trees (*Fraxinus excelsior*; lanes 1 to 3) and the piece of root in question (three replicates; lanes 4 to 6) were amplified with the 10-mer primer OPA-9 (Operon, Alameda), resolved on an agarose gel and stained with ethidium bromide. The RAPD patterns clearly assign the root to tree 3. (B) Microsatellite analysis of two roots and three candidate oak trees (*Quercus robur*) at two loci. Microsatellite markers were resolved on an automated sequencer. The densitograms indicate the assignment of root 1 to tree 1, and of root 2 to tree 2, respectively.

Not only nuclear, but also chloroplast microsatellites (cpSSRs) proved to be helpful in determining the origin of particular plant samples from a number of possible source populations. One example relates to the tracing of the forest from which certain logs were harvested, up to several centuries earlier. Deguilloux et al.[319] developed a protocol for successful extraction and amplification of DNA from wood, and showed that the geographic origin of oak wood can be determined by comparison with a database comprising all cpSSR haplotypes found in oak forests across Europe[321] (see also Chapter 6.5.1). Claims about the origin of wood that is used for, e.g., furniture or wine barrels, can thus be verified or refuted.

6.2.2   Cultivar Identification

The precise, fast, cost-effective, and reliable identification of important plant cultivars is essential in agriculture and horticulture as well as for practical breeding
purposes and related areas such as plant proprietary rights protection. Traditional
methods of cultivar identification frequently are based on the evaluation of sets of
morphological characteristics. Although it is usually cost-effective, morphological
assessments may have their limitations, including (1) insufficient variation among
cultivars (especially if the cultivars to be compared share a closely related pedigree),
(2) subjectivity in the analysis, (3) influence of the environment and management
practice, and (4) expression of some characters only in certain developmental stages.
These considerations triggered the exploration of alternative means of cultivar identification, including allozyme analyses, cytogenetics, analysis of secondary metabolites, and DNA profiling.[209,210,359,789,947]

6.2.2.1   Patenting and Protecting Plant Varieties

A breeder can be granted intellectual property rights for a new plant variety called
plant breeders' rights (PBR). When a breeder has PBR for a certain cultivar and
a certain region, for example, the United Kingdom or Europe, he or she can charge
royalties on sales of the material used for propagation. In Europe, these rights can
be obtained from the Community Plant Variety Office (CPVO). Two other organi-
zations involved with the legal arrangements to protect members from infringement
of their rights are the International Union for the Protection of New Varieties of
Plants (Union Internationale pour la Protection des Obtentions Végétales [UPOV])
and the International Seed Federation (ISF).

The criteria for obtaining PBR include the passing of a so-called **distinctness–uniformity–stability (DUS) test**.[209,210] The new cultivar must be distinct from
all other cultivars already described, it must be uniform so that all propagated
individuals are as similar as possible, and it must be stable so that it stays true to
its specific description. The DUS test is carried out as an observation trial, lasting
one to several years, during which primarily morphological characteristics such as
flower color, height, bushiness, etc., are recorded. The large number of new cultivars
produced requires that a huge number of trials must be undertaken, in which the
new cultivars, as well as older but similar cultivars, are analyzed and compared.[840,1450]
The identification of sufficient distinctness from cultivars in the reference collections
is becoming a problem in the major horticultural and agricultural crop species.
Therefore, the potential use of molecular methods for cultivar description and identification is now being researched, e.g., by the UPOV Working Group on Biochemical
and Molecular Techniques.

One of the main problems with molecular methods is that a minimum genetic
distance between varieties must be defined to protect the breeders of already existing
cultivars. Clearly, a single base pair difference does not warrant the registration of
a new cultivar.[210] Two important ways in which molecular markers could aid DUS
testing are (1) determining distinctness and (2) specifying which existing cultivars
should be included in a reference test set.[782]

New varieties are sometimes derived from already-protected varieties by repeated backcrossing, by genetic engineering, or as a selection from natural or induced mutations or from somaclonal variants. These are called **essentially derived varieties (EDVs)**. EDVs usually look very similar to the protected variety but are sufficiently different to be given a name of their own.[209] Potentially, molecular markers could aid in the recognition of EDVs by testing whether there is a certain level of conformity with the original variety.

Given that transformation and plant regeneration can cause somaclonal variation (see Chapter 6.2.3), expressed as changes in agronomic characters,[162] there is a certain possibility that molecular techniques would produce different DNA profiles for the original cultivar and the EDV. The chances are, however, very small. For example, Zhang et al.[1608] were unable to distinguish an EDV of transgenic sweet potato (*Ipomoea batatas*), containing an intron-β-glucuronidase (GUS) transgene, from the original clone by RAPD markers. On the basis of results from AFLP and microsatellite analyses of inbred lines of maize (*Zea mays*), Heckenberger et al.[589] recommended that specific EDV thresholds are created for marker systems with different degrees of polymorphism. Replicate laboratory assays are needed because even very low percentages of technically induced variation (e.g., due to incomplete digestion of DNA in AFLP analysis) can be difficult to distinguish from true genetic polymorphism. Heckenberger et al.[589] also stressed that precautions should be taken to warrant a high level of homogeneity and reproducibility for DNA markers before applying for plant variety protection (see also Chapter 6.2.4).

Eventually, DNA fingerprints may become part of a passport for well-described crop varieties. However, it is unlikely that they will ever fully replace morphological traits, given that distinctness for DNA markers (whatever their scale and measurement) may not necessarily reflect morphological distinctness. The accuracy of cultivar identification can be quantified by calculating the probability of finding identical fingerprints by chance.[67,635] For this type of estimation, a correct reference group and an associated database are necessary.

6.2.2.2 Choice of DNA Marker Method for Cultivar Identification

A number of comparative investigations have been carried out to explore which technique is most suitable and most reliable for cultivar identification (e.g., Jones et al.,[667] Powell et al.[1095]; see also Chapter 8). Any choice of DNA marker method depends, among other factors, on the scale and purpose of cultivar identification. For small tests comprising only a few samples, all of which are analyzed at the same time, reproducibility and documentation are not very important and any method providing sufficiently variable markers would be acceptable. Numerous studies in this direction have been carried out using multilocus methods such as AFLP, RAPD, and ISSR (e.g., Arnau et al.,[44] Bernet et al.,[123] Charters et al.,[228] Hu and Quiros,[620] Prevost and Wilkinson,[1097] Wolff et al.[1563]).

In one example, Prevost and Wilkinson[1097] analyzed 34 selected potato cultivars with ISSR. Banding patterns were resolved on silver-stained polyacrylamide gels and proved to be highly reproducible. The two most informative primers were each able to distinguish all 34 genotypes. The authors also introduced a new measure to quantify the value of a given PCR primer producing multilocus patterns for cultivar identification. This measure, called **resolving power** (**Rp**), allows the extrapolation of the results from small-scale experiments to large-scale studies. Rp considers both the number of polymorphic bands in a pattern and the informative value of each individual polymorphic band (Ib). Ib is maximal when a band is present in half of the samples examined, and absent in the other half. Rp equals the sum of the Ib values for each band across a multilocus pattern.

If cultivar identification is undertaken for an economically important crop with a large number of registered cultivars, and if DNA profiles for each cultivar will be stored in a database, then repeatability and comparability become very important issues. At present, microsatellites are probably the best tool for this task because they exist in large numbers of polymorphic alleles. The usefulness of microsatellite loci for cultivar identification, especially in vegetatively propagated crops, has been discussed extensively.[94,157,161,349,947,1202,1394] Microsatellites largely fulfill the four criteria originally set up by Bailey[70]: (1) maximal variation between cultivars, (2) minimal variation within cultivars, (3) environmental stability, and (4) experimental reproducibility.

Considering the high allelic diversity of microsatellites, a fifth point should be included: germline stability. Too high mutation rates would make any marker-based cultivar identification system useless. Many factors are known to influence the mutation rate of a microsatellite, including the length and sequence of the repeated motif, the biology of the organism, the chromosomal location, and the length of the allele (e.g., Jin et al.[671]; see Chapter 1.2.2.3). Reported mutation rates of plant, animal, and human microsatellites vary across a wide range, and the frequencies of microsatellite mutations to new length alleles have only rarely been studied in plants.[348,1400,1429] If possible, loci that exhibit relatively low mutation rates should be selected for plant cultivar identification.

Although microsatellites are very useful in general, they also have certain disadvantages, including the relatively high cost of marker development, the need for sophisticated laboratory protocols and equipment (see Chapter 4.8), the occurrence of problems with correct sizing,[439,550] and the occasional occurrence of artefactual amplification products.[161] Known standards should always be included because, for example, the change of the fluorescent tag of the primer, the use of a different detection instrument, or even a different room temperature may influence the exact size of generated fragments.

6.2.2.3 *Propagation and Reproduction of Crop Species*

Successful discrimination among different cultivars to a large extent depends on the commonly used means of propagation, as well as the mating system of the crop species under study. All plants belonging to a particular cultivar of a **vegetatively propagated crop** (e.g., apples, raspberries, carnations, chrysanthemums, roses, bananas, and potatoes) are expected to share identical DNA fingerprints, except for rare mutations. In **sexually propagated crops**, however, some genetic variation may persist also within cultivars, making DNA marker-aided delimitation more difficult. In addition, every seed production cycle inadvertently involves the introduction of genetic variation (e.g., due to recombination or by foreign pollen). This is a common risk in all outcrossing crops such as sugar beet, coconut (*Cocos nucifera*), and oilseed rape (*Brassica napus*). Beets (*Beta vulgaris*) are particularly difficult because they have a fairly high outcrossing rate and therefore require a large number of markers for differentiation of cultivars. In contrast, highly inbreeding crops such as wheat (*Triticum aestivum*) are less problematic because the cultivars are more homogeneous. DNA fingerprinting has proved very useful for discriminating wheat cultivars (reviewed by Gupta and Varshney,[543] Gupta et al.[544]).

Pooling (or **bulking**) of samples is sometimes undertaken for generating DNA profiles of sexually reproducing crops, especially if they are outcrossing. For soybean, Diwan and Cregan[348] suggested the analysis of bulks of 30 to 50 plants. In rye (*Secale cereale*), 10 seedlings per cultivar were pooled for RAPD analysis of 42 cultivars originating from 14 different countries.[864] The cultivars were first grouped into a winter rye and a spring rye group, respectively. Within these groups, cultivars clustered according to their geographic origin. Another example of successful DNA pooling was reported in a RAPD study of genetically heterogeneous rapeseed cultivars.[370] For calculations of the accuracy of parameters obtained from pooled samples, see Kraft and Säll.[745] As an alternative to bulking, several plants from each cultivar can be analyzed individually, and intra- and intercultivar variability are then partitioned statistically, e.g., by AMOVA[421] (see Chapter 5.6.2).

New cultivars can be difficult to discriminate in crops for which the major breeding method involves selection among seedlings originating from a small number of open-pollinated, widespread cultivars.[31,256] In a crop such as peach (*Prunus persica*), which is self-fertile and naturally self-pollinating, new cultivars are sometimes selected that have DNA fingerprints highly similar (or identical) to those of the seed parent.

Another crop with low levels of variation is lemon (*Citrus limon*); new cultivars originate mainly from somatic mutation and nucellar variation. Although they are highly heterozygous, the majority of 57 lemon cultivars analyzed with microsatellite markers remained indistinguishable.[541] Better discrimination was achieved by analyzing the same material with ISSRs. In another study, Bernet et al.[123] found no polymorphism between 13 lemon cultivars with ISSRs, but both RAPD and to an even higher extent, inter-retrotransposon amplified polymorphism (IRAP; see Chapter 2.3.5.4), could discriminate among some of the cultivars studied.

6.2.3 *In Vitro*-Propagated Plant Material and Somaclonal Variation

Tissue culture techniques are commonly used in plants as an efficient way to propagate and store valuable genotypes. Often, some of the regenerants differ from the parental type — a phenomenon called somaclonal variability. This variation is thought to originate either from the release of genetic diversity pre-existing in the explant or from *de novo*-acquired variability during cell line dedifferentiation or callus maintenance *in vitro*. Multiple mechanisms have been discussed, including mutations in chromosome structure and numbers, base substitutions in regulatory genes, alterations in the copy number of repetitive DNA, altered levels of DNA methylation affecting gene regulation patterns, and others (reviewed by Evans,[419]

Karp,[695] Larkin et al.,[780] Lee and Phillips[792]). Somaclonal variation is often expressed as a difference in phenotype, but as with sports (see Chapter 6.2.4), such variation is rarely detectable with DNA fingerprinting techniques.

In rice (*Oryza sativa*), somaclonal variation is well documented. When segregation ratios are analyzed with RAPD, both heterozygous and homozygous mutation events were detected; recessive mutations were more prevalent than dominant ones.[501] Matthes et al.[897] used several AFLP-based approaches to screen for somaclonal variation among morphologically normal and abnormal, tissue culture-derived plants of oil palm (*Elaeis guineensis*). No polymorphisms were found when standard AFLP analysis with *Eco*RI-*Mse*I digestion and 10 different primer combinations was applied. This is in agreement with a previous RAPD study, in which no polymorphisms between normal and abnormal plants could be detected either.[1176] In contrast, low levels of polymorphism were detected when methylation-sensitive restriction enzymes were used for the generation of AFLPs (methylation-sensitive amplified polymorphism [MSAP][1580]; see Chapter 2.3.8.5). The highest number of polymorphic bands (0.3%) was obtained with the combination *Eco*RI-*Hpa*II, followed by 0.04% with each of *Eco*RI-*Msp*I and *Pst*I-*Mse*I. These results showed that tissue culture can cause changes in DNA methylation rather than a change in genome architecture. However, no single polymorphism was consistently different between normal and abnormal clones.

A considerable number of RAPD polymorphisms (34 of 234 products) were found when normal plants were compared with micropropagation-induced dwarf off-types in two Cavendish banana cultivars.[299] One of the polymorphic RAPD fragments was consistently amplified in all 57 normal plants, but was absent in all 59 dwarf plants. Given that these dwarfs had been generated by several independent micropropagation events, this band had a high potential as a diagnostic marker for dwarfs. It was therefore cloned, sequenced, and converted into a sequence characterized amplified region (SCAR; see Chapter 2.3.3.4). PCR and Southern hybridization unexpectedly showed that this marker was chloroplast-encoded, and that normal and dwarf plants were distinguished by some sort of cpDNA rearrangement. Peraza-Echeverria et al.[1067] also investigated micropropagated banana plants, and were able to detect DNA methylation changes using the MSAP approach of Xiong et al.[1580]

In a detailed study on sugarcane, Taylor et al.[1375] found that the extent of somaclonal variation as detectable by RAPDs was strongly dependent on the tissue culture method used. Considerable variation was detected among protoplast-derived calli after prolonged tissue culture (>2 years), but only one single polymorphism was observed among transgenic regenerants derived from embryogenic calli.

PCR with unanchored microsatellite primers was used to screen for tissue culture-induced instability in cauliflower calli (*Brassica oleracea* var. *botrytis*).[799,801] Relatively high levels of marker variation among individual calli were observed, with up to (extrapolated) ~50,000 mutational events during the process of callogenesis.[801] Interestingly, the same technique did not reveal any substantial genetic differences between cauliflower plants regenerated via somatic embryogenesis.[800]

The examples given above show that somaclonal variation may or may not be detected by currently used molecular marker techniques. The numbers of markers associated with somaclonal variation rarely exceed 0.05% of the total number of characters scored. Thus, even if extensive marker analyses are performed, important variations underlying, e.g., dwarfism, leaf variegation, or albinism could easily be missed. Of particular interest are recent methodologies to detect correlations between DNA methylation and particular phenotypic changes.[1580] However, the question about whether specific changes in DNA methylation of certain genes are associated with particular off-types will require sequencing and molecular characterization of the differentially methylated regions. Finding a molecular marker that is associated with a trait of interest remains a difficult task.

6.2.4 Sports and Other Mutants

Spontaneously occurring somatic mutations affecting plant structure and productivity give rise to so-called sports. Some of these constitute an improvement of the original genotype and have therefore been registered and patented as varieties of their own. However, most sports deviate from the original cultivar only in minor characteristics (e.g., flower color), and may thus be very difficult to distinguish with DNA fingerprinting.

In most cases, DNA markers are unable to detect any difference between a sport and its original cultivar. One reason is that only a very small portion of the genome is sampled, even when a large number of markers are employed. Another reason is that the change in phenotype is often a consequence of **chimerism**. This means that there has been a mutation in only one of the three meristematic cell layers that differentiate into the various plant tissues, or that there has been a reversal of cell layers bringing a different layer to the forefront. For DNA fingerprinting, derivatives of all cell layers are extracted as a whole, and a mutation in only one layer would go unnoticed. For example, Fourré et al.[457] showed that RAPD analysis readily distinguishes **among** four different embryogenic clones of Norway spruce (*Picea abies*), but does not detect any polymorphism **within** these clones, despite considerable cytogenetic variation (i.e., trisomy, tetraploidy, mixoploidy, and chimerism).

Debener et al.[316] used fluorescent AFLPs to analyze several rose varieties and their associated sports. Although a large number of polymorphisms separated the different varieties from each other, sports from two cut rose varieties were indistinguishable from their original cultivar, and only five of more than 700 bands were different between a garden rose variety and its sports. This study showed that rose sports can easily be assigned to their parent variety.[316]

Numerous studies have been performed on sporting of grapevine (*Vitis vinifera*), employing most of the currently available DNA fingerprinting methods. Franks et al.[459] performed a detailed microsatellite study on the grapevine cultivar Pinot Meunier. They found that microsatellite loci regularly showed three alleles, although *Vitis vinifera* is a diploid species. This was explained by the occurrence of a mutation to a new allele in only one of the cell layers, so that the extract of the whole plant (all layers) showed the original two alleles present in most cells, plus the novel allele in one of the layers. By regenerating plants from two different cell layers, the authors were able to demonstrate the existence of periclinal chimerism. Plants regenerated from layers L1 and L2 did indeed show different microsatellite alleles as well as different phenotypic characteristics.

Regner et al.[1148] reported that both RAPD and microsatellite markers pick up variation among clones of the grapevine cultivar Riesling. Bellini et al.[106] found that AFLPs are suitable for clonal selection, identification, and certification of *Vitis* clones. However, repeated digestion of DNA samples revealed that some of the observed polymorphisms were of artefactual origin and were based on incomplete digestion of DNA. Imazio et al.[637] applied several DNA marker methods to the analysis of 24 Traminer accessions. AFLP markers distinguished 16 of these clones, even though the average similarity was as high as 97%. In contrast, the use of nine microsatellite loci did not show any polymorphisms. Possible epigenetic differences among the Traminer clones were also examined by MSAP[1580] to quantify the degree of methylation of CCGG target sequences. In this way, two more clones could be identified as having unique fingerprints.[637]

The arbitrary signatures from amplification profiles (ASAP) technique developed by Caetano-Anollés and Gresshoff[199] (see Chapter 4.4.2.1) proved to be an efficient tool to identify somatic mutants and radiation-induced sports in chrysanthemum.[1419] Four genotypes of *Dendranthema grandiflora* were indistinguishable when their genomic DNA was amplified with either of three octamer primers using a standard DNA amplification fingerprinting (DAF) assay (see Chapter 2.3.3). However, secondary PCR of the amplification products using four mini-hairpin decamer primers[198] produced signatures containing about 37% polymorphic loci. Each cultivar was clearly distinguished from the others by a set of unique bands.

To summarize, the chance of a mutation (whether somaclonal, spontaneous, or induced by irradiation or chemicals) being detected by molecular markers will depend on (1) the total number of mutations (which is determined by, e.g., the extent of mutagenic treatment or the duration of tissue culture), (2) the type of marker, (3) the total number of markers screened, and (4) genome size. It decreases with increasing genome size. Preferably, highly sensitive marker systems should be used where many genomic loci are evaluated simultaneously. Suitable markers with high multiplexing rates include RAPDs (including DAF and ASAP), ISSR, AFLPs, and their variants. In addition, the use of MSAP[1580] or a related technique should be considered as a means to estimate the extent of epigenetic variation associated with an altered phenotype.

6.3 GENETIC DIVERSITY

Genetic diversity is important for the survival of wild species and as a source of genetic variation for cultivated plants. This section provides a small set of examples out of the huge number of genetic diversity studies, conducted in both wild and cultivated plants.

6.3.1 Variation and Relatedness among Cultivars

The analysis of plant cultivars with molecular markers often follows a common schedule. In the first phase of the investigation, different types of markers are tested for their ability to identify a cultivar unequivocally, and/or to discriminate between closely related cultivars (see Chapter 6.2.2). The second phase then usually aims at estimating the levels of relatedness among these cultivars (sometimes also including wild progenitors in the study). In some studies in this direction, the only objective is to assign the most likely parents from a selection of candidates (see Chapter 6.3.3 about hybridization). In other cases, the intention is to reveal relationships across the whole set of cultivars available for the study. After the DNA profiles of each cultivar are recorded, a genetic distance matrix is usually calculated. This matrix is then used for a cluster analysis (most commonly based on the unweighted pair group method using arithmetic average [UPGMA] or on the neighbor-joining algorithm), and the results are depicted in a dendrogram (see Chapter 5.5.2). Sometimes an ordination method such as principle coordinates analysis may be more appropriate (see Chapter 5.5.1), especially if intercultivar hybridizations have occurred. The obtained estimates can be used, e.g., to identify cultivars that are most distant to each other in a germplasm collection (see Chapter 6.3.4), or to determine which cultivars should preferentially be selected for a core collection (see Chapter 6.3.5.2).

Whether all available DNA markers or only a genomically well-dispersed subset should be used for such relatedness studies has been discussed.[747] In a study on sugar beet, Kraft et al.[747] found that little was gained by considering the map position of DNA markers in fingerprinting applications. Thus, dendrograms based on a UPGMA analysis of nine sugar beet lines were almost identical, whether they were derived from a carefully chosen subset of 92 mapped markers or from the same number of randomly picked markers. Similarly, Le Clerc et al.[784] found randomly chosen markers to be just as efficient as a set of markers that had been carefully chosen for their even genomic distribution using a linkage map in a study on carrot (*Daucus carota*).

A number of studies examined the relationships between present-day wild populations and domesticated plant material of the same species. The patterns revealed by DNA fingerprinting data can provide information on the origin of the cultigens and the degree of differentiation between wild and domesticated genotypes. On the basis of an AFLP and ISSR data set, Bradeen et al.[159] found that cultivated carrot and wild *Daucus* accessions clustered separately. In Ethiopia, false banana (*Ensete ventricosum*) is an ancient crop plant, and although large-scale plant breeding has never taken place, an RAPD study demonstrated that all analyzed cultivars were strongly differentiated from material collected in the currently very restricted wild populations.[135] Conversely, cultivars of the recently domesticated crop lingonberry (*Vaccinium vitis-idaea*) were completely intermingled with samples from wild populations in an RAPD study.[478]

When available, pedigree information for a set of cultivars can be compared with DNA marker-derived estimates of their relatedness. Sefc et al.[1267] used 32 microsatellite markers to fingerprint grapevine cultivars. Using the computer program Identity, they discovered some genetic relationships that were not anticipated, which led to the reconstruction of pedigrees for several cultivars. In a subsequent study, only nine microsatellite markers were sufficient to group grapevine cultivars according to their region of origin.[1268] However, assigning these cultivars to their region of origin was only effective for the most well-differentiated regions, namely Austria and Portugal.

Using AFLPs and microsatellites, Almanza-Pinzón et al.[24] analyzed a set of 70 spring wheat accessions and calculated the interaccession similarities. A DNA marker-derived distance matrix was significantly correlated with a matrix obtained by calculating the **coefficient of parentage (COP)** from registered pedigrees. The fact that the correlation was higher between AFLP and COP than between microsatellites and COP is probably due mainly to the larger number of AFLP bands evaluated. In addition, overall similarity among cultivars was much higher using the DNA markers than the COP calculations. Probably the marker-derived estimates were more correct, given that the COP calculations depend on some unrealistic assumptions, such as absence of selection.[24]

Most DNA-based studies on variation and relatedness among cultivars are carried out on nuclear DNA, but there have also been some interesting results from analyses of cpDNA. For example, Provan et al.[1110] demonstrated a severe cytoplasmic DNA bottleneck in the history of modern European potato cultivars. When the data were combined from seven polymorphic cpDNA microsatellites, 26 haplotypes resulted among 178 *Solanum tuberosum* accessions, which together represented 95% of the cultivars available in the United Kingdom. One single dominant haplotype was found to be present in 151 of 178 individuals. Interestingly, the apparent lack of diversity among chloroplast haplotypes was not paralleled by decreased levels of nuclear diversity, as evidenced by nuclear microsatellite analyses of the same accessions.

6.3.2 Analysis of Population Genetic Diversity and Its Distribution

The extent of genetic variation in a species and its distribution among and within populations is determined by a large number of factors, such as the breeding system, historical events (regarding, e.g., habitat availability and immigration, population size, migration between populations), and many ecological factors. The influence of a broad range of life history traits on genetic diversity of plant species has been described in a review article by Loveless and Hamrick[846] (1984) and later by Hamrick and Godt.[562-564] These reviews were based on a large number of allozyme studies. More recently, Nybom and Bartish[1001] compiled a total of 106 RAPD-based studies and described the effects of several life history characters and sampling strategies on genetic diversity estimates. An additional article by Nybom,[1000] based on 307 DNA marker studies, complemented the previous reviews by incorporating a wider range of DNA marker methods. The general picture arising from these comparisons is that long-lived, outcrossing, and late successional taxa retain most of their variation within populations, whereas annual, selfing, and early successional taxa allocate more variation among populations. Within-population diversity, in general, is negatively correlated with the level of population differentiation.

6.3.2.1 Choice of Molecular Marker Method for Population Genetics

Given that population genetics can be studied at a wide range of scales and with different questions in mind, the choice of marker system is important (see also Chapter 8). Thus, markers based on slowly evolving DNA sequences are adequate for the analysis of historical events on longer time scales, whereas markers derived from fast-evolving sequences are more suitable for analyzing recently diverged populations. Both dominantly (e.g., AFLP, RAPD, and ISSR) and codominantly inherited markers (e.g., allozymes and microsatellites) have been used to study population structure. The overall patterns regarding the extent and partitioning of genetic variability appear to be quite similar, regardless of marker type, provided that the numbers of analyzed markers are sufficiently high.[1000] One exception is noteworthy: data sets based on allozyme and microsatellite variation often reveal a positive correlation between geographic range and within-population diversity, whereas RAPD-based data sets do not.[1000]

When using microsatellite markers, the appropriate mutational model must be taken into consideration. The theoretical framework for allozyme-based population genetics assumes that any new allele created by mutation is unrelated to the ancestral allele (**infinite alleles model [IAM]**). In microsatellites, however, a majority of mutations may be caused by slipped-strand mispairing during replication,[806] resulting in small gains and losses in repeat copy number, rather than in large changes (see Chapter 1.2.2.3). This type of mutational behavior may be better explained by a **stepwise mutation model (SMM)**. The basic idea behind the SMM is that mutations predominantly differ from their previous state by the change of a single repeat unit. This type of mutational process results in a unimodal distribution of allele sizes. Di Rienzo et al.[338] presented a modification of the model (two-step stepwise mutation) that fitted patterns of variation at microsatellite loci quite well. On the basis of the SMM, Goldstein et al.[508] and Slatkin[1297] independently proposed a method to evaluate genetic distances between microsatellite loci that includes allelic repeat score (see Chapters 5.6.2 and 5.6.3). Goldstein et al.[508] showed that these distances are a linear function of time, whereas allele sharing and Nei's distance level off asymptotically. As a result, distances based on the SMM were considered more appropriate for taxa that are sufficiently diverged, whereas IAM should be more adequate for intraspecific comparisons.[508]

The two mutational models are sometimes compared using the same material. For example, Todokoro et al.[1404] examined the relatedness among Japanese *Arabidopsis* populations. Distance matrices were calculated from pairwise comparisons of microsatellite alleles, using either the proportion of shared alleles (IAM), or the average size difference between alleles (SMM) as a criterion for relatedness. The two distance matrices yielded dendrograms with similar topologies, which partially reflected the geographic origin of the populations. In an extension of this study, a worldwide sample of 42 *Arabidopsis* ecotypes was analyzed.[638]

The observed number of alleles was between the values expected for SMM and IAM. No association between ecotype and geographic origin could be found in the Japanese populations.

The uniparentally inherited chloroplast genome behaves as a single, haploid character, and the effective population size for cpDNA markers is therefore only half that of nuclear (diploid and biparentally inherited) markers. Consequently, differentiation due to genetic drift takes place much faster for cpDNA markers than for nuclear markers. Because of their intraspecific variability, chloroplast microsatellites are a useful tool for studying genetic structure at a species-wide scale. They evolve faster than chloroplast nucleotide sequences and, because of the small effective population size, differentiate faster between populations than do nuclear microsatellites (see Chapters 2.3.4.2 and 6.5.1).

6.3.2.2 Influence of the Breeding System on Genetic Diversity

A broad range of studies from many plant genera indicated that the breeding system is one of the most important determinants of genetic diversity and its distribution (reviewed by Hamrick and Godt,[562–564] Loveless and Hamrick,[846] Nybom,[1000] Nybom and Bartish[1001]) In general, self-fertilizing species allocate a larger proportion of their variation among populations than within populations. In contrast, outcrossing species have a relatively higher within-population component of genetic variation.

Dioecious species have separate female and male plants and are therefore obligate outcrossers. Sea buckthorn (*Hippophae rhamnoides*) is a dioecious, wind-pollinated pioneer tree with a severely fragmented distribution. RAPD analyses were performed on samples from 10 North European populations to estimate the extent and distribution of genetic variability.[82] Within-population gene diversity (H_s) proved to be relatively low for an outcrosser but rather typical for early successional taxa. Only 15% of the variability was allocated among populations, indicating low levels of population differentiation, as expected in outcrossing species. There was a tendency for island populations to be somewhat more differentiated, and to have less within-population diversity than mainland populations, perhaps due to fragmentation.

In the genus *Plantago*, several species with different mating systems have been studied using allozymes, RFLP fingerprinting, RAPDs, and microsatellites.[553,948,1320,1560,1562] Considerable concordance was noted among studies. Thus, the self-incompatible *Plantago lanceolata* generally exhibited a high proportion of genetic variation within populations, whereas the largely selfing *Plantago major* and *Plantago intermedia* had much lower variation within populations, but higher variation among populations. *Plantago coronopus* has a mixed mating system, and took an intermediate position according to both allozymes and RFLP fingerprinting. RAPD markers were used for a more detailed study of the two selfing taxa.[948] *Plantago intermedia* had an average outcrossing rate of only 3 to 6%, which is slightly lower than the 10 to 14% found in its sister species, *P. major*. These relatively small differences were nevertheless accompanied by very different population structures. Thus, population-specific clusters and high population differentiation were observed in *P. intermedia* ($F_{ST} = 0.78$), whereas less well-defined clusters and only moderate differentiation ($F_{ST} = 0.23$) were found in *P. major*.

Awadalla and Ritland[60] first applied microsatellite DNA analysis to closely related taxa with different mating systems. High levels of diversity were found in the monkey flower species complex (*Mimulus guttatus*), which includes both outcrossing and selfing taxa. The selfing taxa, however, did show a more pronounced population differentiation than did the outcrossing taxa. Inbred populations also seemed to suffer from recent bottlenecks, given that they exhibited less variation than expected due solely to the effects of inbreeding.

DNA marker-based evidence of an **isolation-by-distance (IBD)** population structure has been reported for a large number of outcrossing plant species (e.g., Gabrielsen et al.,[468] Graham et al.,[518] Le Corre et al.[785]) but only occasionally in selfing species,[1406] in which a lack of IBD appears to be more common (e.g., Fahima et al.[424]). Great care needs to be taken when analyzing a possible IBD scenario because among-population diversity, whether measured with RAPD (F_{ST} and G_{ST}) or with microsatellites (F_{ST} and R_{ST}), often shows a positive association with maximum geographic distance between the sampled populations.[1000] When the association between collection distance and among-population diversity was analyzed for RAPD data in separate subsets for outcrossing and selfing taxa, a strong positive relationship was detected only for outcrossers.[1001] Obviously, the correlation between genetic and geographic distances is much more pronounced in outcrossing species as a consequence of higher levels of gene flow.

6.3.2.3 Clones and Ramets

To determine the dynamics in vegetatively propagating plant populations, information about the size and relatedness of clones is essential. DNA fingerprinting has revealed the extent of clonal growth in many species.[411,605,678,1088,1331,1349] The type of marker used is largely irrelevant provided that the discriminatory power is sufficiently high (see also Chapter 6.2.4 on sports and other mutants).

In many vegetatively reproducing plant species, DNA marker analyses have demonstrated the existence of clones that are considerably larger (and therefore often also older) than expected from previous data. For example, Steinger et al.[1331] used RAPDs to study *Carex curvula*, a slow-growing rhizomatous sedge found in the European Alps. The plants propagate predominantly through clonal growth. RAPD analysis of 116 tillers from a small patch (2.0×0.4 m) identified a total of 15 multilocus genotypes, each discriminated from other clones by 16 to 39 markers. More than half of the sampled tillers appeared to belong to a single, large clone, which, according to present-day distribution patterns and known growth rate, must be ~2000 years old. In a similar study, Jonsdottír et al.[678] analyzed growth rate and RAPD-based separation of clones and genets in the Siberian sedge *Carex ensifolia* subsp. *arctosibirica*. Two clones were identified — each was well over 3000 years old.

Apparently, one of the most widespread clones worldwide has emerged from the invasive Japanese knotweed (*Fallopia japonica*). RAPD analysis was performed on 150 British samples and 16 samples from the remainder of Europe and the United States.[605] Ten primers produced a total of 108 reproducible fragments, but no variation was observed between any of these samples. In addition, all samples were male-sterile, supporting the conclusion that they are part of the same large clone. Xu et al.[1581] used RAPDs to study genetic variation in another invasive species, namely alligator weed (*Alternanthera philoxeroides*). Samples from eight different sites in China were amplified with 31 primers, producing 196 markers that proved to be monomorphic across all samples. Again, the conclusion was that large-scale vegetative reproduction had taken place.

There are also cases in which DNA marker analyses revealed a considerably larger number of different clones than expected. One such species is *Empetrum hermaphroditum*, a late successional dwarf shrub of boreal forests in northern Sweden. This species propagates vegetatively by layering, which was thought to be its main mode of reproduction. However, an RAPD study comprising one mainland and two island sites that represent different postfire successional ages (145, 375, and 1720 years since the last fire, respectively) revealed that a sexual seed set is much more common than previously assumed.[1357] Using 61 polymorphic RAPD markers, 96 genotypes were identified among a total of 133 samples. All three populations depicted high levels of variation, although some genets were as large as 10 to 40 m in diameter.[1357] For wild garlic (*Allium vineale*), a combination of RAPD and cpDNA-CAPS analysis revealed many multilocus genotypes that originated through sexual reproduction during the expansion of this species across Europe.[221] However, current recruitment seems to take place exclusively by vegetative reproduction, namely through bulbils.

Microsatellites have only rarely been used for clonal discrimination. One such study was carried out in the marine eelgrass *Zostera marina*.[560] Clonal size proved to be positively correlated with heterozygosity. Outbreeding clones were larger and contained more flowering shoots, indicating that inbreeding depression significantly affected vigor and fertility. In a subsequent study, the same authors were able to quantify the genetic neighborhood structures of eelgrass populations using spatial autocorrelation of microsatellite markers.[561] Microsatellites were also used to assess the mode of reproduction in the moss *Polytrichum formosum* in Denmark and the Netherlands.[1446] Low levels of variation suggested that clonal reproduction is predominant in this species, but sexual reproduction and long-distance spore transport also play an important role for shaping the genetic structure.

A number of DNA fingerprinting studies aimed at clarifying the extent of clonal growth in the well-known dandelions (*Taraxacum officinale*).[919,1444,1445] Many *Taraxacum* populations consist solely of triploid individuals that are thought to reproduce mainly through apomixis (i.e., seed set without prior fertilization), and are therefore clonal. An AFLP study showed that some of these clones cover large areas, confirming asexual reproduction.[1444] However, there were also genotypes that resulted from genetic exchange. Sexual reproduction was also demonstrated for a set of apomictic dandelion populations from northern Europe, where only a few genotypes appeared to have originated through somatic mutations.[1445]

6.3.2.4 *Estimating Gene Flow via Pollen and Seeds*

The magnitude and pattern of gene flow is an important factor that influences the effective size and genetic structure of populations. In plants, gene flow is mainly through pollen and seed dispersal. Previous estimates of pollen flow mostly relied on direct observations of the movement of pollinators, usually insects, or by collection of pollen on sticky tape (e.g., from plants in a wind tunnel). DNA fingerprinting has become an important molecular tool for estimating gene flow, both by pollen and seed, for a wide range of species and under a wide range of circumstances (reviewed by Bossart and Prowell,[153] Ouborg et al.,[1019] and Schnabel[1246]).

The most direct way to estimate gene flow is through **parentage analysis**. Usually the maternal genotype is already known, and DNA fingerprinting can serve to pinpoint the offspring and the actual (or most likely) father among all possible fathers. In one of the first studies in this direction, Chase et al.[230,231] analyzed the genetic diversity of the neotropical rain forest tree species *Pithecellobium elegans* (Mimosaceae) at five microsatellite loci. A total of 32 alleles were found (one to 15 alleles per locus) in 52 individual trees from two sampling sites in Costa Rica. Almost all trees could be discriminated by a set of only three marker loci. The authors compared allelic diversity obtained with microsatellites vs. allozymes, and found the former to be much more informative. Individual seeds of a tree were analyzed to estimate the number of different pollen parents. Only one or two fathers contributed to the progeny within a single pod, but many were involved in pollinating a single tree. Pollen parents were found to grow at far distances of each other, and the closest neighbors were usually not the most efficient pollen donators.

Similar studies were carried out by Dawson et al.,[303] who measured pollen-mediated gene transfer in the tropical tree *Gliricidia sepium* (Fabaceae). Using paternity exclusion based on a single microsatellite locus with six alleles, these authors demonstrated that more than 6% of pollen movements in a selected *G. sepium* stand in Guatemala were greater than 75 m, and one extreme example was more than 235 m.

In the South American tree species *Euterpe edulis*, gene flow was also estimated from a direct paternity analysis using microsatellites.[471] First, an exclusion analysis was performed by comparing adult and juvenile genotypes. After that, a so-called paternity index was calculated among those adults that could be the putative parents for a particular juvenile. Some adult trees contributed considerably more to the next generation than did others. Gene flow was over longer distances than expected (up to 22 km), although it was unclear whether this was due to seed or pollen transport. Dick et al.[343] used microsatellite markers and the computer program TWOGENER[55] to estimate pollen dispersal in still another tropical tree species, *Dizinia excelsa*. Dispersal distances were calculated to be 1509 m in open pasture and only 212 m in undisturbed forest. Much longer distances (3.2 km) were found for isolated trees.

The examples outlined above indicate that pollen transport may contribute considerably to gene flow in tropical tree species. They also illustrate that microsatellites provide an excellent marker system for such measurements. Nevertheless, dominant markers may also be used. Gerber et al.[486] compared the efficiency of dominant (AFLP) and codominant (microsatellite) markers as parentage and gene flow estimators in oak trees. High parentage exclusion probabilities were obtained with both types of markers, but microsatellites were more efficient. AFLPs also proved to be adequate for parentage studies, provided that markers were preselected according to their band frequencies in the investigated sample set (optimally between 0.1 and 0.4), and a sufficiently large number of markers (100 to 200) were evaluated.

The extent of gene flow within and among populations is usually inferred from population structure, assuming IBD and absence of other confounding effects. Standard analyses of variation and differentiation may indicate at what scale gene flow occurs in a species. For example, Wolff et al.[1564] used RAPDs to analyze population structure in *Alkanna orientalis*, an insect-pollinated plant in the Sinai Desert. A cluster analysis with samples from four populations showed that individuals from the same population generally clustered together. An AMOVA further revealed that populations separated by a high ridge were the most diverged, indicating that this ridge acted as a barrier for pollen movement by the insects. An RAPD analysis of the South African desert plant *Welwitschia mirabilis* showed that population differentiation was correlated with geographic distance and that gene flow occurred over distances of 6 km, but not more than 18 km.[654]

Indirect estimates of gene flow are most commonly obtained by calculating the **average migration rate, N_m**, from the differences in allele frequency among populations, as expressed by F_{ST} or an analog (see Chapter 5.6.2). An even more detailed measure of gene flow can be obtained when migration rates are estimated from both nuclear (biparentally inherited) and organellar (maternally or, occasionally, paternally inherited) markers.[403,404] Maternally inherited markers (e.g., cpDNA polymorphisms in most angiosperms), are only transported through the seed. According to Ennos[403] and Ennos et al.,[404] the ratio between pollen and seed migration can vary by two orders of magnitude, and is typically much lower for insect- as compared with wind-pollinated plants. Beebee and Rowe[99] compiled a list of molecular marker studies in which the **ratio of pollen to seed migration** ranged from 1.8 in *Eucalyptus nitens* to 500 in *Quercus petraea*. Squirrell et al.[1321] studied both allozymes and chloroplast markers in native European and introduced North American populations of the orchid *Epipactis helleborine*. From nuclear and organellar F_{ST} estimates, a pollen to seed gene flow ratio of only 1.43:1 was calculated. This very low value was not completely unexpected, given that orchid seeds can disperse widely because of their extremely small size.

In general, care should be taken when data sets obtained with different marker types are compared. Comes and Abbott[267] studied gene flow among populations of ragwort (*Senecio gallicus*) on the Iberian Peninsula and in the south of France. Population differentiation was much lower for uniparental cpDNA-CAPS than for biparentally inherited allozymes, indicating a high seed dispersal capacity. However, a subsequent RAPD analysis of the same populations instead showed a moderate level of intraspecific differentiation, more similar to the previous cpDNA results than to the allozyme-derived data.[268] Obviously, differences between the evolutionary rates of the various marker types employed had a stronger influence on the *Senecio* data set than, e.g., high rates of pollen dispersal, slow rates of nuclear lineage sorting, or indirect balancing selection.

Recently, the standard F_{ST} approach has been complemented by fine-scale spatial analyses of gene flow. Based on a small data set derived from five nuclear microsatellite loci and simulation studies, Heuertz et al.[595] estimated pollen and seed dispersal from the spatial genetic structure of ash trees. Another approach is to calculate the **autocorrelation of genotypes** in specific distance classes (see Chapter 5.6.7). The obtained measure, usually expressed in Moran's I, can be used for quantitative estimates of gene flow if the population is in equilibrium and IBD can be assumed.[571] A positive autocorrelation is encountered frequently, especially over shorter spatial distances, even though an overall linear correlation between geographic and genetic distances may be lacking. Using RAPD data, Torres et al.[1410] found autocorrelation in the first distance class (15 m) in populations of the endangered cliff specialist *Antirrhinum microphyllum*, suggesting a patchy distribution of genetic diversity. This is consistent with the territorial behavior of the main pollinator *Rhodanthidium sticticum*, short-distance seed dispersal, and a likewise patchy distribution of suitable habitats.

Cottrell et al.[277] used microsatellite markers to examine the population structure of two oak species (*Quercus robur* and *Quercus petraea*) in a natural forest as well as in a planted and extensively managed forest. Gene flow proved to be relatively high in the natural forest, but had decreased, with a concomitant increase in genetic structuring, in the managed forest.

Combined information on landscape ecology and population genetic structure is needed to determine how historical and temporal gene flow have influenced present-day patterns of variation (see reviews by Manel et al.[872] and Sork et al.[1314]). Detailed knowledge about the landscape, e.g., from geographic information systems (GIS), can then be used in connection with the analysis of plant samples from areas without predefined populations. This is exemplified by a study of Cavers et al.[220] who showed with a combination of DNA fingerprinting methods that the connectivity of the habitat of Spanish cedar (*Cedrela odorata*) was crucial to gene flow: the presence of a high mountain range appeared to have isolated the populations on either side.

6.3.2.5 Effects of Habitat Fragmentation

One important area of population genetics is the study of the effects of human impact on natural environments, which has often led to an increased fragmentation of habitats. The effect of this process on plant and wildlife viability is controversial.[1596] One possible scenario is that the reduced effective population size of isolated populations eventually results in reduced genetic variation, and hence reduced viability of the population. These effects may be most pronounced in species that form small populations, are self-compatible, and have limited seed dispersal abilities. In agreement with this expectation, a microsatellite analysis of naturally fragmented populations of the rare, fire-dependent shrub *Grevillea macleayana* in New South Wales (Australia) revealed high levels of inbreeding and considerable genetic structure.[402]

There is a particular interest in the effect of fragmentation on genetic diversity in forest communities, especially of tropical rain forest areas (for a review, see Aldrich et al.[19] and references therein). Natural, undisturbed tree populations usually show substantial amounts of genetic variation within populations, whereas an enhanced population differentiation due to limited gene flow is typical for tree populations in fragmented forests (e.g., *Eucalyptus albens*[1103]). Several factors may account for such a reduction. First, logging may cause the direct loss of certain genotypes or alleles due to stochastic reasons. Second, gene flow between habitat fragments is expected to decrease due to total or partial loss of the dispersal agent or the inability of the dispersal agent to carry pollen or seeds over large distances. This, in turn, may result in increased inbreeding and genetic drift.

Recently, several studies employed microsatellite data to assess the population genetics of trees in fragmented habitats. The genetic structure of a natural population of 88 trees of *Swietenia humilis* (Meliaceae), an endangered tropical hardwood mahogany species from Central America, was analyzed by White and Powell.[1536] Between four and 23 alleles were identified at 10 microsatellite loci, and the mean observed heterozygosity was 0.415. The extent of subpopulation differentiation at a microgeographical scale was low (mean F_{ST} = 0.036), suggesting an extensive gene flow between the two stands. In a subsequent study, White et al.[1537] compared the parameters of fragmented populations of *S. humilis* with a population in a large continuous forest in Honduras. Genetic variation was still high in all habitat fragments at all microsatellite loci, but the number of low-frequency alleles was reduced in the small fragments, indicating the beginning of genetic erosion. Lowe et al.[848] studied gene flow among *Swietenia macrophylla* populations in a continuous and disturbed forest in Costa Rica. Population differentiation determined with AFLP, microsatellites, and RAPD was reported to be moderate (38, 24, and 20%, respectively). A significant fine-scale structure was found in all populations and gene flow appeared to occur only over short distances, given that most pollinations took place between proximate trees.

Dayanandan et al.[305] investigated fragmented populations of *Carapa guianensis*, another member of the Meliaceae, in Costa Rica. Three microsatellite loci were analyzed, exhibiting four to 28 alleles. No inbreeding was detected in fragmented populations. However, genetic distances and R_{ST} values between populations were greater, and corresponding levels of gene flow were lower among sapling cohorts compared with adult cohorts. The authors conclude that saplings already suffer from a restricted gene flow due to deforestation and habitat fragmentation, whereas the adult cohorts are probably remnants of the times when the area was covered by large continuous forest. This indicates that recent fragmentation events are best monitored by analyzing seedlings and saplings rather than adult trees.

A considerable impact of habitat fragmentation on the genetic structure of a tropical tree species was found by Aldrich et al.,[19] who investigated the population genetics of *Symphonia globulifera* (Clusiaceae), a canopy tree characteristic for primary rain forests of the neotropics, in intact vs. fragmented habitats of southern Costa Rica. The study was designed on a microgeographical scale (all fragmented habitats within a 38.5-ha plot, with two control plots established in an adjacent nature reserve), and from a multistage, demographic perspective. In total, 74 adults, 152 saplings, and 688 seedlings of two size classes were screened for allelic diversity at three microsatellite loci, altogether resulting in 55 alleles. Regarding the numbers of alleles and genotypes per hectare, no differences were observed for continuous primary forest and fragmented forest for all three life stages, except for a larger than expected number of seedling genotypes in fragmented forest. This exception was explained by the massive transportation of seeds into remnant forest patches by bats. In the interpopulation comparison, the most pronounced genetic structure occurred between the two native forest patches treated as one group, and all disturbed patches as the other. In any comparison, significant inbreeding and genetic differentiation (indicated by F_{ST} as well as R_{ST}) were most often associated with seedlings in fragmented forest stands. Principal component analysis of distance matrices confirmed these results and suggested that bottlenecks have occurred through fragmentation, possibly in concert with a pre-existing genetic structure in the adults. It should be emphasized, however, that results from an analysis of only three microsatellite loci need to be interpreted with caution.

In a subsequent study, Aldrich and Hamrick[18] reconstructed a population-level pedigree of *S. globulifera*. Seedlings only occurred in primary and remnant forests, but not in pastures. Surprisingly, however, the majority of seedlings in fragmented forests were derived from a few adult trees located in the open pasture land. Thus the genetic bottleneck experienced by the seedlings in remnant forest patches (see above; Aldrich et al.[19]) was caused by the reproductive dominance of a few spatially isolated trees in pasture land, in conjunction with unusually high levels of selfing in these trees.

As a final example, Collevatti et al.[262] assessed population genetic parameters at 10 microsatellite loci in *Caryocar brasiliense*, an endangered tree of the Brazilian Cerrado. The number of alleles found in 314 individuals from 10 natural populations ranged from 20 to 27. Expected and observed heterozygosities were rather high. Genetic differentiation between populations as measured by pairwise F_{ST} values was correlated with geographical distance, as expected under an IBD model. In this study, no significant differences were detected in fragmented as compared with continuous habitats, probably because Cerrado fragmentation is a relatively recent event.

6.3.3 Hybridization and Introgression

Formation of hybrids by the fusion of gametes from two different entities (species, subspecies, etc.) is a common phenomenon in plants, both in the wild and under cultivation (reviewed by Arnold[47,48] and Rieseberg[1168,1170]). Knowledge about the hybrid origin of a plant can be important for, e.g., cultivar identification, conservation management, or understanding the biology of a species. DNA fingerprinting can aid in identifying the parental species or genotypes that contributed to the hybrid. Nuclear DNA markers generally originate from both parents in roughly equal proportions. Uniparentally inherited organellar markers are useful for tracing the origin of the ovule.

Confirmation of a hybrid origin for first-generation products of a hybridization event is generally very straightforward and can be achieved with almost any type of nuclear DNA markers. If the event involved an unreduced egg cell (which is often the case), we might even determine the direction of the cross solely from nuclear DNA markers, as was demonstrated in an RFLP fingerprinting study of the hybridogenous pentaploid blackberry *Rubus vestervicensis*.[746] Here, a plant of the triploid species *Rubus grabowskii* apparently provided the egg cell and a plant of the tetraploid species *Rubus pedemontanus* provided the pollen cell, as evidenced by the presence of all DNA markers of the former plant in the hybrid but only half of the markers of the latter plant.

6.3.3.1 Hybridization in Wild Populations

In the wild, hybrids usually grow intermingled with one or both of the original species. Introgression may then readily occur if viability and fertility of the hybrids are sufficiently high. If repeated introgression takes place and mainly involves only one of the parental species, more and more of the DNA of the hybrid will be replaced by the DNA of that parental species.

A series of classical studies by Arnold and coworkers[46,49,50,975] combined the use of DNA markers (RAPD, RFLPs of ribosomal RNA genes [rDNA], and cpDNA-CAPS), allozymes, pollination biology, and ecological data to investigate homoploid hybridization among North American *Iris* species. Arnold et al.[49] showed that *Iris fulva* and *Iris hexagona* each had a species-specific rDNA profile. Nason et al.[975] subsequently found that in populations where the two species co-occurred, the DNA profiles indicated interspecific hybridization as well as further introgression in both directions. Diagnostic RAPD and cpDNA-CAPS markers were generated for these two species as well as for *Iris brevicaulis*, and it could be shown that *Iris nelsonii* is derived from hybridization among all three species.[46]

Another set of interesting studies on homoploid interspecific hybridization have been provided for the sunflower genus, *Helianthus*, by Rieseberg and coworkers.[530,1168,1169,1172] RAPD linkage maps were generated for *H. petiolaris* and *Helianthus annuus*.[1168] These maps were then used to study the genome of the recently formed hybrid species, *Helianthus anomalus*, and an artificially generated hybrid.[1168,1169] Apparently, a large amount of genome reorganization had occurred after the formation of the new species. In another example, cpDNA and nuclear microsatellites were used to study the origin of *Helianthus deserticola*.[530] According to the extent of similarity in microsatellite loci between *H. deserticola* and its presumed parents, the hybridization event was estimated to have taken place between 170,000 and 63,000 years ago.

Ayres et al.[62] used RAPD markers to monitor the introgression of genes from one species into another in the aquatic grass genus *Spartina*. Whereas morphological characters did not provide a clear-cut explanation of the hybridization event, RAPD markers demonstrated the introgression of DNA from *Spartina alterniflora*, a species introduced into San Francisco Bay, into *Spartina foliosa*. These markers can now be used to monitor the introgression of *S. alterniflora* into the same or other species in other areas as well.

Maideliza and Okada[870] studied gene flow and hybridization between different cytotypes in the buttercup *Ranunculus silerifolius*. This species is diploid ($2n = 16$) with at least four intraspecific chromosomal races that may reduce gene transfer and, through the resulting reproductive isolation, accelerate speciation. In concordance with this hypothesis, allozyme and ISSR analyses showed that gene flow between allopatric populations of the same cytotype is three to five times higher than that between different cytotypes in the same (parapatric) population.

Palmé and Vendramin[1032] investigated European hazel (*Corylus*) species and populations using cpDNA microsatellites, cpDNA-CAPS, and *mat*K sequencing. They found that haplotype A, which is the dominant haplotype in natural populations of *Corylus avellana* across Europe, is also present in European tree hazel (*Corylus colurna*). In addition, haplotype B, which is rare in *C. avellana*, was identified in *Corylus maxima*. The presence of identical chloroplast haplotypes in different species could either be explained by homoplasy, differential sorting of ancient lineages, or introgression by hybridization. Homoplasy seemed unlikely at least in the case of haplotype B, which was defined by seven independent mutations. The authors favored the hypothesis that these marker patterns are indicative of past hybridization events among the European *Corylus* species. A species-independent geographical distribution of chloroplast haplotypes has been found in other genera such as *Eucalyptus*,[909] *Macaranga*,[1476] and *Quercus*,[373] and may well represent the rule rather than the exception.

6.3.3.2 Hybrid Distances and Diagnostic Markers

Van Raamsdonk et al.[1453] used AFLPs to investigate the relationships among sections of the *Allium* subgenus *Rhizirideum*. The authors introduced a new distance measure called hybrid distance, which indicates the fraction of bands of an accession (e.g., a putative hybrid) that is also found in another accession (e.g., a putative parent). For instance, if all bands of a putative hybrid are also present in a parental accession, the distance is zero, even if there are additional bands in the parent. Because either the visible band of a dominant AFLP marker, or the invisible null allele is passed to the hybrid, the fraction of parental bands found in a hybrid will depend on the level of heterozygosity in the parents. In contrast to other distance measures, the hybrid distance is not reciprocally identical. *Allium roylei* was identified as a putative hybrid between species from the sections *Cepa* and *Rhizirideum*, respectively.

Sometimes a set of diagnostic markers can be identified that occur in widely different frequencies in the parental species. Their occurrence in different hybridogenous populations can then be used as a measure of the degree and direction of introgression. Hybridization events involving the Mexican oak species *Quercus affinis* and *Quercus laurina* could thus be analyzed using only nine diagnostic RAPD markers.[510] Interestingly, the proportion of morphologically intermediate individuals in hybridogenous populations proved to be considerably smaller than the proportion of genetically intermediate individuals in the same populations.

6.3.3.3 Hybridization between Wild and Cultivated Plants

Hybridization can take place between crops and wild plants, with gene flow usually directed from the crop plants to their wild relatives because the former are more numerous. A quantification of this gene flow is important for, e.g., determination of risks involved in growing genetically modified crops.[1544] Gene flow through seed or pollen, followed by hybridization, is one of the possible roads that a transgene can take to become established in the genome of a weedy relative. A recent book by Ellstrand[395] gives a detailed overview of the occurrence of hybridization between crops and their wild relatives. In many cases, DNA fingerprinting has been instrumental and confirmed what the morphology and growth behavior of suspected hybrids already indicated.

One well-studied example is that of weedy, bolting beets that appear within and between rows of sugar beet fields in most European countries. Detailed analyses of chromosome numbers, nuclear, chloroplast, and mitochondrial RFLP fingerprints, and nuclear microsatellites indicated that these weed beets descend from hybridization between sugar beets and wild annual beets (*Beta vulgaris* subsp. *maritima*); the sugar beet was most likely the maternal parent.[155,1468]

6.3.3.4 Hybridization in Gymnosperms

Numerous studies have been carried out on hybridization events in gymnosperms, where both the paternal (cpDNA) and the maternal (mitochondrial DNA [mtDNA]) contributors of hybrid offspring can be characterized. Both nuclear and cytoplasmic markers (cpDNA and mtDNA) were analyzed in a study of two sympatric *Abies* species.[650] Single-strand conformation polymorphism (SSCP) analysis was used to reveal cpDNA and mtDNA polymorphisms, and it could be proven that *Abies homolepsis* was the male parent and *Abies veitchii* was the female parent to all encountered hybrids between these two species.[650] Three informative chloroplast microsatellite markers were used by Bucci et al.[181] to analyze seeds of *Pinus halepensis* and *Pinus brutia* in sympatric stands. A number of putative hybrids could be detected. Interestingly, gene flow proved to be unidirectional also in this case. Whereas 15 of 60 *P. brutia* embryos investigated were found to have been derived by pollination from *P. halepensis*, none of the embryos derived from *P. halepensis* seeds had a *P. brutia* haplotype.

Chloroplast markers (especially if combined with nuclear markers) can also provide some information on auto- vs. allopolyploidy, and single vs. multiple origin of polyploid species. Fady et al.[422] investigated the possibility of interspecific hybridization and cpDNA introgression in three Mediterranean *Cedrus* species. These species normally grow in distinct geographical areas, and have no contact with each other: *Cedrus libani* occurs in Turkey, Syria, and Lebanon; *Cedrus atlantica* occurs in north Africa, and *Cedrus brevifolia* occurs in Cyprus. Seeds from controlled and open-pollinated trees were analyzed by AFLP, RAPD, and cpDNA microsatellites. Interspecific hybridization was revealed by all three types of markers, indicating the absence of reproductive barriers despite geographic separation of the three Mediterranean *Cedrus* species.

6.3.3.5 Polyploidy

Interspecific hybridization in plants may lead to the establishment of new allopolyploid taxa.[1170] Using nuclear multilocus markers, polyploid hybrids can be studied in the same manner as diploid hybrids. However, a positive correlation between ploidy level and number of scored AFLP bands has been observed.[8,693] Nuclear single-locus markers such as microsatellites are more difficult to apply to polyploids because of uncertainty about genomic constitution. If the parental species of an allopolyploid species complex are sufficiently different, genome (and species)-specific microsatellite markers can be developed and have proven very useful (e.g., in analyses of the hexaploid bread wheat and its relatives[547,1315]). Similarly, Lowe et al.[847] developed microsatellite primers that are specific to the A, B, or C genomes in the genus *Brassica*. These microsatellites are not only useful for following the hybridization and polyploidization processes, but also for analyses of gene flow and hybridization between a crop *Brassica* and its wild relatives.

Some allopolyploid species have retained considerable genome integrity as evidenced by polysomic inheritance at microsatellite loci, resulting in complex band profiles that can be quite difficult to score and interpret. There is also a major problem with defining which allele(s) occur in more than one copy, when the number of displayed microsatellite alleles in a sample is fewer than the possible maximum number for that ploidy level. Even unambiguously scored bands are therefore often interpreted as phenotypic banding patterns, and no attempts are made to analyze allelic configurations (e.g., Becher et al.[94]).

The microsatellite DNA allele counting–peak ratios (MAC-PR) approach[413] was developed to solve this problem. It facilitates the analysis of quantitative differences between microsatellite allele peak ratios and thus allows the determination of allelic configuration. MAC-PR was recently applied in studies of dogroses, *Rosa* section *Caninae*.[1003] All dogrose taxa are polyploid and undergo a peculiar meiosis: only seven bivalents are formed, whereas the remaining chromosomes occur as univalents, which are included in viable egg cells but not in viable pollen. Microsatellite analysis was applied to parents and offspring from interspecific crosses involving four pentaploid dogrose species and one tetraploid. The copy numbers of the individual alleles could be determined with MAC-PR. Bivalent formation apparently takes place mainly between a pair of highly homologous genomes, resulting in very restricted sexual recombination and unusually homogeneous offspring groups. The studied taxa showed widely different levels of similarity between bivalent- and univalent-forming genomes, and also among the two to three univalent-forming genomes.

6.3.4 Plant Conservation

Genetic markers play a considerable role in conservation biology. In the 1980s, the importance of genetic factors compared with habitat destruction together with demographic and environmental stochasticity was much debated.[1014] At present, it is widely recognized that integrated demographic and genetic approaches yield the most useful results. A good introduction to conservation genetics is given by Frankham et al.,[458] Falk and Holsinger,[427] and Benson[117] provide comprehensive overviews of what biotechnology (including molecular markers) can offer to plant conservation research.

On their own, molecular markers appear to be insufficient for analyses of the adaptive potential of populations. In one example, Bekessy et al.[102] used RAPD markers and quantitative traits associated with drought tolerance to study levels of genetic variation in the monkey puzzle tree (*Araucaria araucana*). These characters showed different distributions of genetic variation, and RAPDs failed to reveal the prominent differences in ecologically important traits. However, molecular markers have been able to estimate levels and partitioning of genetic variability in numerous threatened plant species. Several DNA-based techniques were used in the relict genus *Borderea* (Dioscoreaceae), in which a previous allozyme study had detected very low levels of genetic variation. Two species were analyzed for their genetic variation with RAPD: the endangered *Borderea choardii* and *Borderea pyrenaica*, endemic to the Pyrenees.[1269] These two species were clearly distinguished by RAPDs, but a high level of variation was also found between and within populations, indicating a recent origin of the present-day population structure. In a subsequent study, Segarra-Moragues et al.[1270] used microsatellites and discovered levels of variation that were similar to those previously obtained with RAPD. In addition, an allopolyploid origin of these rare and endemic *Borderea* taxa was suggested.

Endemic and more widespread congeneric species sometimes show different levels of within-population variation and/or population differentiation.[562,1000,1001] Comparisons of genetic variability in widespread and narrowly distributed species have been carried out using a number of different molecular methods. According to allozymes and RAPD markers, the widespread *Menziesia pentandra* (Ericaceae) was eight and three times, respectively, more polymorphic than the narrow endemic *M. goyozanensis*.[871] Similarly, two relict tree species in the genus *Zelkova* (Ulmaceae) were studied with a range of markers, i.e., internal transcribed spacer sequences (ITS2) of nuclear ribosomal genes, chloroplast sequences (*trn*L), chloroplast microsatellites, and PCR-CAPS.[443] Both *Zelkova abelicea* (from Crete) and *Zelkova sicula* (from Sicily) had diverged from the common species *Zelkova carpinifolia,* and both lacked cpDNA polymorphism altogether.

Many studies have been set up to identify the most diverse or evolutionary significant populations of endangered species, for conservation purposes and/or to use them as source for reintroduction. In a study of Spanish cedar (*Cedrela odorata*; Meliaceae), a tree species that is likely to have undergone genetic erosion,[220] AFLP and cpDNA-CAPS markers were used to study the geographic partitioning of variation. Two different ecotypes, growing in wet and in dry habitats, respectively, were identified, and a strong differentiation due to isolation by distance was found.

Chloroplast microsatellites were used to determine the suitability of particular lady's slipper orchid (*Cypripedium calceolus*) accessions for reintroduction in Britain.[1137] Two accessions that proved unlikely to be of British origin could thus be excluded from the program. Reproduction was an issue in the case of *Spiranthes romanzoffiana*, a British orchid species of conservation priority, which often has a low or no seed set. The reproductive system and its influence on genetic diversity were investigated by AFLPs.[455] Northern and southern populations appeared to differ considerably in their reproductive system, and the two groups also exhibited different chloroplast microsatellites. Northern populations showed only one chloroplast type, but a high level of AFLP variation, suggesting sexual reproduction. The southern populations had a single, but different, chloroplast type, and only 12 unique multilocus AFLP genotypes. These differed by single bands only, which indicates agamospermous or autogamous reproduction.

6.3.5 Germplasm Characterization and Preservation

DNA fingerprinting is an important instrument for the characterization of germplasm, i.e., the total genetic diversity present in the world for a certain crop, encompassing old and newly bred cultivars, land races, and related wild species. One major aim is to determine the extent and distribution of genetic variation, and to understand the geographic and ecological aspects of the processes that have given rise to the observed patterns of variation.[601] Tracing the unknown origin of species or cultivars has been another valuable contribution of DNA fingerprinting to germplasm characterization. For example, Regner et al.[1148] used AFLP, RAPD, ISSR, and microsatellites to study 1200 vines (*Vitis* species) and were able to describe the history of many cultivars still in use, going back to the Middle Ages. Using only microsatellites, Luro et al.[858] successfully studied the origin of different *Citrus* species and their relationships.

6.3.5.1 Gene Banks

DNA markers have provided valuable data for the identification of suitable material for *in situ* preservation, the establishment of *ex situ* gene banks and core collections with maximum diversity, and for the detection of undesirable duplicates. Moreover, the assignment of a permanent bar code to each of the preserved accessions allows unambiguous identification, now and in the future. A few examples illustrate these topics.

An RAPD analysis of clonal structure in rice was instrumental for developing management measures for the *in situ* conservation of wild rice (*Oryza rufipogon*) in China.[1579] Clones were small and levels of sexual reproduction were relatively high in populations that were regularly disturbed or exposed to seasonal drought. In contrast, larger clones and little sexual reproduction were typical for populations with little disturbance and sufficient supply of water. This study also revealed that plants designated for *ex situ* conservation should be collected at distances of more than 12 m to avoid obtaining identical ramets of the same clone.

As one example of DNA-based evaluation of gene banks, Lowe et al.[849] used RAPD analysis to study 56 germplasm accessions of Napier grass (*Pennisetum purpureum*) and its hybrids. This group of cultivars is an important fodder crop in East Africa. There was little or no genetic variation within accessions, probably due mainly to vegetative reproduction. The collections were evaluated and rationalizations were suggested. Genetic distances within and between recognized groups within the crop were large, which suggests that the genetic basis in the germplasm is sufficiently wide.

Preservation of germplasm collections *ex situ* is often very expensive, especially for crops that have to be maintained as vegetatively propagated plants in the field. One such crop is the outbreeding hexaploid *Ipomoea batatas* (sweet potato). The International Potato Center (CIP) hosts the world's largest sweet potato collections, with 5526 cultivated accessions from 57 countries.[1608] Morphological measurements indicated that there are a considerable number of suspected duplications (~1500), among those accessions. It was shown that morphologically different cultivars also differed in their RAPD profiles,[1608] whereas some of the suspected duplicates shared identical RAPD patterns and could therefore be removed from the collections.

In another effort to minimize the amount of preserved plant material, Phippen et al.[1076] showed that a considerable proportion of *Brassica oleracea* var. *capitata* accessions could be omitted from the gene bank collections. About 4.6% of the variation were lost in this way, but at the same time 70% of the cost for maintaining the collection were saved.

6.3.5.2 Core Collections

So-called core collections are sometimes sampled from a larger germplasm collection for more intense characterization. Ideally, a core collection should cover the whole breadth of genetic variation of the crop. Methods and strategies of assembling core collections in different plant species were presented by a number of studies.[226,335,349,798,1249]

One way to choose the most variable plant material for a core collection is to perform a cluster analysis of a pairwise distance matrix generated from all available accessions (see Chapter 5.5.2). Distances can be derived from, e.g., phenotypic characters or molecular marker data. Using stepwise clustering and different sampling regimens (random sampling, preferred sampling, and deviation sampling), Hu et al.[623] showed that core collections in cotton (*Gossypium*) retained a larger amount of genetic variability and included superior representatives when the selection was based on genotypic rather than phenotypic characters.

On the basis of RAPD data, Garkava-Gustavsson et al.[478] compared three different strategies to select plants for a lingonberry (*Vaccinium vitis-idaea*) core collection: (1) a hierarchical sampling strategy based on cluster analyses,[1047] (2) the Maximum Genetic Diversity computer program,[884] and (3) random sampling (as a control). Only the Maximum Genetic Diversity computer program enabled the selection of a core collection that preserved all rare RAPD bands, and with frequencies that had increased over those in the initial plant sample. Similarly, only 1.4 AFLP bands were lost in a Maximum Genetic Diversity program-generated subset representing 32% of the initial plant collection in sweet potato.[426]

6.4 PLANT TAXONOMY AND SYSTEMATICS

6.4.1 Taxonomic Relationships Revealed by Multilocus DNA Methods

Multilocus DNA profiling methods constitute a potential source for phylogenetically informative characters at the level of populations, species, and possibly genera (reviewed by Wolfe and Liston[1558]). Consequently, AFLPs,[77,170,693,740,1449] RAPDs,[5,1138,1139] and ISSRs[643,955,1130,1132] were applied to phylogeny reconstruction in a large number of investigations, only few of which are outlined in more detail below. AFLPs were even considered as the method of choice for analyzing relationships between closely related taxa, in which traditional qualitative characters and/or cpDNA or ITS sequences show little, if any, variation.[331]

Provided that taxon-specific bands are frequent, and intrataxon variation is low, multilocus DNA profiles can also be used for distinguishing taxonomic entities. For example, Anamthawat-Jónsson et al.[28] could easily discriminate between two species of lymegrass, i.e., the tetraploid *Leymus mollis* and the octoploid *L. arenarius*, on the basis of species-specific AFLP bands and a neighbor-joining analysis of Dice distance data. Bartish et al.[83] used a set of 219 polymorphic RAPD markers to analyze species and subspecies in the genus *Hippophae*. They found 16 fixed RAPD markers, i.e., markers that were either present or absent from all plants of a population. Several RAPD bands were informative for the analysis of interspecific relationships (i.e., were present in at least two but not all taxa), whereas others could be considered as taxon-specific markers. Clustering of taxa and populations in a neighbor-joining tree agreed well with some recently suggested taxonomic treatises of *Hippophae*.

When using multilocus markers for taxonomic analyses, the number of scored bands must be sufficiently high.[1138,1139] With AFLPs, acquiring a sufficient number of scored bands is usually not a problem. Kardolus et al.[693] generated AFLP fingerprints for 30 accessions from 19 taxa of *Solanum* sect. *Petota* and three taxa of *Solanum* sect. *Lycopersicum*, representing the closest relatives of potato and tomato, respectively. In total, 551 polymorphic bands were obtained from three primer combinations. The ploidy level was reflected in the profiles, with hexaploids showing more bands than tetraploids and diploids, respectively. Mating system had a considerable impact, with much higher intraspecific polymorphism detected in outcrossing taxa (~40 to 60%) as compared with inbreeders (0 to 2%). Both phenetic and cladistic analyses (see Chapter 5.5.2) were performed at various systematic levels, ranging from individuals to species. The topologies of the resulting phenograms and cladograms were generally similar, and biosystematic classifications based on the AFLP data were generally congruent with those based on traditional characters.

Aggarwal et al.[8] used AFLPs to analyze a total of 77 accessions representing 23 *Oryza* species plus several outgroup genera. Pairwise distances (Dice's index) showed a linear increase depending on the taxonomic level, with 0.02 to 0.21 within species, 0.2 to 0.35 between species sharing the same genome type, and > 0.7 between species carrying different genomes, and between *Oryza* and outgroup genera. Because conspecifics were grouping together with high bootstrap values, some misclassifications were readily identified. The overall dendrogram suggested a monophyletic origin of the genus *Oryza*.

Han et al.[566] used AFLP markers to study the genetic diversity and relatedness of 22 South American *Alstroemeria* species, an interspecific hybrid, and two other genera of the Alstroemeriaceae. Selective primers with 4-bp extensions were used due to the large genome size of these species (see Chapter 4.7.4.2). PCR products from three accessions per species were pooled to obtain a species-specific profile. The authors justified this manipulation with the low intraspecific distances observed (<0.3). Interspecific genetic distances ranged between 0.3 and 0.7 within *Alstroemeria*, and exceeded 0.8 for the intergeneric comparisons. Brazilian and Chilean species formed two clearly distinct groups.

Van Droogenbroeck et al.[1449] used AFLPs to analyze genetic relationships among *Carica papaya* and its wild relatives from Ecuador. Phenetic trees obtained with different distance measures demonstrated the distinctness of the monotypic genus *Carica* from other species of the family Caricaceae. Species and accessions from the three genera investigated each formed well-supported groups.

Bänfer et al.[77] used AFLPs to investigate phylogenetic relationships among 43 species of the paleotropic pioneer tree genus *Macaranga* (Euphorbiaceae). About 30 of these species are obligate ant–plants that are inhabited by specific ant partners. Eight primer combinations produced 426 bands that were analyzed phenetically, cladistically, and by principal coordinates analysis (see Chapter 5.5). The resulting dendrograms were largely congruent with each other and supported the monophyly of several sections and subsectional groups within the genus. Moreover, the AFLP trees provided additional evidence for a multiple evolutionary origin of the ant–plant mutualism within the genus.[77] A typical AFLP pattern and a neighbor-joining dendrogram resulting from an AFLP data set obtained from different *Macaranga* accessions are shown in Figures 6.2 and 6.3, respectively.

The limits of AFLP for taxonomic studies became apparent in two studies on intergeneric and interspecific relationships within the bamboo subtribe[838] and the *Caladieae* tribe of Araceae.[839] Although species were clearly discriminated from each other, Jaccard similarities between genera (and in some instances also between species) were very low (~0.1 to 0.2). In the bamboo study, congeneric species did not form homogeneous clusters. These results may indicate polyphyly of the underlying genera as suggested by the authors, but such conclusions are ambiguous (see below).

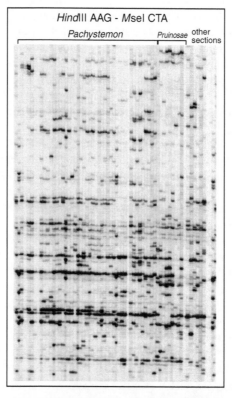

Figure 6.2   Detail of an AFLP[1481,1605] banding pattern of *Macaranga* species (Euphorbiaceae) belonging to various sections.[77] Genomic DNA aliquots were digested with *Hind*II and *Mse*I, ligated to the corresponding adapters, and subjected to two rounds of PCR. Selective primers were *Mse*+CTA and *Hind*+AAG (the latter primer was labeled with the fluorochrome IRDye700). PCR products were separated on a 6% denaturing polyacrylamide gel in an automated LI-COR sequencer.

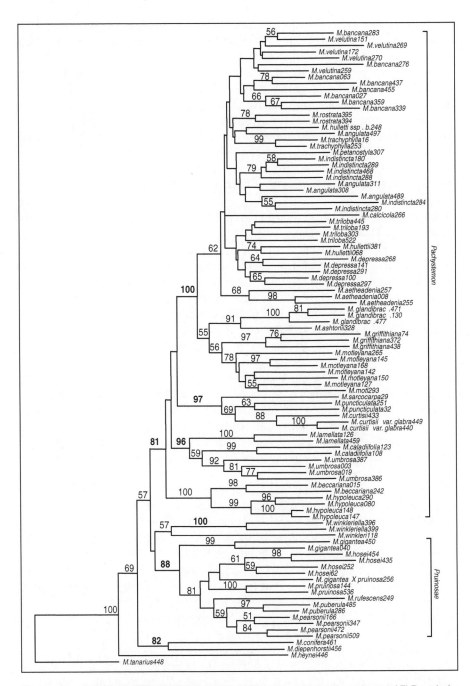

Figure 6.3   Neighbor-joining phenogram of 100 *Macaranga* specimens based on an AFLP analysis with seven primer pair combinations, one of which is exemplified in Figure 6.2. The Nei and Li[981] index of similarity (see Chapter 5.4.1) was used to generate the distance matrix used for tree construction. Horizontal branch lengths correspond to genetic distances. Numbers above branches correspond to bootstrap values.[434] *Macaranga* species belonging to the two sections *Pachystemon* and *Pruinosae* each form a monophyletic group.

In most of the above studies, Jaccard or Dice similarities between congeneric species range from about 0.5 to 0.8. These values often decrease sharply to below 0.2 and then level off when species from different genera are compared,[838,839] indicating random background band sharing. If genetic distances are small between species, and large between genera, AFLP, RAPD, ISSR, and other multilocus techniques may be adequate tools to investigate generic delimitations, and to assign species to certain genera (e.g., Van Drogenbroeck et al.[1449]). However, the low level of band sharing will usually not allow us to draw meaningful conclusions about relationships between genera.

6.4.2 Microsatellite Markers in Taxonomic Studies

There has been much debate about the use of microsatellite markers for phylogenetic studies (e.g., Goldstein and Pollock,[506] Jarne and Lagoda[659]). As opposed to multilocus markers, microsatellite DNA analysis provides a higher chance that specific alleles rather than unrelated bands are compared with each other. In theory, this should increase the probability that shared markers indeed represent orthologous loci. However, uncertainty regarding the taxonomic distance at which microsatellites produce accurate phylogenetic information comes from several sources (see also Chapter 1.2.2.4): (1) mutation rates are unpredictable and often high; (2) the range of possible allele sizes may be constrained; and (3) the frequent occurrence of indels in flanking sequences decreases the accuracy of genealogical information obtained from the microsatellite repeats themselves. As a consequence, alleles of the same length may be homoplasious rather than identical by descent.[30,416,480,1017] SSCP (see Chapter 2.3.9) offers a rapid approach to test whether fragments sharing the same size also share the same sequence. PCR fragments can be analyzed on denaturing gels for length variation, and on native gels for sequence variation. Sunnucks et al.[1345] suggested that SSCP screening for length homoplasy should be a routine part of microsatellite analysis.

Another important factor to consider is the choice of the most appropriate mutational model (IAM, SMM, or some combination of these) for, e.g., estimation of differentiation among taxa (see Chapter 6.3.2.1). Genetically and/or geographically closer populations should probably be analyzed with IAM, and more distant populations should probably be analyzed with SMM. Sampling factors must also be considered, including population size, sample size, number of analyzed loci, and number of alleles per locus. Recent studies now indicate that a two-phase model (TPM) or a generalized stepwise model (GSM) may be the most correct (see review by Estoup et al.[416]), but little data have as yet been published for plants using these models.

Mutation rates of cpDNA microsatellites may be lower than those of nuclear microsatellites. Provan et al.[1109] found that a phylogenetic tree assuming SMM for cpDNA microsatellites fitted well with other molecular classifications within the genus *Zea*. The authors concluded that cpDNA microsatellites (cpSSRs) might be valuable in phylogenetic studies, at least below the genus level. However, size homoplasy is also common at cpSSR loci.[366,555] Doyle et al.[366] compared the size distribution of cpSSR loci amplified from wild perennial relatives of soybean with an independent chloroplast haplotype genealogy hypothesized from restriction mapping studies. Identical cpSSR size classes were found in several separate branches of the haplotype tree. Thus, phylogenetic trees based on cpSSR allele sizes do not faithfully represent relationships among the chloroplast genomes in these organisms. The authors concluded that cpSSRs cannot be used with confidence for broader phylogenetic analyses.

Despite the above reservations, several studies indicate the usefulness of cpSSR data for analyses of relationships between closely related species. For example, genetic distance data derived from cpSSR studies clearly separated the three pine species *Pinus halepensis*, *Pinus brutia*, and *Pinus eldarica*,[181] and have shed some light on phylogenetic relationships between *Pinus cembra*, *Pinus sibirica*, and *Pinus pumila*.[536] In the latter study, both cpSSRs (paternal marker) and the *nad1* intron 2 of mtDNA (maternal marker) indicated a close association between *P. sibirica* and *P. cembra*, and a more distant position of *P. pumila*.

6.4.3 Taxonomic Consequences from DNA Profiling Data

Numerous authors have used DNA fingerprinting to elucidate genetic relationships at various taxonomic levels. In many cases, the reported results confirmed presently accepted taxonomies, but there are also examples in which classical circumscriptions were challenged. Only in a few cases have DNA fingerprint data actually prompted a change in taxonomical treatment. One notable example is the article by Van de Wouw et al.,[1441] who investigated 109 accessions of *Vicia* ser. *Vicia*. This series was previously considered to comprise three species, one of which is represented by the complex *Vicia sativa* aggregate. Continuous morphological and karyological variation within the aggregate made any meaningful subdivision difficult. On the basis of phenetic and principal coordinates analysis of a data set from 83 polymorphic AFLP bands, the authors provided convincing evidence that four species exist in the series, and that the *V. sativa* aggregate consists of six separate groups considered as subspecies.[1441]

Another example is based on cpDNA-CAPS variability detected by PCR with universal cpDNA primers.[84] Two parsimony analyses were carried out of all 15 recognized taxa in *Hippophae* (Elaeagnaceae), one based on cpDNA-CAPS, and one based on a combined data set with the cpDNA data and some morphological traits. The genera *Elaeagnus* and *Shepherdia* were used as outgroups. The results were congruent with those of a previous RAPD study, and in part also with some earlier taxonomic classifications. Three taxa, previously published as *nomina nuda*, could be validated and/or described as *Hippophae neurocarpa* ssp. *stellatopilosa*, *Hippophae goniocarpa*, and *Hippophae litangensis*. The latter two taxa were originally suggested to form one species with two subspecies, but were clearly not monophyletic in the CAPS analyses, and have most likely evolved by two independent hybridization events.[84]

6.5 PHYLOGEOGRAPHY

Phylogeography is a rapidly growing discipline that aims at studying the principles and historical processes governing the geographical distributions of genealogical lineages.[58,59] In contrast to classical population genetics, which is based on allele frequency distributions, phylogeographic procedures are supposed to separate population structure from population history.

In principle, phylogeographic studies can be based on information from nuclear DNA, mtDNA, and cpDNA. In practice, however, uniparentally inherited organellar markers are preferred because they are more likely to retain information about past migration histories than are nuclear markers. Organellar genomes share a number of peculiar properties that distinguish them from nuclear genomes: (1) there is only one circular chromosome present in multiple copies, instead of many linear chromosomes; (2) the organellar genome is haploid and therefore has a smaller effective population size, resulting in stronger differentiation of fragmented populations by genetic drift[404]; (3) there is usually no recombination (but see Marshall et al.[886]) and individual sequence polymorphisms can therefore be combined into haplotypes; and (4) the genome is inherited uniparentally.

In animals, mtDNA has quickly become the molecule of choice for phylogeographic analyses, mainly because the high rate of synonymous base substitutions (~1 to 2% per million years ago [Mya]). In plants, phylogeographic studies have been hampered by the much slower base substitution rates in organellar genomes compared with the nuclear genome[1229] (see Chapter 1.1). Nevertheless, the last decade has seen a rapid increase in the number of explicitly phylogeographic studies also in plants (for reviews, see Avise,[58] Ennos et al.,[404] Newton et al.,[988] Schaal et al.[1229]).

6.5.1   Phylogeography Based on cpDNA

The vast majority of plant phylogeographic studies published so far are based on the chloroplast molecule. Intraspecific polymorphisms in cpDNA are usually searched for by one of three approaches: (1) restriction site polymorphisms (cpDNA-CAPS; see Chapter 2.3.2), (2) microsatellites (cpSSRs; see Chapter 2.3.4.2), or (3) comparative sequencing of one or several stretches of PCR-amplified noncoding cpDNA (see Chapter 2.2.2). Whatever approach is used, the next step is to combine the unique polymorphisms into distinct haplotypes, followed by the analysis of haplotype distribution and frequencies across different geographical regions, quantification of the genetic divergence between haplotypes, and the evaluation of genetic relationships between haplotypes.

Important examples of recent phylogeographic studies include the discovery of glacial refugia in ice-free regions of the Arctic[2] and of southern and southeastern Europe (see reviews by Ennos et al.,[404] Newton et al.,[988] and Taberlet et al.[1359]), the population history of conifers in temperate areas, and the population history of woody angiosperms in the tropics. In the following sections, a few case studies are outlined that illustrate the use of the various types of cpDNA markers listed above.

6.5.1.1 Postglacial Recolonization of Central and Northern Europe

Climatic changes have considerably reshaped the European landscape during the Pleistocene, and plant species were repeatedly forced to retreat from the massive glaciation during the ice ages. Until recently, historical distribution of plant species and the individual fates of thermophilous trees and shrubs after glaciation could only be evaluated by the fossil record, especially of pollen sediments. These findings can now be complemented by molecular studies of extant plant material. Recently, cpDNA markers were used to reconstruct the routes of recolonization of central and northern European plains by woody plant species, including *Quercus petraea* and *Quercus robur*,[373] *Fagus sylvatica*,[327] *Alnus glutinosa*,[719] *Corylus avellana*,[1032] *Carpinus betulus*,[528] *Calluna vulgaris*,[1153] and *Hedera helix*.[527] Additional studies are in the pipeline.

As one example, the phylogeography of common ivy (*Hedera helix* and three related species) throughout its natural range was studied by Grivet and Petit.[527] Chloroplast markers were derived from cpSSRs, cpDNA-CAPS, and direct sequencing of the *trn*L and *trn*K introns. A total of 13 haplotypes were detected among 233 individuals in 27 populations. Contrary to nuclear ribosomal ITS sequences, most haplotypes were shared across species. As was found for most other woody species examined in Europe, haplotype diversity decreased from south to north. *Hedera* apparently survived in Iberian and Balkan refugia during glaciation, with a third refugium in the Alps or in Italy.

Grivet and Petit[528] also studied the phylogeography of European hornbeam (*Carpinus*). They analyzed cpDNA variation in 36 European populations of *C. betulus* and five populations of *Carpinus orientalis*. Again, three different marker types were used (cpDNA-CAPS, cpSSRs, and comparative sequencing of the *trn*K and *trn*L introns). Six haplotypes specific for *C. betulus* were found, and two were found for *C. orientalis*. No haplotypes were shared between species, indicating that there is no ongoing gene flow. One particular haplotype was fixed in all populations from western and northern Europe, but higher diversity was found in the Balkan. The latter region most probably served as glacial refugium, whereas the limited diversity across the remainder of Europe is most likely caused by a bottleneck during recolonization. Haplotypes were fixed in all populations except one, resulting in a very high G_{ST} value (0.972).

The general outcome of the above studies was that (1) recolonization during the last interglacial period was quite rapid (with up to 10 km/generation); (2) haplotypes were often unevenly dispersed, with patches of identical haplotypes covering large areas; (3) highest levels of genetic (haplotype) diversity were retained in southern European refugia; and (4) lowest levels of genetic (haplotype) diversity were encountered in central and northern Europe.

6.5.1.2 Phylogeographic Case Studies in Tropical Trees

An increasing number of phylogeographic studies deal with tropical trees. Collevatti et al.[263] presented a phylogeography of 160 individuals in 10 populations of the endangered neotropical tree species *Caryocar brasiliense* (Caryocaraceae). Networks constructed from 21 cpSSR and 11 sequence haplotypes (*trn*L and *trn*T region), respectively, proved to be incongruent, probably due to homoplasy at the cpSSR loci. Nevertheless, both trees provided evidence for the occurrence of multiple maternal lineages in Brazil.

Lira et al.[827] used seven cpSSR markers to analyze the genetic diversity of the endangered *Caesalpinia echinata* growing in tropical Brazil. With three to 29 individuals from each of seven populations from three regions, the entire extant distribution range was represented. Diversity was extremely low; five of seven populations were fixed for one haplotype. AMOVA revealed a high differentiation between regions (36% of the total variation), and between populations within regions (55%). The current distribution of the species is consistent with the existence of separate glacial refugia rather than being caused by recent anthropogenic influence.

Vogel et al.[1476] used the ccmp primer pairs of Weising and Gardner[1520] to screen for polymorphic regions in the chloroplast genome of *Macaranga*. Five primer pairs revealed 2 to 8 size variants per locus, combining into 17 haplotypes among 29 *Macaranga* accessions from 10 species. Relationships between haplotypes were assessed by phenetic analyses of size variants, and by constructing a parsimony network based on sequence variation. For both types of analysis, the distribution of haplotypes correlated with geographically circumscribed regions rather than with taxonomic boundaries. As mentioned before, such geographically structured, species-independent patterns of chloroplast haplotype distribution have also been observed in other genera,[373,909] suggesting a considerable extent of interspecific exchange of cpDNA in these taxa (see Chapter 6.3.3.1). Phylogeographic studies based on cpDNA variation thus support the concept of Rieseberg and Soltis[1171] that cytoplasmic exchange among congeneric plant species is a common event.

6.5.1.3 Phylogeographic Case Studies in Gymnosperms

Most phylogeographic studies performed in Pinaceae and other conifers have been based on cpSSRs,[379,509,1041,1467] mainly because of the availability of a set of universal cpSSR-flanking primers derived from the fully sequenced cpDNA genome of *Pinus thunbergii*.[1462]

Several studies have been performed in the genus *Abies*. Vendramin and Ziegenhagen[1461] showed that two of 11 primer pairs flanking *P. thunbergii* cpSSRs also revealed variation in silver fir, *Abies alba*. Seven and 15 size variants at these two loci, respectively, combined into 36 haplotypes among 70 *A. alba* plants from seven populations. Vendramin et al.[1464] analyzed genetic variation at the same loci in a larger set of specimens, comprising 714 individuals from 17 populations across Europe. They found eight and 18 different size variants, respectively, which combined into 90 different haplotypes. Thirty-eight haplotypes (i.e., 42%) were population-specific. Genetic distances between most populations were high and significant, genetic differentiation as measured by G_{ST} was 0.17, and the spatial organization of haplotypes suggested a positive correlation between genetic and geographical distances.

Parducci et al.[1041] investigated 29 individuals of the last remaining population of the highly endangered species *Abies nebrodensis* from northern Sicily, together with seven populations of *A. alba*, *Abies cephalonica*, and *Abies numidica*. Altogether, they found 122 haplotypes among 169 individuals. Within-population haplotypic diversity was generally high, but somewhat reduced in *A. nebrodensis* compared with the other *Abies* species. Despite the extreme reduction in population size, the few remaining *A. nebrodensis* individuals still retain a considerable amount of the original variation. AMOVA conducted over all eight populations showed that 19% of total cpSSR variation was allocated between species; 6% was allocated between populations and 74% was allocated within populations.

Several studies have been performed on maritime pine (*Pinus pinaster*). The distribution range of this species is scattered across the western Mediterranean region. Vendramin et al.[1463] found 34 cpSSR haplotypes among 10 populations sampled in different parts of the natural range. Genetic differentiation between populations as measured by the R_{ST} estimate was relatively strong. In contrast, Ribeiro et al.[1158] found almost no genetic differentiation among 12 Portuguese populations of the same species. These authors hypothesized that recent mixing of genetic material in plantations may have caused the lack of genetic structure of *P. pinaster* in Portugal, whereas spatial isolation between larger geographical regions may have prevented gene flow among more distant regions. In a subsequent study, Ribeiro et al.[1159] tested the congruence of genetic diversity and differentiation estimates derived from six cpSSRs (108 haplotypes) and two AFLP primer combinations (62 informative bands), respectively, among 12 French and 12 Portuguese populations of maritime pine. Although cpSSRs revealed much higher genetic diversity than AFLPs (as expected from the higher mutation rates of microsatellites), similar levels of genetic differentiation among populations were found with the two marker types (provided that loci with low-frequency null alleles were removed from the AFLP data set, as suggested by Lynch and Milligan[863] for dominant markers).

As mentioned, organellar markers have a two-fold smaller effective population size than diploid nuclear genomes and are thus more sensitive to genetic drift. Therefore, highly sensitive chloroplast markers are especially useful for the analysis of species that have experienced population bottlenecks. One such example is red pine (*Pinus resinosa*), which has a wide distribution range in North America but is genetically almost monomorphic with most standard molecular markers (e.g., RAPDs[956]). Nevertheless, Echt et al.[379] found substantial variation at nine cpSSR loci, with 23 haplotypes among 179 individuals from the northern part of the red pine distribution range. Most haplotypes were rare, and genetic differentiation was considerable. Somewhat lower levels of variation (six haplotypes) but even stronger geographic differentiation were later reported for the southern part of the distribution range, using an almost identical set of cpSSR markers.[1492] These data indicate substantial isolation and genetic drift within many red pine populations. Interestingly, the area of greatest haplotype diversity cannot be the origin of postglacial populations, because this area was glaciated in the Pleistocene. Walter and Epperson[1492] suggested that the present haplotype diversity in this area could be a consequence of admixture from two or more glacial refugia.

6.5.2 Phylogeography Based on Nuclear Genes

The scarcity of phylogeographic studies in plants compared with those in animals is mainly due to the comparatively low levels of intraspecific variation of plant organellar DNA. An alternative and potentially useful source of intraspecific variation is provided by noncoding regions (e.g., introns) of single-copy nuclear genes.[573,1339] The reconstruction of nuclear phylogeographies is attractive, but it is also technically more demanding compared with studies based on organellar DNA. Three reasons account for this. First, pilot studies involving Southern blot hybridization are required to ensure that the candidate genes are indeed single copy in the investigated taxa. Second, two different alleles will be amplified from heterozygotes, which makes direct sequencing of PCR products difficult. Cloning of PCR products may therefore be required prior to sequencing. Third, intralocus recombination could result in the disruption of haplotypes, which would introduce homoplasy in the data set. Despite these difficulties, intron sequences of low-copy nuclear genes have proved useful in a number of animal phylogeography studies (see review by Hare[573]).

The feasibility of reconstructing nuclear phylogeography in plant species was first demonstrated by Olsen and Schaal,[1011] who sequenced 962 bp of the single-copy nuclear glyceraldehyde 3-phosphate dehydrogenase (*G3pdh*) gene in cultivated cassava (*Manihot esculenta*) and its putative wild relatives, *M. esculenta* subsp. *flabellifolia* and *M. pruinosa*. The sequenced region encompassed four introns and three exons. A total of 64 polymorphic sites were revealed, combining into 28 haplotypes among 212 individuals (424 alleles). Only one haplotype was specifically found in the cultigen. A minisatellite region with up to five repeats of a 25-bp motif was found to be present in the second intron. Variation of the minisatellite repeat number turned out to be homoplasious and was not taken into account in defining haplotypes. However, the variability of the minisatellite often caused a considerable size difference between the two haplotypes present in heterozygotes. Alleles could then be separated by agarose gel electrophoresis, and sequenced individually. For most individuals, the two alleles of a locus could be sequenced together, and heterozygotes were scored by reading double-banded variable sites directly from the autoradiogram. Cloning was not required in either case. A maximum parsimony haplotype network showed a remarkably low level of homoplasy, indicating the absence of intralocus recombination. The distribution of haplotypes across species and regions strongly suggested that cultivated cassava originated solely from wild populations of the conspecific *M. esculenta* ssp. *flabellifolia* along the Southern border of the Amazon basin.

The results from the *G3pdh* phylogeography were corroborated by a study of genetic variation at five microsatellite loci, using the same plant individuals.[1012] The microsatellite alleles present in cultivated cassava proved to be a subset of those found in the wild *M. esculenta* ssp. *flabellifolia* populations, a situation reminiscent of other crops and their ancestors. Moreover, phenetic analysis of allele frequency data grouped the cultigen with wild *M. esculenta* subsp. *flabellifolia* populations from the southern border of the Amazon basin.

Subsequently, Olsen[1010] analyzed the 27 *G3pdh* haplotypes found in *M. esculenta* ssp. *flabellifolia* and *M. pruinosa* in more detail. Geographical distribution of haplotypes indicated IBD across the sampled range of *M. esculenta* ssp. *flabellifolia*. Only northeastern and western populations exhibited higher similarity to each other than expected from the current distribution patterns. A nested clade analysis[1382] suggested that these two regions had been connected by gene flow until recently. Fragmentation could have been the consequence of major habitat shifts in the Amazon basin, which are known to have taken place since the Pleistocene.

7

Linkage Analysis and Genetic Maps

One of the prominent applications of the various molecular markers described in this book is the establishment of genetic linkage maps. The process of genetic mapping can be defined as the determination of the linear order of molecular markers or genes (generally, loci) along a stretch of DNA (e.g., a bacterial artificial chromosome [BAC] clone, a nuclear chromosome, or an organellar genome). The result is a genetic map, which may be described as a graph depicting the relative positions of markers along so-called linkage groups (LGs), based on their frequency of crossovers or recombinations during meiosis. The distance between markers on a genetic map is given as Morgan (M) or centimorgan (cM), where one cM is the distance that separates two markers (or genes), between which a 1% chance of recombination exists (corresponding to one recombination event in 100 meioses). The average extent of recombination is dependent on the genome (e.g., *Arabidopsis thaliana*, 1 cM = 139 kb; human, 1 cM = 1108 kb).

Linkage maps with different marker densities (depths) are now commonplace with most of the more important crop plants (see Chapter 7.1.4), and can be constructed with relative ease and speed. The following steps are prerequisites for a successful genetic mapping of a target genome. First, a careful selection of parent plants precedes their mating to produce a suitable mapping population, then the progeny is individually tested for marker profiles, and pairwise recombination frequencies are calculated, LGs are established, and map distances are estimated, using powerful computer programs such as MapMaker.[773] Finally, the map order is determined.

The methodology of genetic mapping with molecular markers has been the subject of numerous reviews, including Meksem and Kahl,[912] Mohan et al.,[940] O'Malley and Whetten,[1005] Tanksley et al.,[1364] and Young.[1598]

7.1 GENERATING HIGH-DENSITY GENETIC MAPS

7.1.1 Selection of Parent Plants

It is mandatory that genetically divergent parents be chosen that exhibit sufficient polymorphisms, but are not so distant as to cause sterility of the progeny. In the absence of any polymorphism, neither segregation analysis nor linkage mapping is possible. If knowledge about the map position of a certain trait is desired (e.g., resistance vs. susceptibility toward a pathogen), the parents should be polymorphic for that trait. Naturally outcrossing species (e.g., maize) usually exhibit large numbers of DNA polymorphisms. In contrast, the extent of DNA polymorphism may be frustratingly low in inbreeding species. Frequently, the genetic base of cultivated plants is so narrow that hardly any polymorphism can be detected in the progeny of a cross (narrow cross). In such a situation, the polymorphisms in a wild crossable relative of the target cultivar (e.g., *Cicer reticulatum* in the case of *Cicer arietinum* L., the chickpea) is exploited.[1555] Such wide crosses usually produce progeny with an increased number of polymorphisms, which can be detected by any one of the DNA marker techniques described in this book. Yet chromosome pairing and recombination rates can be severely suppressed (disturbed) in wide crosses and generally lead to greatly reduced linkage distances.

7.1.2   Mapping Population

The design of the mapping population is a key and crucial step in the mapping process. Several types of mapping populations may be suitable for a particular project. The simplest one is an F_2 population derived from selfed, true F_1 hybrids. However, such populations are ephemeral; i.e., seed derived from selfing these individuals do not breed true. Of course, this limitation can be circumvented by resorting to cuttings or tissue culture, or bulking of F_3 plants, which altogether complicates the procedure. Still, F_2 populations are not ideal for mapping.[1554]

Permanent sources for mapping are definitely preferred. One such source are recombinant inbred lines (RILs) derived from individual F_2 plants by single-seed descent from sibling F_2 plants through six or more generations.[189,1151] Each RIL contains a different combination of linkage blocks from the original parents, which is the basis for linkage analysis. However, even RILs of the eighth generation are not completely homozygous, as should be expected by theory, but still contain heterozygous genomic regions. F_1 populations also have been used for mapping in, e.g., conifers (such as *Pinus radiata* and *P. taeda*), in which megagametophyte tissue from seeds of open-pollinated trees factually represent a haploid source for mapping purposes[1423] (for review, see O'Malley and Whetten[1005]).

Once a suitable mapping population is chosen, the population size is a point of concern, because the resolution of a map and the ability to determine marker order largely depend on this parameter. Generally, the larger the mapping population, the better. A vague lower threshold that can localize quantitative trait loci (QTL) is a size of 100 individuals. However, high-resolution maps for map-based cloning of target genes ideally require population sizes of more than 500 or even 1000 individuals. For a mere establishment of a genetic map (a rather academic procedure), a mapping population of fewer than 100 individuals suffices. Once the mapping population is finally obtained, DNA from each progeny is isolated and tested for the state of those DNA sequence polymorphisms that distinguish the parents (scoring). The scoring process may pose a heavy burden of time and cost to the breeder, especially when dealing with thousands of individual segregants.

7.1.3   Linkage Analysis

All the data accumulating from scoring the mapping population sequentially with a series of markers are used to construct the linkage map. Linkage analysis is based on the fact that two marker loci that are close to each other on the same chromosome tend to cosegregate; i.e., will be inherited together. This linkage can be broken by recombination between homologous chromosomes in meiosis. The frequency with which recombination occurs (i.e., the recombination fraction q) increases with the physical distance between loci. So, for example, if one microsatellite marker is closely linked to another one, they will cosegregate in the majority of the progeny.

Simple statistical tests such as a χ^2 analysis will test for randomness of segregation and therefore linkage. Two hypotheses have to be discriminated: no linkage ($\theta = \frac{1}{2}$; null hypothesis), and linkage at a recombination fraction $\theta < \frac{1}{2}$ (alternative hypothesis). The statistical criterion for linkage between two loci (traits, markers) is based on an odds ratio L (likelihood odds ratio) provided by the data. This represents a measure to test the null hypothesis that no linkage exists between two loci. The decimal logarithm of L (the so-called Lod score) is conventionally reported. A Lod score of 3 (odds ratio of 1000:1) is normally accepted as a lower threshold to assert linkage because it represents the least acceptable probability that two loci are linked. Higher Lod scores (e.g., 4 or 5) reassure the experimenter that linkage is indeed reliable. As a rule of thumb, the higher the number of progeny, the higher the Lod score.

The mean number of recombination events defines the map distance between two loci. The relationship between map distance and recombination value is characterized by a genetic mapping function (*mf*), which is a formula expressing quantitative relationships between distances in a linkage map using crossover frequency. Several types of mapping functions can be applied; the most common is Haldane's or Kosambi's *mf*. Haldane's *mf* assumes absence of interference, Kosambi's *mf* assumes positive interference (i.e., fewer double recombinants when compared with no interference).

If linkage is indicated by χ^2 analysis of progeny segregation, then the potential for linkage between loci and map distances can be tested mathematically by, e.g., an algorithm such as maximum likelihood or least squares regression methods. Various computer-based linkage programs (e.g., MapMaker[773]) make use of this statistical procedure. MapMaker performs multipoint analyses of many linked loci and has several routines simplifying this analysis, including an algorithm grouping markers rapidly into LGs and another one for suggesting the best possible order of the markers. Once a plausible order is established, another algorithm compares the strength of evidence for that order as compared with possible alternatives in a routine called ripple or ripple function. Ripple confirms the best order in a way that increases only arithmetically with increasing numbers of loci.

7.1.4   The Genetic Map

The linear arrangement of linked loci represents the so-called linkage group (LG) — all of the LGs represent the genetic map. At the start of the mapping procedure, more LGs are usually defined than chromosomes exist. As more and more markers are added to the map, the number of LGs does eventually merge with the number of chromosomes. A so-called **landmark map** illustrates the potential of DNA marker-based genetic mapping, and is exemplified in Figure 7.1 with the legume

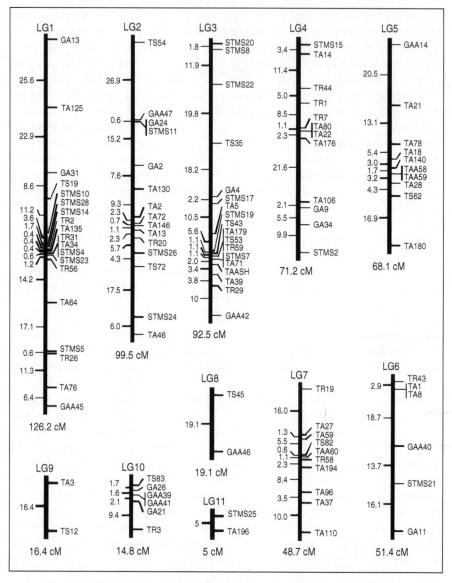

Figure 7.1   Microsatellite-based landmark map of the chickpea genome (*Cicer arietinum*). Genetic distances as computed from recombination frequencies between loci are given in centiMorgans (cM). No recombination is observed between markers connected by a vertical bar. The total size in centiMorgans is indicated below each linkage group (LG).

crop chickpea (*C. arietinum* L.). Principally, it holds for any other genome as well. In Figure 7.1, sequence-tagged microsatellite sites (STMS) and other markers are mapped on seven major LGs (most probably representing seven of the eight chromosomes of chickpea) and a series of yet unlinked smaller LGs. This map, also called a **skeleton map**, provides anchor points for extended mapping, serves to integrate different types of codominant and dominant markers (here, DNA amplification fingerprinting [DAF], sequence characterized amplified region [SCAR], amplified fragment length polymorphism [AFLP], random amplified polymorphic DNA [RAPD], and allozyme markers) into a so-called **integrated map**, and therefore is the starting point for the construction of a map with higher marker density. In such a **saturated map**, markers should optimally be spaced at less than 5-cM intervals over all LGs. Any gene will then probably reside within a few centiMorgans of a neighboring DNA marker.

Maps with hundreds or even thousands of markers now exist for all major crop plants (e.g., wheat,[1181] barley,[1136] rice,[906,1378] chickpea,[1555] cotton,[768] and sunflower,[1363] to name few), and maps have also been created for horticultural species such as roses.[314] There are ongoing efforts to saturate these maps; i.e., to add more and more markers. These markers need not necessarily be developed *de novo*, but at present, are drawn from the pool of already existing markers of a related organism (so-called syntenic markers). For chickpea, an integrated map based on 130 RILs segregating for *Fusarium* resistance was presented by Winter et al.[1555] It encompasses 354 markers including 118 microsatellites, 96 DAFs, 70 AFLPs, 37 intersimple sequence repeats (ISSRs), 17 RAPDs, eight isozymes, three cDNAs, two SCARs, and three *Fusarium* resistance loci. An example of a high-density genetic map developed for chickpea is shown in Figure 7.2.

Genic markers can also be placed on such maps if polymorphisms can be detected in the parental target genes. Because genes are more conserved throughout evolution than are nongenic sequences, polymorphisms in genes are not easily detectable, and mostly represent single-nucleotide polymorphisms (SNPs; see Chapter 9). It is also possible to discover genic polymorphisms by using more sophisticated techniques such as **sequence-specific amplified polymorphism (S-SAP)**[1510] anchored in genes (see Chapter 2.3.8.2), or **resistance gene analog polymorphism (RGAP)**[240] **mapping** (see Chapter 2.3.6). For chickpea, more than 250 different genes from various gene families were tested for restriction site polymorphisms and sequence variants in the vicinity of the genes (e.g., by using a gene-specific forward and a retrotransposon-specific reverse primer for amplification). Only 58 of these gene-anchored fragments proved to be polymorphic and could be mapped.[1075] Using the RGAP approach,[240] degenerate primers targeted at conserved domains in the nucleotide-binding site (NBS) region (and in various combinations with transmembrane domains, leucine-rich repeats [LRRs], and coiled coil domains) were successfully used to define 38 fragments of different RGAs of chickpea.[634] Thirteen of these could be mapped genetically, and only two appeared to cluster.

Still another often quite successful approach to map expressed genes is so-called **expressed sequence tag (EST) mapping**. In an EST analysis, the whole complement of messenger RNAs of a cell, a tissue, an organ, or an organism (in case of single-celled organisms) is converted to cDNAs, which are then directionally cloned into

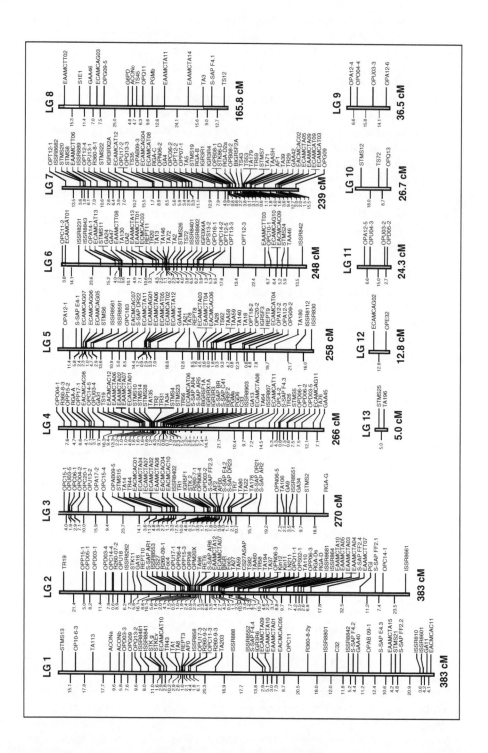

a suitable vector and partially sequenced from either their 5'- or 3'-ends in a single-pass procedure. The resulting 300- to 500-bp sequences allow the identification of the corresponding genes by a similarity search against the public databases. By 2004, the EST database in Genbank (http://www.ncbi.nlm.nih.gov/dbEST/) contains close to 20 million EST entries from almost 500 different organisms.[116] These ESTs can be tested for expressed sequence tag polymorphisms (ESTPs), defined as DNA sequence differences between two (or more) ESTs, that can be detected by restriction analysis or by separation of polymorphic ESTs in denaturing gradient gel electrophoresis.[1060] ESTP markers can be combined in an EST map, which is a graphical depiction of the positions of (preferentially all) ESTs along a chromosome. EST maps are therefore based on genic markers rather than anonymous molecular markers generated by the majority of the techniques detailed in this book. Alternatively, ESTPs can be integrated into existing genetic maps.[1060] In each case, the ESTs label the expressed portion of a genome.

A mapping project is often started with the objective to detect linkage between one or several markers and a trait of interest. Although the density of integrated genetic maps can be extremely high, it may still not be sufficient to tag a specific gene. Gene tagging is the process through which two markers are identified that are closely linked to, and flank the mapped trait. Usually, close linkage is inferred from distances of about 0.1 to 1.0 cM. Thus, a comprehensive map covering all chromosomes evenly might not be dense enough around the locus (trait or gene) of interest. Therefore, **high-density mapping** (fine mapping) is started.

Among several techniques for fine mapping, the so-called **bulked segregant analysis (BSA)** is dominant.[499,928] Two bulked DNA samples from at least 10 individuals of an F_2 population originating from a single cross are drawn. These bulks are homogeneous for a particular trait (e.g., are resistant or susceptible to a specific pathogen, respectively), but heterogeneous at all unlinked regions. The bulks are screened for DNA polymorphisms, and the detected differences compared with a randomized genetic background of unlinked loci. Any differences between these bulks (e.g., presence vs. absence of a band on a gel) represent a candidate for a marker linked closely to the trait in question. Linkage has to be verified in a segregating population.

BSA is superior to other approaches of fine mapping. For example, **near-isogenic lines (NILs)**, generated by repeated backcrossing with selection for the desirable trait at each round of crossing, are essentially identical at all genetic loci except the region bracketing the gene under selection.[890,1600] Thus, any detected polymorphism will most probably reside in close vicinity to the introduced gene. However, NILs require many backcrosses for their development, and, more importantly, may exhibit

---

Figure 7.2 (opposite page) High-density integrated map of the chickpea genome (*Cicer arietinum*), including RAPD, ISSR, AFLP, microsatellite, S-SAP, and RGAP markers.[633,634,1075,1555] Genetic distances as computed from recombination frequencies between loci are given in centiMorgans (cM). The eight largest linkage groups (LG 1 to 8) correspond to the eight chickpea chromosomes. Despite the high map density, several small groups of markers remain unlinked. No recombination is observed between markers connected by a vertical bar. The total size in centiMorgans is indicated below each linkage group.

a linkage drag (i.e., genes incorporated into lines by backcrossing that are flanked by DNA segments from the donor parent). In contrast, all loci detected during BSA will segregate and can be mapped, thus eliminating linkage drag problems (for a review, see Young[1598]). Note that BSA is also widely applied to generate markers for marker-assisted selection, especially for those crops for which classical mapping procedures are complicated by huge genomes and long generation times (e.g., as in many forest trees).

Whatever technique is finally used, a fine map encompassing the gene(s) of interest is created, as exemplified in Figure 7.3. Here, the region of interest, residing on chickpea LG 2 of the above saturated map, harbors several agronomically defined genes conferring resistance toward different races of *Fusarium oxysporum* f.sp. *ciceri*. Fine mapping of this resistance gene cluster with various marker techniques using BSA identified closely linked PCR markers. DNA sequencing of several of these markers revealed high homology to different genes that are potential resistance gene candidates (e.g., a gene encoding an NBS- and LRR-containing receptor kinase, a transcription factor, a chromatin condensation protein, a MutS2-like DNA mismatch

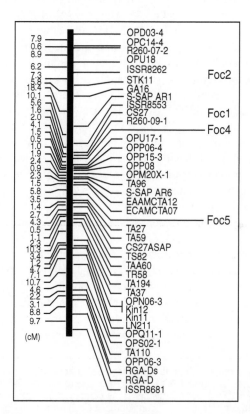

Figure 7.3   Fine mapping of a resistance gene cluster on chickpea linkage group 2, based on a bulked segregant analysis (BSA).[928] This cluster harbors several genes conferring resistance toward different races of *Fusarium oxysporum* f.sp. *ciceri* (Foc loci). See text and legend of Figure 7.2 for additional details.

repair protein, a PR-5 thaumatin-like protein precursor, and an anthranilate *N*-hydroxycinnamoyl/benzoyltransferase, an enzyme of the phytoalexin pathway).[109]

Even if a region of interest is crowded with DNA markers, better techniques are required to resolve marker order. The most straightforward way to improve the genetic resolution is increasing the size of the mapping population. However, huge populations are difficult to manage, especially if resolution as high as 0.1 cM (corresponding to one crossover in every 1000 individuals) is at the aim, which is a prerequisite for map-based cloning of targeted genes (see Chapter 7.4).

If two or more genetic maps share a set of common markers, they can be merged to create a more informative map. Such anchor markers should be transferable between populations, a property that is characteristic of only few marker types (such as the codominant microsatellite and RFLP markers). However, the precision of estimates of recombination frequencies may vary greatly between populations and data sets, so that any merged map will be a compromise only. Computer packages are available that consider the estimates of recombination frequency between a given pair of markers of different origin (data sets or mapping populations), calculate and apply the appropriate weighting, and generate a single recombination value. One such computer program is JoinMap,[1323] now in its fifth version, which assigns weights to all available pairwise combinations and institutes a numerical search for the best-fitting linear arrangement of marker loci.

Genetic maps of plants (and other organisms) are regularly updated on the Internet. Examples of plant genetic maps can be found at http://www.nal.usda.gov/pgdic/Map_proj/ and http://www.ncbi.nlm.nih.gov/genomes/PLANTS/PlantList.html.

7.1.5 Cytogenetic Maps

Once a genetic map is constructed, it is desirable to correlate it to a cytogenetic or chromosome map. To that end, metaphases of the target plant (usually from root tips) are first synchronized, the metaphase chromosomes are isolated by, e.g., fluorescence-assisted chromosome sorting (FACS), and the isolated chromosomes are hybridized to fluorescence-labeled DNA markers diagnostic for the different LGs of the genetic map. For microsatellite markers, the primers complementary to microsatellite-flanking regions are used to amplify the corresponding microsatellite locus in the presence of fluorinated deoxynucleotides. The labeled marker sequences serve as probes in fluorescent *in situ* hybridization (FISH). Preferentially, several markers from a single LG are pooled to increase signal intensity. Eventually, a specific LG is assigned to a specific chromosome. Cytogenetic mapping may be considered a special type of physical mapping. In chickpea, chromosome sorting allowed us to separate all eight chromosomes, and cytogenetic mapping with microsatellite markers proved that the smallest chromosome H corresponds to the smallest LG 8.[1474]

7.1.6 Genetic vs. Physical Maps

A physical map represents the linear arrangement of loci (genes, markers) on a chromosome or part of a chromosome as determined by techniques other than genetic recombination (see, e.g., Klein et al.[722], Röder et al.[1180]). Usually, physical map distances are expressed as numbers of nucleotide pairs (mostly kilobase pairs [kb] or megabase pairs [Mb]) between identifiable genomic sites. Because of the different principles of measurement (genetic map distances are calculated as percentages of genetic recombination; see Chapter 7.1), the map distances of genetic and physical maps generally differ. In addition, the relationship between genetic and physical distances varies dramatically from location to location on a chromosome. For example, in the maize genome, the average ratio is 1500 kb per cM, whereas at the *bronze* gene locus it decreases to 14 kb per cM, but is 217 kb per cM in the *a1/sh2* gene region. The main reason for such effects is that recombination rates vary between different regions of the chromosomes (see below), between male and female sexes, genes and nongenic spacers, and silent genes and expressed genes, and may also be specific for particular genetic backgrounds.[358,791]

In general, recombination is suppressed near centromeres and in heterochromatic and introgressed regions, whereas markers in other regions recombine at a higher level. Therefore, many DNA markers are clustered in linkage maps (see Figures 7.1 and 7.2), the clusters probably corresponding to centromeres or heterochromatic stretches. In other parts of the map, markers are sometimes separated by large gaps that persist even after many hundreds of markers have been placed on the map. Despite the uneven distribution of markers in terms of recombination frequency, most markers are probably more uniformly distributed on a physical map (see Röder et al.[1180,1181]).

For microsatellite markers, the extent of clustering around centromeres seems to have a species-specific component and, e.g., is more pronounced in barley[1136] compared with rice.[1378] If AFLP markers are used for linkage mapping, clustering also strongly depends on the restriction enzyme(s). For example, Young et al.[1601] compared the locations of AFLP markers generated by *Eco*RI-*Mse*I digestion with those obtained by *Pst*I-*Mse*I on the soybean genetic map. As already noted by Keim et al.,[706] the *Eco*RI-*Mse*I-derived markers were quite unevenly distributed, forming at least one dense cluster per LG. Conversely, *Pst*I-*Mse*I-generated fragments showed a more even distribution, and proved to be under-represented in the clusters formed by the *Eco*RI-*Mse*I markers. The common presence of one *Eco*RI-*Mse*I marker cluster per LG indicates their location in heterochromatic regions surrounding the centromere. Such regions are usually also characterized by high levels of cytosine methylation. Given that *Pst*I is methylation-sensitive, it is supposed to cleave only rarely in heterochromatin, and hence *Pst*I-*Mse*I fragments preferentially occupy other map positions than *Eco*RI-*Mse*I fragments. Young et al.[1601] conclude that one single combination of restriction enzymes is not sufficient to provide complete map coverage with AFLP markers, and that an AFLP-based mapping strategy should use both methylation-sensitive and -insensitive enzymes.

The ultimate physical map of a genome is its complete nucleotide sequence (**genome map**). Genome maps are a luxurious source of genetic informations. More than 100 bacterial or archaeal genomes, a dozen fungal genomes, two complete plant genomes (*Arabidopsis thaliana* and rice), several invertebrate and vertebrate genomes including mouse and man are already known base by base; many other genomes from bacteria to mammals are presently being sequenced; and about 100 additional genomes are on target. However, some genomes are huge and filled with repetitive DNA. For example, 5356 Mb of DNA per haploid genome is present in barley, 8300 Mb in rye, 36,000 Mb in *Lilium formosanum*, 140,000 Mb in the lungfish *Protopterus aethiopicus*, and 690,000 Mb in *Amoeba dubia* (see Tables 8.3 and 8.4 in Graur and Li[525]). For comparison, the haploid human genome harbors only about 3300 Mb.

Obviously, genome maps cannot be established right away in all species. This may change soon because novel sequencing strategies (such as whole genome shotgun sequencing [WGS])[1514] compete with the traditional clone-by-clone sequencing strategy, may even replace it in near future, and can also manage gigantic genomes. The WGS approach capitalizes on the mechanical fragmentation of the genomic DNA into overlapping pieces of some 5000 to 10,000 bases each, the sequencing of the ends of these fragments, and their assembly into a genomic sequence using powerful software.

At present, both WGS and clone-by-clone sequencing are used side by side, but more powerful techniques are being designed. For example, single DNA molecule sequencing may become routine, e.g., with the *Staphylococcus aureus* α-hemolysin nanopore detector sequencers, in which hemolysin is part of an artificial membrane immersed in a buffer, across which a voltage is set.[532] Negatively charged double-stranded DNA on the outer side will move toward the positively charged inner side, but can only cross the membrane through the hemolysin channel. Here it is denatured into single strands, which are pulled through the 1.5-nm-wide inner channel of the hemolysin. As each nucleotide crosses the opening, it produces a base-characteristic decrease in current (see Gu et al.[532]). Such systems are capable of an enormous throughput (e.g., 100,000 bases in 1 min), but are still in the early stages of development. Nevertheless, it is expected that future physical maps will be established in extremely short time and at a low cost.

7.2 SYNTENY: THE COMPARATIVE ANALYSIS OF GENOMES

Synteny describes the conserved order, sequence, and orientation of DNA sequences (including genes) in the range from about 500 kb to whole chromosome segments in the genomes of different species, exhibiting various degrees of relationship. The extent of conservation can be visualized on whole chromosomes by FISH (also called **macrosynteny**), but also exists on the level of groups of genes (**microsynteny**). For example, a region containing genes conditioning the absence of ligules is largely conserved among rice, wheat, and maize.[9,944] However, translocations, duplications, and inversions have also created species-specific arrangements of large sequence blocks, so that genetic maps need to be adjusted to detect synteny.

Synteny of colinearly arranged sets of genes or markers is detected in more and more species. As an example, the order of specific blocks of markers was found to be conserved even between *Arabidopsis thaliana* and *Cicer arietinum*,[109] members of different families (Figure 7.4). Although gene sequences and their map order are often highly conserved, neither are the size, sequence, or composition of the DNA in between genes. Therefore, most of the genetic diversity between species resides

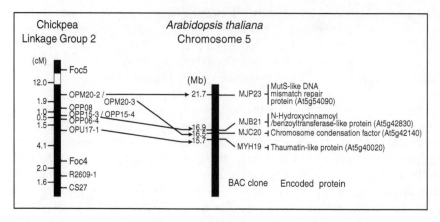

Figure 7.4 An example of microsynteny between *Arabidopsis thaliana* chromosome 5 and *Cicer arietinum* linkage group 2.[109] The sequence of markers in the latter is reversed as compared with Figure 7.3. cM: centiMorgan, Mb: megabase pairs. See text for details.

in intergenic regions. For genetic mapping, recognition of synteny has catalyzed the use of common markers within plant families (e.g., Poaceae[114,334]) as well as across plant families (e.g., between *Arabidopsis* and legumes[109,522,790]).

In many cases, microsynteny assisted in the isolation of specific genes. Once a gene is cloned and sequenced from one plant, the sequence information can be used to isolate the orthologous gene in another plant. For example, the chalcone flavonone isomerase gene in maize was isolated on the basis of sequence information from *Petunia*, snapdragon, and bean.[531] However, a problem arises with the exploitation of syntenic markers because about 15 to 30% of all genes are probably species-specific, and therefore not transferable from species to species, or beyond (e.g., Pereira et al.[1068]).

The pattern of conserved linkage can be extended to include QTLs.[430,726,772,1049] Thus, if a QTL is located in a genetic map of one species, it is highly probable that the same QTL is part of the homoeologous region in a related species (for a review see Bennetzen[111]). As a consequence, comparative mapping (i.e., the establishment of linkage maps in species of unknown genome composition using mapped markers from a related or unrelated species), now is an important component in most mapping projects.[112,1241,1242]

7.3 MARKER-ASSISTED SELECTION

One of the prime uses of DNA markers and genetic maps was, is, and will be marker-assisted selection (MAS; also known as marker-based selection or marker-mediated selection). MAS capitalizes on the identification of individuals in germplasm banks with DNA markers closely linked to an interesting trait, and the further exploitation of these individuals for crosses in breeding programs. Close linkage infers that the markers are flanking the tagged gene(s) at a resolution of at least 2 cM and even better at zero recombination (which is extremely rare). It is expected that the diagnostic DNA marker(s) will tag the gene(s) underlying the trait, so that no circumstantial and time-consuming field tests have to be performed. Thus, instead of selecting for the trait, which can cause undue environmental impact, the breeder selects for a DNA marker — an already successful procedure to tag agronomically relevant genes (see, e.g., Mohan et al.[940]). Such markers can also be identified by, e.g., BSA (see Chapter 7.1.4), and therefore independently of any genetic map. Moreover, such markers can be detected very early in the selection procedure, so that the breeder can significantly reduce the number of seedlings grown and screened, thereby reducing expenses and enhancing efficiency of breeding. MAS applies for monogenic, oligogenic, or qualitative traits, and polygenic or QTL characters, and can be used to pyramid major genes for a trait (e.g., resistance genes) to produce varieties with improved properties. For example, pyramiding of *Xanthomonas* blight-resistance genes *Xa*1, *Xa*3, *Xa*4, *Xa*5, and *Xa*10 in different combinations using molecular marker tags proved to be efficient for developing more resistant rice cultivars.[1594]

MAS is successfully used in some crop breeding schemes and certainly will gradually earn its position in plant breeding methodologies. One of the reasons for its slow introduction into breeding is because of the prevailing separation of marker development and genetic mapping from plant breeding. In addition, the cost of screening germplasm banks for the presence of the tag frequently remains prohibitive, though cost-effective methods have been recommended.[533]

7.4 MOLECULAR MARKERS AND POSITIONAL CLONING

Perhaps the most attractive application of high-density genetic maps is the isolation of tagged genes encoding proteins responsible for the trait in question. The isolation procedure is known as map-based cloning (or positional cloning), and is defined as the molecular cloning of a specific gene in the absence of a transcript or protein, using genetic markers tightly linked to the target gene and a directed or random chromosome walk by linking overlapping clones from a genomic library of bacterial artificial chromosomes (BACs) or yeast artificial chromosomes (YACs). In addition to a dense genetic map and closely linked DNA markers, such a large-insert genomic library is a prerequisite for map-assisted cloning. YAC vectors with a capacity of 400 to 600 kb are still in use, though chimerism of the inserted DNA fragment is a source of concern. During the last decade, BACs with insert sizes of 100 to 150 kb and also cosmid vectors with a capacity of about 40 kb are increasingly being used. Many BAC libraries for animals, plants, and fungi are now available, and many more will soon be added.

In the standard procedure of positional cloning, markers that are tightly linked to the trait of interest are used as probes to isolate clones from a BAC library. The isolated clones are fragmented by, e.g., shearing or restriction, are subcloned, and the various subclones are sequenced from both ends. Overlapping clones are then identified on the basis of their sequence similarity at both ends, and an ordered arrangement of clones by their sequence overlaps, a so-called contig, is established. Eventually, the contig should completely cover the genomic target region where the trait resides.

Now, **chromosome walking** procedures can be employed to "walk" into the gene or genes in question. These can be identified tentatively by sequence comparisons to known genes with similar functions, but candidates will have to be verified by (1) isolation of a corresponding cDNA and (2) the use of this cDNA to complement mutants defective in the target gene function. For example, if a putative resistance gene is isolated, its cDNA has to be retrieved and will be transferred to a susceptible variety by standard gene transfer techniques. The transgenic plant will only show the resistant phenotype if the transferred cDNA was derived from the true resistance gene (gene complementation). Only after these crucial experiments is a gene isolated by map-based cloning procedures considered to be a true candidate, which can then be used to engineer susceptible plants for resistance. The map-assisted cloning of tagged genes, described in detail by Tanksley et al.,[1365] still is the method of choice and exhibits proven potential (exemplified by resistance gene isolation in plants [*pto* gene in tomato,[891] *RPM1* gene in *Arabidopsis*,[523] *RPS2* gene in *A. thaliana*,[935] *Xa*21 gene from rice,[1312] and *Cf*-2 gene in tomato[351]]).

Although the success of map-based cloning can be measured by literally scores of genes, it is still not trivial. For example, a problem *a priori* is the enormous size of eukaryotic genomes, which in higher plants is typically 1000 Mb or more. This suggests that even when 1000 markers are placed on a linkage map, the average physical distance between the markers will be 1 Mb in most plants (and more than 10 Mb in wheat or oats; see Chapter 7.1.6). Such distances are much too large for even sophisticated walking techniques. Another point of concern usually is the relationship between genetic distance (measured as recombination frequency) and physical distance (measured in number of nucleotides), which is not at all uniform throughout eukaryotic genomes (see Chapter 7.1.6). Even markers that appear close to a gene of interest in terms of recombination frequency may still be far away in terms of actual number of bases. Many reports describe this widespread discrepancy, which may exceed ratios of 10,000 or more.[791] Moreover, most eukaryotic genomes contain significant portions of repetitive DNA (see Chapter 1) that altogether hinder facile chromosome walking. Thus, dispersed repetitive DNA sequences frequently cross-hybridize to the overlapping stretch of DNA that identifies the next step of a chromosome walk. Should this happen, contig building and walking beyond the clone may prove impossible.

A comparatively novel approach to exploit genome-wide gene expression data for genetic analysis combines microarray-based expression profiling with genetic or physical mapping.[656] This so-called **genetical genomics** strategy builds on two distinct inbred strains (or parents) that are crossed to produce an F_1 generation, which in turn is crossed to generate the F_2 generation. Each F_2 individual has a random mixture of chromosome blocks originating from either parent. Any character (phenotype) in the F_2 population can be traced to a specific chromosomal location. This is achieved by correlating genetic (genotypic) and phenotypic variation by statistics. Now, the phenotype can also be a variation of gene expression across the individuals of the population. The gene expression levels in the parents and the progeny can be determined by appropriate microarrays, onto which preferentially all genic sequences of the species are spotted as oligonucleotides or cDNA fragments. Such arrays are hybridized to the differentially labeled cDNAs from the progeny to determine what genes are expressed and to what extent. The expression level of each gene is treated as a quantitative trait. Then a statistical analysis correlates DNA variations as determined by microsatellite markers (for example) with gene expression levels. Any statistically significant correlation suggests that the gene (or genes) in the chromosomal region, where the sequence variation is located, accounts for some of the variation in gene expression. Such loci are coined **expressed quantitative trait loci (eQTL)**. Although in its infancy, the concept of eQTLs or expression mapping has already shown its potential in maize, mouse, and man.[1230]

8

Which Marker for What Purpose: A Comparison

In this chapter, we compare different marker techniques for their ability to detect and quantify genetic variation. For additional views on the choice of molecular marker-based methods for various applications in plant genetic research, see Bachmann,[65] Crawford,[283] Hokanson,[604] Mohan et al.,[940] Nybom,[999,1000] Nybom and Bartish,[1001] and Vekemans and Jacquemart.[1458]

8.1 MORPHOLOGICAL CHARACTERS AND ALLOZYMES VS. DNA MARKERS

Morphological characters have long been used to identify species, genera, and families; to evaluate systematic relationships; and to discriminate cultivars, breeding lines, etc. (e.g., in the distinctness–uniformity–stability [DUS] test; see Chapter 6.2.2.1). In contrast to molecular markers, morphological characters are often strongly influenced by the environment, and consequently, special breeding programs and experimental designs are needed to distinguish genotypic from phenotypic variation. In addition, for some small, nonflowering plant species such as mosses and algae, it frequently is difficult to find a sufficient number of morphological characters for a comprehensive study. In contrast, molecular technologies provide almost unlimited numbers of potential markers at the genotypic level.

Correspondence between morphological characters and DNA markers in terms of genetic distances between entities (e.g., populations or cultivars) is sometimes surprisingly high, indicating that both types of data are usually good estimators of genetic relatedness. This is especially true if the morphological descriptors are represented by quantitative rather than qualitative data sets.[893,1201] Morphological variability due to qualitative traits, presumably governed by a single or a few gene mutations, can actually be much more pronounced than the extent of variation indicated by random amplified polymorphic DNAs (RAPDs) or other molecular markers.[804] Consequently, certain groups of plants (e.g., cultivars, subspecies, or species) appear to be considerably better defined through the analysis of qualitative morphological traits rather than by DNA markers.[1013] Sometimes, the evaluated characters have been chosen, on purpose, to reflect plant breeding efforts and may therefore have been subjected to strong selection pressure. Thus, the morphological traits recommended by the International Union for the Protection of New Varieties of Plants (UPOV) for cultivar characterization often show low correspondence to other morphological characters and/or RAPDs.[476,1071]

A frequently used non-DNA method for detecting and estimating genetic variation is allozyme electrophoresis (see Chapter 2.2.1). There are three general advantages of allozymes over many kinds of DNA markers: (1) the low cost for chemicals and labor, (2) the user-friendliness (many individuals can be scored for several allozyme loci within a short time span), and (3) the fact that allozyme markers are codominant, i.e., both alleles in a diploid organism are usually clearly identifiable, and heterozygotes can be discriminated from homozygotes. However, there are also a number of limitations. Thus, a new allele will only be detected if it affects the electrophoretic mobility of the studied molecule. Only about 30% of nucleotide substitutions result in polymorphic fragment patterns, and allozyme analysis therefore underestimates the genetic variability. Another problem is that many plant species are polyploids, which can make the interpretation of allozyme patterns quite difficult. A more practical aspect is that the plant tissue intended for allozyme studies has to be processed shortly after harvest because proteins are usually quite unstable. In contrast, DNA-based methods allow a much longer time span between harvest and processing, and are certainly more suitable for plant material collected, e.g., in remote areas.

In the past, allozymes have been used rather extensively for the discrimination of genotypes, but are now superseded by DNA markers because the latter usually detect much higher levels of polymorphism. Allozyme-derived estimations of genetic distances among pairs of samples or populations often compare quite well with those obtained with DNA markers. However, still higher similarities are generally obtained when different DNA-based methods are compared with each other,[290,642,1473] and allozymes appear to be less efficient than DNA markers when relatively distant populations are analyzed.[268] Reasons for this could be that mutation rates are lower for allozyme loci, or that selection is effective against certain allozyme alleles.

A comparison of a set of 108 RAPD-based studies[1001] with an even larger compilation of earlier allozyme studies[562] yielded quite coherent results. Thus, both (codominant) allozyme and (dominant) RAPD markers indicated that long-lived, outcrossing, late-successional taxa retain most of their genetic variability within populations,[1001] whereas annual, selfing, and/or early successional taxa allocate comparatively more of their genetic variability among populations. When estimates of within-population variation were compared across all studies (with data based exclusively on polymorphic markers), mean values for allozymes and RAPDs were almost identical. In contrast, overall estimates of among-population differentiation were somewhat lower for allozymes than for RAPDs.

8.2 DIFFERENT KINDS OF DNA MARKERS

Chapter 2 illustrates the enormous number of DNA-based marker methods now available for the detection of genetic variation. Hybridization-based methods, involving classical RFLP probes as well as non-species-specific mini- and microsatellite DNA probes, have largely given way to various single- or multilocus, dominant or codominant polymerase chain reaction (PCR) -based methods. Results obtained by the latter are usually well correlated with earlier RFLP data,[360,1202,1398] but the number of polymorphic markers obtained within a given time unit, in general, is considerably higher.

For some applications, it may be sufficient to choose a single, optimally designed marker system. Among the points to consider during the selection process are, e.g., the level of genetic variability in the plant material, financial demands, and the availability of equipment and technical skills.[283,544,1059] For complex research projects endeavoring to analyze, e.g., reticulate evolution, a combination of different marker systems may be more adequate, because it can reveal separate aspects of the involved processes and thus yield more accurate information.[46,268]

8.2.1 Discriminatory Power

Anonymous DNA fragments and PCR amplification products, obtained either as multilocus banding patterns (e.g., amplified fragment length polymorphism [AFLP], RAPD, intersimple sequence repeat [ISSR]) or single-locus allelic data (e.g., restriction fragment length polymorphism [RFLP], sequence characterized amplified regions [SCAR], microsatellite DNA loci) have proven very useful for the identification and discrimination of individuals and cultivars (see Chapter 6.2). Locus-specific microsatellites are often believed to be superior to at least RAPD — and often also to AFLP and ISSR — for cultivar identification, because (1) at least in principle, alleles and genotypes can be assigned unambiguously; (2) primer sequences can easily be distributed among different laboratories; (3) reproducibility is much higher than at least for RAPD[677]; (4) microsatellites are more variable and hence provide higher resolution; and (5) markers are codominant.

Discriminatory power can be measured with various indices and appears to be highly dependent on the number of loci and/or bands analyzed.[414,470,515,907] Because of their multiallelic nature, microsatellite markers are most efficient for the discrimination of genotypes on a per-locus basis. However, they tend to be as informative as RAPDs or even less informative on a per-primer basis, whereas AFLP clearly is the most informative method because of the high multiplex ratio (i.e., the number of markers obtained in a single experiment). In a representative study of olive cultivars (*Olea europaea*), AFLPs scored highest for marker index[1095] (MI), discriminatory power[1384] (D), and effective number of band patterns per assay (P), but quite low for expected heterozygosity averaged over all loci (H_e).[103] Microsatellites had the highest H_e and intermediate values for D and P, and RAPDs yielded intermediate values for MI and low values for H_e, D, and P.[103] For ISSR, efficiency is much higher and approaches that of AFLP when polyacrylamide gels and silver staining are used for fragment separation and visualization, respectively. The efficiency of ISSR is similar to the efficiency of RAPDs, when agarose gels and ethidium bromide staining are used instead.[515,907]

The detection of somaclonal variability generally requires a particularly large number of bands (Chapter 6.2.3). In a comparative analysis of *in vitro*-cultured hop (*Humulus lupulus*) meristems, AFLP analysis (five different combinations of primers) proved to be the most sensitive method, ranking before RAPD (32 primers) and ISSR (seven primers). No variation was detected by microsatellite analysis (10 primer pairs).[1052]

In general, few polymorphisms at the intraspecific level are detectable by analyses of plant organellar DNA. Nevertheless, organellar markers are often used to detect maternal lineages in phylogeographic studies (Chapter 6.5), whereas nuclear markers may differentiate among individual genotypes within these lineages.[168] In olive, the application of three different organellar DNA marker methods (classical RFLP, cleaved amplified polymorphic sequences [CAPS], and chloroplast simple sequence repeat [cpSSR] analyses) produced some contradictory results, with the PCR-based methods yielding more polymorphism and more reliable groupings of the plant material than obtained with the classical RFLP method.[125] Among the various organellar DNA markers available, chloroplast microsatellites (cpSSRs) tend to yield the highest levels of polymorphism[1114] (see Chapter 2.3.4.2). cpSSR markers have become especially popular for studies of gymnosperms, the chloroplasts of which are paternally inherited and therefore suitable for analyses of pollen-mediated gene flow and long-term historical events[289,1159] (see Chapter 6.5.1.2).

8.2.2 Genetic Distances

Patterns of genetic distances or relatedness among pairs of genotypes, cultivars, populations, etc., are usually quite similar when derived from different multilocus methods, provided that enough primers and primer combinations ensure a sufficient minimum number of polymorphic bands and good genome coverage.[79,138,456,470,477,515,516,1434,1473]

Genetic distances revealed by codominantly inherited microsatellite markers are mostly correlated with those obtained by dominant AFLP, RAPD, and ISSR markers, but correlations among the latter methods are generally stronger.[515,1051,1052] In addition, coefficients of variation per band have been reported to be higher for microsatellite analysis than for RAPDs.[1326] Uptmoor et al.[1434] compared genetic distances revealed among a set of *Sorghum* cultivars by several methods. Correlation coefficients were higher between RAPD- and AFLP-derived data than between either one combined with microsatellite data. In other cases, estimates obtained with dominantly inherited DNA markers and microsatellite markers differed considerably.[24,302] However, some of these discrepancies are certainly artefactual and caused by the use of an insufficient number of samples and/or loci.

Simulation studies suggested that four to 10 times more loci must be analyzed when using dominant markers to attain the same accuracy as with codominant microsatellites.[883] Many published studies are based on only five to 10 microsatellite DNA loci, or even fewer. Clearly, this is not sufficient considering that the gain in accuracy is substantial between five and 20 loci, and still improves considerably when 50 loci are used.[883] Obviously, the number of applied markers is crucial for an assessment of the accuracy of the results: the potential precision can be compared by counting the number of polymorphic bands for dominant markers and the total number of alleles minus the number of loci for codominant markers.[570]

8.2.3 Within- and Among-Population Variation

In a large data compilation, overall AFLP-, RAPD-, and ISSR-derived estimates of within-population diversity proved to be quite similar in magnitude, whereas microsatellite-derived estimates were at least twice as high.[1000] In contrast, values of among-population differentiation were somewhat higher when obtained with the three types of dominant markers. Associations between the DNA marker data and various life history traits of the investigated species, however, were quite similar when RAPD and microsatellite analyses were compared.

Estimates of genetic variation obtained with different types of dominant markers in the same plant material are usually also quite similar, both for within-population diversity and for population differentiation[1030,1344,1473,1485,1607] (see also review in Nybom[1000]). In particular, RAPDs and AFLPs often reach very similar estimates, whereas ISSRs tend to produce somewhat higher estimates of within-population variation. ISSRs may also attribute more of the variation to lower levels (e.g., within populations) and less to higher levels (e.g., between populations or regions).[1119] Nonetheless, the opposite has also been reported.[885]

A lack of correlation among estimates for within-population diversity has sometimes been observed when microsatellites and dominant DNA markers were applied to the same set of populations.[882,1343] Such results are probably due to an inadequate number of analyzed microsatellite loci.[302,515] AFLPs are considerably more efficient than microsatellites when assignment of a particular genotype to a particular population is an issue.[211]

In a set of data simulations, Mariette et al.[883] compared levels of discrepancy between diversity estimates obtained with a subset of loci (five, 20, or 50 microsatellite loci and 50, 200, or 500 AFLP loci) and a simulated set of 1000 loci. These levels of discrepancy differed between both the type of marker and the type of evolutionary scenario. Correlations between marker types were especially low in scenarios with (1) low interpopulational heterogeneity (e.g., large populations and high gene flow), (2) high intragenomic heterogeneity (poor genome coverage by using the lowest number of loci), and (3) in recently created populations. AFLP markers were overall more robust than microsatellite DNA markers. The latter proved to be the best predictors when migration rates were high, but performed poorly at low migration rates. In general, about four times as many dominant marker bands are needed to obtain the same efficiency as with codominant marker alleles. If migration rates are very high and genome heterogeneity is low, at least 10 times as many dominant markers are required.

8.2.4 Gene Tagging and Genetic Linkage Mapping

Marker-assisted selection is becoming an increasingly important tool in plant breeding (see Chapter 7.3). Given that cost-effectiveness is a major concern when screening large progenies,[853] simple PCR-based methods such as AFLP, RAPD, and ISSR are generally preferred. In the future, EST markers will become more widely used because they identify a gene of interest with a much higher precision and in a genetically more variable material.[1060] ESTs are often characterized by single-nucleotide polymorphisms (SNPs), which provide biallelic markers that are very amenable to automated analysis systems with either electrophoretic (e.g., by developing CAPS markers) or nonelectrophoretic detection (e.g., DNA microarray) together with fluorometry or colorimetry,[276,1122] (see also Chapter 9).

Genetic linkage maps have been produced for many species, often mainly based on dominant PCR-derived markers, which are especially useful for developing the highly saturated maps required for gene identification and cloning experiments (see Chapter 7). The need for suitable anchor markers, however, is large if genetic maps are designed to be relevant not only for specific crosses but for the whole species, or even for related species. Until now, anchor markers have been provided mainly by codominant allozyme, RFLP, and microsatellite markers (see Chapter 7), but ESTs will become increasingly important.

8.2.5 Costs

Costs of the various marker methods are seldom compared, and can vary considerably among laboratories depending on the availability of automated equipment as well as the level of technical skill. Pooler[1085] reported that the cost per polymorphic band was similar when a RAPD analysis (yielding an average of 2.2 polymorphic bands per primer) was compared with an AFLP analysis (yielding an average of 9.5 polymorphic bands per primer). The higher informativeness of microsatellite and AFLP markers makes them more useful for many studies, but RAPDs will remain attractive when financial investment is limited.[103,1000] AFLPs can be more cost-effective than microsatellite DNA analysis for population assignment studies, especially when the populations are poorly differentiated from each other.[211]

An algorithm for selection of the most cost-effective marker method for cultivar discrimination was suggested by Tessier et al.,[1384] using RAPD and microsatellite data for grapevine as an example. The authors calculated the discriminatory power, D, of each marker on the basis of banding pattern frequencies, as well as the risk of confusion, C, among cultivars, based on the discriminatory power of the markers. As expected, D depended on the number of patterns a marker produces, and on the frequency with which each pattern occurs (maximum when all patterns occur at the same frequency). A combination of eight markers (six RAPD and two microsatellites) was found to be optimal for the discrimination of 224 grapevine varieties.

8.3 CONCLUSIONS

In conclusion, the following points should be taken into account when deciding on the most suitable molecular marker technique for a particular project:

1. Which markers will result in the most appropriate levels of discrimination? Table 8.1 provides some assistance regarding this decision.
2. Do results need to be transferred across laboratories? If so, robust and easily reproducible methods such as allozymes, RFLPs, and microsatellite DNA analyses are the techniques of choice. Allozyme analysis is usually the least expensive alternative, provided that sufficiently high levels of polymorphism are detected and fresh plant material is available.
3. How much time (and funding) is available for the project? Allozymes and RAPD will usually give the quickest results, whereas, e.g., AFLP often requires more work with optimization, and microsatellite DNA analysis may involve time-consuming development of primers[1322] (see Chapter 4.8.5).
4. Is sufficient expertise available? If not, techniques involving cloning and/or sequencing should perhaps be avoided, or be outsourced.
5. What are the specific problems inherent to the organism under study? Availability of only small amounts of DNA and perhaps also partially degraded DNA may necessitate the use of locus-specific, PCR-based methods.

Table 8.1 Levels of Discrimination Provided by the Major Molecular Marker Techniques Currently Used

Method	Above Genus Level	Between Species	Within Species
Allozymes	–	+	+
PCR-based DNA techniques			
Sequencing nuclear DNA			
Genes (e.g., rDNA)	++	+/–	–
Noncoding DNA (e.g., ITS)	+/–	++	+
Multilocus nuclear DNA markers			
AFLP	–	+	++
ISSR	–	+	++
RAPD	–	+	++
SRAP	–	+	+
Single-locus nuclear DNA markers			
Microsatellites	–	+/–	++
SCARs	–	+/–	++
SNPs	+	+	++
Chloroplast markers			
cpDNA-CAPS	+	++	+
cpDNA gene sequences	++	+	–
cpDNA noncoding DNA sequences	+/–	++	+
cpDNA microsatellites	–	+/–	++
Hybridization-based DNA techniques			
Mini- and microsatellite probes	–	+/–	++
RFLP probes	+/–	++	++
RAMPO	–	+	+

Rating: ++, highly useful; +, useful; +/– useful in some cases; –, generally not useful.

9

Future Prospects: SNiPs and Chips for DNA and RNA Profiling

The preceding chapters have described in some detail the principles, methods, and applications of PCR-based DNA fingerprinting techniques currently used in the average plant genetic laboratory. This final chapter aims at introducing recent developments concerning (1) the identification and detection of single-nucleotide polymorphisms (SNPs) as the prototypes of third-generation molecular markers and (2) DNA microarray (so-called chip) technology as a highly sophisticated, nongel approach of DNA and RNA profiling. The treatment of these topics is necessarily brief, given that a comprehensive discussion of the explosive developments in these fields would easily fill another book. For recent reviews on the potential of SNP, microarray and expression profiling technologies in various fields of plant research, see Bhattramakki and Rafalski,[130] Brumfield et al.,[175] Gibson,[491] Gupta et al.,[545] Lemieux,[795] Morin et al.,[953] and Rafalski.[1126,1127]

9.1 SINGLE-NUCLEOTIDE POLYMORPHISMS

9.1.1 What Is a SNiP?

SNPs (pronounced "snips") are single-base pair positions in the genomes of two (or more) individuals, at which different sequence alternatives (alleles) exist in populations. Per definition, the least frequent allele should have an abundance of at least 1%. Among the many types of mutations naturally occurring in genomes, single-nucleotide exchanges (i.e., base substitutions; see Figure 1.1) stand out by their sheer numbers per genome, their relatively low mutation rates (as opposed to, e.g., microsatellites of all types; see Chapter 1.2.2.3), their even distribution across the genomes, and their relative ease of detection. In addition to SNPs, the presence vs. absence of small insertions and deletions (indels) are receiving increasing attention as potential biallelic markers.[131]

SNPs are a consequence of either transition or transversion events. In principle, a SNP locus can have two, three, or four alleles in a population, but biallelic SNPs massively prevail. SNPs fall into several classes, depending on (1) their precise location in a genome and (2) the impact of their location within coding or regulatory regions onto the encoded protein or phenotype. Given that the majority of SNPs are located in noncoding DNA, they are infelicitously called noncoding SNPs (ncSNPs). A subset of these ncSNPs reside in introns. SNPs that reside in exons and the corresponding cDNAs are called coding SNPs, exonic SNPs, or cDNA SNPs, respectively.

Exonic SNPs that do not change the amino acid composition of the encoded domain or protein are called synonymous SNP (synSNP), whereas a nonsynonymous SNP (nsSNP) will change the encoded amino acid. nsSNPs may therefore cause the synthesis of a nonfunctional protein and have an effect on the phenotype. Such so-called diagnostic SNPs may be associated with certain diseases in humans, and with certain agronomic traits in plants. For example, a G/T polymorphism in the leader intron 5′-splice site of the *waxy* gene was found to control the cooking quality of rice,[149] and two nonsynonymous exonic SNPs in the same gene were associated with amylase content and viscosity characteristics.[779] The detection of diagnostic SNPs is a major aim of many SNP discovery projects.

SNPs that reside in promoters or other regulatory regions of the genome are coined promoter SNPs (pSNPs) or regulatory SNPs, respectively. It is obvious that any pSNP can strongly influence the activity of the associated gene. If, for example, a pSNP prevents the binding of a transcription factor to its recognition sequence, the promoter may become nonfunctional. In contrast, intronic SNPs are regarded as more or less inert. However, many researchers value the extragenic SNPs for association studies and whole genome linkage–disequilibrium mapping (see Chapter 9.1.2). Any SNP at a specific site of a genome (or part of a genome, e.g., a bacterial artificial chromosome [BAC] clone) that serves as a reference point for the definition of other SNPs in its neighborhood is called a reference SNP (refSNP, rsID). An refSNP number (tag) is assigned to each rsID at the time of its submission to the databases (e.g., the public dbSNP at the National Center for Biotechnology Information [NCBI]; http://www.ncbi.nlm.nih.gov/SNP). As more and more SNPs are accumulating in the databases, they are labeled with the organism from which they originate (e.g., yeast SNP, human SNP, wheat SNP).[1285]

In general, SNPs are highly abundant, but their density differs substantially in different regions of a genome and from genome to genome in any species, and more so from species to species. For example, the average density of SNPs in the human genome was estimated as about 1 in 1000 bp,[1207] but is considerably larger in some genomic areas such as the noncoding human leukocyte antigen (HLA) regions.[617] In the relatively few plant species analyzed so far, one SNP was usually present per 200 to 500 bp (see Chapter 9.1.4). As may be expected, SNP density is generally higher in intergenic and intronic regions compared with that in exons.

9.1.2 SNP Discovery

SNP discovery usually follows one of two approaches. In the database approach, SNPs are identified by mining sequence databases, and are then coined *in silico* or electronic SNPs (isSNPs, eSNPs). In model organisms and major crops, SNP mining is one of the most promising and efficient SNP discovery strategies.[87,128,1078,1310] However, sequence information from uncharacterized regions of genomes may well contain errors, leading to false-positive results. Given that isSNPs represent virtual polymorphisms, they have to be validated by resequencing the region in which they occur,[247] and special software has been developed to deal with these problems (e.g., POLYBAYES[887]). At present, most of the SNPs are extracted from expressed sequence tag (EST) databases.

In the experimental approach, candidate genes or genome regions of interest are screened for SNPs by a series of techniques such as microchip hybridization,[1494] direct sequencing[1102] or electrophoresis of PCR fragments containing candidate sequences on single-strand conformation polymorphism (SSCP)[1248,1281,1345] or denaturing gradient (DGGE) gels[1060] (see also Chapter 2.3.9). Another option is to convert other types of markers into SNPs, as has been demonstrated for amplified fragment length polymorphism (AFLP) products.[115] Eventually, all common SNPs in the genome of selected model organisms will be discovered and a comprehensive SNP map be established; the first most likely is the complete human SNP map.[1207]

If recombination is low, SNPs that are linearly arranged along a short chromosomal segment form **haplotypes**.[298,467] SNP alleles in these haplotypes are coinherited, and therefore in **linkage disequilibrium** (LD). Research in human populations revealed that regions of high LD extend over relatively long distances (100 kb and more).[35,505] Therefore, genotyping a few carefully chosen SNPs in the target region defined by a certain haplotype provides enough information to predict the constitution of the remainder of SNPs in the region. These selected SNPs are known as tag SNPs. A rough calculation led to the assumption that most of the information about human genetic variation represented by the ~10 million common SNPs in the population is already provided by genotyping only 200,000 to 800,000 tag SNPs across the genome. Haplotype analysis is a logic approach to establish genetic risk profiles, and to predict the clinical reaction of an individual toward pharmaceutically active compounds. Much of the haplotyping effort in humans is bundled in the so-called HapMap Project[1387] (www.hapmap.org), which aims to establish the haplotype map of the human genome, describing the common patterns of variation as well as the association between SNPs.

9.1.3 SNP Genotyping

Numerous assays have been developed for the detection of known and unknown SNPs (for reviews, see Kwok,[764] Landegren et al.,[771] and Syvanen[1356]). Some of these are relatively easy to perform and are low cost, whereas others manage high-volume screening and therefore are very costly. There exists hardly one protocol that meets all needs, so that different protocols may have to be established in a single core genotyping lab to provide flexibility and accurate validation.

The major SNP genotyping techniques fall into at least six groups: (1) direct sequencing, (2) restriction enzyme digestion (cleaved amplified polymorphic sequences [CAPS]), (3) allele-specific PCR, (4) allele-specific primer extension, (5) allele-specific oligonucleotide hybridization, and (6) allele-specific oligonucleotide ligation.[764,771,1356] Additional techniques are being developed continuously and published at surprisingly high rates (see, e.g., the issues of *Nucleic Acids Research* and *Genome Research* since 2000).

Fluorescence-based **sequencing** (or resequencing of SNPs mined from databases) followed by automated slab gel or capillary electrophoresis is the standard method for SNP detection, but also one of the slowest techniques (e.g., Nickerson et al.,[989] Primmer et al.[1102]). For **allele-specific PCR**, primers are designed to amplify one of the two SNP alleles but not the other (e.g., Drenkard et al.[369]; Soleimani et al.[1305]). To achieve this, one primer has a specific base exactly matching the SNP position at its 3′-end. In the **primer extension** technique (e.g., SnaPshot™ Multiplex Kit, Applied Biosystems), only a single, fluorophore-labeled dideoxynucleotide is either incorporated at the SNP position or not, depending on the allelic state. In **allele-specific oligonucleotide hybridization**, fluorescence-labeled PCR fragments are hybridized to immobilized oligonucleotides, each representing a particular SNP allele. After stringent hybridization and washing, fluorescence intensity is measured for each SNP oligonucleotide separately. For **allele-specific oligonucleotide ligation**, the genomic target sequence is first PCR amplified. Then allele-specific oligonucleotides complementary to the target sequence and with the allele-specific bases at the 3′- or 5′-ends are ligated to the DNA adjacent to the polymorphic site. The ligation is possible only in the case of a complete match.

The choice for a particular detection technique depends on (1) the number of SNPs, (2) the number of samples to be screened, (3) the number of simultaneous SNP profiling projects, and (4) the facilities that are available. Usually a decision has to be made as to whether outsourcing is more economical than processing SNPs in-house. In-house SNP discovery requires instrumentation, the value of which ranges from US $25,000 to more than $400,000.

9.1.4   SNPs in Plant Genomes

In plants, SNP research is still in its infancy, and SNPs have been rigorously searched for only in a few species. These include several major crops such as barley,[686,1305] rice,[149,582,779,976] maize,[87,1383] wheat,[1310,1612] and sugar beet,[1248] but also the model plant *Arabidopsis thaliana*[248,369] and some forest trees[488,1060] (see also the reviews by Rafalski[1126,1127]). With approximately one SNP per 200 to 500 bp, the average SNP density in plant genomes appears to be relatively high, but depends on the species investigated. In maize, the analysis of several hundred loci in eight maize inbred lines revealed an extremely high prevalence of SNPs (one SNP per 83 bp), probably a consequence of open pollination in this species.[131] The flanking sequences of microsatellites in maize even contain one SNP per 40 bp, making the estimated total number of SNPs per whole genome 62 million.[384] In wheat, SNP density is only about two SNPs per kilobase pair, but could be much higher in genes encoding enzymes of starch biosynthesis.[131]

More than 25,000 SNPs have become available for *A. thaliana*, based on a comparison of the Landsberg erecta and the Columbia ecotypes, and a first medium-density SNP map has been established.[248] By summer 2004, more than 37.000 publicly available SNPs have been listed in the database of The *Arabidopsis* Information Resource (TAIR) at http://www.arabidopsis.org/Cereon/. SNP genotyping with 12 nuclear and 13 chloroplast interspecific SNPs of different tree species discriminated black spruce (*Picea mariana*) from red (*P. rubens*) and white spruce (*P. glauca*).[488]

9.1.5   Perspective

At present, SNP technologies are mostly used in animal and human genomics. It is therefore obvious that the major advances in so-called SNPology will occur in mammalian systems, be they forensic analyses, comparative and evolutionary genetics, or the use of SNPs in large-scale association studies to identify disease-susceptibility genes for human disorders such as type II diabetes, hypertension, and cancer. Once the culprit genes are identified, the encoded proteins can be targeted by novel therapeutic drugs or diagnostic tests. Comprehensive SNP maps are already available for the human genome, and haplotype blocks for the detection of real associations with candidate disease genes will be identified. SNP detection techniques are already miniaturized and designed for high throughput.[764,771,1356]

Despite an increasing number of reports on the use of SNPs in genotyping and genetic mapping in plants (see Chapter 9.1.4), SNPs are not yet in common practice in the plant sciences. The tremendous cost of developing SNPs, especially the sequencing load, may be one of the reasons that more economic markers such as microsatellites and AFLPs are preferred, at least for exotic plant species that have limited or no economic value. However, ultrafast DNA sequencing, nanopore sequencing, and single-molecule sequencing procedures are being developed and promise lower sequencing and resequencing costs in the future (see *Nature Biotechnology* 21: 1425–1427, 2003; and consult, e.g., www.affymetrix.com, www.illumina.com, www.perlegen.com, and www.sequenom.com). With the availability of specific low-density SNP chips and affordable technologies, SNPs will arrive on the plant (especially crop plant) market as tools for DNA fingerprinting, genomic mapping, and linkage analyses, and will certainly play the role in plant sciences that they already play in human biology.

9.2   DNA MICROARRAYS

Since the first reports on new technological advances to measure the activity of nearly unlimited numbers of genes simultaneously on planar glass supports in the early 1990s, and the seminal paper by Schena et al. in 1995,[1233] DNA arrays have shown enormous potential for almost every aspect of molecular biology. In fact, large-scale parallel expression analyses using cDNA arrays that reflect the whole repertoire of transcripts have revolutionized our understanding of the transcriptome of a cell at a given time (e.g., Ruan et al.[1197]). It succeeded the formerly common but laborious gene-by-gene experimentation, and now allows one to monitor the activity of tens of thousands, and on elaborate arrays, hundreds of thousands of genes at the same time, and nearly quantitatively.

At present, the most widely accepted term for a DNA array is microarray (microscopic array), which stands for any microscale solid support (e.g., nylon membrane, nitrocellulose, glass, quartz, gold, silicon wafer, or other synthetic material) onto which DNA fragments, PCR products, full-length cDNAs or cDNA fragments, oligonucleotides from between 15 to >80 nucleotides, genes or gene fragments, open reading frames, peptides, or proteins (e.g., antibodies) are spotted in an ordered pattern at extremely high density. Such microarrays (in laboratory jargon, chips) are increasingly used for high-throughput expression profiling from bacteria to man. Currently, some hundred different microarrays for different purposes are available, ranging from antibody arrays to cDNA expression arrays, and from transgene arrays to whole genome oligonucleotide arrays, to name but a few. In some areas of microchip manufacture breakthrough advances have already been made, such as the microchannel machining for nanofluidic microarrays (e.g., Cheeks et al.[235]). In addition, detection technologies are being refined and led to the advent of fiberoptic arrays (e.g., Ferguson et al.[436]; Steemers et al.[1328]). Whole genome arrays are increasingly being used (e.g., for comparative transcriptomics) and many complete bacterial genome microarrays already served to decipher transcription differences in different species.[964,1214] These few examples portray a highly fluid field of technology. For recent reviews on various aspects of microarray technology see, e.g., Blohm and Guiseppi-Elie,[142] Gibson,[491] Mantripragada et al.,[878] Quackenbush,[1121] Richmond and Somerville,[1164] and Schena.[1232]

The **technology to manufacture a microarray** is relatively simple, but nevertheless requires experience and care. In short, the carrier for the arrayed targets is first selected and the target molecule(s) defined (see above). Then the mode of spotting of the selected targets is chosen and largely depends on the in-house facilities. Three basic techniques are available: contact printing, noncontact printing, and semiconductor technologies. Contact printing works with a direct contact between the spotting device and the microarray surface, and encircles solid pins, split pins, capillary tubes, and pin-and-ring (PAR) devices. Noncontact printing allows the spots without contact to be addressed, driven by piezoelectric- and microsolenoid-based ink-jet technologies or thermal bubble-jet dispensers. Semiconductor-based spotting represents the most advanced technique and uses photolithographic chrome–glass masks or micromirrors to synthesize oligonucleotides directly on the microarray in a stepwise procedure.

The miniaturization process has been driven to extremes with the development of so-called **nanoarrays**. Nanoarrays are solid supports (e.g., gold-coated glass chips), onto which dots of oligonucleotides or DNAs (also peptides and proteins) are spotted via dip-pen nanolithography (DPN) in arrays of 100 nm (or less) diameter and 100 nm (or less) distance between spots. This dimension is beyond imagination: one spot on a conventional microarray occupies an area of $200 \times 200 \ \mu m^2$, whereas a DPN array easily accommodates 50,000 dots or more on the same area. Interactions between the probes and target molecules on a nanoarray are scanned by the cantilever of an atomic-force microscope and detected by the deflection of the cantilever tip.

The range of spot densities is extremely variable. Some microarrays only accommodate from 100 to 500 targets (so-called poor man's array or low-density microarray), other formats carry from 10,000 to 200,000 (or more) spots (medium-density microarray), and the high-density microarrays may harbor as many as a million spots. The decision for a specific density not only depends on the financial power of the laboratory, but also is made on the specific needs of the experimenter. For example, a genome-wide expression profiling requires at least a medium-density cDNA microarray (for plants, in the density range of 15,000 to 20,000; for mammals, more than 30,000). The expression analysis of genes encoding proteins of a particular metabolic pathway will resort to low-density arrays. Nanoarrays are not yet fully developed for average laboratories.

The next experimental step is the hybridization of probes (usually oligonucleotides or cDNAs labeled with fluorochromes) to the array. For example, the differentially labeled cDNAs from control and test cells are mixed and then hybridized to the array synchronously. The resulting hybrids are then detected after laser excitation of the bound fluorochromes and the signals are analyzed by a computer. The technology is more or less standard in specialized laboratories, but the management of the enormous amounts of data still poses an extraordinary challenge.[1121]

The huge potential of microarrays is mostly exploited for **gene expression analysis**, but also genetic screening and diagnostics (e.g., SNP detection or HLA typing), and the literature abounds with excellent and, in part, breakthrough research articles. However, plant molecular biologists are only reluctantly accepting the promise: the number of articles based on microarray technologies applied to plant or fungal problems is still small (see Schena[1232]). Nearly all of them are devoted to gene expression profiling using cDNA and/or EST arrays (e.g., Ghassemian et al.,[489] Girke et al.,[500] Hertzberg et al.,[594] Reymond et al.,[1156] Richmond and Somerville,[1164] Ruan et al.,[1197] Schaffer et al.,[1231] and Seki et al.[1271]). Microarrays allow the expression profiling of an unprecedented number of genes with relative ease. However, again the costs still prevent their broader use in plant genomics.

Nevertheless, oligonucleotide microarrays can also be designed and used for rapid screening of **DNA polymorphisms** in particular plant genes, and the genetic diversity in a plant population can be determined for these genes. For example, for the identification of the base **X** in the target sequence 5′-TTAGCTATCCCGTXC-CGATGATCGAAT-3′ only the four probes would be sufficient:

3′-ATAGGGCAAGGCTACTA-5′
3′-ATAGGGCAGGGCTACTA-5′
3′-ATAGGGCACGGCTACTA-5′
3′-ATAGGGCATGGCTACTA-5′

If the probes are fluorescently labeled, then the probe with the highest fluorescence intensity would indicate the identity of the base X. This basic concept can be extended to detect polymorphisms (e.g., SNPs) in longer DNA targets relative to a wildtype consensus sequence (see Tillib and Mirzabekov[1401]). If, for example, a target sequence of 1000 bases has to be screened for polymorphisms, then 4000 probes are required. This, of course, presupposes known target gene sequences, the availability of all different oligonucleotides, and appropriate software to call the aberrant bases — again a matter of cost. However, it is to be expected that at least SNP chips (microarrays for the discovery of SNPs in target sequences) will in future be used for SNP fingerprinting in plants, especially because the expenses for such novel fingerprinting will certainly become more modest with the advance of technology.

9.3 EXPRESSION PROFILING AND EXPRESSION MARKERS

A novel concept of markers emerged during the last 2 years that allows one to establish very informative fingerprints of an organism, an organ, a tissue, or a cell. Whereas the molecular markers generated by any of the techniques described in this book consist of DNA sequences of whatever kind, these novel markers derive from messenger RNA, i.e., from transcribed genes. Therefore, they are coined expression markers. The concept is simple. First, the complete genome-wide transcriptome of a target cell is isolated by any of the few high-throughput technologies (e.g., serial analysis of gene expression[1460] or massively parallel signature sequencing[164]), characterized, and quantified. The most abundant transcripts are then taken as indicators and used to generate a transcript profile (**transcript fingerprint**), which is diagnostic for the state of the cell at the time the transcriptome was isolated. The concept of expression markers is here exemplified with a substantially improved version of the conventional **serial analysis of gene expression (SAGE)** technique developed by Velculescu et al.[1460]: the so-called **SuperSAGE** technology.[895]

In short, messenger RNA is first isolated and reverse transcribed into single-stranded cDNA using a reverse transcription primer with the sequence

5′-CTGATCTAGAGGTACCGGATCC**CAGCAG**TTTTTTTTTTTTTTTTTT-3′

containing the 5′-CAGCAG-3′ recognition site for the type III restriction endonuclease *Eco*P15I from *Escherichia coli* strain TG1. The product is converted to double-stranded cDNA, digested with *Nla*III, and the 3′-end fragments of the cDNAs are bound to streptavidin-coated magnetic beads. The bound cDNA is washed, and divided into two portions in separate tubes. Two linkers (linker-1E and linker-2E) are labeled with fluoroisothiocyanate (FITC), and the unblocked 5′-termini of linker-1E and linker-2E are phosphorylated by T4 polynucleotide kinase. Both linker-1E and linker-2E harbor the *Eco*P15I recognition sequence (5′-CAGCAG-3′). Linker-1E or linker-2E, respectively, are then added to the two tubes containing cDNA bound to magnetic beads and ligated to the cDNA ends by T4 DNA ligase. Consequently, each cDNA fragment is flanked by two inverted repeats of 5′-CAGCAG-3′. *Eco*P15I recognizes the asymmetric hexameric sequence 5′-CAGCAG-3′ and cleaves the DNA 25 bp (in one strand) and 27 bp (in the other strand) downstream of the recognition site, leaving a 5′-overhang of two bases. Two unmethylated and inversely oriented recognition sites in head-to-head configuration (5′-CAGCAG-N$_{(i)}$-CTGCTG-3′) are essential for efficient cleavage. Linker-ligated cDNA on the magnetic beads is then digested with *Eco*P15I. Digestion fragments are separated by polyacrylamide gel electrophoresis; the approximately 69-bp linker–tag fragment is visualized by FITC fluorescence under ultraviolet light, and then collected from the gel.

Linker-1E tag and linker-2E tag fragments are mixed, their ends are blunted by filling-in with *Thermococcus kodakaraensis* (KOD) polymerase and subsequently ligated to each other. The resulting so-called ditags are amplified by PCR using two biotinylated primers:

1. 1E: biotin-5'-CTAGGCTTAATACAGCAGCA-3'
2. 2E: biotin-5'-TTCTAACGATGTACGCAGCAGCA-3'

The ditag PCR products are digested with *Nla*III, the resulting fragments are separated on polyacrylamide gels, and the fragment of approximately 54 bp is isolated from the gel. This fragment is concatenated by ligation, cloned into a plasmid vector, transformed into *E. coli* cells, and plated on selective medium. Plasmid inserts are amplified by colony PCR, directly sequenced, and the sequences analyzed by the SAGE2000 software package (extraction of the 22-bp tags adjacent to CATG). The resulting 26- to 27-bp sequence from each cDNA is called a SuperSAGE tag. The main advantage of using *EcoP*15I over conventional enzymes is the longer tag, which allows better identification of the underlying cDNA (or gene) by annotation.

SuperSAGE tags can be used as primers to amplify the 3'-ends of the corresponding cDNAs (small amplified RNA or SAR-SAGE[1470]), the longer cDNAs are annotated, and the most abundantly transcribed sequences are used to establish an **expression fingerprint**. Each of the cDNAs would then be an expression marker. The corresponding expression profile is a complex, context-dependent, and genome-wide pattern of (preferably all) expressed genes at a given time. It is characteristic for a certain cell, tissue, organ, or organism (e.g., a bacterial cell), but changes continuously, depending on the developmental stage and the environment. Comparable to the DNA fingerprints generated by, e.g., DNA markers, the expression fingerprints of two (and more) cells can be compared and differences can be revealed. In addition, and in contrast to the static DNA markers that characterize certain regions of a genome, the dynamic expression markers define the potential of a target cell in a given environment. For example, if a cell is stressed, then expression markers — but not DNA markers — can exactly and quantitatively describe the stressed condition (as compared with the condition without stress).

The concept of expression markers has been described in a novel approach (e.g., Jansen and Nap,[656] Schadt et al.[1230]). Thus, comprehensive screens of plant, mouse, and human transcriptomes identified specific mRNAs, whose abundances correlate with quantitative traits, such as obesity in mice. The corresponding mRNA abundances are treated as quantitative trait loci (so-called gene expression QTLs) that can be mapped in segregating populations.

Plant DNA Isolation Protocols

Appendix 1A   DNA Isolation Protocols Based on CTAB Buffers

Taxa	Remarks	Ref.
Wide range of taxa	The effects of endogenous DNases are examined; grinding of silica gel-dried plant material in ethanol prior to CTAB extraction prevents DNA degradation by DNases	7
Phoenix dactylifera	Modification of the basic protocol	10
Linum usitatissimum	Modifications include the use of high concentrations of β-mercaptoethanol (5%), polysaccharide removal by ethanol precipitation of DNA from 2 M NaCl, and DNA purification by Chelex treatment or gel electrophoresis	17
Saccharum, Lactuca, Fragaria	Variant designed for fresh tissue (meristem cylinders in the case of sugarcane); tissue homogenization with an Ultra-Turrax	22
Hesperis	Proteinase K and potassium acetate are added to the CTAB extraction buffer	32
Sedum telephium	Modification designed for succulent species that are rich in polysaccharides; DNA-CTAB complexes are precipitated by lowering the NaCl concentration, while polysaccharides stay in the supernatant	80
Drosera rotundifolia	Modification of the protocol of Fulton et al.[466]	101
Anthoxanthum, Festuca (and other grasses)	Modification of the basic protocol designed for century-old grass samples	136
Daucus carota	Modification of the basic protocol	166
Various woody species	Modification designed for roots of woody species; high concentrations of spermidine, PVP, PVPP, and mercaptoethanol are included in the extraction buffer	178
Wide range of taxa	DNA is purified via low-melting agarose gel electrophoresis	183
Several tropical plant species	Plant material is desiccated with silica gel; 2% PVP and 4 mM DIECA are included in the extraction buffer. DNA is further purified by CsCl centrifugation	229
Oryza sativa and other species	Miniprep version of the basic protocol; applicability for AFLP analyses tested	236
Citrus	Water-saturated ether is used to remove polysaccharides from the aqueous phase	242
Wide range of taxa	Miniprep version; tissue is ground with ball bearings, liquid nitrogen and a vortex mixer in microfuge tubes	266
Pinus radiata	Modification of the basic protocol	286
Quercus robur, Populus tremula, Ulmus glabra, Abies alba, Pinus sylvestris, Rhododendron luteum, Zea mays	Comparative analysis of the performance of Dellaporta[323] (see Appendix 1B) and CTAB variants; five of the seven plant species tested are considered difficult; best results were obtained with the CTAB protocol including anion exchange chromatography	287

Appendix 1A (continued)   DNA Isolation Protocols Based on CTAB Buffers

Taxa	Remarks	Ref.
Twenty tropical tree species	Combination of CTAB extraction and potassium acetate precipitation; fresh leaves preferred over dried material	309
Cactaceae (85 species)	Variant of the method of De la Cruz et al.,[309] specifically designed for cacti; combination of CTAB extraction and potassium acetate precipitation	310
Adiantum capillus-veneris (and other fern species)	Modification of the method of Porebski et al.[1087] optimized for ferns	328
Magnolia, Quercus	Modification of the basic CTAB protocol of Doyle and Doyle[364]; final purification by ion exchange chromatography	352
Solanum, Glycine	Comparative testing of several field preservation methods for plant tissues; desiccation is recommended	364
Wide range of taxa	Variant specifically designed for fresh tissue	365
Juncus, Luzula	Various modifications of the basic procedure adapted for herbarium specimens	367
Musa	Nuclei are isolated first; then lysed with CTAB buffer	431
Quercus humbolottii	Variant specifically designed for silica gel-dried mature oak leaves that are rich in tannins	438
Lycopersicon esculentum	High-throughput version of the basic protocol designed for several hundred extractions per day	466
Musa acuminata, Ipomoea batatas	Authors report on the influence of different concentrations of β-mercaptoethanol on DNA yield	482
Fragaria, Gladiolus, Allium, Lycopersicon, Malus	Miniprep modification of the basic protocol	586
Wide range of taxa	The initial CTAB extract is centrifuged, and the supernatant loaded onto a Qiagen plasmid isolation column; organic solvents are avoided	591
Saccharum	Nuclei are isolated first, then lysed with CTAB buffer	610
Wide range of taxa	Chaotropic salt and silica particles are added to the aqueous phase after chloroform centrifugation; DNA is selectively bound to silica; particles are washed, dried, and DNA is eluted with low-salt buffer	626
Orobanche	Variant of the technique of Fulton et al.[466] designed for single seeds	673
Nelumbo	High concentrations of β-mercaptoethanol are used (5%)	684
Oryza sativa (and 11 other species)	Variant designed for (half) seeds, treatment of seeds with SDS and proteinase K before adding CTAB buffer	688
Agave	Modification of the basic protocol	705
Wide range of taxa	Variant specifically designed for plants producing large amounts of essential oils and other secondary compounds	711
Two orchid species	Miniprep version of the basic protocol, allowing the simultaneous isolation of RNA and DNA	725
Berberis lycium	Variant designed for dry roots	760
Lycopersicon	Minipreparation; tissue homogenization is performed by means of a viral sap extractor	781
Gossypium	Modification of the basic protocol	807
Pyrrosia (a fern)	Modification of the basic protocol	818
Mimulus, Eichhornia, Aeschynanthus, Lythrum, Antirrhinum	DNA is isolated from flower petals instead of leaf tissue, giving more reliable results in RAPD analysis	822
Various woody species	Modification designed for roots of woody species; the extraction buffer contains 1 M boric acid; gel purification is included to remove PCR inhibitors	825

Appendix 1A (continued) DNA Isolation Protocols Based on CTAB Buffers

Taxa	Remarks	Ref.
Vitis vinifera and other woody plant species	Protocol specifically designed for mature leaves with high polyphenol and polysaccharide content; polyphenols are removed by PVP; polysaccharides are removed by high NaCl concentrations	837
Tageteae and Mutisieae (Asteraceae)	Miniprep version designed for herbarium specimens; plant material is rehydrated in double-distilled water prior to grinding in a glass homogenizer	841
Pennisetum glaucum, Sorghum bicolor, Arachis hypogaea, Cicer arietinum, Cajanus cajan,	High-throughput version based on microtiter plates; CTAB and β-mercaptoethanol concentrations are optimized for each species	866
Proteaceae (10 genera)	Modification of the basic protocol; polysaccharides are removed according to Fang et al.[428]	869
Cichorium, Taraxacum, Lactuca	Variant optimized for latex-containing plants	929
Wide range of taxa	Modification of the basic protocol; numerous variations are reviewed	933
Vigna (mung bean)	DNA-CTAB complexes are precipitated by lowering the NaCl concentration; DNA is purified by CsCl centrifugation	965
Various taxa	Plants are homogenized in CTAB buffer in the field, and filtered extracts are stored at room temperature until returning to the laboratory	990
Picea abies	Modification designed for seedlings and embryogenic cultures of spruce	992
Gossypium hirsutum	Modification of the basic protocol; addition of 0.5 M glucose to the extraction buffer prevents browning	1069
Achillea millefolium, Artemisia dracunculus, Drosera rotundifolia, Aleutherococcus senticosus (ginseng)	Modification specifically designed for medicinal and aromatic plants that are high in secondary metabolites; CTAB extraction buffer is mixed with 8 M LiCl for RNA precipitation	1082
Fragaria	Modification specifically designed for mature leaves with high amounts of polyphenols, tannins, and polysaccharides; polyphenols are removed by PVP; polysaccharides are removed by high NaCl concentrations	1087
Wide range of taxa	Several field preservation methods for plant tissues are compared; desiccation is recommended	1117
Solanum tuberosum	Variant designed for herbarium specimens	1175
Wide range of taxa	Miniprep based on Murray and Thompson,[965] suitable for fresh, herbarium, and mummified specimens	1183
Podophyllum, Polyalthia, Taraxacum	Field-collected tissue is stored in a saturated NaCl–CTAB solution; final DNA purification step involves low-melting agarose gel electrophoresis	1184
Polyalthia glauca, Quercus muehlenbergii, Taraxacum officinale, Tilia americana	DNA is selectively bound to silica particles, which are added to the aqueous phase after chloroform centrifugation; particles are washed, dried, and DNA is eluted with low-salt buffer	1185
Quercus rubra, Castanea sativa	Extracts are treated with pectinase and RNase, DNA is further purified via agarose gel electrophoresis	1187
Vaccinium	DNA is specifically precipitated by PEG 8000	1193
Hordeum vulgare	Authors introduced precipitation of the CTAB-DNA complex by isopropanol instead of lowering the salt concentration	1209

Appendix 1A (continued)   DNA Isolation Protocols Based on CTAB Buffers

Taxa	Remarks	Ref.
Artemisia annua	Modification of the Murray and Thompson[965] protocol, involving DNA purification via ion exchange chromatography on DE-52	1220
Papaver somniferum	Modification of the Murray and Thompson[965] protocol, allowing for the simultaneous isolation of lipids and DNA from seeds	1221
Tropical woody plants (four species)	Modification of the PEG procedure of Rowland and Nguyen,[1193] specifically designed for silica gel-dried woody plant specimens	1234
Physcomitrella patens	Modification of the basic protocol, adapted to moss species; proteinases are included in the extraction buffer	1237
Zea mays	Variant optimized for several-years-old dried corncobs	1247
Seven tropical tree species	Tissue is ground in sand instead of liquid nitrogen; nuclei are isolated first, then lysed with CTAB buffer	1260
Cicer arietinum, Glycine max a.o.	Modification of the basic protocol	1273
Rumen contents	Basic CTAB protocol combined with a purification step on Plant DNeasy columns (Qiagen)	1278
Wide range of taxa	Leaf tissue is fixed with absolute ethanol before grinding	1279
Camellia sinensis	Variant specifically designed for market samples of dry tea leaves pre-washed in water	1293
Pinus radiata	Modification of the procedure of Stewart and Via[1334]	1324
Wide range of taxa	Final DNA purification step involves CsCl centrifugation	1329
Castanea, Vaccinium, Pelargonium, Arachis, Russula	Miniprep version; isolation buffer contains PVP-40, ascorbic acid and DIECA; disposable homogenizers prevent cross-contamination in RAPD analysis	1334
Hieracium	Combination of NaCl–CTAB field preservation method of Rogstad[1184] with grinding in a sorbitol buffer; no liquid nitrogen; nuclear extract is lysed with CTAB buffer	1337
Zingiber, Curcuma	Modification designed for rhizomes	1353
Hylocereus, Selenicereus (climbing cacti)	Using roots as a source material reduces polysaccharide content in the DNA preparation, CTAB extraction buffer has high-salt content (4 M NaCl)	1377
Prunus persica	DNA is isolated from leaves slowly dried at room temperature	1396
Gossypium	DNA is extracted from single seeds; DNA purification involves spun-column chromatography	1496
Vitis amurensis	Silica gel-dried material is ground with solid PVP (final concentration 6%); tissue powder is washed in pre-extraction buffer to remove cytoplasmic contaminants; organelles are lysed with a high-salt (2.5 M NaCl), high β-mercaptoethanol (2%) CTAB buffer	1498
Emblica, Terminalia	Modification of the basic protocol	1503
Cuphea	Modification of the basic protocol (e.g., phenol extraction of CTAB–DNA complexes)	1511
Wide range of taxa	DNA is purified via ammonium acetate treatment, CsCl centrifugation, gel filtration, or ion exchange chromatography	1522
Solanum tuberosum	Modification specifically designed for lyophilized potato tubers	1576

Appendix 1B   DNA Isolation Protocols Based on SDS Buffers and Potassium Acetate–SDS Precipitation of Proteins and Polysaccharides

Taxa	Remarks	Ref.
Dioscorea	Miniprep version of the protocol of Varadarajan et al.[1455]	52
Digitalis obscura	Combination of the procedure of Edwards et al.,[385] with potassium–SDS and PEG precipitation	322
Twenty tropical tree species	Combination of CTAB extraction and potassium acetate precipitation; young leaves preferred over dried material	309
Abelmoschus (okra)	Specifically designed for plant tissues that are rich in viscous polysaccharides; isolation from dark-grown tissue	308
Wide range of taxa	Original description of this strategy of DNA isolation; polysaccharides and proteins are removed by SDS–potassium acetate precipitation	323
Amaranthus	Modification involves PEG precipitation	350
Acer, Magnolia, Elodea, Taxodium, Pinus	Modification of the basic protocol[323]: PVP is included to bind polyphenols, proteinase K is included to remove proteins, RNA is removed by LiCl precipitation	672
Malvaceae, Moraceae, Bombacaceae	Variant specifically designed for plant tissues rich in polyphenols; PVP is included in the isolation buffer	674
Ficus, Citrus, Stenomesson, Caliphruria	Variant specifically designed for plant tissues rich in polysaccharides; crude DNA is purified by passage through a Sephacryl S-1000 column, followed by PEG precipitation	809
Nicotiana tabacum	Miniprep version of the protocol of Pich and Schubert[1077]	823
Pylaiella (brown algae)	DNA is isolated from algal protoplasts to circumvent contamination with cell walls and bacteria; final purification involves CsCl centrifugation	911
Wide range of taxa	Modification of the basic protocol,[323] PVP-360 is included in the isolation buffer	933
Vicia faba, Solanum, Lycopersicon	Variant specifically designed for plant tissues rich in polyphenols; PVP is included in the isolation buffer	1077
Wide range of taxa	Modification of the basic protocol[323]	1200
Arachis hypogaea	Modification of the basic protocol; DNA purification involves ion exchange chromatography on DEAE-cellulose; four techniques are compared	1275
Oryza sativa, Lycopersicon	Plant tissue is dried in a food dehydrator; isolation protocol is based on Dellaporta et al.[323]	1361
Ipomoea	Modification of the basic protocol [323]	1455
Wide range of red and green algae *Spathiphyllum*	Variant specifically designed for red algae; cell lysis at 37°C rather than 65°C reduces the amount of coisolated polysaccharides	1508

Appendix 1C   High-Throughput DNA Isolation Protocols

Taxa	Remarks	Ref.
Nicotiana tabacum, human, lizard, snail	Tissue is extracted in a solution containing commercial laundry detergent	68
Hordeum vulgare, Secale cereale	Crude minipreparation based on the method of Dellaporta et al.[323]	108
Nicotiana tabacum	Leaf and root pieces are directly used for PCR	124
Nicotiana tabacum, Triticum aestivum	RAPDs from single lysed protoplasts or microcolonies; freezing–thawing procedure	174
Brassica napus, Helianthus annuus	DNA samples are isolated within microtiter plates, embedded in agarose, and used for PCR	176
Poaceae, Ipomoea	Single-step procedure involving heating of tissue with microLYSIS, a commercial mixture of detergents; the supernatant is used directly for PCR	190
Wide range of plants, mammals, and insects	High-salt (2 M NaCl) extraction buffer	244
Oryza sativa, Triticum aestivum	Half-seeds (not ground) are treated with a buffer containing proteinase K or 5% Chelex; the supernatant is used directly for PCR	254
Hordeum vulgare	Variant of the alkali method of Klimyuk et al[723]	257
Wide range of taxa	Protoplasts are isolated from small leaf disks, lysed, the DNA precipitated, redissolved, and used for PCR	330
Malus domestica	Small leaf disks are extracted in 96-well plates using a simple extraction buffer and glass beads in an Eppendorf thermomixer; supernatant is used for PCR	347
Brassica napus	DNA is ethanol precipitated from crude leaf extracts, redissolved, and used directly for PCR	385
Lycopersicon esculentum	High-throughput version of the CTAB protocol; designed for several hundred extractions per day	466
Equisetum, wide range of fungi, plants, protists, and animals	Tissue in isolation buffer is heated in a microwave oven instead of homogenization (microwave miniprep)	512
Brassica oleracea	Small pieces of tissue are incubated in a buffer containing proteinase K and RNase; no centrifugation steps are required	537
Porphyra perforata (red algae)	Softening of cell walls by LiCl treatment; crude extracts are precipitated by ethanol	612
Brassica napus	Nondestructive protocol using cotyledon fragments from microspore-derived embryos; based on the method of Dellaporta et al.[323]	616
Oryza sativa	Single-tube procedure involving boiling of tissue in TE buffer, dilution, centrifugation, and use of supernatant for PCR	636
Lycopersicon	Tissue is boiled in alkaline buffer, neutralized, and used for PCR	723
Gossypium	Variant of the method of Benito et al.[108] designed for (half) seeds	751
Glycine max	Method based on the commercial Generation DNA Purification System initially designed for animal tissue; tissue is rubbed onto a collection card, small disks are punched out, and samples are processed for PCR in 96-well microtiter plates	774
Triticum, Trifolium, Nicotiana tabacum	Leaf material is squashed onto a nylon membrane, washed, eluted, and used directly for PCR	775

Appendix 1C (continued)   High-Throughput DNA Isolation Protocols

Taxa	Remarks	Ref.
Lycopersicon esculentum	Direct PCR of pollen grain suspension in distilled water	805
Wide range of taxa	Plant leaves are crushed against FTA paper (a medium usually used to collect blood stains); small disks are collected using a punch, the paper disks are then washed with inorganic reagents and used directly for PCR	821
Nicotiana tabacum, Glycine max, Zea mays	Protocol includes a combination of glass bead homogenization, shock-freezing, and boiling	857
Arabidopsis	Adaption of the method of Edwards et al.[385] to 96-well format	923
Wide range of taxa	Modification of the procedure of Edwards et al.[385]	933
Arum maculatum, Brassica napus	High-throughput method in 96-well format; tissue is lysed and extracted in a mixer mill, using a buffer containing SDS, NaCl, proteinase and RNase; debris and polysaccharides are precipitated by NaCl addition	939
Oryza sativa, Zea mays	Freezing–boiling procedure	1006
Hordeum, Triticum	Alkaline extraction is performed in 96-well plates, using a matrix mixer and dowel pins; the extract is used for PCR after neutralization	1042
Nicotiana tabacum, Zea mays, Beta vulgaris, Beta maritima, Brassica oleracea, Brassica napus, Solanum tuberosum	Comparison of six small-scale methods (Edwards et al.,[385] Cheung et al.,[244] Oard and Dronavalli,[1006] Chunwongse et al.,[254] Wang et al.,[1496] and Guidet[537]) for their performance in seven plant species	1182
Hordeum, Arabidopsis, (and bacteria, fungi, algae, vertebrates)	Commercial DNA isolation kit; DNA is selectively bound to magnetic beads added to the homogenized tissue; after several washing steps, DNA is eluted and used for PCR	1199
Lotus corniculatus	Leaf tissue is dried, homogenized in a shaking-mill, extracted by heat treatment in a buffer with high EDTA concentration; one-tube procedure; extract is diluted for PCR	1330
Wide range of taxa	Single step DNA isolation; variation of the boiling procedure; salt and EDTA concentration, pH, incubation time, and temperature are optimized	1397
Arabidopsis	Leaf tissue is boiled in alkali, neutralized, and used for microsatellite PCR	1472
Hordeum vulgare	Small samples drilled out of single seeds are treated with alkali, heated in a microwave oven, neutralized, and used directly for PCR	1479
Gossypium	DNA is extracted from single seeds; purification involves spun-column chromatography	1496
Brassica napus, Arabidopsis	Single step DNA isolation; alkaline extraction	1497
Wide range of taxa	Miniprep version of the potassium ethylxanthogenate protocol of Jhingan[669,670]; tissue homogenization is not required	1545
Oryza sativa	Crude extracts of seedlings are used directly for PCR	1593

Appendix 1D   **DNA Isolation Protocols Involving the Isolation of Nuclei**

Taxa	Remarks	Ref.
Gossypium	Combination of the methods of Paterson et al.[1050] and Lassner et al.,[781] pre-extraction leaves nuclei intact, which are then lysed with a CTAB–sarkosyl buffer	232
Vitis vinifera	Tissue is homogenized in reaction tubes using a motor-driven metal homogenizer	265
Theobroma cacao	Protocol is specifically designed for plant tissues rich in polyphenols; PVP, BSA, and DIECA are included in the isolation buffer	278
Gossypium, Cenchrus	Protocol is specifically designed for plant tissues rich in polyphenols; nuclei are isolated using a glucose-containing buffer, and lysed by proteinase K–SDS–EDTA; DNA purification involves CsCl centrifugation	294
Wide range of taxa	Nuclei minipreparation via protoplasts; specifically designed for RAPDs and other PCR analyses	330
Musa	Nuclei isolation involves sucrose step gradient centrifugation followed by a variant of the CTAB method	431
Saccharum	Isolated nuclei are lysed with CTAB–SDS	610
Gossypium	Combination of the methods of Paterson et al.[1050] and Fulton et al.[466]; pre-extraction leaves nuclei intact, which are then lysed with a CTAB–sarkosyl buffer	807
Kelp (Laminariales)	Protocol is specifically designed for brown algae; final purification via gel filtration on Sepharose spun columns	901
Zea mays, Brassica napus, Gossypium, Helianthus annuus	Tissue is stored and preincubated in reagent grade ethanol; nuclei are isolated in a hexylene glycol buffer and lysed with SDS–proteinase K	966
Gossypium	Protocol is specifically designed for tissues rich in polyphenols and polysaccharides; PVP, ascorbic acid, and DIECA are included in the nuclear isolation buffer; nuclei are lysed with CTAB	1050
Theobroma cacao	Crude nuclei are isolated first, nuclear DNA is then extracted by a variant of the CTAB method	1070
Lycopersicon esculentum	Large-scale protocol specifically designed for plant tissues rich in polyphenols, based on steps and buffers of Watson and Thompson[1505] and Couch and Fritz[278]	1072
Wide range of taxa, ferns	Nuclei are isolated along with chloroplasts; DNA is purified via CsCl centrifugation; alternative protocols are reviewed	1329
Pisum sativum	Tissue is treated with ether; nuclei are stabilized by hexylene glycol and purified via Percoll step gradient centrifugation; DNA is purified via CsCl centrifugation	1505
Nicotiana tabacum	Nuclei are stabilized by polyamines and purified via Percoll step gradient centrifugation (see Chapter 4.2.6 for details)	1549

Appendix 1E   Isolation Protocols for Megabase DNA

Taxa	Remarks	Ref.
Arabidopsis, *Nicotiana*	Liquid isolation procedure gives higher yields than isolation via protoplasts	75
Triticum, Secale cereale, Hordeum vulgare	DNA is isolated via protoplasts embedded in agarose plugs	243
Triticum	DNA is isolated from liquid nitrogen-powdered tissue embedded in agarose microbeads; an additional gel purification step is included	494
Poaceae, Fabaceae	Extension of the protocol of Guidet et al.[539]	538
Triticum, Secale cereale, Nicotiana tabacum	Liquid nitrogen-powdered tissue is embedded into agarose plugs, and DNA isolation is performed within the agarose	539
Oryza sativa	Nuclei are isolated from rice germ via several Percoll step gradients, and embedded into agarose plugs; DNA isolation is performed within the agarose	578
Glycine max	DNA is isolated via protoplasts derived from suspension-cultured cells, embedded in agarose plugs	611
Arabidopsis	DNA is isolated via nuclei embedded in agarose beads; endogenous DNase activity is inhibited by treatment with 160 mM L-lysine plus 4 mM EGTA	832
Arabidopsis, *Oryza sativa*	DNA is isolated via nuclei embedded in agarose plugs or microbeads	834
Citrus sinensis	DNA is isolated via nuclei isolated from liquid nitrogen-powdered tissue embedded in agarose plugs; efficiency of the method is compared with that of Guidet et al.[539]	858
Helianthus annuus	DNA either is isolated from protoplasts or from nuclei (purified from liquid nitrogen-powdered tissue) embedded in agarose plugs	1021
Lycopersicon	DNA is isolated via protoplasts embedded in agarose plugs	1439
Lycopersicon	DNA is isolated via protoplasts embedded into agarose microbeads that provide an increased surface area, facilitating enzymatic treatments	1553
Sorghum bicolor	DNA is isolated via protoplasts embedded in either microbeads or agarose plugs	1567
Wide range of taxa	DNA is isolated via nuclei (obtained from liquid nitrogen-powdered or homogenized fresh tissue), embedded into agarose plugs or microbeads	1609
Gossypium	DNA is isolated via nuclei embedded in agarose microbeads	1614

Appendix 1F   Miscellaneous DNA Isolation Protocols

Taxa	Remarks	Ref.
Wide range of plant, fungal and animal tissues	Tissue is homogenized in 0.4 M NaCl–TE buffer; followed by proteinase K–SDS treatment and salt extraction of DNA with high concentrations of NaCl	22
Nicotiana tabacum, human, lizard, snail	Isolation buffer contains commercial laundry detergent	68
Pinus strobus, Gossypium hirsutum	Simultaneous isolation of RNA and DNA from recalcitrant tissues; DNA purification involves CsCl centrifugation	72
Cattleya	Tissue homogenization takes place in a Mini-bead-beater; polysaccharides are precipitated with 0.1% ethanol	110
Anthurium andreanum	DNA is specifically precipitated by spermine	182
Gossypium	A guanidine-hydrochloride buffer is used; DNA is purified via ion exchange chromatography	207
Hymenaea herbarium specimen and fossil insects	General method for fossil, herbarium and museum specimens; DNA is specifically bound to glassmilk in the presence of guanidine isothiocyanate	212
Phaseolus and other taxa	Chloroform is displaced by dichloromethane in organic extractions (cheeper and less hazardous than chloroform), DNA quality is indistinguishable	233
Stachys and various nonplant material	Method initially developed for the simultaneous isolation of DNA and RNA from biopsy material. Involves urea–SDS lysis of cells and CsCl centrifugation	353
Cupressus sempervirens	Method based on Qiagen Plant DNeasy kit, involving several modifications needed for cupress needles	361
Transgenic Zea mays residues in soil	Protocol is designed for soil; PCR-inhibiting humic acids are removed by calcium precipitation	410
Several taxa	Polysaccharides are removed by differential precipitation in the presence of 2 M NaCl	428
Dysosma	Method designed for dried roots and rhizomes; extraction buffer contains urea and SDS	465
Picea abies and a wide range of other taxa	Specifically designed for tissues rich in terpenoids and polyphenols; acidic extraction medium; PVP; cysteine; DNA purification on RPC-5 columns	540
Wide range of taxa	Extensive phenol, PVP and PEG treatment of extract; DEAE Sephacel column chromatography	579
Betula	Specifically designed for plant tissues rich in polyphenols; high molarity urea-phosphate buffer; inclusion of DIECA and PVP	619
Wide range of taxa	Cell walls are solubilized by inclusion of potassium or sodium salts of ethyl xanthogenate in the extraction buffer; small amounts of fresh tissue can be processed without homogenization	669, 670
Triticum aestivum, Hordeum vulgare	Simultaneous grinding of 16 samples using ball bearings; lysis in a buffer containing sarkosyl and PVPP	692
Various taxa with a high content of polyphenols	Tissue is initially ground in 1% β-mercaptoethanol, SDS lysis buffer is supplemented with 6% PVP and 3.75 M ammonium acetate	716
Begonia	An initial washing step with a low-salt buffer removes organic acids that otherwise render DNA insoluble	741
Abies alba, Picea abies	Minipreparation from 5 mg of dormant buds	749

Appendix 1F (continued)   Miscellaneous DNA Isolation Protocols

Taxa	Remarks	Ref.
Multinucleate green algae	Total nucleic acids are first extracted with SDS–proteinase K; RNA and DNA are purified by differential LiCl precipitation and CsCl centrifugation, respectively	765
Davidia involucrata	Tissue is lysed in high SDS–mercaptoethanol buffer	813
Fragaria	Specifically designed for (DNA and RNA) isolation from tissues rich in polysaccharides and polyphenols; differential solubility of these compounds as compared to DNA or RNA in 2-butoxyethanol is exploited	875
Wide range of taxa	The method of Guillemaut and Marechal-Drouard[540] is combined with ion exchange chromatography on DEAE cellulose	879
Nicotiana tabacum, Zea mays, Helianthus annuus	Protocol specifically designed for protoplasts and tissue-cultured cells, also suitable for other tissues. Polysaccharides are removed by precipitation with 0.1 vol ethanol	921
Arabidopsis	Polysaccharides are removed by precipitation with 0.35 vol of ethanol in low salt (0.25 M NaCl)	924
Phoenix dactylifera	Method designed for fresh tissue; extraction buffer derived from an isolation protocol for plant mitochondria, contains mannitol and PEG 6000	1020
Wide range of taxa	Polysaccharides are removed by a mixture of glycoside hydrolases	1154
Rhodophyta (red algae)	Miniprep for fresh and dried algal materials; gel purification of crude DNA	1227
Various woody plant species	Specifically designed for pollen; pollen coat is removed by mechanical grinding in a bead mill; DNA is isolated with various lysis buffers	1290
Vicia faba	Small scale 5-h procedure involving CsCl centrifugation in a tabletop centrifuge; applicable for plants, algae, yeast, mammals, insects, and bacteria	1515
Wide range of taxa	Benzyl chloride is used in the extraction medium, since it reacts with –OH residues in polysaccharides	1618
Abies alba	Miniprep version of the protocol of Guillemaut and Maréchal-Drouard[540]	1619

APPENDIX 2

Commercial Companies

Appendix 2A   Suppliers and Sellers of Reagents and Equipment

The companies mentioned below are a selection of those regularly used. It must not be considered as a complete list and the mention of a particular company does not imply a recommendation by the authors.

Company Name and Website	Address and Contacts	Products and Services
Ab Peptides Inc. www.abpeps.com	8224 Manchester Road, Ste. 101 St. Louis, MO 63144 Phone: +1 314 968 4944 Fax:+1 314 968 8988	Klentaq, *Taq* polymerase
Aldrich (see Sigma-Aldrich)		
Amersham Biosciences www1.amershambiosciences.com	SE-751 84 Uppsala, Sweden Phone: +46 18 612 00 00 Fax: +46 18 612 12 00	Radiochemicals; general DNA techniques, chemicals, and equipment; chromatography
Applied Biosystems www.appliedbiosystems.com	850 Lincoln Centre Drive Foster City, CA 94404 Phone: +1 650 638 5800 Fax: +1 650 638 5884	DNA sequencing and fragment analysis systems, services, thermocyclers, genomics and proteomics
Beckman Coulter Inc. www.beckmancoulter.com	Oakley Court, Kingsmead Business Park, London Road High Wycombe Buckinghamshire HP11 1JU, U.K. Phone: +44 1494 441181 Fax: +44 1494 463843	Biomedical instruments, DNA sequencing, genomics
BIO 101, Inc. (see Qbiogene)		

Company Name and Website	Address and Contacts	Products and Services
Bioline www.bioline.com	PMB 311 28 South Main Street Randolph, MA 02368-4800 Phone: +1 781 830 0360 Fax: +1 781 830 0205	Molecular biology reagents
Bio-Rad www.bio-rad.com	1000 Alfred Nobel Drive Hercules, CA 94547 Phone: +1 510 724 7000 Fax: +1 510 741 5817	Life sciences, diagnostics
Calbiochem (includes Merck for Europe, Novagen) www.calbiochem.com	See Merck	Biochemicals, genomics, proteins
Cambrex (includes FMC Bioproducts) www.cambrex.com	One Meadowlands Plaza East Rutherford, NJ 07073 Phone: 201-804-3000	Agarose, SYBR Green, biochemicals
Cellmark Diagnostics www.cellmark.co.uk	PO Box 265, Abingdon Oxfordshire OX14 1YX, U.K. Phone: +44 1235 528000	Paternity testing
Elchrom www.elchrom.com	Gewerbestrasse 8 6330 Cham, Switzerland Phone: +41 41 747 25 50 Fax: +41 41 743 25 36	Electrophoresis, gels
Eppendorf www.eppendorf.com	Barkhausenweg 1 22339 Hamburg, Germany Phone: +49 40 53 8010 Fax: +49 40 53 801 556	Liquid handling, separation, molecular biology products, plastics
Finnzymes www.finnzymes.com	Keilaranta 16 A 02150 Espoo, Finland Phone: +358 9 584 121 Fax +358 9 5841 2200	Molecular biology enzymes
Fisher Scientific www.fishersci.com	2000 Park Lane Pittsburgh PA 15275 Phone: +1 800 766 7000 Fax: +1 800 926 1166	Distributor
Gibco-BRL (see Invitrogen)		
Gilson www.gilson.com	3000 W. Beltline Hwy. P.O. Box 620027 Middleton, WI 53562-0027 Phone: +1 608 836 1551 Fax: +1 608 831 4451	Pipets, liquid handling, chromatography
Grant/Boekel Instruments www.grant.co.uk	Shepreth Cambridgeshire SG8 6GB, U.K. Tel: +44 1763 260811 Fax: +441763 262410	Laboratory equipment
Heraeus, see Kendro		
Hettich www.hettichlab.com	Gartenstrasse 100 D-78532 Tuttlingen, Germany Phone: + 49 7461 705 201	Centrifuges
Hybaid (see Thermo Electron)		

Company Name and Website	Address and Contacts	Products and Services
Invitrogen (includes Gibco, Molecular Probes) www.invitrogen.com	PO Box 3326 4800 DH Breda, the Netherlands Phone: 0800 099 8882 Fax: 0800 023 4212	Molecular biology, cloning, oligos, genomics
Kendro www.kendro.com	Stortford Hall Park Bishop's Stortford Hertfordshire CM23 5GZ, U.K. Phone: +44 1279 82 77 00 Fax: +44 1279 82 77 50	Equipment, incubators, centrifuges
Merck Biosciences, Ltd. www.merckbiosciences.co.uk	Boulevard Industrial Park Padge Road, Beeston Nottingham NG9 2JR, U.K. Phone: +44 115 943 0840 Fax: +44 115 943 0951	Biochemicals, distributor
Microsynth GmbH www.microsynth.ch	Schützenstrasse 15 9436 Balgach, Switzerland Phone: +41 71 722 8333 Fax: +41 71 722 87 58	Oligo synthesis, sequencing service
Millipore www.millipore.com	290 Concord Rd. Billerica, MA 01821 Phone: +1 978 7154321	Filters
MJ Research www.mjr.com	590 Lincoln Street Waltham, MA 02451 Phone: +1 617 972 8180 Fax: +1 617 923 8080	Thermocyclers
MWG www.mwgbiotech.com	Anzinger Strasse 7a D-85560 Ebersberg, Germany Phone: +49 8092 82890 Fax: +49 8092 21084	Oligos, sequencing, genomics
National Diagnostics www.nationaldiagnostics.com	305 Patton Drive Atlanta, GA 30336 Phone: +1 404 699 2121 Fax: +1 404 699 2077	Electrophoresis, histology, solvents
New England Biolabs www.neb.com	32 Tozer Road Beverly, MA 01915-5599 Phone: +1 978 927 5054 Fax: +1 978 921 1350	Molecular biology enzymes
Perkin Elmer www.perkinelmer.com	45 William Street Wellesley, MA 02481-4078 Phone: +1 781 237 5100	Imaging, biochemicals
Pharmacia LKB (see Amersham)		
Promega www.promega.com	2800 Woods Hollow Road Madison WI 53711 Phone: +1 608 274 4330	General molecular biology products
Qiagen www.qiagen.com	28159 Avenue Stanford Valencia, CA 91355 Phone: +1 800 426 8157 Fax: +1 800 718 2056	Molecular biology, kits, genomics
Qbiogene www.qbiogene.com	Parc d'Innovation, BP 50067 67402 Illkirch Cedex, France Phone: +33 3 88 67 54 25 Fax: +33 3 88 67 19 45	Consumables and biochemicals

Company Name and Website	Address and Contacts	Products and Services
Sarstedt www.sarstedt.com	Rommelsdorfer Straße Postfach 1220 51582 Nümbrecht, Germany Phone: +49 2293 305 0 Fax: +49 2293 305 122	Consumables, plasticware
Schleicher and Schuell GmbH www.schleicher-schuell.com	Hahnestraße 3 D-37586 Dassel, Germany Phone: +49 5561 791 0 Fax: +49 5564 230 9	Filter paper, membranes, filters
Sigma-Aldrich Family (includes Sigma, Aldrich, Fluka, Supelco) www.sigmaaldrich.com	3050 Spruce St. St. Louis, MO 63103	Oligos, chemicals and biochemicals
Sorvall, see Kendro		
Stratagene www.stratagene.com	11011 N. Torrey Pines Road La Jolla, CA 92037 Phone: +1 858 535 5400	General molecular biology products
Syngene www.syngene.com	Beacon House, Nuffield Rd. Cambridge CB4 1TF, U.K. Phone: +44 1223 727123 Fax: +44 1223 727101	Gel documentation and analysis systems
Thermo Electron Corporation www.thermo.com	Hemel Hempstead, P2 7SH, U.K. Phone: +44 870 609 9223 Fax: +44 870 609 9222	Life and laboratory sciences, equipment, consumables,
University of British Columbia www.biotech.ubc.ca	NAPS Unit University of British Columbia 6174 University Boulevard Vancouver, BC V6T 1Z3, Canada	Primers
Uvitec www.uvitec.co.uk	Avebury House 36a Union Lane Cambridge. CB4 1QB, U.K. Phone: +44 1223 568060 Fax: +44 1223 306198	UV transillumunators, documentation systems
Ultra-Violet Products Ltd, UVP Inc www.uvp.com	Unit 1, Trinity Hall Farm Estate Nuffield Road Cambridge CB4 1TG, U.K. Phone: +44 1223 420022 Fax: +44 1223 420561	UV transillumunators, documentation systems
Whatman Lab Products www.whatman.co.uk	Whatman House St Leonard's Road 20/20 Maidstone Kent ME16 0LS, U.K.	Separation techniques, filters
VWR scientific www.vwr.com	Goshen Corporate Park West 1310 Goshen Parkway West Chester, PA 19380 Phone: +1 610 429 2850 Fax: +1 610 429 9340	Laboratory equipment, chemicals etc

Appendix 2B   Companies That Offer Development of Microsatellite Libraries and Genotyping

Company Name and Web Site	Contacts
Bioprofiles, Ltd. www.bioprofiles.co.uk	1 Ryelea Longhoughton NE66 3DE, U.K. information@bioprofiles.co.uk
Biopsytech www.biopsytec.de	Rheinbach, Germany gerhards@biopsytec.com
BC Research www.bcresearch.com	Canada (noncommercial applications) cnewton@bcresearch.com
CIRAD www.cirad.fr	CIRAD Montpellier, France norbert.billotte@cirad.fr
Genetic Identification Services (GIS) www.genetic-id-services.com	Chatsworth, CA gisemail@genetic-id-services.com
Genome Express www.genomex.com	Grenoble, France bogden@genomex.com
Amplicon Express www.genomex.com	1610 NE Eastgate Blvd Suite Pullman, WA 99163 bogden@genomex.com
Northern Bioidentification Service, Ltd. www.biobank.co.kr/maker/nnn/northern-bio.shtml	403-63 Albert Street Winnipeg, MB R3B 1G4, Canada northern@mts.net
Traitgenetics GmbH www.traitgenetics.com	Am Schwabeplan 1b D-06466 Gatersleben, Germany contact@traitgenetics.de
Ecogenics GmbH www.ecogenics.ch	Winterthurerstrasse 190 8057 Zuerich, Switzerland
GENterprise GmbH www.genterprise.de	J.-J.-Becherweg 34-36 D-55128 Mainz, Germany kraemer@genterprise.de

Computer Programs Dealing with the Evaluation of DNA Sequence Variation and Molecular Marker Data

The programs mentioned below are a selection of those regularly used. It must not be considered as a complete list and the mention of a particular program does not imply a recommendation by the authors. We apologize to those whose programs are not included below and for missing references or data.

Appendix 3A   Data Sorting and Checking

Program	Web Site	Operating System	Description
4Peaks	www.mekentosj.com/4peaks	MacOS	Shows and edits sequences
BioEdit	www.mbio.ncsu.edu/BioEdit/ bioedit.html	Windows	Handles and aligns DNA sequences
ClustalX (Thompson et al.[1395])	www.icgeb.org/netsrv/ clustalx.html	Windows, MacOS, UNIX	Provides multiple alignment of DNA sequences
EMBOSS	www.emboss.org	LINUX and MacOS	Analyzes and aligns sequences
Excel Microsatellite Toolkit	oscar.gen.tcd.ie/%7Esdepark/ ms-toolkit/	Windows	Checks and formats microsatellite data
GeneScanView	bmr.cribi.unipd.it	Windows	Reads files (ABI and some other brands) for analyzing fragments for AFLP or microsatellite analysis
Genographer	hordeum.oscs.montana.edu/ genographer/	Windows	Reads files (ABI and some other brands) for analyzing fragments for AFLP or microsatellite analysis

Appendix 3A (continued) Data Sorting and Checking

Program	Web Site	Operating System	Description
Micro-Checker	www.microchecker.hull.ac.uk	Windows	Identifies scoring errors due to stuttering, large allele dropout, and null alleles
ProSeq v2.9 (Filatov[442])	helios.bto.ed.ac.uk/evolgen/ filatov/proseq.html	Windows	Visualizes and edits ABI chromatograms, aligns DNA sequences
STRand	www.vgl.ucdavis.edu/STRand/	Windows	Reads files (ABI and some other brands) for analyzing fragments for AFLP or microsatellite analysis
CodonCode Aligner	www.codoncode.com	Commercial	Edits a variety of input files, aligns and analyzes sequences
Sequencher	www.genecodes.com	Commercial	Aligns and analyzes sequences, provides restriction mapping and physical mapping options

Appendix 3B   Allele Frequencies, Population Structure, Population Assignment

Program	Web Site	Operating System	Description
ADE-4	pbil.univ-lyon1.fr/ADE-4/ADE-4.html	Windows	Performs PCA, FCA, on ecological data
AFLP-SURV	www.ulb.ac.be/sciences/lagev/aflp-surv.html	Web based	Analyzes dominant data, calculates distances between populations and between individuals, with specified mating system
API-CALC 1.0 (Ayres and Overall[63])	www.rdg.ac.uk/statistics/genetics/	Windows	Calculates probability of identity
Arlequin	lgb.unige.ch/arlequin/	Windows, MacOS, Linux	Analyzes population genetic data
Assignment calculator Doh	www2.biology.ualberta.ca/jbrzusto/Doh.php	Web based	Performs assignment and migration tests
BOTTLENECK	www.montpellier.inra.fr/URLB/bottleneck/bottleneck.html	Windows	Detects bottlenecks from allele frequency data
DISPAN	mep.bio.psu.edu/	DOS	Calculates heterozygosities, G_{ST}, and phylogenetic trees
DNaSP	www.ub.es/dnasp	Windows, MacOS	Calculates population genetic parameters from sequence data
FSTAT (Goudet[514])	www.unil.ch/izea/softwares/fstat.html	Windows	Calculates population genetic parameter
GDA (Weir[1518])	hydrodictyon.eeb.uconn.edu/people/plewis/software.php	Windows	Analyzes genetic data
GenAlEx	www.anu.edu.au/BoZo/GenAlEx	Windows, MacOS	Analyzes population structure, PCO, Mantel test
GENECLASS	www.montpellier.inra.fr/URLB/index.html	Windows	Performs assignment tests
GENEPOP (Raymond and Rousset[1145])	wbiomed.curtin.edu.au/genepop	DOS and web based	Performs population genetics calculations
Genetix	www.univ-montp2.fr/~genetix/genetix/genetix.htm	Windows	Performs population genetics calculations
GenoType/GenoDive	staff.science.uva.nl/~meirmans/	Windows, MacOS	Analyzes clonal structure
GeoDis (Posada et al.[1091])	darwin.uvigo.es/software/geodis.html	Windows, Linux and MacOS	Analyzes phylogeographic data and population structure

Appendix 3B (continued)   Allele Frequencies, Population Structure, Population Assignment

Program	Web Site	Operating System	Description
Hickory	darwin.eeb.uconn.edu/hickory/hickory.html	Windows Linux	Analyzes geographic structure from dominant and codominant markers
LCDMV	www.cimmyt.org/ABC/manual/contents.htm	Windows, Unix (SAS required)	Identifies cultivars
LIAN	adenine.biz.fh-weihenstephan.de/lian/	Web based	Analyzes linkage equilibrium for multilocus data
MICROSAT	hpgl.stanford.edu/projects/microsat/	Window, MacOS	Analyzes genetic distances
Microsatellite Analyzer (MSA) (Dieringer and Schlötterer[345])	i122server.vu-wien.ac.at/MSA/MSA_download.html	Web based	Analyzes population genetic structure
Migrate	evolution.genetics.washington.edu/lamarc/migrate.html	Windows. MacOS, Linux	Analyzes population size and migration rate
MLGsim	www.molbiol.umu.se/forskning/saura/software.htm	DOS	Predicts multilocus identity, clonal structure
MLNE	www.zoo.cam.ac.uk/ioz/software.htm	Windows	Predicts effective population size and migration
PASSAGE	lsweb.la.asu.edu/rosenberg/Passage/	Windows	Performs spatial analysis, analyzes ecological data
PCAGEN	www2.unil.ch/popgen/softwares/	Windows	Analyzes codominant data, PCA and F_{ST} values
POPGENE	www.ualberta.ca/~fyeh/	Windows	Analyzes population genetic data
POPTREE	mep.bio.psu.edu/	DOS, and Linux	Analyzes heterozygosity, phylogenetic trees
PSAwinD	homepage3.nifty.com/makotot_ftbc/PSAwinD100E.htm	Windows	Calculates autocorrelation and population structure
RAPDistance	www.anu.edu.au/BoZo/software	DOS	Analyzes population structure for dominant markers
RSTCALC	helios.bto.ed.ac.uk/evolgen/rst/rst.html	Windows	Analyzes microsatellites, genetic variance, population structure
Spatial Genetic Software (SGS)	kourou.cirad.fr/genetique/software.html	Windows	Analyzes population structure for any type of marker
Sites	lifesci.rutgers.edu/~heylab/ProgramsandData/Programs/SITES/SITES_Documentation.htm	Windows, MacOS	Performs population analysis of DNA sequence data

Appendix 3B (continued)   Allele Frequencies, Population Structure, Population Assignment

Program	Web Site	Operating System	Description
SPAGeDi (Hardy and Vekemans[572])	www.ulb.ac.be/sciences/lagev/ spagedi.html	Windows	Calculates autocorrelation and population structure
Structure	pritch.bsd.uchicago.edu	Windows, UNIX	Analyzes population structure, migration, assignment, hybrid zones
test_h_diff	www.ucl.ac.uk/tcga/software/ index.html	Windows	Tests gene diversity differences
TFPGA	bioweb.usu.edu/mpmbio/	Windows	Analyzes population genetics, dominant/ codominant data
NTSYS	www.exetersoftware.com/cat/nts yspc/ntsyspc.html	Commercial	Performs PCO, Mantel test; also analyzes quantitative data

Appendix 3C   **Parentage and Relatedness[a]**

Program	Web Site	Operating System	Description
CERVUS	helios.bto.ed.ac.uk/evolgen/cervus/cervus.html	Windows	Determines paternity, using codominant loci
Delirious (Stone and Björklund[1336])	www.zoo.utoronto.ca/stone/DELRIOUS/delrious.htm	Windows, Linux and MacOS	Determines relatedness
FaMoZ (Gerber et al.[487])	www.pierroton.inra.fr/genetics/labo/Software	Web based	Determines parentage
Identity	www.boku.ac.at/zag/forsch/		Determines parents and offspring
IDENTIX (Belkhir et al.[103a])	www.univ-montp2.fr/%7Egenetix/#programs	Windows	Calculates relatedness
MER	www.zoo.cam.ac.uk/ioz/software.htm	Windows	Calculates relatedness
MLTR	genetics.forestry.ubc.ca/ritland/programs.html	Windows	Analyzes mating systems, relatedness
Newhybrids	ib.berkeley.edu/labs/slatkin/eriq/software/software.htm	Windows, MacOS	Detects hybrids from multilocus data
PAPA	www.bio.ulaval.ca/louisbernatchez/downloads_fr.htm	Windows	Determines relatedness, kinship, parentage
PARENTE (Cercueil et al.[222])	www2.ujf-grenoble.fr/leca/membres/manel.html		Determines relatedness, kinship, parentage
POPAIRS	chkuo.name/software/POpairs.html	MacOS	Determines parentage
Relatedness and Kinship	www.gsoftnet.us/GSoft.html	MacOS	Determines relatedness, kinship

[a] See also Jones and Ardren.[675]

Appendix 3D   **Mapping and Linkage**

Program	Web Site	Operating System
GMENDEL	cropandsoil.oregonstate.edu/G-mendel/Default.htm	PC and UNIX
MAPMAKER (Lander et al.[773])	www.broad.mit.edu/genome_software	
QTL Express (Seaton et al.[1265a])	qtl.cap.ed.ac.uk/	Web based
QTL Cartographer	statgen.ncsu.edu/qtlcart/index.php	Windows
JOINMAP, (Stam[1323]) Mapchart, MAPQTL	www.kyazma.nl/index1.php	Commercial

Appendix 3E   **Clustering and Phylogenetic Analysis**

Program	Web Site	Operating System	Description
CITE	adenine.biz.fh-weihenstephan.de/cite/	Web based	Provides confidence intervals for divergence time estimates
DAMBE (Xia and Xie[1578])	aix1.uottawa.ca/%7Exxia/software/ software.htm	Windows	Analyzes phylogenetic data
FastME	www.ncbi.nlm.nih.gov/CBBresearch/ Desper/FastME.html	DOS	Analyzes phylogenetic data
MEGA2	www.megasoftware.net	Windows	Analyzes phylogenetic data
Network	www.fluxus-engineering.com/ sharenet.htm	Windows, DOS	Lists phylogenetic networks
PHYLIP (Felsenstein [435])	evolution.gs.washington.edu/phylip.html	Windows, PacOS, Dos	Analyzes phylogenetic data
Tree-PUZZLE	www.tree-puzzle.de	Windows, MacOS, UNIX	Analyzes phylogenetic data
TREEVIEW	taxonomy.zoology.gla.ac.uk/rod/ treeview.html	Windows and MacOS	Provides visualization of trees, from PAUP*, PHYLIP, or ClustalW
PAUP*	paup.csit.fsu.edu/, www.sinauer.com	Commercial	Analyzes phylogenetic data

Appendix 3F   Primer Development

Program	Web Site	Operating System	Description and Comments
Various	www.hgmp.mrc.ac.uk/GenomeWeb/nuc-primer.html		Offers a range of programs
Amplify	engels.genetics.wisc.edu/amplify	MacOS	Tests primers for dimer formation
OligoAnalyzer	biotools.idtdna.com/analyzer	Web based	Tests primers
Oligoperfect Designer	www.invitrogen.com/content.cfm?pageid=9716	Web based	Designs primers
Oligowiz	www.cbs.dtu.dk/services/OligoWiz	Web based	Designs any type of oligo
Primer3 (Rozen and Skaletsky[1196])	frodo.wi.mit.edu/cgi-bin/primer3/primer3_www.cgi/	Web based	Designs primers
Primo	www.changbioscience.com/primo/primo.html	Web based	Designs primers
LASERGENE	www.dnastar.com	Commercial	Analyzes oligos and DNA
MacVector software	www.accelrys.com/products/macvector	Commercial	Performs complete DNA analysis
Oligo (Rychlik and Rhoads[1203])	www.oligo.net	Commercial	Designs primers
Primer Designer	www.scied.com/ses_pd5.htm	Commercial	Designs primers

Web Pages of Interest

The web pages mentioned below are a selection of those regularly used. It must not be considered as a complete list and the mention of a particular web page does not imply a recommendation by the authors. We apologize to those whose web pages are not mentioned.

Web sites with overview of and links to a large variety of data analysis programs:

http://www.pierroton.inra.fr/genetics/labo/Software/
http://www.nceas.ucsb.edu/papers/geneflow/software/index.html
http://evolution.genetics.washington.edu/phylip/software.htm
http://lewis.eeb.uconn.edu/lewishome/software.html
http://linkage.rockefeller.edu/soft/
http://taxonomy.zoology.gla.ac.uk/software/software.html
http://www.gsoftnet.us/GSoft.html
http://courses.washington.edu/fish543/Software.htm
http://www.biology.lsu.edu/general/software.html
http://www.cellbiol.com/soft.htm
http://iubio.bio.indiana.edu/IUBio-Software+Data/molbio/Listings.html
http://mep.bio.psu.edu/
http://www.bio.psu.edu/People/Faculty/Nei/Lab/
http://www.univ-montp2.fr/%7Egenetix/#programs
http://www2.biology.ualberta.ca/jbrzusto/
http://uwadmnweb.uwyo.edu/zoology/mcdonald/molmark/Data/WebSoft.html

Web sites with programs useful for teaching and simulations:

http://www.anu.edu.au/BoZo/GenAlEx
http://evol.biology.mcmaster.ca/paulo/winpop.php
http://darwin.eeb.uconn.edu/simulations/simulations.html
http://www.cbs.umn.edu/populus/
ftp://evolution.gs.washington.edu/pub/popgen/popg.html
http://faculty.washington.edu/~herronjc/SoftwareFolder/software.html
http://www.evotutor.org/Software.html
http://cc.oulu.fi/~jaspi/popgen/popgen.htm

Green Plant Phylogeny, Research Coordination Group, DEEP GREEN:
http://ucjeps.berkeley.edu/bryolab/greenplantpage.html

GRIN taxonomy, National Plant Germplasm System:
http://www.ars-grin.gov/cgi-bin/npgs/html/index.pl

Molecular Ecology Notes Primer Database:
http://tomato.bio.trinity.edu/home.html

Web Resources in Molecular Evolution and Systematics:
http://darwin.eeb.uconn.edu/molecular-evolution.html

Societies:

Ecological Society of Australia, http://www.ecolsoc.org.au/
European Society for the Study of Evolution, http://www.eseb.org/
Society for Molecular Biology and Evolution, http://www.smbe.org/

Links to wide range of evolution sites:
http://dorakmt.tripod.com/evolution/link.html

Programs, tools, and contacts:

http://www.bioexchange.com/index.cfm
http://webdoc.sub.gwdg.de/ebook/y/1999/whichmarker/index.htm

References

1. Aagard, J.E., Vollmer, S.S., Sorensen, F.C., and Strauss, S.H., (1995) Mitochondrial DNA products among RAPD profiles are frequent and strongly differentiated between races of Douglas-fir, *Mol. Ecol.* 4: 441–447.
2. Abbott, R.J., Smith, L.C., Milne, R.I., Crawford, R.M.M., Wolff, K., and Balfour, J., (2000) Molecular analysis of plant migration and refugia in the Arctic, *Science* 289: 1343–1346.
3. Abe, J., Xu, D.H., Suzuki, Y., Kanazawa, A., and Shimamoto, Y., (2003) Soybean germplasm pools in Asia revealed by nuclear SSRs, *Theor. Appl. Genet.* 106: 445–453.
4. Adams, M.D., and 196 coauthors, (2000) The genome sequence of *Drosophila melanogaster*, *Science* 287: 2185–2195.
5. Adams, R.P., and Demeke, T., (1993) Systematic relationships in *Juniperus* based on random amplified polymorphic DNAs (RAPDs), *Taxon* 42: 553–571.
6. Adams, R.P., and Rieseberg, L.H., (1998) The effects of non-homology in RAPD bands on similarity and multivariate statistical ordination in *Brassica* and *Helianthus*, *Theor. Appl. Genet.* 97: 323–326.
7. Adams, R.P., Zhong, M., and Fei, Y., (1999) Preservation of DNA in plant specimens: inactivation and re-activation of DNases in field specimens, *Mol. Ecol.* 8: 681–684.
8. Aggarwal, R.K., Brar, D.S., Nandi, S., Huang, N., and Khush, G.S., (1999) Phylogenetic relationships among *Oryza* species revealed by AFLP markers, *Theor. Appl. Genet.* 98: 1320–1328.
9. Ahn, S., Anderson, J.A., Sorrels, M.E., and Tanksley, S.D., (1993) Homoeologous relationships of rice, wheat and maize chromosomes, *Mol. Gen. Genet.* 241: 483–490.
10. Aitchitt, M., Ainsworth, C.C., and Thangavelu, M., (1993) A rapid and efficient method for the extraction of total DNA from mature leaves of the date palm (*Phoenix dactylifera* L.), *Plant Mol. Biol. Rep.* 11: 317–319.
11. Aitman, T.J., Hearne, C.M., McAleer, M.A., and Todd, J.A., (1991) Mononucleotide repeats are an abundant source of length variants in mouse genomic DNA, *Mamm. Genome* 1: 206–210.
12. Akagi, H., Yokozeki, Y., Inagaki, A., and Fujimura, T., (1996) Microsatellite DNA markers for rice chromosomes, *Theor. Appl. Genet.* 93: 1071–1077.
13. Akagi, H., Yokozeki, Y., Inagaki, A., Mori, K., and Fujimura, T., (2001) *Micron*, a microsatellite-targeting transposable element in the rice genome, *Mol. Genet. Genomics* 266: 471–480.
14. Akkaya, M.S., Bhagwat, A.A., and Cregan, P.B., (1992) Length polymorphisms of simple sequence repeat DNA in soybean, *Genetics* 132: 1131–1139.

15. Akkaya, M.S., Shoemaker, R.C., Specht, J.E., Bhagwat, A.A., and Cregan, P.B., (1995) Integration of simple sequence repeat DNA markers into a soybean linkage map, *Crop Sci.* 35: 1439–1445.

16. Albertini, E., Porceddu, A., Marconi, G., Barcaccia, G., Pallottini, L., and Falcinelli, M., (2003) Microsatellite-AFLP for genetic mapping of complex polyploids, *Genome* 46: 824–832.

17. Aldrich, J., and Cullis, C.A., (1993) RAPD analysis in flax: optimization of yield and reproducibility using *KlenTaq*1 DNA polymerase, Chelex 100, and gel purification of genomic DNA, *Plant Mol. Biol. Rep.* 11: 128–141.

18. Aldrich, P.R., and Hamrick, J.L., (1998) Reproductive dominance of pasture trees in a fragmented tropical forest mosaic, *Science* 281: 103–105.

19. Aldrich, P.R., Hamrick, J.L., Chavarriaga, P., and Kochert, G., (1998) Microsatellite analysis of demographic genetic structure in fragmented populations of the tropical tree *Symphonia globulifera*, *Mol. Ecol.* 7, 933–944.

20. Ali, S., Müller, C.R., and Epplen, J.T., (1986) DNA finger printing by oligonucleotide probes specific for simple repeats, *Hum. Genet.* 74: 239–243.

21. Alix, K., Paulet, F., Glaszmann, J.C., and D'Hont, A., (1999) Inter-*Alu*-like species-specific sequences in the *Saccharum* complex, *Theor. Appl. Genet.* 99: 962–968.

22. Aljanabi, S.M., and Martinez, I., (1997) Universal and rapid salt-extraction of high quality genomic DNA for PCR-based techniques, *Nucleic Acids Res.* 25: 4692–4693.

23. Aljanabi, S.M., Forget, L., and Dookun, A., (1999) An improved and rapid protocol for the isolation of polysaccharide- and polyphenol-free sugarcane DNA, *Plant Mol. Biol. Rep.* 17: 281 (1–8).

24. Almanza-Pinzón, M.I., Khairallah, M., Fox, P.N., and Warburton, M.L., (2003) Comparison of molecular markers and coefficients of parentage for the analysis of genetic diversity among spring bread wheat accessions, *Euphytica* 130: 77–86.

25. Alphey, L., (1997) *DNA Sequencing*, Bios Scientific Publ., Oxford, U.K.

26. Amarger, V., Gaugier, D., Yerle, M., Apiou, F., Pinton, P., Giraudeau, F., Monfouilloux, S., Lathrop, M., Dutrillaux, B., Buard, J., and Vergnaud, G., (1998) Analysis of distribution in the human, pig, and rat genomes points toward a general subtelomeric origin of minisatellite structures, *Genomics* 52: 62–71.

27. Amos, W., and Rubinsztein, D.C., (1996) Microsatellites are subject to directional evolution, *Nat. Genet.* 12: 13–14.

28. Anamthawat-Jónsson, K., Bragason, B.T., Bödvarsdóttir, S.K., and Koebner, R.M.D., (1999) Molecular variation in *Leymus* species and populations, *Mol. Ecol.* 8: 309–315.

29. Andersen, T.H., and Nilsson-Tillgren, T., (1997) A fungal minisatellite, *Nature* 386: 771.

30. Angers, B., and Bernatchez, L., (1997) Complex evolution of a salmonid microsatellite locus and its consequences in inferring allelic divergence from size information, *Mol. Biol. Evol.* 14: 230–238.

31. Aranza, M.J., Carbó, J., and Arús, P., (2003) Microsatellite variability in peach [*Prunus persica* (L.) Batsch]: cultivar identification, marker mutation, pedigree inferences and population structure, *Theor. Appl. Genet.* 106: 1341–1352.

32. Aras, S., Duran, A., and Yenilmez, G., (2003) Isolation of DNA for RAPD analysis from dry leaf material of some *Hesperis* L. specimens, *Plant Mol. Biol. Rep.* 21: 461a–461f.

33. Archibald, J.K., Mort, M.E., and Crawford, D.J., (2003) Bayesian inference of phylogeny: a non-technical primer, *Taxon* 52: 187–191.

34. Arcot, S.S., Wang, Z., Weber, J.L., Deininger, P.L., and Batzer, M.A., (1995) *Alu* repeats: a source for the genesis of primate microsatellites, *Genomics* 29: 136–144.

35. Ardlie, K.G., Kruglyak, L., and Seielstad, M., (2002) Patterns of linkage disequilibrium in the human genome, *Nat. Rev. Genet.* 3: 299–309.

36. Arens, P., Odinot, P., Van Heusden, A.W., Lindhout, P., and Vosman, B., (1995) GATA- and GACA-repeats are not evenly distributed throughout the tomato genome, *Genome* 38: 84–90.

37. Arens, P., Coops, H., Janse, J., and Vosman, B., (1998) Molecular genetic analysis of black poplar (*Populus nigra* L.) along Dutch rivers, *Mol. Ecol.* 7: 11–18.

38. Areshchenkova, T., and Ganal, M.W., (1999) Long tomato microsatellites are predominantly associated with centromeric regions, *Genome* 42: 536–544.

39. Areshchenkova, T., and Ganal, M.W., (2002) Comparative analysis of polymorphism and chromosomal location of tomato microsatellite markers isolated from different sources, *Theor. Appl. Genet.* 104: 229–235.

40. Armour, J.A.L., Wong, Z., Wilson, V., Royle, N.J., and Jeffreys, A.J., (1989) Sequences flanking the repeat arrays of human minisatellites: association with tandem and dispersed repeat elements, *Nucleic Acids Res.* 13: 4925–4935.

41. Armour, J.A.L., Povey, S., Jeremiah, S., and Jeffreys, A.J., (1990) Systematic cloning of human minisatellites from ordered charomid libraries, *Genomics* 8: 501–512.

42. Armour, J.A.L., Neumann, R., Gobert, S., and Jeffreys, A.J., (1994) Isolation of human simple repeat loci by hybridization selection, *Hum. Mol. Genet.* 3: 599–605.

43. Armour, J.A.L., Alegre, S.A., Miles, S., Williams, L.J., and Badge, R.M., (1999) Minisatellites and mutation processes in tandemly repetitive DNA. In *Microsatellites: Evolution and Applications*, Goldstein, D.B., and Schlötterer, C., Eds., Oxford University Press, Oxford, U.K., pp. 24–33.

44. Arnau, G., Lallemand, J., and Bourgoin, M., (2002) Fast and reliable strawberry cultivar identification using inter simple sequence repeat (ISSR) amplification, *Euphytica* 129: 69–79.

45. Arnold, C., Rossetto, M., McNally, J., and Henry, R.J., (2002) The application of SSRs characterized for grape (*Vitis vinifera*) to conservation studies in Vitaceae, *Am. J. Bot.* 89: 22–28.

46. Arnold, M.L., (1993) *Iris nelsonii* (Iridaceae): origin and genetic composition of a homoploid hybrid species, *Am. J. Bot.* 80: 577–583.

47. Arnold, M.L., (1997) *Natural Hybridisation and Evolution*, Oxford University Press, Oxford, U.K.

48. Arnold, M.L., (2004) Natural hybridisation and the evolution of domesticated, pest and disease organisms, *Mol. Ecol.* 13: 997–1007.

49. Arnold, M.L., Bennett, B.D., and Zimmer, E.A., (1990) Natural hybridisation between *Iris fulva* and *Iris hexagona*: pattern of ribosomal DNA variation, *Evolution* 44: 1512–1521.

50. Arnold, M.L., Buckner, C.M., and Robinson, J.J., (1991) Pollen-mediated introgression and hybrid speciation in Louisiana irises, *Proc. Natl. Acad. Sci. U.S.A.,* 88: 1398–1402.

51. Arroyo-García, R., Lefort, F., de Andrés, M.T., Ibáñez, J., Borrego, J., Jouve, N., Cabello, F., and Martínez-Zapater, J.M., (2002) Chloroplast microsatellite polymorphisms in *Vitis* species, *Genome* 45: 1142–1149.

52. Asemota, H.N., (1995) A fast, simple, and efficient miniscale method for the preparation of DNA from tissues of yam (*Dioscorea* spp.), *Plant Mol. Biol. Rep.* 13: 214–218.

53. Ashikawa, I., (2001) Surveying CpG methylation at 5′-CCGG in the genomes of rice cultivars, *Plant Mol. Biol.* 45: 31–39.

54. Atienzar, F., Evenden, A., Jha, A., Savva, D., and Depledge, M., (2000) Optimized RAPD analysis generates high-quality genomic DNA profiles at high annealing temperature, *BioTechniques* 28: 52–54.

55. Austerlitz, F., and Smouse, P.E., (2001) Two-generation analysis of pollen flow across a landscape. II. Relation between Φ_{ft}, pollen dispersal and interfemale distance, *Genetics,* 157: 851–857.

56. Ausubel, F.M., Brent, R., Kingston, R.E., Moore, D.D., Seidman, J.G., Smith, J.A., and Struhl, K., (1999) *Short Protocols in Molecular Biology*, Wiley, New York.

57. Avise, J.C., (1994) *Molecular Markers, Natural History, and Evolution*, Chapman & Hall, New York and London.

58. Avise, J.C., (2000) *Phylogeography: The History and Formation of Species*, Harvard University Press, Cambridge, MA.

59. Avise, J.C., Arnold, J., Ball, R.M., Jr., Bermingham, T., Lamb, T., Neigel, J.E., Reeb, C.A., and Saunders, N.C., (1987) Intraspecific phylogeography: the mitochondrial bridge between population genetics and systematics, *Annu. Rev. Ecol. Syst.* 18: 489–522.

60. Awadalla, P., and Ritland, K., (1997) Microsatellite variation and evolution in the *Mimulus guttatus* species complex with contrasting mating systems, *Mol. Biol. Evol.* 14: 1023–1034.

61. Ayliffe, M.A., Lawrence, G.J., Ellis, J.G., and Pryor, A.J., (1994) Heteroduplex molecules formed between allelic sequences cause nonparental RAPD bands, *Nucleic Acids Res.* 22: 1632–1636.

62. Ayres, D.R., Garcia-Rossi, D., Davis, H., and Strong, D.R., (1999) Extent and degree of hybridisation between exotic (*Spartina aterniflora*) and native (*S. foliosa*) cordgrass (Poaceae) in California, USA determined by random amplified polymorphic DNA (RAPDs), *Mol. Ecol.* 8: 1179–1186.

63. Ayres, K., and Overall, D.J., (2004) API-CALC 1.0: a computer program for calculating the average probability of identity allowing for substructure, inbreeding and the presence of close relatives, *Mol. Ecol. Notes* 4: 315–318.

64. Bachem, C.W.B., van der Hoeven, R.S., de Bruijn, S.M., Vreugdenhil, D., Zabeau, M., and Visser, R.G.F., (1996) Visualization of differential gene expression using a novel method of RNA fingerprinting based on AFLP: analysis of gene expression during potato tuber development, *Plant J.* 9: 745–753.

65. Bachmann, K., (1997) Nuclear DNA markers in plant biosystematic research, *Opera Bot.* 132: 137–148.

66. Backert, S., Nielsen, B.L., and Börner, T., (1997) The mystery of the rings: structure and replication of mitochondrial genomes from higher plants, *Trends Plant Sci.* 2: 477–483.

67. Badouin, L., and Lebrun, L., (2001) An operational Bayesian approach for the identification of sexually reproduced cross-fertilized populations using molecular markers, *Acta Hortic.* 546: 81–93.

68. Bahl, A., and Pfenninger, M., (1996) A rapid method of DNA isolation using laundry detergent, *Nucleic Acids Res.* 24: 1587–1588.

69. Bai, G., Ayele, M., Tefera, H., and Nguyen, H.T., (1999) Amplified fragment length polymorphism analysis of tef [*Eragrostis tef* (Zucc.) Trotter], *Crop Sci.* 39: 819–824.

70. Bailey, D.C., (1983) Isozymic variation and plant breeders' rights. In *Isozymes in Plant Genetics and Breeding, Part A*, Tanksley, S.D., and Orton, J.T., Eds., Elsevier, Amsterdam, pp. 425–441.

71. Baker, A.J., Ed., (2000) *Molecular Methods in Ecology*, Blackwell Science Ltd., Oxford, U.K.

72. Baker, S.S., Rugh, C.L., and Kamalay, J.C., (1990) RNA and DNA isolation from recalcitrant plant tissues, *BioTechniques* 9: 268–272.

73. Baldwin, B.G., (1992) Phylogenetic utility of the internal transcribed spacers of nuclear ribosomal DNA in plants: an example from the Compositae, *Mol. Phylogenet. Evol.* 1: 3–16.

74. Baldwin, B.G., Sanderson, M.J., Porter, J.M., Wojciechowski, M.F., Campbell, C.S., and Donoghue, M.J., (1995) The ITS region of nuclear ribosomal DNA: a valuable source of evidence on angiosperm phylogeny, *Ann. Missouri Bot. Gard.* 82: 247–277.

75. Balestrazzi, A., Bernaccchia, G., Cella, R., Ferretti, L., and Sora, S., (1991) Preparation of high molecular weight plant DNA and its use for artificial chromosome construction, *Plant Cell Rep.* 10: 315–320.

76. Balloux, F., and Lugon-Moulin, N., (2002) The estimation of population differentiation with microsatellite markers, *Mol. Ecol.* 11: 155–165.

77. Bänfer, G., Fiala, B., and Weising, K., (2004) AFLP analysis of phylogenetic relationships among myrmecophytic species of *Macaranga* (Euphorbiaceae) and their allies, *Plant Syst. Evol.*, 249: 213–231.

78. Baril, C.P., Verhaegen, D., Vigneron, P., Bouvet, J.M., and Kremer, A., (1997) Structure of the specific combining ability between two species of *Eucalyptus*, I. RAPD data. *Theor. Appl. Genet.* 94: 796–803.

79. Barker, J.H.A., Matthes, M., Arnold, G.M., Edwards, K.J., Åhman, I., Larsson, S., and Karp, A., (1999) Characterisation of genetic diversity in potential biomass willows (*Salix* spp.) by RAPD and AFLP analyses, *Genome* 42: 173–183.

80. Barnwell, P., Blanchard, A.N., Bryant, J.A., Smirnoff, N.S., and Weir, A.F., (1998) Isolation of DNA from the highly mucilagenous succulent plant *Sedum telephium*, *Plant Mol. Biol. Rep.* 16: 133–138.

81. Barrier, M., Friar, E., Robichaux, R., and Purugganan, M. (2000) Interspecific evolution in plant microsatellite structure, *Gene* 241: 101–105.

82. Bartish, I.V., Jeppsson, N., and Nybom, H., (1999) Population genetic structure in the dioecious pioneer plant species *Hippophae rhamnoides* investigated by random amplified polymorphic DNA (RAPD) markers, *Mol. Ecol.* 8: 791–802.

83. Bartish, I.V., Garkava, L.P., Rumpunen, K., and Nybom, H., (2000) Phylogenetic relationships and differentiation among and within populations of *Chaenomeles* Lindl. (Rosaceae), estimated with RAPDs and isozymes, *Theor. Appl. Genet.* 101: 554–563.

84. Bartish, I.V., Jeppsson, N., Swenson, U., and Nybom, H., (2002) Phylogeny of *Hippophae* (Elaeagnaceae) inferred from parsimony analysis of chloroplast DNA and morphology, *Syst. Bot.* 27: 41–54.

85. Bassam, B.J., and Caetano-Anollés, G., (1993) Automated "hot start" PCR using mineral oil and paraffin wax, *BioTechniques* 14: 32–34.

86. Bassam, B.J., Caetano-Anollés, G., and Gresshoff, P.M., (1991) Fast and sensitive silver staining of DNA in polyacrylamide gels, *Anal. Biochem.* 196: 80–83.

87. Batley, J., Barker, G., O'Sullivan, H., Edwards, K.J., and Edwards, D., (2003) Mining for single nucleotide polymorphisms and insertions/deletions in maize expressed sequence tag data, *Plant Physiol.* 132: 84–91.

88. Baurens, F.-C., Noyer, J.-L., Lanaud, C., and Lagoda, P.J.L., (1998) Inter-*Alu* PCR like genomic profiling in banana, *Euphytica* 99: 137–142.

89. Baurens, F.-C., Bonnot, F., Bienvenu, D., Causse, S., and Legavre, T., (2003) Using SD-AFLP and MSAP to assess CCGG methylation in the banana genome, *Plant Mol. Biol. Rep* 21: 339–348.

90. Baverstock, P.R., and Moritz, C., (1996) Project design. In *Molecular Systematics, 2nd Edition*, Hillis, D.M., Moritz, C., and Mable, B.K., Eds., Sinauer Associates, Sunderland, MA, pp. 17–27.

91. Beaumont, M.A., (1999) Detecting population expansion and decline using microsatellites, *Genetics* 153: 2013–2029.

92. Beaumont, M.A., and Rannala, B., (2004) The Bayesian revolution in genetics, *Nat. Rev. Genet.* 5: 251–261.

93. Bebeli, P.J., Zhou, Z., Somers, D.J., and Gustafson, J.P., (1997) PCR primed with minisatellite core sequences yields DNA fingerprinting probes in wheat, *Theor. Appl. Genet.* 95: 276–283.

94. Becher, S.A., Steinmetz, K., Weising, K., Boury, S., Peltier, D., Renou, J.-P., Kahl, G., and Wolff, K., (2000) Microsatellites for cultivar identification in *Pelargonium*, *Theor. Appl. Genet.* 101: 643–651.

95. Becker, J., and Heun, M., (1995) Mapping of digested and undigested random amplified microsatellite polymorphisms in barley, *Genome* 38: 991–998.

96. Beckmann, J.S., and Soller, M., (1990) Toward a unified approach to genetic mapping of eukaryotes based on sequence tagged microsatellite sites, *Bio/Technology* 8: 930–932.

97. Beckmann, J.S., and Weber, J.L., (1992) Survey of human and rat microsatellites, *Genomics* 12: 627–631.

98. Deleted in proof.

99. Beebee, T., and Rowe, G., (2004) *An Introduction to Molecular Ecology*, Oxford University Press, Oxford.

100. Beismann, H., Barker, J.H.A., Karp, A., and Speck, T., (1997) AFLP analysis sheds light on distribution of two *Salix* species and their hybrid along a natural gradient, *Mol. Ecol.* 6: 989–993.

101. Bekesiova, I., Nap, J.-P., and Mlynarova, L., (1999) Isolation of high quality DNA and RNA from leaves of the carnivorous plant *Drosera rotundifolia*, *Plant Mol. Biol. Rep.* 17: 269–277.

102. Bekessy, S.A., Allnutt, T.R., Premoli, A.C., Lara, A., Ennos, R.A., Burgman, M.A., Cortes, M., and Newton, A.C., (2002) Genetic variation in the vulnerable and endemic monkey puzzle tree, detected using RAPDs, *Heredity* 88: 243–249.

103. Belaj, A., Satovic, Z., Cipriani, G., Baldoni, L., Testolin, R., Rallo, L., and Trujillo, I., (2003) Comparative study of the discriminating capacity of RAPD, AFLP and SSR markers and of their effectiveness in establishing genetic relationships in olive, *Theor. Appl. Genet.* 107: 736–744.

103a. Belkhir, K., Castric, V., and Bonhomme, F., (2002) IDENTIX, a software to test for relatedness in a population using permutation methods. *Mol. Ecol. Notes* 2: 611–614.

104. Bell, C.J., and Ecker, J.R., (1994) Assignment of 30 microsatellite loci to the linkage map of *Arabidopsis*. *Genomics* 19: 137–144.

105. Bell, G.I., Selby, M.J., and Rutter, W.J., (1982) The highly polymorphic region near the human insulin gene is composed of simple tandemly repeating sequences, *Nature* 295: 31–35.

106. Bellini D., Velasco, R., and Grando, M.S., (2001) Intravarietal DNA polymorphisms in grapevine (*Vitis vinifera* L.), *Acta Hortic.* 546: 343–349.

107. Bellstedt, D.U., Linder, H.P., and Harley, E.H., (2001) Phylogenetic relationships in *Disa* based on non-coding *trn*L-*trn*F chloroplast sequences: evidence of numerous repeat regions, *Am. J. Bot.* 88: 2088–2100.

108. Benito, C., Figueiras, A.M., Zaragoza, C., Gallego, F.J., and De la Pena, A., (1993) Rapid identification of Triticeae genotypes from single seeds using the polymerase chain reaction, *Plant Mol. Biol.* 21: 181–183.

109. Benko-Iseppon, A.-M., Winter, P., Hüttel, B., Staginnus, C., Muehlbauer, F.J., and Kahl, G., (2003) Molecular markers closely linked to *Fusarium* resistance genes in chickpea show significant alignments to pathogenesis-related genes located on *Arabidopsis* chromosomes 1 and 5, *Theor. Appl. Genet.* 107: 379–386.

110. Benner, M.S., Braunstein, M.D., and Weisberg, M.U., (1995) Detection of DNA polymorphisms within the genus *Cattleya* (Orchidaceae), *Plant Mol. Biol. Rep.* 13: 147–155.

111. Bennetzen, J.L., (1996) The use of comparative genome mapping in the identification, cloning and manipulation of important plant genes, In *The Impact of Plant Molecular Genetics*, Sobral, B.W.S., Ed., Birkhäuser, Boston, pp. 71–85.

112. Bennetzen, J.L., (2000) Comparative sequence analysis of plant nuclear genomes: microcolinearity and its many exceptions, *Plant Cell* 12: 1021–1029.

113. Bennetzen, J.L., (2000) Transposable element contributions to plant gene and genome evolution, *Plant Mol. Biol.* 42: 251–269.

114. Bennetzen, J.L., and Ma, J., (2003) The genetic colinearity of rice and other cereals on the basis of genome sequence analysis, *Curr. Opin. Plant Biol.* 6: 128–133.

115. Bensch, S., Åkesson, S., and Irwin, D.E., (2002) The use of AFLPs to find an informative SNP: genetic differences across a migratory divide in willow warblers, *Mol. Ecol.* 11: 2359–2366.

116. Benson, D.A., Karsch-Mizrachi, I., Lipman, D.J., Ostell, J., and Wheeler, D.L., (2004) GenBank: update, *Nucleic Acids Res.* 32: D23–D26.

117. Benson, E.E., (1999) *Plant Conservation Biotechnology*, Taylor and Francis, London.

118. Benter, T., Papadopoulos, S., Pape, M., Manns, M., and Poliwoda, H., (1995) Optimization and reproducibility of random amplified polymorphic DNA in human, *Anal. Biochem.* 230: 92–100.

119. Bentzen, P., and Wright, J.M., (1993) Nucleotide sequence and evolutionary conservation of a minisatellite variable number tandem repeat cloned from Atlantic salmon, *Salmo salar*, *Genome* 36: 271–277.

120. Benzecri, J.P., (1992) *Correspondence Analysis Handbook*, Marcel Dekker, New York.

121. Berenyi, M., Gichucki, S.T., Schmidt, J., and Burg, K., (2002) Ty1-*copia* retrotransposon-based S-SAP (sequence-specific amplified polymorphism) for genetic analysis of sweetpotato, *Theor. Appl. Genet.* 105: 862–869.

122. Berg, E., and Hamrick, J., (1995) Fine-scale genetic structure of a Turkey oak forest, *Evolution* 49: 110–120.

123. Bernet, G.P., Mestre, P.F., Pina, J.A., and Asins, M.J., (2004) Molecular discrimination of lemon cultivars, *HortScience* 39: 165–169.

124. Berthomieu, P., and Meyer, C., (1991) Direct amplification of plant genomic DNA from leaf and root pieces using PCR, *Plant Mol. Biol.* 17: 555–557.

125. Besnard, G., and Bervillé, A., (2002) On chloroplast DNA variations in the olive (*Olea europaea* L.) complex: comparison of RFLP and PCR polymorphisms, *Theor. Appl. Genet.* 104: 1157–1163.

126. Besnard, G., Khadari, B., Baradat, P., and Bervillé, A., (2002) *Olea europaea* (Oleaceae) phylogeography based on chloroplast DNA polymorphism, *Theor. Appl. Genet.* 104: 1353–1361.

127. Besse, P., Taylor, G., Carroll, B., Berding, N., Burner, D., and McIntyre, C.L., (1998) Assessing genetic diversity in a sugarcane germplasm collection using an automated AFLP analysis, *Genetica* 104: 143–153.

128. Beutow, K.H., Edmonson, M.N., and Cassidy, A.B., (1999) Reliable identification of large numbers of candidate SNPs from public EST data, *Nat. Genet.* 21: 323–325.

129. Beyermann, B., Nürnberg, P., Weihe, A., Meixner, M., Epplen, J.T., and Börner, T., (1992) Fingerprinting plant genomes with oligonucleotide probes specific for simple repetitive DNA sequences, *Theor. Appl. Genet.* 83: 691–694.

130. Bhattramakki, D., and Rafalski, A., (2001) Discovery and application of single nucleotide polymorphism markers in plants. In *Plant Genotyping: The DNA Fingerprinting of Plants*, Henry, R.J., Ed., CABI Publishing, Wallingford, U.K., pp. 179–192.

131. Bhattramakki, D., Dolan, M., Hanafey, M., Wineland, R., Vaske, D., Register J.C., III, Tingey, S.V., and Rafalski, A., (2002) Insertion-deletion polymorphisms in 3′-regions of maize genes occur frequently and can be used as highly informative genetic markers, *Plant Mol. Biol.* 48: 539–547.

132. Bielawski, J.P., Noack, K., and Pumo, D.E., (1995) Reproducible amplification of RAPD markers from vertebrate DNA, *BioTechniques* 18: 856–860.

133. Bierwerth, S., Kahl, G., Weigand, F., and Weising, K., (1992) Oligonucleotide fingerprinting of plant and fungal genomes: a comparison of radioactive, colorigenic and chemiluminescent detection methods, *Electrophoresis* 13: 115–122.

134. Birky, C.W., (1995) Uniparental inheritance of mitochondrial and chloroplast genes: mechanisms and evolution, *Proc. Natl. Acad. Sci. U.S.A.* 92: 11331–11338.

135. Birmeta, G., Nybom, H., and Bekele, E., (2004) Distinction between wild and cultivated enset (*Ensete ventricosum*) gene pools in Ethiopia using RAPD markers, *Hereditas* 140: 139–148.

136. Biss, P., Freeland, J., Silvertown, J., McConway, K., and Lutman, P., (2003) Successful amplification of rice chloroplast microsatellites from century-old grass samples from the park grass experiment, *Plant Mol. Biol. Rep.* 21: 249–257.

137. Blackwell, M., and Chapman, R.L., (1993) Collection and storage of fungal and algal samples, *Meth. Enzymol.* 224: 65–77.

138. Blair, M.W., Panaud, O., and McCouch, S.R., (1999) Application of ISSR amplification for the fingerprinting of rice cultivars and for the analysis of microsatellite motif frequencies in the rice genome, *Theor. Appl. Genet.* 98: 780–792.

139. Blair, M.W., Hedetale, V., and McCouch, S.R., (2002) Fluorescent-labelled microsatellite panels useful for detecting allelic diversity in cultivated rice (*Oryza sativa* L.), *Theor. Appl. Genet.* 105: 449–457.

140. Blanck, A., Glück, B., Wartbichler, R., Bender, S., Pöll, M., and Brandl, A., (1997) Activity of restriction enzymes in a PCR mix, *Biochemica* 3: 25.

141. Blanquer-Maumont, A., and Crouau-Roy, B., (1995) Polymorphism, monomorphism, and sequences in conserved microsatellites in primate species, *J. Mol. Evol.* 41: 492–497.

142. Blohm, D.H., and Guiseppi-Elie, A., (2001) New developments in microarray technology, *Curr. Opin. Biotechnol.* 12: 41–47.

143. Blouin, M.S., (2003) DNA-based methods for pedigree reconstruction and kinship analysis in natural populations, *Trends Ecol. Evol.* 18: 503–511.

144. Blum, H., Beier, H., and Gross, H.J., (1987) Improved silver staining of plant proteins, RNA and DNA in polyacrylamide gels, *Electrophoresis* 8: 93–99.

145. Böttger, E.C., (1990) Frequent contamination of *Taq* polymerase with DNA, *Clin. Chem.* 36: 1258–1259.

146. Bois, P., and Jeffreys, A.J., (1999) Minisatellite instability and germline mutation, *Cell Mol. Life Sci.* 55: 1636–1648.

147. Bois, P., Stead, J.D.H., Bakshi, S., Williamson, J., Neumann, R., Moghadaszadeh, B., and Jeffreys, A., (1998) Isolation and characterization of mouse minisatellites, *Genomics* 50: 317–330.

148. Bonnin, I., Huguet, T., Gherardi, M., Prosperi, J.-M., and Olivieri, I., (1996) High level of polymorphism and spatial structure in a selfing plant species, *Medicago truncatula* (Leguminosae), shown using RAPD markers, *Am. J. Bot.* 83: 843–855.

149. Bormans, C.A., Rhodes, R.B., Kephart, D.D., McClung, A.M., and Park, W.D., (2002) Analysis of a single nucleotide polymorphism that controls the cooking quality of rice using a non-gel based assay, *Euphytica* 128: 261–267.

150. Bornet, B., and Branchard, M., (2001) Nonanchored inter simple sequence repeat (ISSR) markers: reproducible and specific tools for genome fingerprinting, *Plant Mol. Biol. Rep.* 19: 209–215.

151. Bornet, B., Muller, C., Paulus, F., and Branchard, M., (2002) Highly informative nature of inter simple sequence repeat (ISSR) sequences amplified using tri- and tetra-nucleotide primers from DNA of cauliflower (*Brassica oleracea* var. *botrytis* L.), *Genome* 45: 890–896.

152. Borstnik, B., and Pumpernik, D., (2002) Tandem repeats in protein coding regions of primate genes, *Genome Res.* 12: 909–915.

153. Bossart, J.L., and Prowell, D.P., (1998) Genetic estimates of population structure and gene flow: limitations, lessons and new directions, *Trends Ecol. Evol.* 13: 202–206.

154. Botstein, D., White, R.L., Skolnick, M., and Davis, R.W., (1980) Construction of a genetic linkage map in man using restriction fragment length polymorphism, *Am. J. Hum. Genet.* 32: 314–331

155. Boudry, P., Wieber, R., Samitou-Laprade, P., Pillen, K., Van Dijk, H., and Jung, C., (1994) Identification of RFLP markers closely linked to the bolting gene B and their significance for the study of the annual habit in beets (*Beta vulgaris* L.), *Theor. Appl. Genet.* 88: 852–858.

156. Bowditch, B., Albright, D.G., Williams, J.G.K., and Braun, M.J., (1993) Use of randomly amplified polymorphic DNA markers in comparative genome studies, *Meth. Enzymol.* 224: 294–309.

157. Bowers, J.E., Dangl, G.S., Vignani, R., and Meredith, C.P., (1996) Isolation and characterization of new polymorphic simple sequence repeat loci in grape (*Vitis vinifera* L.), *Genome* 39: 628–633.

158. Brachet, S., Jubier, M.F., Richard, M., Jung-Muller, B., and Frascaria-Lacoste, N., (1999) Rapid identification of microsatellite loci using 5′-anchored PCR in the common ash *Fraxinus excelsior*, *Mol. Ecol.* 8: 160–163.

159. Bradeen, J.M., Bach, I.C., Briard, M., Le Clerc, V., Grzebelus, D., Senalik, D.A., and Simon, P.W., (2002) Molecular diversity analysis of cultivated carrot (*Daucus carota* L.) and wild *Daucus* populations reveals a genetically nonstructured composition, *J. Am. Soc. Hortic. Sci.* 127: 383–391.

160. Braun, A., Little, D.P., Reuter, D., Müller-Mysok, B., and Köster, H., (1997) Improved analysis of microsatellites using mass spectrometry, *Genomics* 46: 18–23.

161. Bredemeijer, G.M.M., Arens, P., Wouters, D., Visser, D., and Vosman, B., (1998) The use of semi-automated fluorescent microsatellite analysis for tomato cultivar identification, *Theor. Appl. Genet.* 97: 584–590.

162. Bregitzer, P., Halbert, S.E., and Lemaux, P.G., (1998) Somaclonal variation in the progeny of transgenic barley, *Theor. Appl. Genet.* 96: 421–425.

163. Bremer, K., (1988) The limits of amino acid sequence data in angiosperm phylogenetic reconstruction, *Evolution* 42: 795–803.

164. Brenner, K., and 23 coauthors, (2000) Gene expression analysis by massively parallel signature sequencing (MPSS) on microbead arrays, *Nat. Biotechnol.* 18: 630–634.

165. Breyne, P., Rombaut, D., Van Gysel, A., Van Montagu, M., and Gerats, T., (1999) AFLP analysis of genetic diversity within and between *Arabidopsis thaliana* ecotypes, *Mol. Gen. Genet.* 261: 627–634.

166. Briard, M., Le Clerc, V., Grzebelius, D., Senalik, D., and Simon, P.W., (2000) Modified protocols for rapid carrot genomic DNA extraction and AFLP analysis using silver stain or radioisotopes, *Plant Mol. Biol. Rep.* 18: 235–241.

167. Brohede, J., Primmer, C.R., Möller, A., and Ellegren, H., (2002) Heterogeneity in the rate and pattern of germline mutation at individual microsatellite loci, *Nucleic Acids Res.* 30: 1997–2003.

168. Bronzini de Caraffa, V., Maury, J., Gambotti, C., Breton, C., Bervillé, A., and Giannettini, J., (2002) Mitochondrial and RAPD mark oleasters, olive and feral olive from western and eastern Mediterranean, *Theor. Appl. Genet.* 104: 1209–1216.

169. Brookfield, J.F.Y., (1992) DNA fingerprinting in clonal organisms, *Mol. Ecol.* 1: 21–26.

170. Brouat, C., McKey, D., and Douzery, J.P., (2004) Differentiation in a geographical mosaic of plants coevolving with ants: phylogeny of the *Leonardoxa africana* complex (Fabaceae: Caesalpinioideae) using amplified fragment length polymorpism markers, *Mol. Ecol.* 13: 1157–1171.

171. Broun, P., and Tanksley, S.D., (1993) Characterization of tomato clones with sequence similarity to human minisatellites 33.6 and 33.15, *Plant Mol. Biol.* 23: 231–242.

172. Broun, P., and Tanksley, S.D., (1996) Characterization and genetic mapping of simple repeat sequences in the tomato genome, *Mol. Gen. Genet.* 250: 39–49.

173. Brown, L.Y., and Brown, S.A., (2004) Alanine tracts: the expanding story of human illness and trinucleotide repeats, *Trends Genet.* 20: 51–58.

174. Brown, P.T.H., Lange, F.D., Kranz, E., and Lörz, H., (1993) Analysis of single protoplasts and regenerated plants by PCR and RAPD technology, *Mol. Gen. Genet.* 237: 311–317.

175. Brumfield, R.T., Beerli, P., Nickerson, D.A., and Edwards, S.V., (2003) The utility of single nucleotide polymorphisms in inferences of population history, *Trends Ecol. Evol.* 18: 249–256.

176. Brunel, D., (1992) An alternative, rapid method of plant DNA extraction for PCR analyses, *Nucleic Acids Res.* 20: 4676.

177. Brunell, M.S., and Whitkus, R., (1997) RAPD marker variation in *Eriastrum densifolium* (Polemoniaceae): implications for subspecific delimitations and conservation, *Syst. Bot.* 22: 543–553.

178. Brunner, I., Brodbeck, S., Büchler, U., and Sperisen, C., (2001) Molecular identification of fine roots of trees from the alps: reliable and fast DNA extraction and PCR-RFLP analyses of plastid DNA, *Mol. Ecol.* 10: 2079–2087.

179. Deleted in proof.

180. Bryan, G.J., McNicoll, J., Ramsay, G., Meyer, R.C., and De Jong, W.S., (1999) Polymorphic simple sequence repeat markers in chloroplast genomes of Solanaceous plants, *Theor. Appl. Genet.* 99: 859–867.

181. Bucci, G., Anzidei, M., Madaghiele, A., and Vendramin, G.G., (1998) Detection of haplotypic variation and natural hybridization in *halepensis*-complex pine species using chloroplast simple sequence repeat (SSR) markers, *Mol. Ecol.* 7: 1633–1643.

182. Buldewo, S., and Jaufeerally-Fakim, Y.F., (2002) Isolation of clean and PCR-amplifiable DNA from *Anthurium andreanum*, *Plant Mol. Biol. Rep.* 20: 71a–71g.

183. Bult, C., Källersjö, M., and Suh, Y., (1992) Amplification and sequencing of 16S/18S rDNA from gel-purified total plant DNA, *Plant Mol. Biol. Rep.* 10: 273–284.

184. Bureau, T.E., and Wessler, S.R., (1992) *Tourist*: a large family of small inverted repeat elements frequently associated with maize genes, *Plant Cell* 4: 1283–1294.

185. Bureau, T.E., and Wessler, S.R., (1994) Mobile inverted-repeat elements of the *Tourist* family are associated with the genes of many cereal grasses, *Proc. Natl. Acad. Sci. U.S.A.* 91: 1411–1415.

186. Bureau, T.E., and Wessler, S.R., (1994) Stowaway: a new family of inverted repeat elements associated with the genes of both monocotyledonous and dicotyledonous plants, *Plant Cell* 6: 907–916.

187. Bureau, T.E., Ronald, P.C., and Wessler, S.R., (1996) A computer-based systematic survey reveals the predominance of small inverted-repeat elements in wild-type rice genes, *Proc. Natl. Acad. Sci. U.S.A.* 93: 8524–8529.

188. Burke, T., Dolf, G., Jeffreys, A.J., and Wolff, R., Eds., (1991) *DNA Fingerprinting: Approaches and Applications*, Birkhäuser, Basel, Switzerland.

189. Burr, B., and Burr, F.A., (1991) Recombinant inbred lines for molecular mapping in maize. Theoretical and practical considerations, *Trends Genet.* 7: 55–60.

190. Burr, K., Harper, R., and Linacre, A., (2001) One-step isolation of plant DNA suitable for PCR amplification, *Plant Mol. Biol. Rep.* 19: 367–371.

191. Bush, S.P., and Mulcahy, D.L., (1999) The effects of regeneration by fragmentation upon clonal diversity in the tropical forest shrub *Poikilacanthus macranthus*: random amplified polymorphic DNA (RAPD) results, *Mol. Ecol.* 8: 865–870.

192. Bussell, J.D., (1999) The distribution of random amplified polymorphic DNA (RAPD) diversity amongst populations of *Isotoma petraea* (Lobeliaceae), *Mol. Ecol.* 8: 775–789.

193. Buzas, Z., and Varga, L., (1995) Rapid method for separation of microsatellite alleles by the PhastSystem, *PCR Methods Appl* 4: 380–381.

194. Caetano-Anollés, G., (1998) DAF optimization using Taguchi methods and the effect of thermal cycling parameters on DNA amplification, *BioTechniques* 25: 472–480.

195. Caetano-Anollés, G., (1998) Genetic instability of bermudagrass (*Cynodon*) cultivars "Tifgreen" and "Tifdwarf" detected by DAF and ASAP analysis of accessions and offtypes, *Euphytica* 101: 165–173.

196. Caetano-Anollés, G., (1999) High genome-wide mutation rates in vegetatively propagated bermudagrass, *Mol. Ecol.* 8: 1211–1221.

197. Caetano-Anollés, G., (2001) Plant genotyping using arbitrarily amplified DNA. In *Plant Genotyping. The DNA Fingerprinting of Plants*, Henry, R.J., Ed., CABI Publishing, CAB International, Wallingford, U.K., pp. 29–46.

198. Caetano-Anollés, G., and Gresshoff, P.M., (1994) DNA amplification fingerprinting using arbitrary mini-hairpin oligonucleotide primers, *Bio/Technology* 12: 619–623.

199. Caetano-Anollés, G., and Gresshoff, P.M., (1996) Generation of sequence signatures from DNA amplification fingerprints with mini-hairpin and microsatellite primers, *BioTechniques* 20: 1044–1056.

200. Caetano-Anollés, G., and Gresshoff, P.M., (1998) *DNA Markers. Protocols, Applications, and Overviews*, Wiley-VCH, New York.

201. Caetano-Anollés, G., Bassam, B.J., and Gresshoff, P.M., (1991) High resolution DNA amplification fingerprinting using very short arbitrary oligonucleotide primers, *Bio/Technology* 9, 553–557.

202. Caetano-Anollés, G., Bassam, B.J., and Gresshoff, P.M., (1991) DNA amplification fingerprinting: a strategy for genome analysis, *Plant Mol. Biol. Rep.* 9: 294–307.

203. Caetano-Anollés, G., Bassam, B.J., and Gresshoff, P.M. (1992) DNA fingerprinting: MAAPing out a RAPD redefinition? *Bio/Technology* 10: 937.

204. Caetano-Anollés, G., Bassam, B.J., and Gresshoff, P.M., (1992) Primer-template interactions during DNA amplification fingerprinting with single arbitrary oligonucleotides, *Mol. Gen. Genet.* 235: 157–165.

205. Caetano-Anollés, G., Bassam, B.J., and Gresshoff, P.M., (1993) Enhanced detection of polymorphic DNA by multiple arbitrary amplicon profiling of endonuclease-digested DNA: identification of markers tightly linked to the supernodulation locus in soybean, *Mol. Gen. Genet.* 241: 57–64.

206. Cafasso, D., Pellegrino, G., Musacchio, A., Widmer, A., and Cozzolino, S., (2001) Characterization of a minisatellite repeat locus in the chloroplast genome of *Orchis palustris. Curr. Genet.* 39: 394–398.

207. Callahan, F.E., and Mehta, A.M., (1991) Alternative approach for consistent yields of total genomic DNA from cotton (*Gossypium hirsutum* L.), *Plant Mol. Biol. Rep.* 9: 252–261.

208. Callen, D.F., Thompson, A.D., Shen, Y., Phillips, H.A., Richards, R.I., Mulley, J.C., and Sutherland, G.R., (1993) Incidence and origin of "null" alleles in the $(AC)_n$ microsatellite markers, *Am. J. Hum. Genet.* 52: 922–927.

209. Camlin, M.S., (2001) Possible future roles for molecular techniques in the identification and registration of new plant cultivars, *Acta Hortic.* 546: 289–296.

210. Camlin, M.S., (2003) Plant cultivar identification and registration — the role for molecular techniques, *Acta Hortic.* 625: 37–47.

211. Campbell, D., Duchesne, P., and Bernatchez, L., (2003) AFLP utility for population assignment studies: analytical investigation and empirical comparison with microsatellites, *Mol. Ecol.* 12: 1993–1998.

212. Cano, R.J., and Poinar, H., (1993) Rapid isolation of DNA from fossil and museum specimens suitable for PCR, *BioTechniques* 15: 432–435.

213. Capy, P., Gasperi, G., Biémont, C., and Bazin, C., (2000) Stress and transposable elements: co-evolution or useful parasites? *Heredity* 85: 101–106.

214. Cardle, L., Ramsay, L., Milbourne, D., Macaulay, M., Marshall, D., and Waugh, R., (2000) Computational and experimental characterization of physically clustered simple sequence repeats in plants, *Genetics* 156: 847–854.

215. Deleted in proof.

216. Casa, A., Brouwer, C., Nagel, A., Wang, L., Zhang, Q., Kresovich, S., and Wessler, S.R., (2000) The MITE family *Heartbreaker* (*Hbr*): molecular markers in maize, *Proc. Natl. Acad. Sci. U.S.A.* 97: 10083–10089.

217. Cascuberta, E., Cascuberta, J.M., Puigdomènech, P., and Monfort, A., (1998) Presence of miniature inverted-repeat transposable elements (MITEs) in the genome of *Arabidopsis thaliana*: characterisation of the *Emigrant* family of elements, *Plant J.* 16: 79–85.

218. Cato, S.A., and Richardson, T.E., (1996) Inter- and intraspecific polymorphism at chloroplast SSR loci and the inheritance of plastids in *Pinus radiata* D. Don, *Theor. Appl. Genet.* 93: 587–592.

219. Cavalli-Sforza, L.L., and Edwards, A.W., (1967) Phylogenetic analysis: models and estimation procedures. *Am. J. Hum. Genet.* 19: 233–257.

220. Cavers, S., Navarro, C., and Lowe, A.J., (2003) A combination of molecular markers identifies evolutionarily significant units in *Cedrela odorata* L. (Meliaceae) in Costa Rica, *Conserv. Genet.* 4: 571–580.

221. Ceplitis, A., (2001) The importance of sexual and asexual reproduction in the recent evolution of *Allium vineale, Evolution* 55: 1581–1591.

222. Cercueil, A., Bellemain, E., and Manel, S., (2002) PARENTE: computer program for parentage analysis. *J. Hered.* 93: 458–459.

223. Cervera, M.-T., Ruiz-Garcia, L., and Martinez-Zapater, J.M., (2002) Analysis of DNA methylation in *Arabidopsis thaliana* based on methylation-sensitive AFLP markers, *Mol. Genet. Genomics* 268: 543–552.

224. Chakraborty, R., Kimmel, M., Stivers, L., Davison, L.J., and Deka, R., (1997) Relative mutation rates at di-, tri- and tetranucleotide microsatellite loci, *Proc. Natl. Acad. Sci. U.S.A.* 94: 1041–1046.

225. Chalhoub, B.A., Thibault, S., Laucou, V., Rameau, C., Höfte, H., and Cousin, R., (1997) Silver staining and recovery of AFLP™ amplification products on large denaturing polyacrylamide gels, *BioTechniques* 22: 216–220.

226. Chandra, S., Huaman, Z., Hari Krishna, S., and Ortiz, R., (2002) Optimal sampling strategy and core collection size of Andean tetraploid potato based on isozyme data — a simulation study, *Theor. Appl. Genet.* 104: 1325–1334.

227. Chang, R.-Y., O'Donoughue, L.S., and Bureau, T.E., (2001) Inter-MITE polymorphisms (IMP): a high throughput transposon-based genome mapping and fingerprinting approach, *Theor. Appl. Genet.* 102: 773–781.

228. Charters, Y.M., Robertson, A., Wilkinson, M.J., and Ramsay, G., (1996) PCR analysis of oilseed rape cultivars (*Brassica napus* L. ssp. *oleifera*) using 5′-anchored simple sequence repeat (SSR) primers, *Theor. Appl. Genet.* 92: 442–447.

229. Chase, M.W., and Hills, H.H., (1991) Silica gel: an ideal material for field preservation of leaf samples for DNA studies, *Taxon* 40: 215–220.

230. Chase, M., Kesseli, R., and Bawa, K., (1996) Microsatellite markers for population and conservation genetics of tropical trees, *Am. J. Bot.* 83: 51–57.

231. Chase, M., Moller, C., Kesseli, R., and Bawa, K.S., (1996) Distant gene flow in tropical trees, *Nature* 383: 398–399.

232. Chaudhry, B., Yasmeen, A., Husnain, T., and Riazuddin, S., (1999) Mini-scale genomic DNA extraction from cotton, *Plant Mol. Biol. Rep.* 17: 280 (1–7).

233. Chaves, A.L., Vergara, C.E., and Mayer, J.E., (1995) Dichloromethane as an economic alternative to chloroform in the extraction of DNA from plant tissues, *Plant Mol. Biol. Rep.* 13: 18–25.

234. Chaw, S.-M., Parkinson, C.L., Cheng, Y., Vincent, T.M., and Palmer, J.D., (2000) Seed plant phylogeny inferred from all three plant genomes: monophyly of extant gymnosperms and origin of Gnetales from conifers, *Proc. Natl. Acad. Sci. U.S.A.* 97: 4086–4091.

235. Cheek, B.J., Steel, A.B., Torres, M.P., Yu, Y.Y., and Yang, H. (2001) Chemiluminescence detection for hybridization assays on the flow-through chip, a three-dimensional microchannel biochip, *Anal. Chem.* 73: 5777–5783.

236. Chen, D.-H., and Ronald, P.C., (1999) A rapid DNA minipreparation method suitable for AFLP and other PCR applications, *Plant Mol. Biol. Rep.* 17: 53–57.

237. Chen, H., and Sun, M., (1998) Consensus multiplex PCR-restriction fragment length polymorphism (RFLP) for rapid detection of plant mitochondrial DNA polymorphism, *Mol. Ecol.* 7: 1553–1556.

238. Chen, L.-F.O., Kuo, H.-Y., Chen, M.-H., Lai, K.-N., and Chen, S.-C.G., (1997) Reproducibility of the differential amplification between leaf and root DNAs in soybean revealed by RAPD markers, *Theor. Appl. Genet.* 95: 1033–1043.

239. Chen, X., Temnykh, S., Xu, Y., Cho, Y.G., and McCouch, S.R., (1997) Development of a microsatellite framework map providing genome-wide coverage in rice (*Oryza sativa*), *Theor. Appl. Genet.* 95: 553–567.

240. Chen, X.M., Line, R.F., and Leung, H., (1998) Genome scanning for resistance gene analogs in rice, barley, and wheat by high resolution electrophoresis, *Theor. Appl. Genet.* 97: 345–355.

241. Cheng, S., Chang, S.-Y., Gravitt, P., and Respess, R., (1994) Long PCR, *Nature* 369: 684–685.

242. Cheng, Y.-J., Guo, W.-W., Yi, H.-L., Pang, X.-M., and Deng, X., (2003) An efficient protocol for genomic DNA extraction from *Citrus* species, *Plant Mol. Biol. Rep.* 21: 177a–177g.

243. Cheung, W.Y., and Gale, M.D., (1990) The isolation of high molecular weight DNA from wheat, barley and rye for analysis by pulse-field gel electrophoresis, *Plant Mol. Biol.* 14: 881–888.

244. Cheung, W.Y., Hubert, N., and Landry, B.S., (1993) A simple and rapid DNA microextraction method for plant, animal, and insect suitable for RAPD and other PCR analyses, *PCR Methods Appl.* 3: 69–70.

245. Chevet, E., Lemaitre, G., and Katinka, M.D., (1995) Low concentrations of tetramethylammonium chloride increase yield and specificity of PCR, *Nucleic Acids Res.* 23: 3343–3344.

246. Chin, E.C.L., Senior, M.L., Shu, H., and Smith, J.S.C., (1996) Maize simple repetitive DNA sequences: abundance and allele variation, *Genome* 39: 866–873.

247. Ching, A., and Rafalski, A., (2002) Rapid genetic mapping of ESTs using SNP pyrosequencing and indel analysis, *Cell. Mol. Biol. Lett.* 7: 803–810.

248. Cho, R.J., Mindrinos, M., Richards, D.R., Sapolsky, R.J., Anderson M., Drenkard, E., Dewdney, J., Reuber, T.L., Stammers, M., Federspiel, N., Theologis, A., Yang, W.-H., Hubbell, E., Au, M., Chung, E.Y., Lashkari, D., Lemieux, B., Dean, C., Lipshutz, R.J., Ausubel, F.M., Davis, R.W., and Oefner, P.J., (1999) Genome-wide mapping with biallelic markers in *Arabidopsis thaliana*, *Nat. Genet.* 23: 203–207.

249. Cho, Y.G., Blair, M.W., Panaud, O., and McCouch, S.R., (1996) Cloning and mapping of variety-specific rice genomic DNA sequences: amplified fragment length polymorphisms (AFLP) from silver-stained polyacrylamide gels, *Genome* 39: 373–378.

250. Cho, Y.G., Ishii, T., Temnykh, S., Chen, X., Lipovich, L., McCouch, S.R., Park, W.D., Ayres, N., and Cartinhour, S., (2000) Diversity of microsatellites derived from genomic libraries and GenBank sequences in rice (*Oryza sativa* L.), *Theor. Appl. Genet.* 100: 713–722.

251. Chomczynski, P., Mackey, K., Drews, R., and Wilfinger, W., (1997) DNAzol®: A reagent for the rapid isolation of genomic DNA, *BioTechniques* 22: 550–553.

252. Chou, Q., Russell, M., Birch, D.E., Raymond, J., and Bloch, W., (1992) Prevention of pre-PCR mis-priming and primer dimerization improves low-copy-number amplifications, *Nucleic Acids Res.* 20: 1717–1723.

253. Chung, S.-M., and Staub, J.E., (2003) The development and evaluation of consensus chloroplast primer pairs that possess highly variable sequence regions in a diverse array of plant taxa, *Theor. Appl. Genet.* 107: 757–767.

254. Chunwongse, J., Martin, G.B., and Tanksley, S.D., (1993) Pre-germination genotypic screening using PCR amplification of half-seeds, *Theor. Appl. Genet.* 86: 694–698.

255. Cifarelli, R.A., Gallitelli, M., and Cellini, F., (1995) Random amplified hybridization microsatellites (RAHM): isolation of a new class of microsatellite-containing DNA clones, *Nucleic Acids Res.* 23: 3802–3803.

256. Cipriani, G., Marazzo, M.T., Di Gaspero, G., and Testolin, R., (2001) DNA microsatellite in fruit crops: isolation, length polymorphism, inheritance, somatic stability and cross-species conservation, *Acta Hortic.* 546: 145–150.

257. Clancy, J.A., Jitkov, V.A., Han, F., and Ullrich, S.E., (1996) Barley tissue as direct template for PCR: a practical breeding tool, *Mol. Breed.* 2: 181–183.

258. Clauss, M.J., Cobban, H., and Mitchell-Olds, T., (2002) Cross-species microsatellite markers for elucidating population genetic structure in *Arabidopsis* (Brassicaceae), *Mol. Ecol.* 11: 591–602.

259. Clegg, M.T., (1993) Chloroplast gene sequences and the study of plant evolution, *Proc. Natl. Acad. Sci. U.S.A.* 90: 363–367.

260. Clement, M., Posada, D., and Crandall, K.A., (2000) TCS: a computer program to estimate gene genealogies, *Mol. Ecol.* 9: 1657–1659.

261. Cnops, G., den Boer, B., Gerats, A., Van Montagu, M., and Van Lisjebettens, M., (1996) Chromosome landing at the *Arabidopsis* TORNADO1 locus using an AFLP-based strategy, *Mol. Gen. Genet.* 253: 32–41.

262. Collevatti, R.G., Grattapaglia, D., and Hay, J.D., (2001) Population genetic structure of the endangered tropical tree species *Caryocar brasiliense*, based on variability at microsatellite loci, *Mol. Ecol.* 10: 349–356.

263. Collevatti, R.G., Grattapaglia, D., and Hay, J.D., (2003) Evidences for multiple maternal lineages of *Caryocar brasiliense* populations in the Brazilian Cerrado based on the analysis of chloroplast DNA sequences and microsatellite haplotype variation, *Mol. Ecol.* 12: 105–115.

264. Collick, A., and Jeffreys, A.J., (1990) Detection of a novel minisatellite-specific DNA-binding protein, *Nucleic Acids Res.* 18: 625–629.

265. Collins, G.G., and Symons, H., (1992) Extraction of nuclear DNA from grape vine leaves by a modified procedure, *Plant Mol. Biol. Rep.* 10: 233–235.

266. Colosi, J.C., and Schaal, B.A., (1993) Tissue grinding with ball bearings and vortex mixer for DNA extraction, *Nucleic Acids Res.* 21: 1051–1052.

267. Comes, H.P., and Abbott, R.J., (1998) The relative importance of historical events and gene flow on the population structure of a Mediterranean ragwort, *Senecio gallicus* (Asteraceae), *Evolution* 52: 355–367.

268. Comes, H.P., and Abbott, R.J., (2000) Random amplified polymorphic DNA (RAPD) and quantitative trait analyses across a major phylogeographical break in the Mediterranean ragwort *Senecio gallicus* Vill. (Asteraceae), *Mol. Ecol.* 9: 61–76.

269. Condit, R., and Hubbell, S.P., (1991) Abundance and DNA sequence of two-base repeat regions in tropical tree genomes, *Genome* 34: 66–71.

270. Congiu, L., Chicca, M., Cella, R., Rossi, R., and Bernacchia, G., (2000) The use of random amplified polymorphic DNA (RAPD) markers to identify strawberry varieties: a forensic application, *Mol. Ecol.* 9: 229–232.

271. Connell, J.P., Pammi, S., Iqbal, M.J., Huizinga, T., and Reddy, A.S., (1998) A high-throughput procedure for capturing microsatellites from complex plant genomes, *Plant Mol. Biol. Rep.* 16: 341–349.

272. Cooper, A., and Poinar, H.N., (2000) Ancient DNA: do it right, or not at all, *Science* 289: 1139.

273. Cordeiro, G.M., Maguire, T.L., Edwards, K.J., and Henry, R.J. (1999) Optimisation of a microsatellite enrichment technique in *Saccharum* spp., *Plant Mol. Biol. Rep.* 17: 225–229.

274. Cordeiro, G.M., Casu, R., McIntyre, C.L., Manners, J.M., and Henry, R.J., (2001) Microsatellite markers from sugarcane (*Saccharum* spp.) ESTs cross-transferable to erianthus and sorghum, *Plant Sci.* 160: 1115–1123.

275. Corley-Smith, G.E., Lim, C.J., Kalmar, G.B., and Brandhorst, B.P., (1997) Efficient detection of DNA polymorphisms by fluorescent RAPD analysis, *BioTechniques* 22: 690–699.

276. Coryell, V.H., Jessen, H., Schupp, J.M., Webb, D., and Keim, P., (1999) Allele-specific hybridization markers for soybean, *Theor. Appl. Genet.* 98: 690–696.

277. Cottrell, J.E., Munro, R.C., Tabbener, H.E., Milner, A.D., Forrest, G.I., and Lowe, A.J., (2003) Comparison of fine-scale genetic structure using nuclear microsatellites within two British oak woods differing in population history, *For. Ecol. Management* 176: 287–303.

278. Couch, J.A., and Fritz, P.J., (1990) Isolation of DNA from plants high in polyphenolics, *Plant Mol. Biol. Rep.* 8: 8–12.

279. Coyle, H.M., Ladd, C., Palmbach, T., and Lee H.C., (2001) The green revolution: botanical contributions to forensic and drug reinforcement, *Croatian Med. J.* 42: 340–345.

280. Cozzolino, S., Cafasso, D., Pellegrino, G., Musacchio, A., and Widmer, A., (2003) Fine-scale phylogeographical analysis of Mediterranean *Anacamptis palustris* (Orchidaceae) populations based on chloroplast minisatellite and microsatellite variation, *Mol. Ecol.* 12: 2783–2792.

281. Cozzolino, S., Cafasso, D., Pellegrino, G., Musacchio, A., and Widmer, A., (2003) Molecular evolution of a plastid tandem repeat locus in an orchid lineage, *J. Mol. Evol.* 57: S41–S49.

282. Craft, K.J., Ashley, M.V., and Koenig, W.D., (2002) Limited hybridization between *Quercus lobata* and *Quercus douglasii* (Fagaceae) in a mixed stand in Central coastal California, *Am. J. Bot.* 89: 1792–1798.

283. Crawford, D.J., (1997) Molecular markers for the study of genetic variation within and between populations of rare plants, *Opera Bot.* 132: 149–157.

284. Cregan, P.B., and Quigley, C.V., (1998) Simple sequence repeat DNA marker analysis. In *DNA Markers. Protocols, Applications and Overviews*, Caetano-Anollés, G., and Gresshoff, P.M., Eds., Wiley-VCH, New York, pp. 173–185.

285. Creste, S., Neto, A.T., and Figueira, A., (2001) Detection of simple sequence repeat polymorphisms in denaturing polyacrylamide sequencing gels by silver staining, *Plant Mol. Biol. Rep.* 19: 299–306.

286. Crowley, T.M., Muralitharan, M.S., and Stevenson, T.W., (2003) Isolating conifer DNA: a superior polysaccharide elimination method, *Plant Mol. Biol. Rep* 21: 97a–97d.

287. Csaikl, U.M., Bastian, H., Brettschneider, R., Gauch, S., Meir, A., Schauerte, M., Scholz, F., Sperisen, C., Vornam, B., and Ziegenhagen, B., (1998) Comparative analysis of different DNA extraction protocols: a fast, universal maxi-preparation of high quality plant DNA for genetic evaluation and phylogenetic studies, *Plant Mol. Biol. Rep.* 16: 69–86.

288. Cuadrado, A., and Schwarzacher, T., (1998) The chromosomal organization of simple sequence repeats in wheat and rye genomes, *Chromosoma* 107: 587–594.

289. Cuenca, A., Escalante, A.E., and Pinero, D., (2003) Long-distance colonization, isolation by distance, and historical demography in a relictual Mexican pinyon pine (*Pinus nelsonii* Shaw) as revealed by paternally inherited genetic markers (cpSSRs), *Mol. Ecol.* 12: 2087–2097.

290. Culley, T.M., and Wolfe, A.D., (2001) Population genetic structure of the cleistogamous plant species *Viola pubescens* Aiton (Violaceae), as indicated by allozyme and ISSR molecular markers, *Heredity* 86: 545–556.

291. Culley, T.M., Wallace, L.E., Gengler-Nowak, K.M., and Crawford, D.J., (2002) A comparison of two methods of calculating G_{st}, a genetic measure of population differentiation, *Am. J. Bot.* 89: 460–465.

292. Cummings, C.J., and Zoghbi, H.Y., (2000) Fourteen and counting: unraveling trinucleotide repeat diseases, *Hum. Mol. Genet.* 9: 909–916.

293. D'Aquila, R.T., Bechtel, L.J., Videler, J.A., Eron, J.J., Gorczyca, P., and Kaplan, J.C., (1991) Maximizing sensitivity and specificity of PCR by preamplification heating, *Nucleic Acids Res.* 19: 3749.

294. Dabo, S.M., Mitchell, E.D., and Melcher, U., (1993) A method for the isolation of nuclear DNA from cotton (*Gossypium*) leaves, *Anal. Biochem.* 210: 34–38.

295. Dallas, J.F., (1988) Detection of DNA "fingerprints" of cultivated rice by hybridization with a human minisatellite DNA probe, *Proc. Natl. Acad. Sci. U.S.A.* 85: 6831–6835.

296. Dallas, J.F., (1992) Estimation of microsatellite mutation rates in recombinant inbred strains of mouse, *Mamm. Genome* 3: 452–456.

297. Dally, A.M., and Second, G., (1989) Chloroplast DNA isolation from higher plants: an improved non-aqueous method, *Plant Mol. Biol. Rep.* 7: 135–143.

298. Daly, M.J., Rioux, J.D., Schaffner, S.F., Hudson, T.J., and Lander, E.S., (2001) High-resolution haplotype structure in the human genome, *Nat. Genet.* 29: 229–232.

299. Damasco, O.P., Graham, G.C., Henry, R.J., Adkins, S.W., Smith, M.K., and Godwin, I.D., (1996) Random amplified polymorphic DNA (RAPD) detection of dwarf off-types in micropropagated Cavendish (*Musa* spp. AAA) bananas, *Plant Cell Rep.* 16: 118–123.

300. Dávila, J.A., Sánchez de la Hoz, M.P., Loarce, Y., and Ferrer, E., (1998) The use of random amplified microsatellite polymorphic DNA and coefficients of parentage to determine genetic relationships in barley, *Genome* 41: 477–486.

301. Dávila, J.A., Loarce, Y., and Ferrer, E., (1999) Molecular characterization and genetic mapping of random amplified microsatellite polymorphism in barley, *Theor. Appl. Genet.* 98: 265–273.

302. Dávila, J.A., Loarce, Y., Ramsay, L., Waugh, R., and Ferrer, E., (1999) Comparison of RAMP and SSR markers for the study of wild barley genetic diversity, *Hereditas* 131: 5–13.

303. Dawson, I.K., Waugh, R., Simons, A.J., and Powell, W., (1997) Simple sequence repeats provide a direct estimate of pollen-mediated gene dispersal in the tropical tree *Gliricidia sepium*. *Mol. Ecol.* 6: 179–183.

304. Dayanandan, S., Bawa, K.S., and Kesseli, R., (1997) Conservation of microsatellites among tropical trees (Leguminosae), *Am. J. Bot.* 84: 1658–1663.

305. Dayanandan, S., Dole, J., Bawa, K., and Kesseli, R., (1999) Population structure delineated with microsatellite markers in fragmented populations of a tropical tree, *Carapa guianensis* (Meliaceae), *Mol. Ecol.* 8: 1585–1592.

306. De Boer, S.H., Ward, L.J., Li, X., and Chittaranjan, S., (1995) Attenuation of PCR inhibition in the presence of plant compounds by addition of BLOTTO, *Nucleic Acids Res.* 23: 2567–2568.

307. De Castro, O., and Menale, B., (2004) PCR amplification of Michele Tenore's historical specimens and facility to utilize an alternative approach to resolve taxonomic problems, *Taxon* 53: 147–151.

308. De Kochko, A., and Hamon, S., (1990) A rapid and efficient method for the isolation of restrictable total DNA from plants of the genus *Abelmoschus*, *Plant Mol. Biol. Rep.* 8: 3–7.

309. De la Cruz, M., Whitkus, R., and Mota-Bravo, L., (1995) Tropical tree DNA isolation and amplification, *Mol. Ecol.* 4: 787–789.

310. De la Cruz, M., Ramirez, F., and Hernandez, H., (1997) DNA isolation and amplification from cacti, *Plant Mol. Biol. Rep.* 15: 319–325.

311. De la Rosa, R., James, C.M., and Tobutt, K.R., (2002) Isolation and characterization of polymorphic microsatellites in olive (*Olea europaea* L.) and their transferability to other genera in the Oleaceae, *Mol. Ecol. Notes* 2: 265–267.

312. De Riek, J., Dendauw, J., Mertens, M., De Loose, M., Heursel, J., and van Blockstaele, E., (1999) Validation of criteria for the selection of AFLP markers to assess the genetic variation of a breeders' collection of evergreen azaleas, *Theor. Appl. Genet.* 99: 1155–1165.

313. Debener, T., and Mattiesch, L., (1998) Effective pairwise combination of long primers for RAPD analyses in roses, *Plant Breed.* 117: 147–151.

314. Debener, T., and Mattiesch, L., (1999) Construction of a genetic linkage map for roses using RAPD and AFLP markers, *Theor. Appl. Genet.* 99: 891–899.

315. Debener, T., Salamini, F., and Gebhardt, C., (1990) Phylogeny of wild and cultivated *Solanum* species based on nuclear restriction fragment length polymorphism (RFLPs), *Theor. Appl. Genet.* 79: 360–368.

316. Debener, T., Janakiram, T., and Mattiesch, L., (2000) Sports and seedlings of rose varieties analysed with molecular markers, *Plant Breed.* 119: 71–74.

317. Decrocq, V., Favé, M.G., Hagen, L., Bordenave, L., and Decrocq, S., (2003) Development and transferability of apricot and grape EST microsatellite markers across taxa, *Theor. Appl. Genet.* 106: 912–922.

318. Decrocq, V., Hagen, L.S., Favé, M.-G., Eyquard, J.P., and Pierronet, A., (2004) Microsatellite markers in the hexaploid *Prunus domestica* species and parentage lineage of three European plum cultivars using nuclear and chloroplast simple-sequence repeats, *Mol. Breed.* 13: 135–142.

319. Deguilloux, M.F., Pemonge, M.H., and Petit, R.J., (2002) Novel perspectives in wood certification and forensics: dry wood as a source of DNA, *Proc. R. Soc. London Ser. B* 269: 1039–1046.

320. Degouilloux, M.-F., Dumolin-Lapègue, S., Gielly, L., Grivet, D., and Petit, R.J., (2003) A set of primers for the amplification of chloroplast microsatellites in *Quercus*, *Mol. Ecol. Notes*: 24–27.

321. Deguilloux, M.F., Pemonge, M.H., Bertel, L., Kremer, A., and Petit R.J., (2003) Checking the geographical origin of oak wood: molecular and statistical tools. *Mol. Ecol.* 12: 1629–1636.

322. Del Castillo Agudo, L., Gavidia, I., Pérez-Bermudez, P., and Segura, J., (1995) PEG precipitation, a required step for PCR amplification of DNA from wild plants of *Digitalis obscura* L., *BioTechniques* 18: 766–768.

323. Dellaporta, S.L., Wood, J., and Hicks, J.B., (1983) A plant DNA minipreparation: version II. *Plant Mol. Biol. Rep.* 1: 19–21.

324. Demeke, T., and Adams, R.P., (1992) The effects of plant polysaccharides and buffer additives on PCR, *BioTechniques* 12: 332–334.

325. Demeke, T., Adams, R.P., and Chibbar, R., (1992) Potential taxonomic use of random amplified polymorphic DNA (RAPD): a case study in *Brassica*, *Theor. Appl. Genet.* 84: 990–994.

326. Demesure, B., Sozi, N., and Petit, R.J., (1995) A set of universal primers for amplification of polymorphic non-coding regions of mitochondrial and chloroplast DNA in plants, *Mol. Ecol.* 4: 129–131.

327. Demesure, B., Comps, B., and Petit, R.J., (1996) Chloroplast DNA phylogeography of the common beech (*Fagus sylvatica* L.) in Europe, *Evolution* 50: 2515–2520.

328. Dempster, E.L., Pryor, K.V., Francis, D., Young, J.E., and Rogers, H.J., (1999) Rapid DNA extraction from ferns for PCR-based analyses, *BioTechniques* 27: 66–68.

329. Depeiges, A., Goubely, C., Lenoir, A., Cocherel, S., Picard, G., Raynal, M., Grellet, F., and Delseny, M., (1995) Identification of the most represented repeated motifs in *Arabidopsis thaliana* microsatellite loci, *Theor. Appl. Genet.* 91: 160–168.

330. Deragon, J.-M., and Landry, B.S., (1992) RAPD and other PCR-based analyses of plant genomes using DNA extracted from small leaf discs, *PCR Methods Appl.* 1: 175–180.

331. Despres, L., Gielly, L., Redoutet, B., and Taberlet, P., (2003) Using AFLP to resolve phylogenetic relationships in a morphologically diversified plant species complex when nuclear and chloroplast sequences fail to reveal variability, *Mol. Phylogenet. Evol.* 27: 185–196.

332. Dessauer, H.C., Cole, C.J., and Hafner, M.S., (1996) Collection and storage of tissues. In *Molecular Systematics, 2nd Edition*, Hillis, D.M., Moritz, C., and Mable, B.K., Eds., Sinauer Associates, Sunderland, MA, pp. 29–47.

333. Devos, K.M., and Gale, M.D., (1992) The use of random amplified polymorphic DNA markers in wheat, *Theor. Appl. Genet.* 84: 567–572.

334. Devos, K.M., and Gale, M.D., (2000) Genome relationships: the grass model in current research, *Plant Cell* 12: 637–646.

335. Dhanaraj, A.L., Rao, E.V.V.B., Swamy, K.R.M., Bhat, M.G., Prasad, D.T., and Sondur, S.N., (2002) Using RAPDs to assess the diversity in Indian cashew (*Anacardium occidentale* L.) germplasm, *J. Hortic. Sci. Biotechnol.* 77: 41–47.

336. Di Gaspero, G., and Cipriani, G., (2002) Resistance gene analogs are candidate markers for disease-resistance genes in grape (*Vitis* spp.), *Theor. Appl. Genet.* 106: 163–172.

337. Di Gaspero, G., Peterlunger, E., Testolin, R., Edwards, K.J., and Cipriani, G., (2000) Conservation of microsatellite loci within the genus *Vitis*, *Theor. Appl. Genet.* 101: 301–308.

338. Di Rienzo, A., Peterson, A.C., Garza, J.C., Valdes, A.M., Slatkin, M., and Freimer, N.B., (1994) Mutational processes of simple-sequence repeat loci in human populations, *Proc. Natl. Acad. Sci. U.S.A.* 91: 3166–3170.

339. Diadema, K., Baumel, A., Lebris, M., and Affre, L., (2003) Genomic DNA isolation and amplification from callus culture in succulent plants, *Carpobrotus species* (Aizoaceae), *Plant Mol. Biol. Rep.* 21: 173a–173e.

340. Diaz, O., Sun, G.L., Salomon, B., and von Bothmer, R., (1999) Levels and distribution of allozyme and RAPD variation in populations of *Elymus fibrosus* (Schrenk) Tzvel. (Poaceae). *Genet. Res. Crop Evol.* 47: 11–24.

341. Diaz, V., and Ferrer, E., (2003) Genetic variation of populations of *Pinus oocarpa* revealed by resistance gene analog polymorphism, *Genome* 46: 404–410.

342. Dib, C., Fauré, S., Fizames, C., Samson, D., Drouot, N., Vignal, A., Millasseau, P., Marc, S., Hazan, J., Seboun, E., Lathrop, M., Gyapay, G., Morissette, J., and Weissenbach, J., (1996) A comprehensive genetic map of the human genome based on 5,264 microsatellites, *Nature* 380: 152–154.

343. Dick, C.W., Etchelecu, G., and Austerlitz, F., (2003) Pollen dispersal of tropical trees (*Dinizia excelsa*: Fabaceae) by native insects and African honeybees in pristine and fragmented Amazonian rainforest, *Mol. Ecol.* 2: 753–764.

344. Dieringer, D., and Schlötterer, C., (2003) Two distinct modes of microsatellite muta-
     tion processes: evidence from the complete genomic sequences of nine species,
     Genome Res. 13: 2242–2251.

345. Dieringer, D., and Schlötterer, C., (2003) Microsatellite analyser (MSA): a platform
     independent analysis tool for large microsatellite data sets, *Mol. Ecol. Notes* 3: 167–169.

346. Dietrich, W.F., Miller, J., Steen, R., Merchant, M.A., Damron-Boles, D., Husain, Z.,
     Dredge, R., Daly, M., Ingalis, K.A., O'Connor, T.J., Evans, C.A., DeAngelis, M.M.,
     Levinson, D.M., Kruglyak, L., Goodman, N., Copeland, N.G., Jenkins, N.A., Haw-
     kins, T.L., Stein, L., Page, D.C., and Lander, E.S., (1996) A comprehensive genetic
     map of the mouse genome, *Nature* 380: 149–152.

347. Dilworth, E., and Frey, J.E., (2000) A rapid method for high throughput DNA extrac-
     tion from plant material for PCR amplification, *Plant Mol. Biol. Rep.* 18: 61–64.

348. Diwan, N., and Cregan, P.B., (1997) Automated sizing of fluorescent-labeled simple
     sequence repeat (SSR) markers to assay genetic variation in soybean, *Theor. Appl.
     Genet.* 95: 723–733.

349. Diwan, N., McIntosh, M.S., and Bauchan, G.R., (1995) Methods of developing a
     core collection of annual *Medicago* species, *Theor. Appl. Genet.* 90: 755–761.

350. Dixit, A., (1998) A simple and rapid procedure for isolation of *Amaranthus* DNA
     suitable for fingerprint analysis, *Plant Mol. Biol. Rep.* 16: 91 (1–8).

351. Dixon, M.S., Jones, D.A., Keddie, J.S., Thomas, C.M., Harrison, K., and Jones,
     J.D.G., (1996) The tomato *cf-2* disease resistance locus comprises two functional
     genes encoding leucine-rich repeat proteins, *Cell* 84: 451–459.

352. Do, N., and Adams, R.P., (1991) A simple technique for removing plant polysaccha-
     ride contaminants from DNA., *BioTechniques* 10: 162–166.

353. Döbbeling, U., Böni, R., Häffner, A., Dummer, R., and Burg, G., (1997) Method for
     simultaneous RNA and DNA isolation from biopsy material, culture cells, plants and
     bacteria, *BioTechniques* 22: 88–90.

354. Don, R.H., Cox, P.T., Wainwright, B.J., Baker, K., and Mattick, J.S., (1991) 'Touch-
     down' PCR to circumvent spurious priming during gene amplification, *Nucleic Acids
     Res.* 19: 4008.

355. Dong, J., and Wagner, D.B., (1993) Taxonomic and population differentiation of
     mitochondrial diversity in *Pinus banksiana* and *Pinus contorta*, *Theor. Appl. Genet.*
     86: 573–578.

356. Donini, P., Elias, M.L., Bougourd, S.M., and Koebner, R.M.D., (1997) AFLP finger-
     printing reveals pattern differences between template DNA extracted from different
     plant organs, *Genome* 40: 521–526.

357. Doolittle, W.F., and Sapienza, C., (1980) Selfish genes, the phenotype paradigm and
     genome evolution, *Nature* 284: 601–603.

358. Dooner, H.K., (1994) Genetic fine structure from testcross progeny analysis. In *The
     Maize Handbook*, Freeling, M., and Walbot, V., Eds., Springer, New York, pp. 303–306.

359. Doré, C., Dosba, F., and Baril, C. (Eds.), (2001) *Proceedings of the International
     Symposium on Molecular Markers for Characterizing Genotypes and Identifying
     Cultivars in Horticulture, Acta Horticult.,* Vol. 546, ISHS Comission Biotechnology,
     Montpellier, France.

360. Dos Santos, J.B., Nienhuis, J., Skroch, P., Tivang, J., and Slocum, M.K., (1994)
     Comparison of RAPD and RFLP genetic markers in determining genetic similarity
     among *Brassica oleracea* L. genotypes, *Theor. Appl. Genet.* 87: 909–915.

361. Doulis, A.G., Harfouche, A.L., and Aravanopoulos, F.A., (2000) Rapid, high quality
     DNA isolation from cypress (*Cupressus sempervirens* L.) needles and optimization
     of the RAPD marker technique, *Plant Mol. Biol. Rep.* 17: 1–14.

362. Dowling, T.E., Moritz, C., Palmer, J.D., and Rieseberg, L.H., (1996) Nucleic acids III: analysis of fragments and restriction sites. In *Molecular Systematics, 2nd Edition*, Hillis, D.M., Moritz, C., and Mable, B.K., Eds., Sinauer Associates, Sunderland, MA, pp. 249–320.

363. Doyle, J.J., (1992) Gene trees and species trees: molecular systematics as one-character taxonomy, *Syst. Bot.* 17: 144–163.

364. Doyle, J.J., and Dickson, E.E., (1987) Preservation of plant samples for DNA restriction endonuclease analysis, *Taxon* 36: 715–722.

365. Doyle, J.J., and Doyle, J.L., (1990) Isolation of plant DNA from fresh tissue, *Focus* 12: 13–15.

366. Doyle, J.J., Morgante, M., Tingey, S.V., and Powell, W., (1998) Size homoplasy in chloroplast microsatellites of wild perennial relatives of soybean (*Glycine* subgenus *Glycine*), *Mol. Biol. Evol.* 15: 215–218.

367. Drábková, L., Kirschner, J., and Vlcek, C., (2002) Comparison of seven DNA extraction and amplification protocols in historical herbarium specimens of Juncaceae, *Plant Mol. Biol. Rep.* 20: 161–175.

368. Drenkard, E., Glazebrook, J., Preuss, D., and Ausubel, F.M. (1998) Use of cleaved amplified polymorphic sequences (CAPS) for genetic mapping and typing. In *DNA Markers. Protocols, Applications, and Overviews*, Caetano-Anollés, G., and Gresshoff, P.M., Eds., Wiley-VCH, New York, pp. 187–197.

369. Drenkard, E., Richter, B.G., Rozen, S., Stutius, L.M., Angell, N.A., Mindrinos, M., Cho, R.J., Oefner, P.J., Davis, R.W., and Ausubel, F.M., (2000) A simple procedure for the analysis of single nucleotide polymorphisms facilitates map-based cloning in *Arabidopsis*, *Plant Physiol.* 124: 1483–1492.

370. Dulson, J., Kott, L.S., and Ripley, V.L., (1998) Efficacy of bulked DNA samples for RAPD DNA fingerprinting of genetically complex *Brassica napus* cultivars, *Euphytica* 102: 65–70.

371. Duminil, J., Pemonge, M.-H., and Petit, R.J., (2002) A set of 35 consensus primer pairs amplifying genes and introns of plant mitochondrial DNA, *Mol. Ecol. Notes* 2: 428–430.

372. Dumolin-Lapègue, S., Pemonge, M.-H., and Petit, R.J., (1997) An enlarged set of consensus primers for the study of organelle DNA in plants, *Mol. Ecol.* 6: 393–397.

373. Dumolin-Lapègue, S., Demesure, B., Fineschi, S., Le Corre, V., and Petit, R.J., (1997) Phylogeographic structure of white oaks throughout the European continent, *Genetics* 146: 1475–1487.

374. Dumolin-Lapègue, S., Pemonge, M.H., Gielly, L., Taberlet, P., and Petit, R.J., (1999) Amplification of oak DNA from ancient and modern wood, *Mol. Ecol.* 8: 2137–2140.

375. Dweikat, I.M., and Mackenzie, S.A., (1998) RAPD-DGGE: applications to pedigree assessment, cultivar identification and mapping. In *DNA Markers. Protocols, Applications, and Overviews*, Caetano-Anollés, G., Gresshoff, P., Eds., Wiley-VCH, New York, pp. 85–90.

376. Dweikat, I., Mackenzie, S., Levy, M., and Ohm, H., (1993) Pedigree assessment using RAPD-DGGE in cereal crop species, *Theor. Appl. Genet.* 85: 497–505.

377. Echt, C.S., and May-Marquardt, P., (1997) Survey of microsatellite DNA in pine, *Genome* 40: 9–17.

378. Echt, C.S., Erdahl, L.A., and McCoy, T.J., (1992) Genetic segregation of random amplified polymorphic DNA in diploid cultivated alfalfa, *Genome* 35: 84–87.

379. Echt, C.S., DeVerno, L.L., Anzidei, M., and Vendramin, G.G., (1998) Chloroplast microsatellites reveal population genetic diversity in red pine, *Pinus resinosa* Ait, *Mol. Ecol.* 7: 307–316.

380. Eckert, R.L., and Green, H., (1986) Structure and evolution of the human involucrin gene, *Cell* 46: 583–589.

381. Economou, E.P., Bergen, A.W., Warren, A.C., and Antonorakis, S.E., (1990) The polydeoxyadenylate tract of *Alu* repetitive elements is polymorphic in the human genome, *Proc. Natl. Acad. Sci. U.S.A.* 87: 2951–2954.

382. Edwards, A., Civitello, A., Hammond, H.A., and Caskey, C.T., (1991) DNA typing and genetic mapping with trimeric and tetrameric tandem repeats, *Am. J. Hum. Genet.* 49: 746–756.

383. Edwards, D., Coghill, J., Batley, J., Holdsworth, M., and Edwards, K.J., (2002) Amplification and detection of transposon insertion flanking sequences using fluorescent MuAFLP, *BioTechniques* 32: 1090–1097.

384. Edwards, K., and Mogg R., (2001) Plant genotyping by analysis of single nucleotide polymorphisms. In *Plant Genotyping: The DNA Fingerprinting of Plants*, Henry, R.J., Ed., CABI Publishing, Wallingford, U.K., pp. 1–13.

385. Edwards, K., Johnstone, C., and Thompson, C., (1991) A simple and rapid method for the preparation of plant genomic DNA for PCR analysis, *Nucleic Acids Res.* 19: 1349.

386. Edwards, K.J., Barker, J.H.A., Daly, A., Jones, C., and Karp, A., (1996) Microsatellite libraries enriched for several microsatellite sequences in plants, *BioTechniques* 20: 758–760.

387. Eichler, E.E., Holden, J.J.A., Popovich, B.W., Reiss, A.L., Snow, K., Thibodeau, S.N., Richards, S., Ward, P.A., and Nelson, D.L., (1994) Length of uninterrupted CGG repeats determines instability in the FMR1 gene, *Nat. Genet.* 8: 88–94.

388. Ellegren, H., (1995) Mutation rates at porcine microsatellite loci, *Mamm. Genome* 6: 376–377.

389. Ellegren, H., (2004) Microsatellites: simple sequences with complex evolution, *Nat. Rev. Genet.* 5: 435–445.

390. Ellegren, H., Primmer, C.R., and Sheldon, B.C., (1995) Microsatellite 'evolution': directionality or bias? *Nat. Genet.* 11: 360–362.

391. Ellegren, H., Moore, S., Robinson, N., Byrne, K., Ward, W., and Sheldon, B.C., (1997) Microsatellite evolution — a reciprocal study of repeat lengths at homologous loci in cattle and sheep, *Mol. Biol. Evol.* 14: 854–860.

392. Ellinghaus, P., Badehorn, D., Blümer, R., Becker, K., and Seedorf, U., (1999) Increased efficiency of arbirarily primed PCR by prolongued ramp times, *BioTechniques* 26: 626–630.

393. Ellis, J., Dodds, P., and Pryor, T., (2000) Structure, function and evolution of plant disease resistance genes, *Curr. Opin. Plant Biol.* 3: 278–284.

394. Ellis, T.H.N., Poyser, S.J., Knox, M.R., Vershinin, A.V., and Ambrose, M.J., (1998) Polymorphism of insertion sites of Ty1-*copia* class retrotransposons and its use for linkage and diversity analysis in pea, *Mol. Gen. Genet.* 260: 9–19.

395. Ellstrand, N.C., (2003) *Dangerous Liaisons? When Cultivated Plants Mate with Their Wild Relatives,* Johns Hopkins University Press, Baltimore, MD.

396. Ellstrand, N.C., and Roose, M.L., (1987) Patterns of genotypic diversity in clonal plant species, *Am. J. Bot.* 74: 123–131.

397. Ellsworth, D.L., Rittenhouse, K.D., and Honeycutt, R.L., (1993) Artifactual variation in randomly amplified polymorphic DNA banding patterns, *BioTechniques* 14: 214–217.

398. Elphinstone, M.S., Hinten, G.N., Anderson, M.J., and Nock, C.J., (2003) An inexpensive and high-throughput procedure to extract and purify total genomic DNA for population studies, *Mol. Ecol. Notes* 3: 317–320.

399. Elsik, C.G., and Williams, C.G., (2001) Low-copy microsatellite recovery from a conifer genome, *Theor. Appl. Genet.* 103: 1189–1195.

400. Elsik, C.G., Minihan, V.T., Hall, S.E., Scarpa, A.M., and Williams, C.G., (2000) Low-copy microsatellite markers for *Pinus taeda* L., *Genome* 43: 550–555.

401. Ender, A., Schwenk, K., Städler, T., Streit, B., and Schierwater, B., (1996) RAPD identification of microsatellites in *Daphnia*, *Mol. Ecol.* 5: 437–441.

402. England, P.R., Usher, A.V., Whelan, R.J., and Ayre, D.J., (2002) Microsatellite diversity and genetic structure of fragmented populations of the rare, fire-dependent shrub *Grevillea macleayana*, *Mol. Ecol.* 11, 967–977.

403. Ennos, R.A., (1994) Estimating the relative rates of pollen and seed migration among plant populations, *Heredity* 72: 250–259.

404. Ennos, R.A., Sinclair, W.T., Hu, X.-S., and Langdon, A., (1999) Using organelle markers to elucidate the history, ecology and evolution of plant populations. In *Molecular Systematics and Plant Evolution*, Hollingsworth, P.M., Bateman, R.M., and Gornall. R.J., Eds., Taylor & Francis, London, pp. 1–19.

405. Epplen, J.T., (1988) On simple repeated GATA/GACA sequences in animal genomes: a critical reappraisal, *J. Hered.* 79: 409–417.

406. Epplen, J.T., and Lubjuhn, T., Eds. (1999) *DNA Profiling and DNA Fingerprinting*. Birkhäuser, Basel, Switzerland.

407. Epplen, C., Melmer, G., Siedlaczck, I., Schwaiger, F.W., Mäueler, W., and Epplen, J.T., (1993) On the essence of "meaningless" simple repetitive DNA in eukaryote genomes. In *DNA Fingerprinting: State of the Science*, Pena, S.D.J., Chakraborty, R., Epplen, J.T., and Jeffreys, A.J., Eds., Birkhäuser, Basel, Switzerland, pp. 29–45.

408. Epplen, J.T., Kyas, A., and Mäueler, W., (1996) Genomic simple repetitive DNAs are targets for differential binding of nuclear proteins, *FEBS Lett.* 389: 92–95.

409. Epplen, J.T., Mäueler, W., and Santos, E.J.M., (1998) On GATAGATA and other "junk" in the barren stretch of genomic desert, *Cytogenet. Cell Genet.* 80: 75–82.

410. Ernst, D., Kiefer, E., Drouet, A., and Sandermann, H., Jr., (1996) A simple method of DNA extraction from soil of composite transgenic plants by PCR, *Plant Mol. Biol. Rep.* 14: 143–148.

411. Escaravage, N., Questiau, S., Pornon, A., Doche, B., and Taberlet, P., (1998) Clonal diversity in a *Rhododendron ferrugineum* L. (Ericaceae) population inferred from AFLP markers, *Mol. Ecol.* 7: 975–982.

412. Escudero, A., Iriondo, J.M., and Torres, M.E., (2003) Spatial analysis of genetic diversity as a tool for plant conservation, *Biol. Conserv.* 113: 351–365.

413. Esselink, G.D., Nybom, H., and Vosman, B., (2004) Assignment of allelic configuration in polyploids using the MAC-PR (microsatellite DNA allele counting-peak ratios) method, *Theor. Appl. Genet.* 109: 402–408.

414. Esselman, E.J., Jianqiang, L., Crawford, D.J., Windus, J.L., and Wolfe, A.D., (1999) Clonal diversity in the rare *Calamagrostis porteri* ssp. *insperata* (Poaceae): comparative results for allozymes and random amplified polymorphic DNA (RAPD) and intersimple sequence repeat (ISSR) markers, *Mol. Ecol.* 8: 443–451.

415. Estoup, A., Tailliez, C., Cornuet, J.-M., and Solignac, M., (1995) Size homoplasy and mutational processes of interrupted microsatellites in two bee species, *Apis mellifera* and *Bombus terrestris* (Apidae), *Mol. Biol. Evol.* 12; 1074–1084.

416. Estoup, A., Jarne, P., and Cornuet, J.-M., (2002) Homoplasy and mutation model at microsatellite loci and their consequences for population genetics analysis, *Mol. Ecol.* 11: 1591–1604.

417. Eujayl, I., Sorrells, M.E., Baum, M., Wolters, P., and Powell, W., (2002) Isolation of EST-derived microsatellite markers for genotyping the A and B genomes of wheat, *Theor. Appl. Genet.* 104: 399–407.

418. Eujayl, I., Sledge, M.K., Wang, L., May, G.D., Chekhovskiy, K., Zwonitzer, J.C., and Mian, M.A.R., (2004) *Medicago truncatula* EST-SSRs reveal cross-species genetic markers for *Medicago* spp., *Theor. Appl. Genet.* 108: 414–422.

419. Evans, D.A., (1989) Somaclonal variation — genetic basis and breeding applications, *Trends Genet.* 5: 46–50.

420. Evett, I.W., and Weir, B.S., (1998) *Interpreting DNA Evidence*, Sinauer Associates, Sunderland, Mass.

421. Excoffier, L., Smouse, P.E., and Quattro, J.M., (1992) Analysis of molecular variance inferred from metric distances among DNA haplotypes: applications to human mitochondrial DNA restriction data, *Genetics* 131: 479–491.

422. Fady, B., Lefèvre, F., Reynaud, M., Vendramin, G.G., Dagher-Kharrat, M.B., Anzidei, M., Pastorelli, R., Savouré, A., and Bariteau, M., (2003) Gene flow among different taxonomic units: evidence from nuclear and cytoplasmic markers in *Cedrus* plantation forests, *Theor. Appl. Genet.* 107: 1132–1138.

423. Fager, E.W., (1972) Diversity: a sampling study, *Am. Nat.* 106: 293–310.

424. Fahima, T., Sun, G.L., Beharav, A., Krugman, T., Beiles, A., and Nevo, E., (1999) RAPD polymorphism in wild emmer wheat populations, *Triticum dicoccoides*, in Israel, *Theor. Appl. Genet.* 98: 434–447.

425. Fahima, T., Röder, M.S., Wendehake, K., Kirzhner, V.M., and Nevo, E., (2002) Microsatellite polymorphism in natural populations of wild emmer wheat, *Triticum dicoccoides*, in Israel, *Theor. Appl. Genet.* 104: 17–29.

426. Fajardo, D.S., La Bonte, D.R., and Jarret, R.L., (2002) Identifying and selecting for genetic diversity in Papua New Guinea sweetpotato *Ipomoea batatas* (L.) Lam. germplasm collected as botanical seed, *Genet. Res. Crop Evol.* 49: 463–470.

427. Falk, D.A, and Holsinger, K.E., (1991) *Genetics and Conservation of Rare Plants*, Oxford University Press, Oxford, U.K.

428. Fang, G., Hammar, S., and Grumet, R., (1992) A quick and inexpensive method for removing polysaccharides from plant genomic DNA, *BioTechniques* 13: 52–55.

429. Farris, J.S., Albert, V.A., Källersjö, M., Lipscomb, D., and Kluge, A.G., (1996) Parsimony jackknifing outperforms neighbour joining, *Cladistics* 11: 99–124.

430. Fatokun, C.A., Menancio-Hautea, D.I., Danesh, D., and Young N.D., (1992) Evidence for orthologous seed weight genes in cowpea and mung bean based on RFLP mapping, *Genetics* 132: 841–846.

431. Fauré, S., Noyer, J.L., Horry, J.P., Bakry, F., Lanaud, C., and González de Léon, C., (1993) A molecular marker-based linkage map of diploid bananas (*Musa acuminata*), *Theor. Appl. Genet.* 87: 517–526.

432. Fedoroff, N., (2000) Transposons and genome evolution in plants, *Proc. Natl. Acad. Sci. U.S.A.* 97: 7002–7007.

433. Feinberg, A.P., and Vogelstein, B., (1983) A technique for radiolabelling DNA restriction endonuclease fragments to high specific activity, *Anal. Biochem.* 137: 6–13.

434. Felsenstein, J., (1985) Confidence limits on phylogenetics: an approach using the bootstrap, *Evolution* 39: 783–791.

435. Felsenstein, J., (2004) *Inferring Phylogenies*, Sinauer Associates, Sunderland, MA.

436. Ferguson, J.A., Steemers, F.J., and Walt, D.R. (2000) High-density fiber optic DNA random microsphere array, *Anal. Chem.* 72: 5618–5624.

437. Ferguson, M.E., Newbury, H.J., Maxted, N., Ford-Lloyd, B.V., and Robertson, L.D., (1998) Population genetic structure in *Lens* taxa revealed by isozyme and RAPD analysis, *Genet. Res. Crop Evol.* 45: 549–559.

438. Fernandez, J.F., Sork, V.L., Gallego, G., López, J., Bohorques, A., and Tohme, J., (2000) Cross-amplification of microsatellite loci in a neotropical *Quercus* species and standardization of DNA extraction from mature leaves dried in silica gel, *Plant Mol. Biol. Rep.* 18: 397a–397e.

439. Fernando, P., Evans, B.J., Morales, J.C., and Melnick, D.J., (2001) Electrophoresis artefacts — a previously unrecognized cause of error in microsatellite analysis, *Mol. Ecol. Notes* 1: 325–328.

440. Ferriol, M., Picó, M.B., and Nuez, F., (2003) Genetic diversity of some accessions of *Cucurbita maxima* from Spain using RAPD and SBAP markers, *Genet. Res. Crop Evol.* 50: 227–238.

441. Field, D., and Wills, C., (1996) Long, polymorphic microsatellites in simple organisms, *Proc. R. Soc. London Ser. B* 263: 209–215.

442. Filatov, D.A., (2002) PROSEQ: a software for preparation and evolutionary analysis of DNA sequence data sets, *Mol. Ecol. Notes* 2: 621–624.

443. Fineschi, S., Anzidei, M., Cafasso, D., Cozzolino, S., Garfì, G., Pastorelli, R., Salvini, D., Taurchini, D., and Vendramin, G.G., (2002) Molecular markers reveal a strong genetic differentiation between two European relic tree species: *Zelkova abelicea* (Lam.) Boissier and *Z. sicula* Di Pasquale, Garfì & Quézel (Ulmaceae), *Conserv. Genet.* 3: 145–153.

444. Finnegan, D.J., (1997) How non-LTR retrotransposons do it, *Curr. Biol.* 7: R245–R248.

445. Fischer, D., and Bachmann, K., (1998) Microsatellite enrichment in organisms with large genomes (*Allium cepa*), *BioTechniques* 24: 796–802.

446. Fisher, P., Gardner, R.C., and Richardson, T.E., (1996) Single locus microsatellites isolated using 5′-anchored PCR, *Nucleic Acids Res.* 24: 4369–4371.

447. FitzSimmons, N.N., Moritz, C., and Moore, S.S., (1995) Conservation and dynamics of microsatellite loci over 300 million years of marine turtle evolution, *Mol. Biol. Evol.* 12: 432–440.

448. Flagstad, Ø., Røed, K., Stayc, J.E., and Jakobsen, K.S., (1999) Reliable noninvasive genotyping based on excremental PCR of nuclear DNA purified with a magnetic bead protocol, *Mol. Ecol.* 8: 879–883.

449. Flavell, A.J., (1999) Long terminal repeat retrotransposons jump between species, *Proc. Natl. Acad. Sci. U.S.A.* 96: 12211–12212.

450. Flavell, A.J., Dunbar, E., Anderson, R., Pearce, S.R., Hartley, R., and Kumar, A., (1992) Ty1-*copia* group retrotransposons are ubiquitous and heterogeneous in higher plants, *Nucleic Acids Res.* 20: 3639–3644.

451. Flavell, A.J., Smith, D.B., and Kumar, A., (1992) Extreme heterogeneity of Ty1-*copia* group retrotransposons in plants, *Mol. Gen. Genet.* 231: 233–242.

452. Flavell, A.J., Pearce, S.R., and Kumar, A., (1994) Plant transposable elements and the genome, *Curr. Opin. Genet. Dev.* 4: 838–844.

453. Flavell, A.J., Knox, M.R., Pearce, S.R., Ellis, T.H.N., (1998) Retrotransposon-based insertion polymorphisms for high throughput marker analysis, *Plant J.* 16: 643–650.

454. Flournoy, L.E., Adams, R.P., and Pandey, R.N., (1996) Interim and archival preservation of plant specimens in alcohols for DNA studies, *BioTechniques* 20: 657–660.

455. Forrest, A.D., Hollingsworth, M.L., Hollingsworth, P.M., Sydes, C., and Bateman, R.M., (2004) Population genetic structure in European populations of *Spiranthes romanzoffiana* set in the context of other genetic studies on orchids, *Heredity* 92: 218–227.

456. Fossati, T., Labra, M., Castiglione, S., Failla, O., Scienza, A., and Sala, F., (2001) The use of AFLP and SSR markers to decipher homonyms and synonyms in grapevine cultivars: the case of the varietal group known as "Schiave," *Theor. Appl. Genet.* 102: 200–205.

457. Fourré, J.-L., Berger, P., Niquet, L., and André, P., (1997) Somatic embryogenesis and somaclonal variation in Norway spruce: morphogenetic, cytogenetic and molecular approaches, *Theor. Appl. Genet.* 94: 159–169.

458. Frankham, R., Ballou, J.D., and Broscoe, J.D., (2002) *Introduction to Conservation Genetics,* Cambridge University Press, Cambridge, U.K.

459. Franks, T., Botta, R., and Thomas, M.R., (2002) Chimerism in grapevines: implications for cultivar identity, ancestry and genetic improvement, *Theor. Appl. Genet.* 104: 192–199.

460. Fraser, L., Harvey, C.F., Crowhurst, R.N., and De Silva H.N., (2004) EST-derived microsatellites from *Actinidia* species and their potential for mapping, *Theor. Appl. Genet.* 108: 1010–1016.

461. Frey, M., Stettner, C., and Gierl, A., (1998) A general method for gene isolation in tagging approaches: amplification of insertion mutagenised sites (AIMS), *Plant J.* 13: 717–721.

462. Deleted in proof.

463. Fritsch, P., and Rieseberg, L.H., (1992) High outcrossing rates maintain male and hermaphrodite individuals in populations of the flowering plant *Datisca glomerata,* *Nature* 359: 633–636.

464. Fritsch, P., Hanson, M.A., Spore, C.D., Pack, P.E., and Rieseberg, L.H., (1993) Constancy of RAPD primer amplification strength among distantly related taxa of flowering plants, *Plant Mol. Biol. Rep.* 11: 10–20.

465. Fu, R.-Z., Wang, J., Sun, Y.-R., and Shaw, P.-C., (1998) Extraction of genomic DNA suitable for PCR analysis from dried plant rhizomes/roots, *BioTechniques* 25: 796–801.

466. Fulton, T.M., Chunwongse, J., and Tanksley, S.D., (1995) Microprep protocol for extraction of DNA from tomato and other herbaceous plants, *Plant Mol. Biol. Rep.* 13: 207–209.

467. Gabriel, S.B., Schaffner, S.F., Nguyen, H., Moore, J.M., Roy, J., Blumenstiel, B., Higgins, J., DeFelice, M., Lochner, A., and Faggart, M. (2002) The structure of haplotype blocks in the human genome, *Science* 296: 2225–2229.

468. Gabrielsen, T.M., Bachmann, K., Jakobsen, K.S., and Brochmann, C., (1997) Glacial survival does not matter: RAPD phylogeography of Nordic *Saxifraga oppositifolia,* *Mol. Ecol.* 6: 831–842.

469. Gaggiotti, O.E., Lange, O., Rassmann, K., and Gliddon, C., (1999) A comparison of two indirect methods for estimating average levels of gene flow using microsatellite data, *Mol. Ecol.* 8: 1513–1520.

470. Gagne, G., Roeckel-Drevet, P., Grezes-Besset, B., Shindrova, P., Ivanov, P., Grand-Ravel, C., Vear, F., Charmet, G., and Nicolas, P., (2000) Amplified fragment length polymorphism (AFLP) as suitable markers to study *Orobanche cumana* genetic diversity, *J. Phytopathol.* 148: 457–459.

471. Gaiotto, F.A., Grattapaglia, D., and Vencovsky, R., (2003) Genetic structure, mating system, and long distance gene flow in heart of palm (*Euterpe edulis* Mart.), *J. Hered.* 94: 399–406.

472. Gallego, F.J., and Martínez, I., (1997) Method to improve reliability of random-amplified polymorphic DNA markers, *BioTechniques* 23: 663–664.

473. Galperin, M.Y., (2004) The molecular biology database collection: 2004 update, *Nucleic Acids Res.* 32: D3–D22.

474. Gao, L., Tang, J., Li, H., and Jia, J. (2003) Analysis of microsatellites in major crops assessed by computational and experimental approaches, *Mol. Breed.* 12: 245–261.

475. Gao, L.F., Jing, R.L., Huo, N.X., Li, Y., Li, X.P., Zhou, R.H., Chang, X.P., Tang, J.F., Ma, Z.Y., and Jia, J.Z., (2004) One hundred and one new microsatellite loci derived from ESTs (EST-SSRs) in bread wheat, *Theor. Appl. Genet.* 108: 1392–1400.

476. Garcia, M.G., Ontivero, M., Diaz Ricci, J.C., and Castagnaro, A., (2002) Morphological traits and high resolution RAPD markers for the identification of the main strawberry varieties cultivated in Argentina, *Plant Breed.* 121: 76–80.

477. Garcia-Mas, J., Oliver, M., Gómez-Paniagua, H., and de Vicente, M.C., (2001) Comparing AFLP, RAPD and RFLP markers for measuring genetic diversity in melon, *Theor. Appl. Genet.* 101: 860–864.

478. Garkava-Gustavsson, L., Persson, H.A., Nybom, H., Rumpunen, K., Gustavsson, B.A., and Bartish, I.V., (2005) RAPD-based analysis of genetic diversity and selection of lingonberry (*Vaccinium vitis-idaea* L.) material for *ex situ* conservation. *Genet. Res. Crop. Evol.,* in press.

479. Garrett, P.E., Tao, F., Lawrence, N., Ji, J., Schumacher, R.T., and Manak, M., (2002) Tired of the same old grind in the new generation genomics and proteomics era? *Targets* 1: 156–162.

480. Garza, J.C., and Freimer, N.B., (1996) Homoplasy for size at microsatellite loci in humans and chimpanzees, *Genome Res.* 6: 211–217.

481. Gaudeul, M., Taberlet, P., and Till-Bottraud, I., (2000) Genetic diversity in an endangered alpine plant, *Eryngium alpinum* L., (Apiaceae) inferred from amplified fragment length polymorphism markers, *Mol. Ecol.* 9, 1625–1637.

482. Gawel, N.J., and Jarret, R.L., (1991) A modified CTAB DNA extraction procedure for *Musa* and *Ipomoea*, *Plant Mol. Biol. Rep.* 9: 262–266.

483. Gemmell, N.J., Allen, P.J., Goodman, S.J., and Reed, J.Z., (1997) Interspecific microsatellite markers for the study of pinniped populations, *Mol. Ecol.* 6: 661–666.

484. Georges, M., Gunawardana, A., Threadgill, D.W., Lathrop, M., Olsaker, I., Mishra, A., Sargeant, L.L., Schoeberlein, A., Steele, M.R., Terry, C., Threadgill, D.S., Zhao, X., Holm, T., Fries, R., and Womack, J.E., (1991) Characterization of a set of variable number of tandem repeat markers conserved in Bovidae, *Genomics* 11: 24–32.

485. Gerber, H.P., Seipel, K., Georgiev, O., Höfferer, M., Hug, M., Rusconi, S., and Schaffner, W., (1994) Transcriptional activation modulated by homopolymeric glutamine and proline stretches, *Science* 263: 808–811.

486. Gerber, S., Streiff, R., Bodénès, C., Mariette, S., and Kremer, A., (2000) Comparison of microsatellites and amplified fragment length polymorphism markers for parentage analysis, *Mol. Ecol.* 9: 1037–1048.

487. Gerber, S., Chabrier, P., and Kremer, A., (2003) FAMOZ: a software for parentage analysis using dominant, codominant and uniparentally inherited markers, *Mol. Ecol. Notes* 3: 479–481.

488. Germano, J., and Klein, A.S., (1999) Species-specific nuclear and chloroplast single nucleotide polymorphisms to distinguish *Picea glauca, P. mariana* and *P. rubens*, *Theor. Appl. Genet.* 99: 37–49.

489. Ghassemian, M., Waner, D., Tchieu, J., Gribskov, M., and Schroeder, J.I., (2001) An integrated *Arabidopsis* annotation database for Affymetrix Genechip® data analysis, and tools for regulatory motif searches, *Trends Plant Sci.* 6: 448–449.

490. Ghérardi, M., Mangin, B., Goffinet, B., Bonnet, D., and Huguet, T., (1998) A method to measure genetic distance between allogamous populations of alfalfa (*Medicago sativa*) using RAPD molecular markers, *Theor. Appl. Genet.* 96: 406–412.

491. Gibson, G., (2002) Microarrays in ecology and evolution: a preview, *Mol. Ecol.* 11: 17–24.

492. Gierl, A., and Saedler, H., (1992) Plant-transposable elements and gene tagging, *Plant Mol. Biol.* 19: 39–49.

493. Giese, H., Holm-Jensen, A.G., Mathiassen, H., Kjaer, B., Rasmussen, S.K., Bay, H., and Jensen, J., (1994) Distribution of RAPD markers on a linkage map of barley, *Hereditas* 120: 267–273.

494. Gill, K.S., Nasuda, S., and Gill, B.S., (1996) Isolation, cloning and gel blot analysis of high molecular weight wheat DNA, *BioTechniques* 21: 572–576.

495. Gill, P., Evett, I.W., Woodroffe, S., Lygo, J.E., Millican, E., and Webster, M., (1991) Databases, quality control and interpretation of DNA profiling in the home office forensic science service, *Electrophoresis* 12: 204–209.

496. Gillings, M., and Holley, M., (1997) Amplification of anonymous DNA fragments using pairs of long primers generates reproducible DNA fingerprints that are sensitive to genetic variation, *Electrophoresis* 18: 1512–1518.

497. Gilmore, S., Peakall, R., and Robertson, J., (2003) Short tandem repeats (STR) markers are hypervariable and informative in *Cannabis sativa*: implications for forensic investigations, *Forensic. Sci. Int.* 131: 65–74.

498. Gilmour, D.S., Thomas, G.H., and Elgin, S.C.R., (1989) *Drosophila* nuclear proteins bind to regions of alternating C and T residues in gene promoters, *Science* 245: 1487–1490.

499. Giovanonni, J., Wing, R., and Tanksley, S.D. (1992) Isolation of molecular markers from specific chromosomal intervals using DNA pools from existing mapping populations, *Nucleic Acids Res.* 19: 6553–6558.

500. Girke, T., Todd, J., Ruuska, S., White, J., Benning, C., and Ohlrogge, J., (2000) Microarray analysis of developing *Arabidopsis* seeds, *Plant Physiol.* 124: 1570–1581.

501. Godwin, I.D., Sangduen, N., Kunanuvatchaidach, R., Piperidis, G., and Adkins, S.W., (1997) RAPD polymorphisms among variant and phenotypically normal rice (*Oryza sativa* var. *indica*) somaclonal progenies, *Plant Cell Rep.* 16: 320–324.

502. Godwin, I.D., Aitken, E.A.B., and Smith, L.W., (1997) Application of intersimple sequence repeat (ISSR) markers to plant genetics, *Electrophoresis* 18: 1524–1528.

503. Goff, S.A., and 55 co-authors, (2002) A draft sequence of the rice genome (*Oryza sativa* L. ssp. *japonica*), *Science* 296: 92–100.

504. Goffeau, A., Barrell, B.G., Bussey, H., Davis, R.W., Dujon, B., Feldmann, H., Galibert, F., Hoheisel, J.D., Jacq, C., Johnston, M., Louis, E.J., Mewes, H.W., Murakami, Y., Philippsen, P., Tettelin, H., and Oliver, S.G., (1996) Life with 6000 genes, *Science* 274: 546–567.

505. Goldstein, D.B., (2001) Islands of linkage disequilibrium, *Nat. Genet.* 29: 109–111.

506. Goldstein, D.B., and Pollock, D.D., (1997) Launching microsatellites: a review of mutation processes and methods of phylogenetic inference, *J. Hered.* 88: 335–342.

507. Goldstein, D.B., and Schlötterer, C., Eds., (1999) *Microsatellites: Evolution and Applications*, Oxford University Press, Oxford, U.K.

508. Goldstein, D.B., Linares, A.R., Cavalli-Sforza, L.L., and Feldman, M.W., (1995) An evaluation of genetic distances for use with microsatellite loci, *Genetics* 139: 463–471.

509. Gómez, A., González-Martínez, S.C., Collada, C., Climent, J., and Gil, L., (2003) Complex population genetic structure in the endemic Canary Island pine revealed using chloroplast microsatellite markers, *Theor. Appl. Genet.* 107: 1123–1131.

510. González-Rodríguez, A., Arias, D.M., Valencia, S., and Oyama, K., (2004) Morphological and RAPD analysis of hybridization between *Quercus affinis* and *Q. laurina* (Fagaceae), two Mexican red oaks, *Am. J. Bot.* 91: 401–409.

511. Goodman, S.J., (1997) R_{ST} Calc: a collection of computer programs for calculating estimates of genetic differentiation from microsatellite data and determining their significance, *Mol. Ecol.* 6: 881–885.

512. Goodwin, D.C., and Lee, S.B., (1993) Microwave miniprep of total genomic DNA from fungi, plants, protists and animals for PCR, *BioTechniques* 15: 438–444.

513. Gortner, G., Nenno, M., Weising, K., Zink, D., Nagl, W., and Kahl, G., (1998) Chromosomal localization and distribution of simple sequence repeats and the *Arabidopsis*-type telomere sequence in the genome of *Cicer arietinum* L., *Chromosome Res.* 6: 97–104.

514. Goudet, J., (1995) FSTAT (version 1.2): A computer program to calculate F-statistics, *J. Hered.* 86: 485–486.

515. Goulao, L, and Oliveira, C.M., (2001) Molecular characterisation of cultivars of apple (*Malus* x *domestica* Borkh.) using microsatellite (SSR and ISSR) markers, *Euphytica* 122: 81–89.

516. Goulao, L., Cabrita, L., Oliveira, C.M., and Leitao, J.M., (2001) Comparing RAPD and AFLP™ analysis in discrimination and estimation of genetic similarities among apple (*Malus domestica* Borkh.) cultivars, *Euphytica* 119: 259–270.

517. Goulding, S.E., Olmstead, R.G., Morden, C.W., and Wolfe, K.H., (1996) Ebb and flow of the chloroplast inverted repeat, *Mol. Gen. Genet.* 252: 195–206.

518. Graham, J., Squire, G.R., Marshall, B., and Harrison, R.E., (1997) Spatially dependent genetic diversity within and between colonies of wild raspberry *Rubus idaeus* detected using RAPD markers, *Mol. Ecol.* 6: 1001–1008.

519. Graham, S.W., and Olmstead, R.G., (2000) Evolutionary significance of an unusual chloroplast DNA inversion found in two basal angiosperm lineages, *Curr. Genet.* 37: 183–188.

520. Grandbastien, M.-A., (1992) Retroelements in higher plants, *Trends Genet.* 8: 103–108.

521. Granok, H., Leibovitch, B.A., Shaffer, C.D., and Elgin, S.C.R. (1995) Ga-ga over GAGA factor, *Curr. Biol.* 5: 238–241.

522. Grant, D., Cregan, P., and Shoemaker, R.C., (2000) Genome organization in dicots: genome duplication in *Arabidopsis* and synteny between soybean and *Arabidopsis*, *Proc. Natl. Acad. Sci. U.S.A.* 97: 4168–4173.

523. Grant, M.R., Godiard, L., Straube, E., Ashfield, T., Lewald, J., Sattler, A., Innes, R.W., and Dang, J.L. (1995) Structure of the *Arabidopsis* RPM1 gene enabling dual specificity disease resistance, *Science* 269: 843–846.

524. Grassi, F., Imazio, S., Failla, O., Scienza, A., Ocete Rubio, R., Lopez, M.A., Sala, F., and Labra, M., (2003) Genetic isolation and diffusion of wild grapevine Italian and Spanish populations as estimated by nuclear and chloroplast SSR analysis, *Plant Biol.* 5: 608–614.

525. Graur, D., and Li, W.-H., (2000) *Fundamentals of Molecular Evolution, 2nd Edition*, Sinauer Associates, Sunderland, MA.

526. Grimaldi, M.-C., and Crouau-Roy, B., (1997) Microsatellite allele homoplasy due to variable flanking sequences, *J. Mol. Evol.* 44: 336–340.

527. Grivet, D., and Petit, R.J., (2002) Phylogeography of the common ivy (*Hedera* sp.) in Europe: genetic differentiation through space and time, *Mol. Ecol.* 11: 1351–1362.

528. Grivet, D., and Petit, R.J., (2003) Chloroplast DNA phylogeography of the hornbeam in Europe: evidence for a bottleneck at the outset of postglacial colonization, *Conserv. Genet.* 4: 47–56.

529. Grivet, D., Heinze, B., Vendramin, G.G., and Petit, R.J., (2001) Genome walking with consensus primers: application to the large single copy region of chloroplast DNA, *Mol. Ecol. Notes* 1: 345–349.

530. Gross, B.L., Schwarzbach, E., and Rieseberg, L.H., (2003) Origin(s) of the diploid hybrid species *Helianthus deserticola* (Asteraceae), *Am. J. Bot.* 90: 1708–1719.

531. Grotewold, E., and Peterson, T. (1994) Isolation and characterization of a maize gene encoding chalcone flavonone isomerase, *Mol. Gen. Genet.* 242: 1–8.

532. Gu, L.-Q., Cheley S., and Bayley H., (2001) Capture of a single molecule in a nanocavity, *Science* 291: 636–640.

533. Gu, W.K., Weeden, N.F., Yu, J., and Wallace, D.H. (1995) Large-scale, cost-effective screening of PCR products in marker-assisted selection applications, *Theor. Appl. Genet.* 91: 465–470.

534. Guarino, L., Rao, V.R., and Reid, R., Eds., (1995) *Collecting Plant Genetic Diversity: Technical Guidelines*, CAB International, Wallingford, U.K.

535. Guerra-Sanz, J.M., (2002) *Citrullus* simple sequence repeats markers from sequence databases, *Mol. Ecol. Notes* 2: 223–225.

536. Gugerli, F., Senn, J., Anzidei, M., Madaghiele, A., Büchler, U., Sperisen, C., and Vendramin, G.G., (2001) Chloroplast microsatellites and mitochondrial *nad*1 intron 2 sequences indicate congruent phylogenetic relationships among Swiss stone pine (*Pinus cembra*), Siberian stone pine (*Pinus sibirica*) and Siberian dwarf pine (*Pinus pumila*), *Mol. Ecol.* 10: 1489–1497.

537. Guidet, F., (1994) A powerful new technique to quickly prepare hundreds of plant extracts for PCR and RAPD analyses, *Nucleic Acids Res.* 22: 1772–1773.

538. Guidet, F., and Langridge, P., (1992) Megabase DNA preparation from plant tissue, *Meth. Enzymol.* 216: 3–12.

539. Guidet, F., Rogowsky, P., and Langridge, P., (1990) A rapid method of preparing megabase plant DNA, *Nucleic Acids Res.* 18: 4955.

540. Guillemaut, P., and Maréchal-Drouard, L., (1992) Isolation of plant DNA: a fast, inexpensive, and reliable method, *Plant Mol. Biol. Rep.* 10: 60–65.

541. Gulsen, O., and Roose, M.L., (2001) Lemons: diversity and relationships with selected *Citrus* genotypes as measured with nuclear genome markers, *J. Am. Soc. Hortic. Sci.* 126: 309–317.

542. Gupta, M., Chyi, Y.-S., Romero-Severson, J., and Owen, J.L., (1994) Amplification of DNA markers from evolutionary diverse genomes using single primers of simple-sequence repeats, *Theor. Appl. Genet.* 89: 998–1006.

543. Gupta, P.K., and Varshney, R.K., (2000) The development and use of microsatellite markers for genetic analysis and plant breeding with emphasis on bread wheat, *Euphytica* 113: 163–185.

544. Gupta, P.K., Varshney, R.K., Sharma, P.C., and Ramesh, B., (1999) Molecular markers and their application in wheat breeding, *Plant Breed.* 118: 369–390.

545. Gupta, P.K., Roy, J.K., and Prasad, M., (2001) Single nucleotide polymorphisms: a new paradigm for molecular marker technology and DNA polymorphism detection with emphasis on their use in plants, *Curr. Sci.* 80: 524–535.

546. Gur-Arie, R., Cohen, C.J., Eitan, Y., Shelef, L., Hallerman, E.M., and Kashi, Y., (2000) Simple sequence repeats in *Escherichia coli*: abundance, distribution, composition, and polymorphism, *Genome Res.* 10: 62–71.

547. Guyomarc'h, H., Sourdille, P., Charmet, G., Edwards, K.J., and Bernard, M., (2002) Characterisation of polymorphic microsatellite markers from *Aegilops tauschii* and transferability to the D-genome of bread wheat, *Theor. Appl. Genet.* 104: 1164–1172.

548. Gyllensten, U.B., Jakobsson, S., Temrin, H., and Wilson, A.C., (1989) Nucleotide sequence and genomic organization of bird minisatellites, *Nucleic Acids Res.* 17: 2203–2214.

549. Haber, J.E., and Louis, E.J., (1998) Minisatellite origins in yeast and humans, *Genomics* 48: 132–135.

550. Haberl, M., and Tautz, D., (1999) Comparative allele sizing can produce inaccurate allele size differences for microsatellites, *Mol. Ecol.* 8: 1347–1350.

551. Hadrys, H., Balick, M., and Schierwater, B., (1992) Applications of random amplified polymorphic DNA (RAPD) in molecular ecology, *Mol. Ecol.* 1: 55–63.

552. Hakki, E.E., and Akkaya, M.S., (2000) Microsatellite isolation using amplified fragment length polymorphism markers: no cloning, no screening, *Mol. Ecol.* 9: 2152–2154.

553. Hale, M.L., and Wolff, K., (2003) Polymorphic microsatellite loci in *Plantago lanceolata*, *Mol. Ecol. Notes* 3: 134–135.

554. Hale, M.L., Squirrell, J., Borland, A.M., and Wolff, K., (2002) Isolation of polymorphic microsatellite loci in the genus *Clusia* (Clusiaceae), *Mol. Ecol. Notes* 2: 506–508.

555. Hale, M.L., Borland, A.M., Gustafsson, M.H.G., and Wolff, K., (2004) Causes of size homoplasy among chloroplast microsatellites in closely related *Clusia* species, *J. Mol. Evol.* 58: 182–190.

556. Hall, B.G., (2001) *Phylogenetic Trees Made Easy*, Sinauer Associates, Sunderland, MA.

557. Halldén, C., Hansen, M., Nilsson, N.-O., Hjerdin, A., and Säll, T., (1996) Competition as a source of errors in RAPD analysis, *Theor. Appl. Genet.* 93: 1185–1192.

558. Hamilton, M.B., (1999) Four primer pairs for the amplification of chloroplast intergenic regions with intraspecific variation, *Mol. Ecol.* 8: 521–523.

559. Hamilton, M.B., Pincus, E.L., DiFiore, A., and Fleischer, R.C., (1999) Universal linker and ligation procedures for construction of genomic DNA libraries enriched for microsatellites, *BioTechniques* 27: 500–507.

560. Hämmerli, A., and Reusch, T.B.H., (2003) Inbreeding depression influences genet size distribution in a marine angiosperm, *Mol. Ecol.* 12: 619–629.

561. Hämmerli, A., and Reusch, T.B.H., (2003) Genetic neighbourhood of clone structures in eelgrass meadows quantified by spatial autocorrelation of microsatellite markers, *Heredity* 91: 448–455.

562. Hamrick, J.L., and Godt, M.J.W., (1989) Allozyme diversity in plant species, in *Plant Population Genetics, Breeding and Genetic Resources*, Brown, A.H.D., Clegg, M.T., Kahler, A.L., and Weir, B.S., Eds., Sinauer Associates, Sunderland, MA, pp. 43–63.

563. Hamrick, J.L., and Godt, M.J.W., (1996) Effects of life history traits on genetic diversity in plant species. *Phil. Trans. R. Soc. London, Ser. B* 351: 1291–1298.

564. Hamrick, J.L., and Godt, M.J.W., (1997) Allozyme diversity in cultivated crops, *Crop Sci.* 37: 26–30.

565. Han, T.H., Van Eck, H.J., De Jeu, M.J., and Jacobsen, E., (1999) Optimization of AFLP fingerprinting of organisms with a large-sized genome: a study on *Alstroemeria* spp., *Theor. Appl. Genet.* 98: 465–471.

566. Han, T.-H., De Jeu, M., Van Eck, H., and Jacobsen, E., (2000) Genetic diversity of Chilean and Brazilian *Alstromeria* species assessed by AFLP analysis, *Heredity* 84: 564–569.

567. Hanley, S., Edwards, D., Stevenson, D., Haines, S., Hegarty, M., Schuch, W., and Edwards, K.J., (2000) Identification of transposon-tagged genes by the random sequencing of *Mutator*-tagged DNA fragments from *Zea mays*, *Plant J.* 22: 557–566.

568. Hansen, M., Halldén, C., and Säll, T., (1998) Error rates and polymorphism frequencies for three RAPD protocols, *Plant Mol. Biol. Rep.* 16: 139–146.

569. Hansen, M., Kraft, T., Christansson, M., and Nilsson, N.O., (1999) Evaluation of AFLP in *Beta*, *Theor. Appl. Genet.* 98: 845–852.

570. Hardy, O.J., (2003) Estimation of pairwise relatedness between individuals and characterization of isolation-by-distance processes using dominant genetic markers, *Mol. Ecol.* 12: 1577–1588.

571. Hardy, O.J., and Vekemans, X., (1999) Isolation by distance in a continuous population: reconciliation between spatial autocorrelation analysis and population genetics models, *Heredity* 83: 145–154.

572. Hardy, O.J., and Vekemans, X., (2002) SPAGeDi: a versatile computer program to analyse spatial genetic structure at the individual or population levels, *Mol. Ecol. Notes* 2: 618–620.

573. Hare, M.P., (2001) Prospects for nuclear gene phylogeography, *Trends Ecol. Evol.* 16: 700–706.

574. Harr, B., and Schlötterer, C., (2000) Long microsatellite alleles in *Drosophila melanogaster* have a downward mutation bias and short persistence times, which cause their genome-wide underrepresentation, *Genetics* 155: 1213–1220.

575. Harris, H., (1966) Enzyme polymorphism in man, *Proc. R. Soc. London Ser. B* 164: 298–310.

576. Hartl, D.L., and Clark, A.G., (1997) *Principles of Population Genetics, 3rd Edition*, Sinauer Associates, Sunderland, Mass.

577. Hatada, I., Hayashizaki, Y., Hirotsune, S., Komatsubara, H., and Mukai, T., (1991) A genomic scanning method for higher organisms using restriction sites as landmarks, *Proc. Natl. Acad. Sci. U.S.A.* 88: 9523–9527.

578. Hatano, S., Yamaguchi, J., and Hirai, A., (1992) The preparation of high-molecular weight DNA from rice and its analysis by pulsed-field gel electrophoresis, *Plant Sci.* 83: 55–64.

579. Hattori, J., Gottlob-McHugh, S.G., and Johnson, D.A., (1987) The isolation of high-molecular weight DNA from plants, *Anal. Biochem.* 165: 70–74.

580. Hauge, X.Y., and Litt, M., (1993) A study of the origin of "shadow bands" seen when typing dinucleotide repeat polymorphisms by the PCR, *Hum. Mol. Genet.* 2: 411–415.

581. Hayashi, K., (1991) PCR-SSCP: a simple and sensitive method for detection of mutations in the genomic DNA, *PCR Methods Appl.* 1: 34–38.

582. Hayashi, K., Hashimoto, N., Daigen, M., and Ashikawa, I., (2004) Development of PCR-based SNP markers for rice blast resistance genes at the *Piz* locus, *Theor. Appl. Genet.* 108: 1212–1220.

583. Hayden, M.J., and Sharp, P.J., (2001) Sequence-tagged microsatellite profiling (STMP): a rapid technique for developing SSR markers, *Nucleic Acids Res.* 29: e43.

584. Hayden, M.J., and Sharp, P.J., (2001) Targeted development of informative microsatellite (SSR) markers, *Nucleic Acids Res.* 29: e44

585. Hayden, M.J., Good, G., and Sharp, P.J., (2002) Sequence-tagged microsatellite profiling (STMP): improved isolation of DNA sequence flanking target SSRs, *Nucleic Acids Res.* 30: e129.

586. Haymes, K.M., (1996) Mini-prep method suitable for a plant breeding program, *Plant Mol. Biol. Rep.* 14: 280–284.

587. He, Q., Viljanen, M.K., and Mertsola, J., (1994) Effects of thermocyclers and primers on the reproducibility of banding patterns in randomly amplified polymorphic DNA analysis, *Mol. Cell. Probes* 8: 155–160.

588. Heath, D.D., Iwama, G.K., and Devlin, R.H., (1993) PCR primed with the VNTR core sequences yields species specific patterns and hypervariable probes, *Nucleic Acids Res.* 21: 5782–5785.

589. Heckenberger, M., Rouppe van der Voort, J., Melchinger, A.E., Peleman, J., and Bohn, M., (2003) Variation in DNA fingerprints among accessions within maize inbred lines and implications for identification of essentially derived varieties: II. Genetic and technical sources of variation in AFLP data and comparison with SSR data, *Mol. Breed.* 12: 97–106.

590. Hedrick, P.W., (1999) Perspective: highly variable loci and their interpretation in evolution and conservation, *Evolution* 53: 313–318.

591. Hellwig, F.H., Nolte, M., Ochsmann, J., and Wissemann, V., (1999) Rapid isolation of total cell DNA from milligram plant tissue, *Haussknechtia* 7: 29–34.

592. Henegariu, O., Heerema, N.A., Dlouhy, S.R., Vance, G.H., and Vogt, P.H., (1997) Multiplex PCR: critical parameters and step-by-step protocol, *BioTechniques* 23: 504–511.

593. Henry, R.J., (2001) *Plant Genotyping: The DNA Fingerprinting of Plants*, CAB International Publishing, Wallingford, U.K.

594. Hertzberg, M., Aspeborg, H., Schrader, J., Andersson, A., Erlandsson, R., Blomqvist, K., Bhalerao, R., Uhlen, M., Teeri, T.T., Lundeberg, J., Sundberg, B., Nilsson, P., and Sandberg, G., (2001) A transcriptional road map to wood formation, *Proc. Natl. Acad. Sci. U.S.A.* 98: 14732–14737.

595. Heuertz, M., Vekemans, X., Hausman, F., Palada, M., and Hardy, O.J., (2003) Estimating seeds vs. pollen dispersal from spatial genetic structure in the common ash, *Mol. Ecol.* 12: 2483–2495.

596. Heun, M., and Helentjaris, T., (1993) Inheritance of RAPDs in F_1 hybrids of corn, *Theor. Appl. Genet.* 85: 961–968.

597. Hillis, D.M., Moritz, C., and Mable, B.K. (1996) *Molecular Systematics, 2nd Edition.* Sinauer Associates, Sunderland, MA.

598. Hiratsuka, J., Shimada, H., Whittier, R., Ishibashi, T., Sakamoto, M., Mori, M., Kondo, C., Honji, Y., Sun, C.-R., Meng, B.-Y., Li, Y.-Q., Kanno, A., Nishizawa, Y., Hirai, A., Shinozaki, K., and Sugiura, M., (1989) The complete sequence of the rice (*Oryza sativa*) chloroplast genome: intermolecular recombination between distinct tRNA genes accounts for a major plastid DNA inversion during the evolution of cereals, *Mol. Gen. Genet.* 217: 185–194.

599. Hirochika, H., Fukuchi, A., and Kikuchi, F., (1992) Retrotransposon families in rice, *Mol. Gen. Genet.* 233: 209–216.

600. Hisatomi, Y., Hanada, K., and Iida, S., (1997) The retrotransposon *RTip1* is integrated into a novel type of minisatellite, *MiniSip1*, in the genome of the common morning glory and carries another new type of minisatellite, *MiniSip2, Theor. Appl. Genet.* 95: 1049–1056.

601. Hodgkin, T., Rovigliono, R., De Vicente, M.C., and Dudnik, N., (2001) Molecular methods in the conservation and use of plant genetic resources, *Acta Hortic.* 546: 107–118.

602. Hodkinson, T.R., Chase, M.W., and Renvoize, S.A., (2002) Characterization of a genetic resource collection for *Miscanthus* (Saccharinae, Andropogoneae, Poaceae) using AFLP and ISSR PCR, *Ann. Bot.* 89: 627–636.

603. Hoelzel, A.R., (1998) *Molecular Genetic Analysis of Populations: A Practical Approach, 2nd Edition*, IRL Press, Oxford, U.K.

604. Hokanson, S.C., (2001) SNiPs, Chips, BACs, and YACs: are small fruits part of the party mix?, *HortScience* 36: 859–871.

605. Hollingsworth, M.L., and Bailey, J.P., (2000) Evidence for massive clonal growth in the invasive weed *Fallopia japonica* (Japanese knotweed), *Bot. J. Linn. Soc.* 133: 463–472.

606. Hollingsworth, P.M., and Ennos, R.A., (2004) Neighbour joining trees, dominant markers and population genetic structure, *Heredity* 92: 490–498.

607. Holsinger, K.E., Lewis, P.O., and Dey, D.K., (2002) A Bayesian approach to inferring population structure from dominant markers, *Mol. Ecol.* 11: 1157–1164.

608. Holton, T.A., (2001) Plant genotyping by analysis of microsatellites. In *Plant Genotyping: The DNA Fingerprinting of Plants*, Henry, R.J., Ed. CABI Publishing, CAB International, Wallingford, U.K., pp. 15–27.

609. Holton, T.A., Christopher, J.T., McClure, L., Harker, N., and Henry, R.J., (2002) Identification and mapping of polymorphic SSR markers from expressed gene sequences of barley and wheat, *Mol. Breed.* 9: 63–71.

610. Honeycutt, R.J., Sobral, B.W.S., Keim, P., and Irvine, J.E., (1992) A rapid DNA extraction method for sugarcane and its relatives, *Plant Mol. Biol. Rep.* 10: 66–72.

611. Honeycutt, R.J., Sobral, B.W.S., McClelland, M., and Atherly, A.G., (1992) Analysis of large DNA from soybean (*Glycine max* L. Merr.) by pulsed-field gel electrophoresis, *Plant J.* 2: 133–135.

612. Hong, Y.-K., Coury, D.A., Polne-Fuller, M., and Gibor, A., (1992) Lithium chloride extraction of DNA from the seaweed *Porphyra perforata* (Rhodophyta). *J. Phycol.* 28: 717–720.

613. Hoopes, B.C., and McClure, W.R., (1981) Studies on the selectivity of DNA precipitation by spermine, *Nucleic Acids Res.* 9: 5493–5505.

614. Hopkins, K.L., and Hilton, A.C., (2001) Use of multiple primers in RAPD analysis of clonal organisms provides limited improvement in discrimination, *BioTechniques* 30: 1262–1267.

615. Horn, G.T., Richards, B., and Klinger, K.W., (1989) Amplification of a highly polymorphic VNTR segment by the polymerase chain reaction, *Nucleic Acids Res.* 17: 2140.

616. Horn, P., and Rafalski, A., (1992) Non-destructive RAPD genetic diagnostics of microspore-derived *Brassica* embryos, *Plant Mol. Biol. Rep.* 10: 285–293.

617. Horton, R., Niblett, D., Milne, S., Palmer, S., Tubby, B., Trowsdale, J., and Beck, S., (1998) Large-scale sequence comparisons reveal unusually high levels of variation in the HLA-DQB1 locus in the class II region of the human MHC, *J. Mol. Biol.* 282: 71–97.

618. Höss, M., and Pääbo, S., (1993) DNA extraction from Pleistocene bones by a silica-based purification method, *Nucleic Acids Res.* 21: 3913–3914.

619. Howland, D.E., Oliver, R.P., and Davy, A.J., (1991) A method of extraction of DNA from birch, *Plant Mol. Biol. Rep.* 9: 340–344.

620. Hu, J., and Quiros, C.F., (1991) Identification of broccoli and cauliflower cultivars with RAPD markers, *Plant Cell Rep.* 10: 505–511.

621. Hu, J., and Vick, B.A., (2003) Target region amplification polymorphism: a novel marker technique for plant genotyping, *Plant Mol. Biol. Rep.* 21: 289–294.

622. Hu, J., van Eysden, J., and Quiros, C.F., (1995) Generation of DNA-based markers in specific genome regions by two-primer RAPD reactions, *PCR Methods Appl.* 4: 346–351.

623. Hu, J., Zhu, J., and Xu, H.M., (2000) Methods of constructing core collections by stepwise clustering with three sampling strategies based on genotypic values of crops, *Theor. Appl. Genet.* 101: 264–268.

624. Huang, J., and Sun, M., (1999) A modified AFLP with fluorescence-labelled primers and automated DNA sequencer detection for efficient fingerprinting analysis in plants, *Biotechnol. Tech.* 13: 277–278.

625. Huang, J., and Sun, M., (2000) Fluorescein PAGE analysis of microsatellite-primed PCR: a fast and efficient approach for genomic fingerprinting, *BioTechniques* 28: 1968–1072.

626. Huang, J., Ge, X., and Sun, M., (2000) Modified CTAB protocol using a silica matrix for isolation of plant genomic DNA, *BioTechniques* 28: 432–434.

627. Huang, W.-G., Cipriani, G., Morgante, M., and Testolin, R., (1998) Microsatellite DNA in *Actinidia chinensis*: isolation, characterisation, and homology in related species, *Theor. Appl. Genet.* 97: 1269–1278.

628. Huelsenbeck, J.P., and Crandall, K.A., (1997) Phylogeny estimation and hypothesis testing using maximum likelihood, *Annu Rev. Ecol. Syst.* 28: 437–466.

629. Huelsenbeck, J.P., Ronquist, F., Nielsen, R., and Bollback, J.P., (2001) Bayesian inference of phylogeny and its impact on evolutionary biology, *Science* 294: 2310–2314.

630. Hughes, A.E., (1993) Optimization of microsatellite analysis for genetic mapping, *Genomics* 15: 433–434.

631. Hung, T., Mak, K., and Fong, K., (1990) A specificity enhancer for polymerase chain reaction, *Nucleic Acids Res.* 18: 4953.

632. Hunt, G.J., and Page, R.E., (1992) Patterns of inheritance with RAPD molecular markers reveal novel types of polymorphism in the honey bee, *Theor. Appl. Genet.* 85: 15–20.

633. Hüttel, B., Winter, P., Weising, K., Choumane, W., Weigand, F., and Kahl, G., (1999) Sequence-tagged microsatellite site markers for chickpea (*Cicer arietinum* L.), *Genome* 42: 210–217.

634. Hüttel, B., Santra, D., Muehlbauer, F.J., and Kahl, G. (2002) Resistance gene analogues of chickpea (*Cicer arietinum* L.): Isolation, genetic mapping and association with a *Fusarium* resistance gene cluster, *Theor. Appl. Genet.* 105: 479–490.

635. Ibañez, J., (2001) Mathematical analysis of RAPD data to establish reliability of varietal assignment in vegetatively propagated species, *Acta Hortic.* 546: 73–79.

636. Ikeda, N., Bautista, N.S., Yamada, T., Kamijama, O., and Ishii, T., (2001) Ultra-simple DNA extraction method for marker-assisted selection using microsatellite markers in rice, *Plant Mol. Biol. Rep.* 19: 27–32.

637. Imazio, S., Labra, M., Grassi, F., Winfield, M., Bardini, M., and Scienza, A., (2002) Molecular tools for clone identification: the case of the grapevine cultivar "Traminer," *Plant Breed.* 121: 531–535.

638. Innan, H., Terauchi, R., and Miyashita, N.T., (1997) Microsatellite polymorphism in natural populations of the wild plant *Arabidopsis thaliana*, *Genetics* 146: 1441–1452.

639. Innis, M.A., Gelfand, D.H., Sninsky, J.J., and White, T.J., (1990) *PCR Protocols: A Guide to Methods and Applications*, Academic Press, San Diego, CA.

640. Inoue, S., Takahashi, K., and Ohta, M., (1999) Sequence analysis of genomic regions containing trinucleotide repeats isolated by a novel cloning method, *Genomics* 57: 169–172.

641. Inukai, T., and Sano, Y., (2002) Sequence rearrangement in the AT-rich minisatellite of the novel rice transposable element *Basho*, *Genome* 45: 493–502.

642. Ipek, M., Ipek, A., and Simon, P.W., (2003) Comparison of AFLPs, RAPD markers, and isozymes for diversity assessment of garlic and detection of putative duplicates in germplasm collections, *J. Am. Soc. Hortic. Sci.* 128: 246–252.

643. Iruela, M., Rubio, J., Cubero, J.I., Gil, J., and Millán, T., (2002) Phylogenetic analysis in the genus *Cicer* and cultivated chickpea using RAPD and ISSR markers, *Theor. Appl. Genet.* 104: 643–651.

644. Isabel, N., Tremblay, L., Michaud, M., Tremblay, F.M., and Bousquet, J., (1993) RAPDs as an aid to evaluate the genetic integrity of somatic embryogenesis-derived populations of *Picea mariana* (Mill.) B.S.P., *Theor. Appl. Genet.* 86: 81–87.

645. Ishibashi, Y., Saitoh, T., Abe, S., and Yoshida, M.C., (1996) Null microsatellite alleles due to nucleotide sequence variation in the grey-sided vole *Clethrionomys rufocanus*, *Mol. Ecol.* 5: 589–590.

646. Ishii, T., and McCouch, S.R., (2000) Microsatellites and microsynteny in the chloroplast genomes of *Oryza* and eight other Gramineae species, *Theor. Appl. Genet.* 100: 1257–1266.

647. Ishii, T., Mori, N., and Ogihara, Y., (2001) Evaluation of allelic diversity at chloroplast microsatellite loci among common wheat and its ancestral species, *Theor. Appl. Genet.* 103: 896–904.

648. Ishii, T., Xu, Y., and McCouch, S.R., (2001) Nuclear- and chloroplast-microsatellite variation in A-genome species of rice, *Genome* 44: 658–666.

649. Islam, M.S., Lian, C., Kameyama, N., Wu, B., and Hogetsu, T., (2004) Development of microsatellite markers in *Rhizophora stylosa* using a dual-suppression-polymerase chain reaction technique, *Mol. Ecol. Notes* 4: 110–112.

650. Isoda, K., Shiraishi, S., Watanabe, S., and Kitamura, K. (2000) Molecular evidence of natural hybridisation between *Abies veitchii* and *A. homolepis* (Pinaceae) revealed by chloroplast, mitochondrial and nuclear markers, *Mol. Ecol.* 9: 1965–1974.

651. Ito, T., Smith, C.L., and Cantor, C.R., (1992) Sequence-specific DNA purification by triplex affinity capture, *Proc. Natl. Acad. Sci. U.S.A.* 89: 495–498.

652. Ivandic, V., Hackett, C.A., Nevo, E., Keith, R., Thomas, W.T.B., and Forster, B.P., (2002) Analysis of simple sequence repeats (SSRs) in wild barley from the Fertile Crescent: associations with ecology, geography and flowering time, *Plant Mol. Biol.* 48: 511–527.

653. Iwahana, H., Adzuma, K., Takahashi, Y., Katashima, R., Yoshimoto, K., and Itakura, M., (1995) Multiple fluorescence-based PCR-SSCP analysis with postlabeling, *PCR Methods Appl.* 4: 275–282.

654. Jacobsen, K.M., and Lester, E., (2003) A first assessment of genetic variation in *Welwitschia mirabilis* Hook, *J. Hered.* 94: 212–217.

655. Jakše, J., and Javornik, B., (2001) High throughput isolation of microsatellites in hop (*Humulus lupulus* L.), *Plant Mol. Biol. Rep.* 19: 217–226.

656. Jansen, R.C., and Nap, J.P.H., (2001) Genetical genomics: the added value from segregation, *Trends Genet.* 17: 388–391.

657. Jansen, R.K., Wee, J.L., and Millie, D. (1998) Comparative utility of chloroplast DNA restriction site and DNA sequence data for phylogenetic studies in plants. In *Molecular Systematics of Plants II: DNA Sequencing*, Soltis, D.E., Soltis, P.S., and Doyle, J.J., Eds., Kluwer Academic, Dordrecht, pp. 87–100.

658. Jarman, A.P., and Wells, R.A., (1989) Hypervariable minisatellites: recombinators or innocent bystanders? *Trends Genet.* 5: 367–371.

659. Jarne, P., and Lagoda, P.J.L., (1996) Microsatellites, from molecules to populations and back, *Trends Genet.* 11: 424–429.

660. Jeandroz, S., Bastien, D., Chandelier, A., Du Jardin, P., and Favre, J.M., (2002) A set of primers for amplification of mitochondrial DNA in *Picea abies* and other conifer species, *Mol. Ecol. Notes* 2: 389–392.

661. Jeffreys, A.J., and Morton, D.B., (1987) DNA fingerprinting of dogs and cats, *Anim. Genet.* 18: 1–15.

662. Jeffreys, A.J., Wilson, V., and Thein, S.L., (1985) Hypervariable "minisatellite" regions in human DNA, *Nature* 314: 67–73.

663. Jeffreys, A.J., Wilson, V., and Thein, S.L., (1985) Individual-specific "fingerprints" of human DNA, *Nature* 316: 76–79.

664. Jeffreys, A.J., Wilson, V., Neumann, R., and Keyte, J., (1988) Amplification of human minisatellites by the polymerase chain reaction: towards DNA fingerprinting of single cells, *Nucleic Acids Res.* 16: 10053–10971.

665. Jeffreys, A.J., Neumann, R., and Wilson, V., (1990) Repeat unit sequence variation in minisatellites: a novel source of DNA polymorphism for studying variation and mutation by single molecule analysis, *Cell* 60: 473–485.

666. Jeffreys, A.J., Turner, M., and Debenham, P., (1991) The efficiency of multilocus DNA fingerprint probes for the individualization and establishment of family relationships, determined from extensive casework, *Am. J. Hum. Genet,* 48: 824–840.

667. Jeffreys, A.J., MacLeod, A., Tamaki, K., Neil, D.L., and Monckton, D.G., (1991) Minisatellite repeat coding as a digital approach to DNA typing, *Nature* 354: 204–209.

668. Jeffreys, A.J., Tamaki, K., MacLeod, A., Monckton, D.G., Neil, D.L., and Armour, J.L.L., (1994) Complex gene conversion events in germline mutation at human minisatellites, *Nat. Genet.* 6: 136–145.

669. Jhingan, A.K, (1992) A novel technology for DNA isolation, *Meth. Mol. Cell. Biol.* 3: 15–22.

670. Jhingan, A.K., (1992) Efficient procedure for DNA extraction from lyophilized plant material, *Meth. Mol. Cell Biol.* 3: 185–187.

671. Jin, L., Macaubas, C., Hallmayer, J., Kimura, A., and Mignot, E., (1996) Mutation rate varies among alleles at a microsatellite locus: phylogenetic evidence, *Proc. Natl. Acad. Sci. U.S.A.* 93: 15285–15288.

672. Jobes, D.V., Hurley, D.L., and Thien, L.B., (1995) Plant DNA isolation: a method to efficiently remove polyphenolics, polysaccharides, and RNA, *Taxon* 44: 379–386.

673. Joel, D.M., Portnoy, V., and Katzir, N., (1996) Identification of single tiny seeds of *Orobanche* using RAPD analysis, *Plant Mol. Biol. Rep.* 14: 243–248.

674. John, M.E., (1992) An efficient method for isolation of RNA and DNA from plants containing polyphenolics, *Nucleic Acids Res.* 20: 2381.

675. Jones, A.G., and Ardren, W.R., (2003) Methods of parentage analysis in natural populations, *Mol. Ecol.* 12: 2511–2523.

676. Jones, A.G., Stockwell, C.A., Walker, D., and Avise, J.C., (1998) The molecular basis of a microsatellite null allele from the white sands pupfish, *J. Hered.* 89: 339–342.

677. Jones, C.J., Edwards, K.J., Castaglione, S., Winfield, M.O., Sala, F., Van de Wiel, C., Bredemeijer, G., Vosman, B., Matthes, M., Daly, A., Brettschneider, R., Bettini, P., Buiatti, M., Maestri, E., Malcevschi, A., Marmiroli, N., Aert, R., Volckaert, G., Rueda, J., Linacero, R., Vazquez, A., and Karp, A., (1997) Reproducibility testing of RAPD, AFLP and SSR markers in plants by a network of European laboratories, *Mol. Breed.* 3: 381–390.

678. Jonsdottír, I.S., Augner, M., Fagerström, T., Persson, H., and Stenström, A., (2000) Genet age in marginal populations of two clonal *Carex* species in the Siberian Arctic, *Ecography* 23: 402–412.

679. Jost, J.P., and Saluz, H.P., (1993) *DNA Methylation: Molecular Biology and Biological Significance*. Springer, Basel, Switzerland.

680. Jurka, J., and Pethiyagoda, C., (1995) Simple repetitive DNA sequences from primates: compilation and analysis, *J. Mol. Evol.* 40: 120–126.

681. Kalendar, R., Grob, T., Regina, M., Suoniemi, A., and Schulmann, A., (1999) IRAP and REMAP: two new retrotransposon-based DNA fingerprinting techniques, *Theor. Appl. Genet.* 98: 704–711.

682. Kalendar, R., Tanskanen, J., Immonen, S., Nevo, E., and Schulman, A.H., (2000) Genome evolution of wild barley (*Hordeum spontaneum*) by *Bare*-1 retrotransposon dynamics in response to sharp microclimatic divergence, *Proc. Natl. Acad. Sci. U.S.A.* 97: 6603–6607.

683. Kalinowski, S.T., (2002) Evolutionary and statistical properties of three genetic distances, *Mol. Ecol.* 11: 1263–1273.

684. Kanazawa, A., and Tsutsumi, N., (1992) Extraction of restrictable DNA from plants of the genus *Nelumbo*, *Plant Mol. Biol. Rep.* 10: 316–318.

685. Kanazin, V., Marek, L.F., and Shoemaker, R.C., (1996) Resistance gene analogs are conserved and clustered in soybean, *Proc. Natl. Acad. Sci. U.S.A.* 93: 11746–11750.

686. Kanazin, V., Talbert, H., See, D., DeCamp, P., Nevo, E., and Blake, T., (2002) Discovery and assay of single-nucleotide polymorphisms in barley (*Hordeum vulgare*), *Plant Mol. Biol.* 48: 529–537.

687. Kandpal, R.P., Kandpal, G., and Weissman, S.M., (1994) Construction of libraries enriched for sequence repeats and jumping clones, and hybridization selection for region-specific markers, *Proc. Natl. Acad. Sci. U.S.A.* 91: 88–92.

688. Kang, H.W., Cho, Y.G., Yoon, U.H., and Eun, M.Y., (1998) A rapid DNA extraction method for RFLP and PCR analysis from a single dry seed, *Plant Mol. Biol. Rep.* 16: 90 (1–9).

689. Kantety, R.V., Zeng, X., Bennetzen, J.L., and Zehr, B.E. (1995) Assessment of genetic diversity in dent and popcorn (*Zea mays* L.) inbred lines using inter-simple sequence repeat (ISSR) amplification, *Mol. Breed.* 1: 365–373.

690. Kantety, R.V., La Rota, M., Matthews, D.E., and Sorrells, M.E., (2002) Data mining for simple sequence repeats in expressed sequence tags from barley, maize, rice, sorghum and wheat, *Plant Mol. Biol.* 48: 501–510.

691. Karagyozov, L., Kalcheva, I.D., and Chapman, V.M., (1993) Construction of random small-insert genomic libraries highly enriched for simple sequence repeats, *Nucleic Acids Res.* 21: 3911–3912.

692. Karakousis, A., and Langridge, P., (2003) A high-throughput plant DNA extraction method for marker analysis, *Plant Mol. Biol. Rep.* 21: 95a–95f.

693. Kardolus, J.P., Van Eck, H.J., and Van den Berg, R.G., (1998) The potential of AFLPs in biosystematics: a first application in *Solanum* taxonomy (Solanaceae), *Plant Syst. Evol.* 210: 87–103.

694. Karhu, A., Dietrich, J.-H., and Savolainen, O., (2000) Rapid expansion of microsatellite sequences in pines, *Mol. Biol. Evol.* 17: 259–265.

695. Karp, A., (1991) On the current understanding of somaclonal variation, *Oxford Surv. Plant Mol. Cell Biol.* 7: 1–58.

696. Karp, A., Kresovich, S., Bhat, K.V., Ayad, W.G., and Hodgkin, T., (1997) *Molecular Tools in Plant Genetic Resources Conservation: A Guide to the Technologies*. IPGRI Technical Bulletin No. 2. International Plant Genetic Resources Institute, Rome, Italy.

697. Kashi, Y., and Soller, M., (1999) Functional roles of microsatellites and minisatellites. In *Microsatellites: Evolution and Applications*, Goldstein, D.B., and Schlötterer, C., Eds., Oxford University Press, pp. 10–23.

698. Kashi, Y., King, D., and Soller, M., (1997) Simple sequence repeats as a source of quantitative genetic variation, *Trends Genet.* 13: 74–78.

699. Katti, M.V., Ranjekar, P.K., and Gupta, V.S, (2001) Differential distribution of simple sequence repeats in eukaryotic genome sequences, *Mol. Biol. Evol.* 18: 1161–1167.

700. Katzir, N., Danin-Poleg, Y., Tzuri, G., Karchi, Z., Lavi, U., and Cregan, P.B., (1996) Length polymorphism and homologies of microsatellites in several Cucurbitaceae species, *Theor. Appl. Genet.* 93: 1282–1290.

701. Kaukinen, J., and Varvio, S.-L., (1992) Artiodactyl retroposons: association with microsatellites and use in SINEmorph detection by PCR, *Nucleic Acids Res.* 20: 2955–2958.

702. Kaundun, S.S., and Matsumoto, S., (2002) Heterologous nuclear and chloroplast microsatellite amplification and variation in tea, *Camellia sinensis, Genome* 45: 1041–1048.

703. Kawamoto, S., Ohnishi, T., Kita, H., Chisaka, O., and Okubo, K., (1999) Expression profiling by iAFLP: a PCR-based method for genome-wide gene expression profiling, *Genome Res.* 9: 1305–1312.

704. Kawase, M., (1994) Application of the restriction landmark genomic scanning (RLGS) method to rice cultivars as a new fingerprinting technique, *Theor. Appl. Genet.* 89: 861–864.

705. Keb-Llanes, M., González, G., Chi-Manzanero, B., and Infante, D., (2002) A rapid and simple method for small-scale DNA extraction in Agavaceae and other tropical plants, *Plant Mol. Biol. Rep.* 20: 299a–299e.

706. Keim, P., Schupp, J.M., Travis, S.E., Clayton, K., Zhu, T., Shi, L., Ferreria, A., and Webb, D.M., (1997) A high-density soybean genetic map based on AFLP markers, *Crop Sci.* 37: 537–543.

707. Kelchner, S.A., and Wendel, J.F., (1996) Hairpins create minute inversions in non-coding regions of chloroplast DNA, *Curr. Genet.* 30: 259–262.

708. Kennedy, G.C., German, M.S., and Rutter, W.J., (1995) The minisatellite in the diabetes susceptibility locus IDDM2 regulates insulin transcription, *Nat. Genet.* 9: 293–298.

709. Kesseli, R.V., Paran, I., and Michelmore, R.W., (1994) Analysis of a detailed genetic linkage map of *Lactuca sativa* (lettuce) constructed from RFLP and RAPD markers, *Genetics* 136: 1435–1446.

710. Khadari, B., Breton, C., Moutier, N., Roger, J.P., Besnard, G., Bervillé, A., and Dosba, F., (2003) The use of molecular markers for germplasm management in a French olive collection, *Theor. Appl. Genet.* 106: 521–529.

711. Khanuja, S.P.S., Shasany, A.K., Darokar, M.P., and Kumar, S., (1999) Rapid isolation of DNA from dry and fresh samples of plants producing large amounts of secondary metabolites and essential oils, *Plant Mol. Biol. Rep.* 17: 74 (1–7).

712. Kidwell, M.G., and Lisch, D.R., (2000) Transposable elements and host genome evolution, *Trends Ecol. Evol.* 15: 95–99.

713. Kijas, J.M.H., Fowler, J.C.S., Garbett, C.A., and Thomas, M.R., (1994) Enrichment of microsatellites from the *Citrus* genome using biotinylated oligonucleotide sequences bound to streptavidin-coated magnetic particles, *BioTechniques* 16: 658–662.

714. Kijas, J.M.H., Fowler, J.C.S., and Thomas, M.R., (1995) An evaluation of sequence tagged microsatellite site markers for genetic analysis within *Citrus* and related species, *Genome* 38: 349–355.

715. Kijas, J.M.H., Thomas, M.R., Fowler, J.C.S., and Roose, M.L., (1997) Integration of trinucleotide microsatellites into a linkage map of *Citrus, Theor. Appl. Genet.* 94: 701–706.

716. Kim, C.S., Lee, C.H., Shin, J.S., Chung, Y.S., and Hyung, N.I., (1997) A simple and rapid method for isolation of high quality genomic DNA from fruit trees and conifers using PVP, *Nucleic Acids Res.* 25: 1085–1086.

717. Kim, K.J., and Mabry, T.J., (1991) Phylogenetic and evolutionary implications of nuclear ribosomal DNA variation in dwarf dandelions (*Krigia*, Lactuceae, Asteraceae), *Plant Syst. Evol.* 177: 53–69.

718. Kimpton, C.P., Gill, P., Walton, A., Urquhart, A., Millican, E.S., and Adams, M., (1993) Automated DNA profiling employing multiplex amplification of short tandem repeat loci, *PCR Methods Appl.* 3: 13–22.

719. King, R.A., and Ferris, C., (2000) Chloroplast DNA and nuclear DNA variation in the sympatric alder species, *Alnus cordata* (Lois.) Duby and *A. glutinosa* (L.) Gaertn, *Biol. J. Linn. Soc.* 70: 147–160.

720. King, R.A., and Ferris, C., (2002) A variable minisatellite sequence in the chloroplast genome of *Sorbus* L. (Rosaceae: Maloideae), *Genome* 45: 570–576.

721. Kishore, V.K., Velasco, P., Shintani, S.K., Rowe, J., Rosato, C., Adair, N., Slabaugh, M.B., and Knapp, S.J., (2004) Conserved simple sequence repeats for the Limnanthaceae (Brassicales), *Theor. Appl. Genet.* 108: 450–457.

722. Klein, P.E., Klein, R.R., Cartinhour, S.W., Ulanch, P.E., Dong, J., Obert, J.A., Morishige, D.T., Schlueter, S.D., Childs, K.L., Ale, M., and Mullet, J.E., (2000) A high-throughput AFLP-based method for constructing integrated genetic and physical maps: progress toward a sorghum genome map, *Genome Res.* 10: 789–807.

723. Klimyuk, V.I., Carroll, B.J., Thomas, C.M., and Jones, J.D.G., (1993) Alkali treatment for rapid preparation of plant material for reliable PCR analysis, *Plant J.* 3: 493–494.

724. Klinkicht, M., and Tautz, D., (1992) Detection of simple sequence length polymorphisms by silver staining, *Mol. Ecol.* 1: 133–134.

725. Knapp, J.E., and Chandlee, J.M., (1996) RNA/DNA mini-prep from a single sample of orchid tissue, *BioTechniques* 21: 54–56.

726. Knapp, S.J., Bridges, W.C., and Birkes, D. (1990) Mapping quantitative trait loci using molecular marker linkage maps, *Theor. Appl. Genet.* 79: 585–592.

727. Knoop, V., Unseld, M., Marienfeld, J., Brandt, P., Sünkel, S., Ullrich, H., and Brennicke, A., (1996) *copia*-, *gypsy*- and LINE-like retrotransposon fragments in the mitochondrial genome of *Arabidopsis thaliana*, *Genetics* 142: 579–585.

728. Knowles, L.L., and Maddison, W.P., (2002) Statistical phylogeography, *Mol. Ecol.* 11: 2623–2635.

729. Knox, M.R., and Ellis, T.H.N., (2001) Stability and inheritance of methylation states at *Pst*I sites in *Pisum*, *Mol. Genet. Genomics* 265: 497–507.

730. Koblizkova, A., Dolezel, J., and Macas, J., (1998) Subtraction with 3′-modified oligonucleotides eliminates amplification artifacts in DNA libraries enriched for microsatellites, *BioTechniques* 25: 32–38.

731. Koebner, R.M.D., (1995) Predigestion of DNA template improves the level of polymorphism of random amplified polymorphic DNAs in wheat, *Genet. Anal. Biomol. Eng.* 12: 63–67.

732. Kokoska, R.J., Stefanovic, L., Tran, H.T., Resnick, M.A., Gordenin, D.A., and Petes, T.D., (1998) Destabilization of yeast micro- and minisatellite DNA sequences by mutations affecting a nuclease involved in Okazaki fragment processing (*rad27*) and DNA polymerase delta (*pol3t*), *Mol. Cell. Biol.* 18: 2779–2788.

733. Kölliker, R., Jones, E.S., Jahufer, M.Z.Z., and Forster, J.W., (2001) Bulked AFLP analysis for the assessment of genetic diversity in white clover (*Trifolium repens* L.), *Euphytica* 121: 305–315.

734. Kolchinsky, A., Kolesnikova, M., and Ananiev, E., (1991) "Portraying" of plant genomes using polymerase chain reaction amplification of ribosomal 5S genes, *Genome* 34: 1028–1031.

735. Koller, B., Lehmann, B., McDermott, J.M., and Gessler, C. (1993) Identification of apple cultivars using RAPD markers, *Theor. Appl. Genet.* 85: 901–904.

736. Komarnitsky, I.K., Samoylov, A.M., Redko, V.V., Peretyayko, V.G., and Gleba, Y.Y., (1990) Interspecific diversity of sugar beet (*Beta vulgaris*) mitochondrial DNA, *Theor. Appl. Genet.* 80: 253–257.

737. Kominami, R., Mitani, K., and Muramatsu, M., (1988) Nucleotide sequence of a mouse minisatellite DNA, *Nucleic Acids Res.* 16: 1197.

738. Konieczny, A., and Ausubel, F.M., (1993) A procedure for mapping *Arabidopsis* *mutations* using co-dominant ecotype-specific PCR-based markers, *Plant J.* 4: 403–410.

739. Konieczny, A., Voytas, D.F., Cummings, M.P., and Ausubel, F.M., (1991) A superfamily of *Arabidopsis thaliana* retrotransposons, *Genetics* 127: 801–809.

740. Koopman, W.J.M., Zevenbergen, M.J., and Van den Berg, R.G., (2001) Species relationships in *Lactuca s.l.* (Lactuceae, Asteraceae) inferred from AFLP fingerprints, *Am. J. Bot.* 88: 1881–1887.

741. Kopperud, C., and Einset, J.W., (1995) DNA isolation from *Begonia* leaves, *Plant Mol. Biol. Rep.* 13: 129–130.

742. Korpelainen, H., and Virtanen, V., (2003) DNA fingerprinting of mosses, *J. Forensic Sci.* 48: 804–807.

743. Kostia, S., Varvio, S.-L., Vakkari, P., and Pulkkinen, P., (1995) Microsatellite sequences in a conifer, *Pinus sylvestris*, *Genome* 38: 1244–1248.

744. Kovárova, M., and Dráber, P., (2000) New specificity and yield enhancer of polymerase chain reactions, *Nucleic Acids Res.* 28: e70.

745. Kraft, T., and Säll, T., (1999) An evaluation of the use of pooled samples in studies of genetic variation, *Heredity* 82: 488–494.

746. Kraft, T., Nybom, H., and Werlemark, G., (1995) *Rubus vestervicensis* (Rosaceae) — its hybrid origin revealed by DNA fingerprinting, *Nord. J. Bot.* 15: 237–242.

747. Kraft, T., Hansen, M., and Nilsson, N.O., (2000) Linkage disequilibrium and fingerprinting in sugar beet, *Theor. Appl. Genet.* 101: 323–326.

748. Krauss, S.L., (2000) Patterns of mating in *Persoonia mollis* (Proteaceae) revealed by an analysis of paternity using AFLP: implications for conservation, *Aust. J. Bot.* 48: 349–356.

749. Kreike, J., (1990) Genetic analysis of forest tree populations: isolation of DNA from spruce and fir apices, *Plant Mol. Biol.* 14: 877–879.

750. Kresovich, S., Szewc-McFadden, A.K., Bliek, S.M., and McFerson, J.R., (1995) Abundance and characterization of simple-sequence repeats (SSRs) isolated from a size-fractionated genomic library of *Brassica napus* L. (rapeseed), *Theor. Appl. Genet.* 91: 206–211.

751. Krishna, T.G., and Jawali, N., (1997) DNA isolation from single or half seeds suitable for random amplified polymorphic DNA analyses, *Anal. Biochem.* 250: 125–127.

752. Kristensen, V.N., and Børresen-Dale, A.-L., (1997) Improved electrophoretic separation of polymorphic short tandem repeats in agarose gels using bis-benzimide, *BioTechniques* 23: 634–636.

753. Kruglyak, S., Durrett, R.T., Schug, M., and Aquadro, C.F., (1998) Equilibrium distributions of microsatellite repeat length resulting from a balance between slippage events and point mutations, *Proc. Natl. Acad. Sci. U.S.A.* 95: 10774–10778.

754. Kubelik, A.R., and Szabo, L.J., (1995) High-GC primers are useful in RAPD analysis of fungi, *Curr. Genet.* 28: 384–389.

755. Kubis, S.E., Heslop-Harrison, J.S., Desel, C., and Schmidt, T., (1998) The genomic organization of non-LTR retrotransposons (LINEs) from three *Beta* species and five other angiosperms, *Plant Mol. Biol.* 36: 821–831.

756. Kubo, T., Nishizawa, S., Sugawara, A., Itchoda, N., Estiati, A., and Mikami, T., (2000) The complete nucleotide sequence of the mitochondrial genome of sugar beet (*Beta vulgaris* L.) reveals a novel gene for tRNA^{Cys} (GCA), *Nucleic Acids Res.* 28: 2571–2576.

757. Kuhn, D.N., Heath, M., Wisser, R.J., Meerow, A., Brown, J.S., Lopes, U., and Schnell, R.J., (2003) Resistance gene homologues in *Theobroma cacao* as useful genetic markers, *Theor. Appl. Genet.* 107: 191–202.

758. Kuleung, C., Baenziger, P.S., and Dweikat, I., (2004) Transferability of SSR markers among wheat, rye, and triticale, *Theor. Appl. Genet.* 108: 1147–1150.

759. Kumar, A., and Bennetzen, J.L., (1999) Plant retrotransposons, *Annu. Rev. Genet.* 33, 479–532.

760. Kumar, A., Pushpangadan, P., and Mehrotra, S., (2003) Extraction of high-molecular-weight DNA from dry root tissue of *Berberis lycium* suitable for RAPD, *Plant Mol. Biol. Rep.* 21: 309a–309d.

761. Kutil, B.L., and Williams, C.G., (2001) Triplet-repeat microsatellites shared among hard and soft pines, *J. Hered.* 92: 327–332.

762. Kuzoff, R.K., and Gasser, C.S., (2000) Recent progress in reconstructing angiosperm phylogeny, *Trends Plant Sci.* 5: 330–336.

763. Kuzoff, R.K., Sweere, J.A., Soltis, D.E., Soltis, P.S., and Zimmer, E.A., (1998) The phylogenetic potential of entire 26S rDNA sequences in plants, *Mol. Biol. Evol.* 15: 251–263.

764. Kwok, P.-Y., (2001) Methods for genotyping single nucleotide polymorphisms, *Annu. Rev. Genomics Hum. Genet.* 2: 235–258.

765. La Claire, J.W., II, and Herrin, D.L., (1997) Co-isolation of high-quality DNA and RNA from coenocytic green algae, *Plant Mol. Biol. Rep.* 15: 263–272.

766. Labarca, C., and Paigen, K., (1980) A simple, rapid and sensitive DNA assay procedure, *Anal. Biochem.* 102: 344–352.

767. Labate, J.A., (2000) Software for population genetic analyses of molecular marker data, *Crop Sci.* 40: 1521–1528.

768. Lacape, J.-M., Nguyen, T.-B., Thibivilliers, S., Bojinov, B., Courtois, B., Cantrell, R.G., Burr, B., and Hau, B. (2003) A combined RFLP-SSR-AFLP map of tetraploid cotton based on a *Gossypium hirsutum* x *Gossypium barbadense* backcross population, *Genome* 46: 612–626.

769. Lagercrantz, U., Ellegren, H., and Andersson, L., (1993) The abundance of various polymorphic microsatellite motifs differs between plants and vertebrates, *Nucleic Acids Res.* 21: 1111–1115.

770. Lan, R., and Reeves, P.R., (2000) Unique adaptor design for AFLP fingerprinting, *BioTechniques* 29: 745–750.

771. Landegren, U., Nilsson, M., and Kwok, P.Y., (1998) Reading bits of genetic information: methods for single-nucleotide polymorphism analysis, *Genome Res.* 8: 769–776.

772. Lander, E.S., and Botstein D., (1989) Mapping Mendelian factors underlying quantitative traits using RFLP linkage maps, *Genetics* 121: 185–199.

773. Lander E.S., Green P., Abrahamson J., Barlow A., Daly M.J., Lincoln S.E., Newburg L. (1987) MAPMAKER: an interactive computer package for constructing primary genetic linkage maps of experimental and natural populations. *Genomics* 1: 174–181.

774. Lange, D.A., Penuela, S., Denny, R.L., Mudge, J., Concibido, V.C., Orf, J.H., and Young, N.D., (1998) A plant DNA isolation protocol suitable for polymerase chain reaction based marker-assisted breeding, *Crop Sci.* 38: 217–220.

775. Langridge, U., Schwall, M., and Langridge, P., (1991) Squashes of plant tissue as substrate for PCR, *Nucleic Acids Res.* 19: 6954.

776. Lanham, P.G., Fennell, S., Moss, J.P., and Powell, W. (1992) Detection of polymorphic loci in *Arachis* germplasm using random amplified polymorphic DNAs, *Genome* 35: 885–889.

777. Lannér, C., Bryngelsson, T., and Gustafsson, M., (1996) Genetic validity of RAPD markers at the intra- and interspecific level in wild *Brassica* species with n=9, *Theor. Appl. Genet.* 93: 9–14.

778. Lannér-Herrera, C., Gustafsson, M., Fält, A.-S., and Bryngelsson, T., (1996) Diversity in natural populations of wild *Brassica oleracea* as estimated by isozyme and RAPD analysis, *Genet. Res. Crop Evol.* 43: 13–23.

779. Larkin, P.D., and Park, W.D., (2003) Association of *waxy* gene single nucleotide polymorphisms with starch characteristics in rice (*Oryza sativa* L.), *Mol. Breed.* 12: 335–339.

780. Larkin, P.J., Banks, P.M., Bhati, R., Brettell, R.I.S., Davies, P.A., Ryan, S.A., Scowcroft, W.R., Spindler, L.H., and Tanner, G.J., (1989) From somatic variation to variant plants: mechanisms and applications, *Genome* 31: 705–711.

781. Lassner, M.W., Peterson, P., and Yoder, J.I., (1989) Simultaneous amplification of multiple DNA fragments by polymerase chain reaction in the analysis of transgenic plants and their progeny, *Plant Mol. Biol. Rep.* 7: 116–128.

782. Law, J.R., Reeves, J.C., Jackson, J., Donini, P., Matthews, D., Smith, J.S.C., and Cooke, R.J., (2001) Most similar variety comparisons — a grouping tool for use in distinctness, uniformity and stability (DUS) testing, *Acta Hortic.* 546: 96–100.

783. Lazzaro, B.P., Sceurman, B.K., Carney, S.L., and Clark, A.G., (2002) fRFLP and fAFLP: medium-throughput genotyping by fluorescently post-labeling restriction digestion, *BioTechniques* 33: 539–546.

784. Le Clerc, V., Briard, M., and Peltier, D., (2001) Evaluation of carrot genetic substructure: comparison of the efficiency of mapped molecular markers with randomly chosen markers, *Acta Hortic.* 546: 127–134.

785. Le Corre, V., Dumolin-Lapègue, S., and Kremer, A., (1997) Genetic variation at allozyme and RAPD loci in sessile oak *Quercus petraea* (Matt.) Liebl.: the role of history and geography, *Mol. Ecol.* 6: 519–529.

786. Le Thierry D'Ennequin, M.L., Panaud, O., Robert, T., and Ricroch, A. (1997) Assessment of genetic relationships among sexual and asexual forms of *Allium cepa* using morphological traits and RAPD markers, *Heredity* 78: 403–409.

787. Le, Q.H., Wright, S., Yu, Z., and Bureau, T. (2000) Transposon diversity in *Arabidopsis thaliana*, *Proc. Natl. Acad. Sci. U.S.A.* 97: 7376–7381.

788. Leamon, J.H., Moiseff, A., and Crivello, J.F., (2000) Development of a high-throughput process for detection and screening of genetic polymorphisms, *BioTechniques* 28: 994–1005.

789. Lee, D., Reeves, J.C., and Cooke, R.J., (1996) DNA profiling and plant variety registration: 1. The use of random amplified DNA polymorphisms to discriminate between varieties of oilseed rape, *Electrophoresis* 17: 261–265.

790. Lee, J.M., Grant, D., Vallejos, C.E., and Shoemaker R.C., (2001) Genome organization in dicots. II. *Arabidopsis* as a "bridging species" to resolve genome evolution events among legumes, *Theor. Appl. Genet.* 103: 765–773.

791. Lee, M., (1995) DNA markers and plant breeding programs, *Adv. Agronomy* 55: 265–344.

792. Lee, M., and Phillips, R.L., (1988) The chromosomal basis of somaclonal variation, *Annu. Rev. Plant Physiol. Plant Mol. Biol.* 39: 413–437.

793. Lee, M., and Nicholson, P., (1997) Isolation of genomic DNA from plant tissues, *Nat. Biotechnol.* 15: 805–806.

794. Leister, D., Ballvora, A., Salamini, F., and Gebhardt, C., (1996) A PCR-based approach for isolating pathogen resistance genes from potato with potential for wide application in plants, *Nat. Genet.* 14: 421–429.

795. Lemieux, B., (2001) Plant genotyping based on analysis of single nucleotide polymorphisms using microarrays. In *Plant Genotyping: The DNA Fingerprinting of Plants*, Henry, R. J., Ed., CABI Publishing, Wallingford, U.K., pp. 47–57.

796. Lenoir, A., Cournoyer, B., Warwick, S., Picard, G., and Deragon, J.-M., (1997) Evolution of SINE S1 retroposons in Cruciferae plant species, *Mol. Biol. Evol.* 14: 934–941.

797. Leonard, J.A., Wayne, R.K., and Cooper, A., (2000) Population genetics of Ice Age brown bears, *Proc. Natl. Acad. Sci. U.S.A.* 97: 1651–1654.

798. Lerceteau, E., Robert, T., Pétiard, V., and Crouzillat, D., (1997) Evaluation of the extent of genetic variability among *Theobroma cacao* accessions using RAPD and RFLP markers, *Theor. Appl. Genet.* 95: 10–19.

799. Leroy, X.J., and Leon, K., (2000) A rapid method for detection of plant genomic instability using unanchored-microsatellite primers, *Plant Mol. Biol. Rep.* 18: 283a–283g.

800. Leroy, X.J., Leon, K., Charles, G., and Branchard, M., (2000) Cauliflower somatic embryogenesis and analysis of regenerant stability by ISSRs, *Plant Cell Rep.* 19: 1102–1107.

801. Leroy, X.J., Leon, K., Hily, J.M., Chaumeil, P., and Branchard, M., (2001) Detection of *in vitro* culture-induced instability through inter-simple sequence repeat analysis, *Theor. Appl. Genet.* 102: 885–891.

802. Lessa, E.P., and Applebaum, G., (1993) Screening techniques for detecting allelic variation in DNA sequences, *Mol. Ecol.* 2: 119–129.

803. Levi, A., Rowland, L.J., and Hartung, J.S., (1993) Production of reliable randomly amplified polymorphic DNA (RAPD) markers from DNA of woody plants, *HortScience* 28: 1188–1190.

804. Levi, A., Thomas, C.E., Wehner, T., and Zhang, X., (2001) Low genetic diversity indicates the need to broaden the genetic base of cultivated watermelon, *HortScience* 36: 1096–1101.

805. Levin, I., and Gilboa, N., (1997) Direct PCR using tomato pollen grain suspensions, *BioTechniques* 23: 986–990.

806. Levinson, G., and Gutman, G.A., (1987) Slipped-strand mispairing: a major mechanism for DNA sequence evolution, *Mol. Biol. Evol.* 4: 203–221.

807. Li, H., Luo, J., Hemphill, J.K., Wang, J.-T., and Gould, J.H., (2001) A rapid and high yielding DNA miniprep for cotton (*Gossypium* spp.), *Plant Mol. Biol. Rep.* 19: 183a–183e.

808. Li, J.Z., Sjakste, T.G., Röder, M.S., and Ganal, M.W., (2003) Development and genetic mapping of 127 new microsatellite markers in barley, *Theor. Appl. Genet.* 107: 1021–1027.

809. Li, Q.-B., Cai, Q., and Guy, C.L., (1994) A DNA extraction method for RAPD analysis from plants rich in soluble polysaccharides, *Plant Mol. Biol. Rep.* 12: 215–220.

810. Li, Y.C., Fahima, T., Peng, J.H., Röder, M.S., Kirzhner, V.M., Beiles, A., Korol, A.B., and Nevo, E., (2000) Edaphic microsatellite DNA divergence in wild emmer wheat, *Triticum dicoccoides*, at a microsite: Tabigha, Israel, *Theor. Appl. Genet.* 101: 1029–1038.

811. Li, Y.C., Korol, A.B., Fahima, T., Beiles, A., and Nevo, E., (2002) Microsatellites: genomic distribution, putative functions and mutational mechanisms: a review, *Mol. Ecol.* 11: 2453–2465.

812. Li, Y.C., Röder, M.S., Fahima, T., Kirzhner, V.M., Beiles, A., Korol, A.B., and Nevo, E., (2002) Climatic effects on microsatellite diversity in wild emmer wheat (*Triticum dicoccoides*) at the Yehudiya microsite, Israel, *Heredity* 89: 127–132.

813. Li, Y.X., Su, Z.X., and Chen, F., (2002) Rapid extraction of genomic DNA from leaves and bracts of dove tree (*Davidia involucrata*), *Plant Mol. Biol. Rep.* 20: 185a–185e.

814. Lian, C., and Hogetsu, T., (2002) Development of microsatellite markers in black locust (*Robinia pseudoacacia*) using a dual-suppression-PCR technique, *Mol. Ecol. Notes* 2: 211–213.

815. Lian, C., Zhou, Z., and Hogetsu, T., (2001) A simple method for developing microsatellite markers using amplified fragments of inter-simple sequence repeat (ISSR), *J. Plant Res.* 114: 381–385.

816. Liang, P., (2002) A decade of differential display, *BioTechniques* 33: 338–346.

817. Liang, P., and Pardee, A.B., (1992) Differential display of eukaryotic messenger RNA by means of the polymerase chain reaction, *Science* 257: 967–971.

818. Lim, S.H., Looi, L.K.C., Ong, B.L., and Wee, Y.C., (1997) A method of DNA isolation from epiphytic CAM ferns for use in random amplified polymorphic DNA analysis, *Biol. Plant.* 39: 637–639.

819. Lin, J.-J., Kuo, J., and Ma, J., (1996) A PCR-based DNA fingerprinting technique: AFLP for molecular typing of bacteria, *Nucleic Acids Res.* 24: 3649–3650.

820. Lin, J.-J., Ma, J., and Kuo, J., (1999) Chemiluminescent detection of AFLP markers, *BioTechniques* 26: 344–348.

821. Lin, J.-J., Fleming, R., Kuo, J., Matthews, B.F., and Saunders, J.A., (2000) Detection of plant genes using a rapid, nonorganic DNA purification method, *BioTechniques* 28: 346–350.

822. Lin, J.-Z., and Ritland, K., (1995) Flower petals allow simpler and better isolation of DNA for plant RAPD analyses, *Plant Mol. Biol. Rep.* 13: 210–213.

823. Lin, R.-C., Ding, Z.-S., Li, L.-B., and Kuang, T.-Y., (2001) A rapid and efficient DNA minipreparation suitable for screening transgenic plants, *Plant Mol. Biol. Rep.* 19: 379a–379e.

824. Lindahl, T., (1997) Facts and artifacts of ancient DNA, *Cell* 90: 1–3.

825. Linder, C.R., Moore, L.A., and Jackson, R.B., (2000) A universal molecular method for identifying underground plant parts to species, *Mol. Ecol.* 9: 1549–1559.

826. Linz, U., (1990) Thermocycler temperature variation invalidates PCR results, *BioTechniques* 9: 286–292.

827. Lira, C.F., Cardoso, S.R.S., Ferreira, P.C.G., Cardoso, M.A., and Provan, J., (2003) Long-term population isolation in the endangered tropical tree species *Caesalpinia echinata* Lam. revealed by chloroplast microsatellites, *Mol. Ecol.* 12: 3219–3225.

828. Liston, A., Rieseberg, L.H., Adams, R.P., Do, N., and Zhu, G., (1990) A method for collecting dried plant specimens for DNA and isozyme analyses, and the results of a field experiment in Xinjiang, China, *Ann. Missouri Bot. Gard.* 77: 859–863.

829. Litt, M., and Luty, J.A., (1989) A hypervariable microsatellite revealed by *in vitro* amplification of a dinucleotide repeat within the cardiac muscle actin gene, *Am. J. Hum. Genet.* 44: 397–401.

830. Litt, M., Hauge, X., and Sharma, V., (1993) Shadow bands seen when typing poly-morphic dinucleotide repeats: some causes and cures, *BioTechniques* 15: 280–284.

831. Liu, B., and Wendel, J.F., (2001) Intersimple sequence repeat (ISSR) polymorphisms as a genetic marker system in cotton, *Mol. Ecol. Notes* 1: 205–208.

832. Liu, D., and Wu, R., (1999) Protection of megabase-sized chromosomal DNA from breakage by DNase activity in plant nuclei, *BioTechniques* 26: 258–261.

833. Liu, G., and Quiros, C.F., (2001) Sequence-related amplified polymorphism (SRAP), a new marker system based on a simple PCR reaction: its application to mapping and gene tagging in *Brassica*, *Theor. Appl. Genet.* 103: 455–461.

834. Liu, Y.-G., and Whittier, R.F., (1994) Rapid preparation of megabase plant DNA from nuclei in agarose plugs and microbeads, *Nucleic Acids Res.* 22: 2168–2169.

835. Liu, Y.G., Mitsukawa, N., and Whittier, R.F., (1993) Rapid sequencing of unpurified PCR products by thermal asymmetric PCR cycle sequencing using unlabeled sequencing primers, *Nucleic Acids Res.* 21: 3333–3334.

836. Liu, Z.-W., Biyashev, R.M., and Saghai-Maroof, M.A., (1996) Development of simple sequence repeat DNA markers and their integration into a barley linkage map, *Theor. Appl. Genet.* 93: 869–876.

837. Lodhi, M.A., Ye, G.-N., Weeden, N.F., and Reisch, B.I., (1994) A simple and efficient method for DNA extraction from grapevine cultivars and *Vitis* species. *Plant Mol. Biol. Rep.* 12: 6–13.

838. Loh, J.P., Kiew, R., Hay, A., Kee, A., Gan, L.H., and Gan, Y.-Y., (2000) Intergeneric and interspecific relationships in Araceae tribe Caladiae and development of molecular mark-ers using amplified fragment length polymorphisms (AFLP), *Ann. Bot.* 85: 371–378.

839. Loh, J.P., Kiew, R., Set, O., Gan, L.H., and Gan, Y.-Y., (2000) A study of genetic variation and relationships within the bamboo subtribe Bambusinae using amplified fragment length polymorphism, *Ann. Bot.* 85: 607–612.

840. Lombard, V., Dubreuil, P., Dillmann, C., and Varil, C., (2001) Genetic distance estimators based on the molecular data for plant registration and protection: a review, *Acta Hortic.* 546: 55–63.

841. Loockerman, D.J., and Jansen, R.K., (1996) The use of herbarium material for DNA studies. In *Sampling the Green World*, Stuessy, T.F., and Sohmer, S.H., Eds., Columbia University Press, New York, pp. 205–220.

842. Loomis, M.D., (1974) Overcoming problems of phenolics and quinines in the isola-tion of plant enzymes and organelles, *Meth. Enzymol.* 31: 528–544.

843. Lorenz, M., Weihe, A., and Börner, T., (1994) DNA fragments of organellar origin in random amplified polymorphic DNA (RAPD) patterns of sugar beet (*Beta vulgaris* L.), *Theor. Appl. Genet.* 88: 775–779.

844. Lorenz, M., Weihe, A., and Börner, T., (1997) Cloning and sequencing of RAPD fragments amplified from mitochondrial DNA of male-sterile and male-fertile cyto-plasm of sugar beet (*Beta vulgaris*), *Theor. Appl. Genet.* 94: 273–278.

845. Loridon, K., and Saumitou-Laprade, P., (2002) Detection of gene-anchored amplifi-cation polymorphism (GAAP) in the vicinity of plant mitochondrial genes, *Mol. Genet. Genomics* 267: 329–337.

846. Loveless, M.D., and Hamrick, J.L., (1984) Determinants of genetic structure in plant populations, *Ann. Rev. Ecol. Syst.* 15: 65–95.

847. Lowe, A.J., Jones, A.J., Raybould, A.E., Trick, M., Moule, C.L., and Edwards, K.J., (2002) Transferability and genome specificity of a new set of microsatellite primers among *Brassica* species of the U triangle, *Mol. Ecol. Notes* 2: 7–11.

848. Lowe, A.J., Jourde, B., Breyne, P., Coplaert, N., Navarro, C., Wilson, J., and Cavers, S., (2003) Fine-scale genetic structure and gene flow within Costa Rican populations of mahogany (*Swietenia macrophylla*), *Heredity* 90: 268–275.

849. Lowe, A.J., Thorpe, W., Teale, A., and Hanson, J. (2003) Characterisation of germplasm accessions of napier grass (*Pennisetum purpureum* and *P. purpureum x P. glaucum* hybrids) and comparison with farm clones using RAPD, *Genet. Res. Crop Evol.* 50: 121–132.

850. Lowe, A.J., Harris, S., and Ashton, P., (2004) *Ecological Genetics: Design, Analysis and Application,* Blackwell Publishing, Malden, Mass.

851. Lowe, A.J., Moule, C., Trick, M., and Edwards, K.J., (2004) Efficient large-scale development of microsatellites for marker and mapping applications in *Brassica* crop species, *Theor. Appl. Genet.* 108: 1103–1112.

852. Lu, J., Knox, M.R., Ambrose, M.J., Brown, J.K.M., and Ellis, T.H.N., (1996) Comparative analysis of genetic diversity in pea assessed by RFLP- and PCR-based methods, *Theor. Appl. Genet.* 93: 1103–1111.

853. Luby, J.J., and Shaw, D.V., (2001) Does marker-assisted breeding make dollars and sense in a fruit breeding program?, *HortScience* 36: 872–879.

854. Luikart, G., and England, P.R., (1999) Statistical analysis of microsatellite data, *Trends Ecol. Evol.* 14: 253–256.

855. Lunt, D.H., Whipple, L.E., and Hyman, B.C., (1998) Mitochondrial DNA variable number tandem repeats (VNTRs): utility and problems in molecular ecology, *Mol. Ecol.* 7: 1441–1455.

856. Lunt, D.H., Hutchinson, W.F., and Carvalho, G.R., (1999) An efficient method for PCR-based isolation of microsatellite arrays, *Mol. Ecol.* 8: 891–894.

857. Luo, G., Hepburn, A.G., and Widholm, J.M., (1992) Preparation of plant DNA for PCR analysis: a fast, general and reliable procedure, *Plant Mol. Biol. Rep.* 10: 319–323.

858. Luro, F., and Laigret, F., (1995) Preparation of high molecular weight genomic DNA from nuclei of woody plants, *BioTechniques* 19: 388–392.

859. Luro, F., Rist, D., and Ollitraut, P., (2001) Evaluation of genetic relationships in *Citrus* genus by means of sequence tagged microsatellites, *Acta Hortic.* 546: 237–242.

860. Lyall, J.E.W., Brown, G.M., Furlong, R.A., Ferguson-Smith, M.A., and Affara, N.A., (1993) A method for creating chromosome-specific plasmid libraries enriched in clones containing $(CA)_n$ microsatellite repeat sequences directly from flow-sorted chromosomes, *Nucleic Acids Res.* 21: 4641–4642.

861. Lynch, M., (1990) The similarity index and DNA fingerprinting, *Mol. Biol. Evol.* 7: 478–484.

862. Lynch, M., (1991) Analysis of population genetic structure by DNA fingerprinting, in *DNA Fingerprinting: Approaches and Applications,* Burke, T., Dolf, G., Jeffreys, A.J., and Wolff, R. Eds., Birkhäuser, Basel, Switzerland, pp. 113–126.

863. Lynch, M., and Milligan, B.G., (1994) Analysis of population genetic structure with RAPD markers, *Mol. Ecol,* 3: 91–99.

864. Ma, R., Yli-Mattila, T., and Pulli, S., (2004) Phylogenetic relationships among genotypes of worldwide collection of spring and winter ryes (*Secale cereale* L.) determined by RAPD-PCR markers, *Hereditas* 140: 210–221.

865. Macaulay, M., Ramsay, L., Powell, W., and Waugh, R., (2001) A representative, highly informative "genotyping set" of barley SSRs, *Theor. Appl. Genet.* 102: 801–809.

866. Mace, E.S., Buhariwalla, H.K., and Crouch, J.H., (2003) A high-throughput DNA extraction protocol for tropical molecular breeding programs, *Plant Mol. Biol. Rep.* 21: 459a–459h.

867. MacPherson, J.M., Eckstein, P.E., Scoles, G.J., and Gajadhar, A.A., (1993) Variability of the random amplified polymorphic DNA assay among thermal cyclers, and effects of primer and DNA concentration, *Mol. Cell. Probes* 7: 293–299.

868. Maguire, T.L., and Sedgley, M., (1997) Genetic diversity in *Banksia* and *Dryandra* (Proteaceae) with emphasis on *Banksia cuneata*, a rare and endangered species, *Heredity* 79: 394–401.

869. Maguire, T.L., Collins, G.G., and Sedgley, M., (1994) A modified CTAB DNA extraction procedure for plants belonging to the family Proteaceae, *Plant Mol. Biol. Rep.* 12: 106–109.

870. Maideliza, T., and Okada, H., (2004) Evidence of reduction of gene flow between two cytotypes of *Ranunculus silerifolius* Lév. (Ranunculaceae) revealed with allozyme and intersimple sequence repeat polymorphisms, *Plant Species Biol.* 19: 23–31.

871. Maki, M., Horie, S., and Yokoyama, J., (2002) Comparison of genetic diversity between narrowly endemic shrub *Menziesia goyozanensis* and its widespread congener *M. pentandra* (Ericaceae), *Conserv. Genet.* 3: 421–425.

872. Manel, S., Schwartz, M.K., Luikart, G., and Taberlet, P., (2003) Landscape genetics: combining landscape ecology and population genetics, *Trends Ecol. Evol.* 18: 189–197.

873. Manly, B.F.J., (1992) *Multivariate Statistical Methods: A Primer*, Chapman and Hall, London.

874. Manninen, I., and Schulman, A.H., (1993) BARE-1, a *copia*-like retroelement in barley (*Hordeum vulgare*), *Plant Mol. Biol.* 22: 829–864.

875. Manning, K., (1991) Isolation of nucleic acids from plants by differential solvent precipitation, *Anal. Biochem.* 195: 45–50.

876. Mansfield, E.S., Vainer, M., Enad, S., Barker, D.L., Harris, D., Rappaport, E., and Fortina, P., (1996) Sensitivity, reproducibility, and accuracy in short tandem repeat genotyping using capillary array electrophoresis, *Genome Res.* 6: 893–903.

877. Mantel, N., (1967) The detection of disease clustering and a generalized regression approach, *Cancer Res.* 27: 209–220.

878. Mantripragada, K.K., Buckley, P.G., de Stahl, T.D., and Dumanski, J.P., (2004) Genomic microarrays in the spotlight, *Trends Genet.* 20: 87–94.

879. Maréchal-Drouard, L., and Guillemaut, P., (1995) A powerful but simple technique to prepare polysaccharide-free DNA quickly and without phenol extraction, *Plant Mol. Biol. Rep.* 13: 26–30.

880. Mariac, C., Trouslot, P., Poteaux, C., Bezancon, G., and Renno, J.-F., (2000) Chloroplast DNA extraction from herbaceous and woody plants for direct restriction fragment length polymorphism analysis, *BioTechniques* 28: 110–113.

881. Marienfeld, J., Unseld, M., and Brennicke, A., (1999) The mitochondrial genome of *Arabidopsis* is composed of both native and immigrant information, *Trends Plant Sci.* 4: 495–502.

882. Mariette, S., Chagné, D., Lézier, C., Pastuszka, P., Raffin, A., Plomion, C., and Kremer, A., (2001) Genetic diversity within and among *Pinus pinaster* populations: comparisons between AFLP and microsatellite markers, *Heredity* 86: 469–479.

883. Mariette, S., Le Corre, V., Austerlitz, F., and Kremer, A., (2002) Sampling within the genome for measuring within-population diversity: trade-offs between markers, *Mol. Ecol.* 11: 1145–1156.

884. Marita, J.M., Rodrigues, J.M., and Nienhuis, J., (2000) Development of an algorithm identifying maximally diverse core collections, *Genet. Res. Crop. Evol.* 47: 515–526.

885. Marsh, G., and Ayres, D.R., (2002) Genetic structure of *Senecio layneae* (Composital): a rare plant of the Chaparral, *Madrono* 49: 150–157.

886. Marshall, H.D., Newton, C., and Ritland, K., (2001) Sequence-repeat polymorphisms exhibit the signature of recombination in lodgepole pine chloroplast DNA, *Mol. Biol. Evol.* 18: 2136–2138.

887. Marth, G.T., Korf, I., Yandell, M.D., Yeh, R.T., Gu, Z., Zakeri, H., Stitziel, N.O., Hillier, L., Kwok, P.Y., and Gish, W.R., (1999) A general approach to single nucleotide polymorphism discovery, *Nat. Genet.* 23: 452–456.

888. Martienssen, R.A., and Baulcombe, D.C., (1989) An unusual wheat insertion sequence (WIS1) lies upstream of an α-amylase gene in hexaploid wheat, and carries a "minisatellite" array, *Mol. Gen. Genet.* 217: 401–410.

889. Martienssen, R.A., and Colot, V., (2001) DNA methylation and epigenetic inheritance in plants and filamentous fungi, *Science* 293: 1070–1074.

890. Martin, G.B., Williams, J.G.K., and Tanksley, S.D., (1991) Rapid identification of markers linked to a *Pseudomonas* resistance gene in tomato by using random primers and near isogenic lines, *Proc. Natl. Acad. Sci. U.S.A.* 88: 2336–2340.

891. Martin, G.B., Brommonschenkel, S.H., Chungwongse, J., Frary, A., Ganal, M.W., Spivey, R., Wu, T., Earle, E.D., and Tanksley, S.D. (1993) Map-based cloning of a protein kinase gene confering disease resistance in tomato, *Science* 262: 1432–1436.

892. Martin, W., Stoebe, B., Goremykin, V., Hansmann, S., Hasegawa, M., and Kowallik, K.V., (1998) Gene transfer to the nucleus and the evolution of chloroplasts, *Nature* 393: 162–165.

893. Martinello, G.E., Leal, N.R., Amaral Júnior, A.T., Pereira, M.G., and Daher, R.F., (2001) Comparison of morphological characteristics and RAPD for estimating genetic diversity in *Abelmoschus* spp, *Acta Hortic.* 546: 101–104.

894. Matioli, S.R., and de Brito, R.A., (1995) Obtaining genetic markers by using double-stringency PCR with microsatellites and arbitrary primers, *BioTechniques* 19: 752–758.

895. Matsumura, H., Reich, S., Ito, A., Saitoh, H., Winter, P., Kahl, G., Reuter, M., Krüger D., and Terauchi, R., (2003) Gene expression analysis of plant host-pathogen interactions by SuperSAGE, *Proc. Natl. Acad. Sci. U.S.A.* 100: 15718–15723.

896. Matsuyama, T., Abe, T., Bae, C.-H., Takahashi, Y., Kiuchi, R., Nakano, T., Asami, T., and Yoshida, S., (2000) Adaptation of restriction landmark genomic scanning (RLGS) to plant genome analysis, *Plant Mol. Biol. Rep.* 18: 331–338.

897. Matthes, M., Singh, R., Cheah, S.-C., and Karp, A., (2001) Variation in oil palm (*Elaeis guineensis* Jacq.) tissue culture-derived regenerants revealed by AFLPs with methylation-sensitive enzymes, *Theor. Appl. Genet.* 102: 971–979.

898. Deleted in proof.

899. Maxam, A.M., and Gilbert, W., (1977) A new method for sequencing DNA, *Proc. Natl. Acad. Sci. U.S.A.* 74: 560–564.

900. May, B., (1998) Starch gel electrophoresis of allozymes. In *Molecular Genetic Analysis of Populations: A Practical Approach, 2nd Edition,* Hoelzel, A.R., Ed., IRL Press, Oxford, U.K., pp. 1–28.

901. Mayes, C., Saunders, G.W., Tan, I.H., and Druehl, L.D., (1992) DNA extraction methods for kelp (Laminariales) tissue, *J. Phycol.* 28: 712–716.

902. McCarthy, P.L., Hansen, J.L., Zemetra, R.S., and Berger, P.H., (2002) Rapid identification of transformed wheat using a half-seed PCR assay, *BioTechniques* 32: 560–564.

903. McClelland, M., Arensdorf, H., Cheng, R., and Welsh, J., (1994) Arbitrarily primed PCR fingerprints resolved on SSCP gels, *Nucleic Acids Res.* 22: 1770–1771.

904. McClelland, M., Nelson, M., and Raschke, E., (1994) Effect of site-specific modification on restriction endonucleases and DNA modification methyltransferases, *Nucleic Acids Res.* 22: 3640–3659.

905. McClelland, M., Mathieu-Daude, F., and Welsh, J., (1995) RNA fingerprinting and differential display using arbitrarily primed PCR, *Trends Genet.* 11: 242–246.

906. McCouch, S.R., Chen, X., Panaud, O., Temnykh, S., Xu, Y., Cho, Y.G., Huang, N., Ishii, T., and Blair, M., (1997) Microsatellite marker development, mapping and applications in rice genetics and breeding, *Plant Mol. Biol.* 35: 89–99.

907. McGregor, C.E., Lambert, C.A., Greyling, M.M., Louw, J.H., and Warnick, L., (2000) A comparative assessment of DNA fingerprinting techniques (RAPD, ISSR, AFLP and SSR) in tetraploid potato (*Solanum tuberosum* L.) germplasm, *Euphytica* 113: 135–144.

908. McGregor, C.E., Van Treuren, R., Hoekstra, R., and Van Hintum, T.J.L., (2002) Analysis of the wild potato germplasm of the series *Acaulia* with AFLPs: implications for *ex situ* conservation, *Theor. Appl. Genet.* 104: 146–156.

909. McKinnon, G.E., Vaillancourt, R.E., Jackson, H.D., and Potts, B.M., (2001) Chloroplast sharing in the Tasmanian eucalypts, *Evolution* 55: 703–711.

910. Meekins, J.F., Ballard, H.E., Jr., and McCarthy, B.C., (2001) Genetic variation and molecular biogeography of a North American invasive plant species (*Alliaria petiolata*, Brassicaceae), *Int. J. Plant Sci.* 162: 161–169.

911. Mejjad, M., Vedel, F., and Ducreux, G., (1994) Improvement of DNA preparation and of PCR cycling in RAPD analysis of marine macroalgae, *Plant Mol. Biol. Rep.* 12: 101–105.

912. Meksem, K., and Kahl, G., (2004) *The Handbook of Genome Mapping*, Wiley-VCH, New York.

913. Mellersh, C., and Sampson, J., (1993) Simplifying detection of microsatellite length polymorphisms, *BioTechniques* 15: 582–584.

914. Mengoni, A., Gori, A., and Bazzicalupo, M., (2000) Use of RAPD and microsatellite (SSR) variation to assess genetic relationships among populations of tetraploid alfalfa, *Medicago sativa*, *Plant Breed.* 119: 311–317

915. Mengoni, A., Ruggini, C., Vendramin, G.G., and Bazzicalupo, M., (2000) Chloroplast microsatellite variations in tetraploid alfalfa, *Plant Breed.* 119: 509–512

916. Mengoni, A., Barabesi, C., Gonnelli, C., Galardi, F., Gabbrielli, R., and Bazzicalupo, M., (2001) Genetic diversity of heavy metal-tolerant populations in *Silene paradoxa* L. (Caryophyllaceae): a chloroplast microsatellite analysis, *Mol. Ecol.* 10: 1909–1916.

917. Menke, U., and Mueller-Roeber, B., (2001) RNA fingerprinting of specific plant cell types: adaptation to plants and optimization of RNA arbitrarily primed PCR (RAP-PCR), *Plant Mol. Biol. Rep.* 19: 33–48.

918. Menkir, A., Goldsbrough, P., and Ejeta, G., (1997) RAPD based assessment of genetic diversity in cultivated races of *Sorghum*, *Crop Sci.* 37: 564–569.

919. Mes, T.H.M., Kuperus, P., Kirschner, J., Stepanek, J., Storchova, H., Oosterveld, P., Den Nijs, J.C.M., (2002) Detection of genetically divergent clone mates in apomictic dandelions, *Mol. Ecol.* 11: 253–265.

920. Messier, W., Li, S.-H., and Stewart, C.-B., (1996) The birth of microsatellites, *Nature* 381: 483.

921. Mettler, I.J., (1987) A simple and rapid method for minipreparation of DNA from tissue cultured plant cells, *Plant Mol. Biol. Rep.* 5: 346–349.

922. Meyer, W., Mitchell, T.G., Freedman, E.Z., and Vilgalys, R., (1993) Hybridization probes for conventional DNA fingerprinting used as single primers in the polymerase chain reaction to distinguish strains of *Cryptococcus neoformans*, *J. Clin. Microbiol.* 31: 2274–2280.

923. Michaels, S.D., and Amasino, R.M., (2001) High throughput isolation of DNA and RNA in 96-well format using a paint shaker, *Plant Mol. Biol. Rep.* 19: 227–233.

924. Michaels, S.D., John, M.C., and Amasino, R.M., (1994) Removal of polysaccharides from plant DNA by ethanol precipitation, *BioTechniques* 17: 274–276.

925. Michalalakis, Y., and Excoffier, L., (1996) A generic estimation of population subdivision using distances between alleles with special reference for microsatellite loci, *Genetics* 142: 1061–1064.

926. Micheli, M.R., Bova, R., Calissano, P., and D'Ambrosio, E., (1993) Randomly amplified polymorphic DNA fingerprinting using combinations of oligonucleotide primers, *BioTechniques* 15: 388–390.

927. Micheli, M.R., Bova, R., Pascale, E., and D'Ambrosio, E., (1994) Reproducible DNA fingerprinting with the random amplified polymorphic DNA (RAPD) method, *Nucleic Acids Res.* 22: 1921–1922.

928. Michelmore, R.W., Paran, I., and Kesseli, R.V., (1991) Identification of markers linked to disease-resistance genes by bulked segregant analysis: a rapid method to detect markers in specific genomic regions by using segregating populations, *Proc. Natl. Acad. Sci. U.S.A.* 88: 9828–9832.

929. Michiels, A., Van den Ende, W., Tucker, M., Van Riet, L., and Van Laere, A., (2003) Extraction of high-quality genomic DNA from latex-containing plants, *Anal. Biochem.* 315: 85–89.

930. Milbourne, D., Meyer, R.C., Collins, A.J., Ramsay, L.-D., Gebhardt, C., and Waugh, R., (1998) Isolation, characterization and mapping of simple sequence repeat loci in potato, *Mol. Gen. Genet.* 259: 233–245.

931. Miller, J., and Schmidt, H., (1998) The Missouri Botanical Garden's DNA bank. In *Conservation of Plant Genes III: Conservation and Utilization of African Plants*, Adams, R.P., and Adams, J.E., Eds., Missouri Botanical Garden Press, St. Louis, MO, pp. 175–182.

932. Miller, K.M., Ming, T.J., Schulze, A.D., and Withler, R.E., (1999) Denaturing gradient gel electrophoresis (DGGE): a rapid and sensitive technique to screen nucleotide sequence variation in populations, *BioTechniques* 27: 1016–1030.

932a. Miller Coyle, H., Ed., (2004) *Forensic Botany: Principles and Applications to Criminal Casework*, CRC Press, Boca Raton, FL.

933. Milligan, B.G., (1998) Total DNA isolation. In *Molecular Genetic Analysis of Populations: A Practical Approach, 2nd Edition*, Hoelzel, A.R., Ed., IRL Press, Oxford, U.K., pp. 29–64.

934. Milligan, B.G., and McMurry, C.K., (1993) Maximum likelihood analysis of male fertility using dominant and codominant genetic markers, *Mol. Ecol.* 2: 257–283.

935. Mindrinos, M., Katagiri, F., Yu, G.L., and Ausubel, F.M., (1994) The *Arabidopsis thaliana* disease resistance gene *RPS*2 encodes a protein containing a nucleotide-binding site and leucine-rich repeats, *Cell* 78: 1089–1099.

936. Mitchell, S.E., Kresovich, S., Jester, C.A., Hernandez, C.J., and Szewcz-McFadden, A.K., (1997) Application of multiplex PCR and fluorescence-based, semi-automated allele sizing technology for genotyping plant resources, *Crop Sci.* 37: 617–624.

937. Miyada, C.G., and Wallace, R.B., (1987) Oligonucleotide hybridization techniques, *Meth. Enzymol.* 154: 94–107.

938. Mogg, R., Batley, J., Hanley, S., Edwards, D., O'Sullivan, H., and Edwards, K.J., (2002) Characterization of the flanking regions of *Zea mays* microsatellites reveals a large number of useful sequence polymorphisms, *Theor. Appl. Genet.* 105: 532–543.

939. Mogg, R.J., and Bond, J.M., (2003) A cheap, reliable and rapid method of extracting high-quality DNA from plants, *Mol. Ecol. Notes* 3: 666–668.

940. Mohan, M., Nair, S., Bhagwat, A., Krishna, T.G., Yano, M., Bhatia, C.R., and Sasaki, T., (1997) Genome mapping, molecular markers and marker-assisted selection in crop plants. *Mol. Breed.* 3: 87–103.

941. Monckton, D.G., and Jeffreys, A.J., (1991) Minisatellite "isoallele" discrimination in pseudohomozygotes by single molecule PCR and variant repeat mapping, *Genomics* 11: 465–467.

942. Monckton, D.G., Neumann, R., Guram, T., Fretwell, N., Tamaki, K., MacLeod, A., and Jeffreys, A.J., (1994) Minisatellite mutation rate variation associated with a flanking DNA sequence polymorphism, *Nat. Genet.* 8: 162–170.

943. Money, T., Reader, S., Qu, L.J., Dunford, R.P., and Moore, G., (1996) AFLP-based mRNA fingerprinting, *Nucleic Acids Res.* 24: 2616–2617.

944. Moore, G., (2000) Cereal chromosome structure, evolution and pairing, *Annu. Rev. Plant Physiol. Plant Mol. Biol.* 51: 195–222.

945. Moore, S.S., Sargeant, L.L., King, T.J., Mattick, J.S., Georges, M., and Hetzel, D.J.S., (1991) The conservation of dinucleotide microsatellites among mammalian genomes allows the use of heterologous PCR primer pairs in closely related species, *Genomics* 10: 654–660.

946. Deleted in proof.

947. Morell, M.K., Peakall, R., Appels, R., Preston, L.R., and Lloyd, H.L., (1995) DNA profiling techniques for plant variety identification, *Aust. J. Exp. Agric.* 35: 807–819.

948. Morgan-Richards, M., and Wolff, K., (1999) Genetic structure and differentiation of *Plantago major* reveals a pair of sympatric sister species, *Mol. Ecol.* 8: 1027–1036.

949. Morgante, M., and Olivieri, A.M., (1993) PCR-amplified microsatellites as markers in plant genetics, *Plant J.* 3: 175–182.

950. Morgante, M., and Vogel, J., (1994) Compound microsatellite primers for the detection of genetic polymorphisms, U.S. Patent Application 08/326456.

951. Morgante, M., Hanafey, M., and Powell, W., (2002) Microsatellites are preferentially associated with nonrepetitive DNA in plant genomes, *Nat. Genet.* 30: 194–200.

952. Morin, P.A., and Smith, D.G., (1995) Nonradioactive detection of hypervariable simple sequence repeats in short polyacrylamide gels, *BioTechniques* 19: 223–228.

953. Morin, P.A., Luikart, G., Wayne, R.K., and the SNP workshop group, (2004) SNPs in ecology, evolution and conservation, *Trends Ecol. Evol.* 19: 208–216.

954. Mort, M.E., Soltis, P.S., Soltis, D.E., and Mabry, M.L., (2000) Comparison of three methods for estimating internal support on phylogenetic trees, *Syst. Biol.* 49: 160–170.

955. Mort, M.E., Crawford, D.J., Santos-Guerra, A., Francisco-Ortega, J., Esselman, E.J., and Wolfe, A.D., (2003) Relationships among the Macaronesian members of *Tolpis* (Asteraceae: Lactuceae) based upon analyses of inter simple sequence repeat (ISSR) markers, *Taxon* 52: 511–518.

956. Mosseler, A., Egger, K.N., and Hughes, G.A., (1992) Low levels of genetic diversity in red pine confirmed by random amplified polymorphic DNA markers, *Can. J. For. Res.* 22: 1332–1337.

957. Motohashi, R., Mochizuki, K., Ohtsubo, H., and Ohtsubo, E., (1997) Structures and distribution of pSINE1 members in rice genomes, *Theor. Appl. Genet.* 95: 359–368.

958. Moxon, E.R., and Wills, C., (1999) DNA microsatellites: agents of evolution? *Sci. Am.* 280(1): 94–99.

959. Mueller, U.G., and Wolfenbarger, L.L., (1999) AFLP genotyping and fingerprinting, *Trends Ecol. Evol.* 14: 389–394.

960. Mullis, K.B., Ferré, F., and Gibbs, R.A., Eds. (1994) *The Polymerase Chain Reaction*, Birkhäuser, Basel, Switzerland.

961. Munthali, M., Ford-Lloyd, B.V., and Newbury, H.J., (1992) The random amplification of polymorphic DNA for fingerprinting plants, *PCR Method Appl.* 1: 274–276.

962. Muralidharan, K., and Wakeland, E.K., (1993) Concentration of primer and template qualitatively affects products in random-amplified polymorphic DNA PCR, *BioTechniques* 14: 362–364.

963. Murphy, R.W., Sites, J.W., Jr., Buth, D.G., and Haufler, C.H., (1996) Proteins: isozyme electrophoresis. In *Molecular Systematics, 2nd Edition*, Hillis, D.M., Moritz, C., and Mable, B.K., Eds., Sinauer Associates, Sunderland, MA, pp. 51–120.

964. Murray, A.E., Lies, D., Li, G., Nealson, K., Zhou, J., and Tiedje, J.M., (2001) DNA/DNA hybridization to microarrays reveals gene-specific differences between closely related microbial genomes, *Proc. Natl. Acad. Sci. U.S.A.* 98: 9853–9858.

965. Murray, M.G., and Thompson, W.F., (1980) Rapid isolation of high molecular weight plant DNA, *Nucleic Acids Res.* 8: 4321–4325.

966. Murray, M.G., and Pitas, J.W., (1996) Plant DNA from alcohol-preserved samples, *Plant Mol. Biol. Rep.* 14: 261–265.

967. Murray, V., (1989) Improved double-stranded DNA sequencing using the linear polymerase chain reaction, *Nucleic Acids Res.* 17: 8889.

968. Murray, V., Monchawin, C., and England, P.R., (1993) The determination of the sequences present in the shadow bands of a dinucleotide repeat PCR, *Nucleic Acids Res.* 21: 2395–2398.

969. Myburg, A.A., Remington, D.L., O'Malley, D.M., Sederoff, R.R., and Whetten, R.W., (2001) High-throughput AFLP analysis using infrared dye-labeled primers and an automated DNA sequencer, *BioTechniques* 30: 348–357.

970. Myers, R.M., Maniatis, T., and Lerman, L.S., (1986) Detection and localization of single base changes by denaturing gradient gel electrophoresis, *Meth. Enzymol.* 155: 501–527.

971. N'Goran, J.A.K., Laurent, V., Risterucci, A.M., and Lanaud, C., (1994) Comparative genetic diversity studies of *Theobroma cacao* L. using RFLP and RAPD markers, *Heredity* 73: 589–597.

972. Nadir, E., Margalit, H., Gallily, T., and Ben-Sasson, S.A., (1996) Microsatellite spreading in the human genome: evolutionary mechanisms and structural implications, *Proc. Natl. Acad. Sci. U.S.A.* 93: 6470–6475.

973. Nakamura, Y., Leppert, M., O'Connell, P., Wolff, R., Holm, T., Culver, M., Martin, C., Fujimoto, E., Hoff, M., Kumlin, E., and White, R., (1987) Variable number of tandem repeat (VNTR) markers for human gene mapping, *Science* 235: 516–522.

974. Narvel, J.M., Chu, W.-C., Fehr, W.R., Cregan, P.B., and Shoemaker, R.C., (2000) Development of multiplex sets of simple sequence repeat DNA markers covering the soybean genome, *Mol. Breed.* 6: 175–183.

975. Nason, J.D., Ellstrand, N.C., and Arnold, M.L., (1992) Patterns of hybridisation and introgression in populations of oaks, manzanitas and irises, *Am. J. Bot.* 79: 101–111.

976. Nasu, S., Suzuki, J., Ohta, R., Hasegawa, K., Yui, R., Kitazawa, N., Monna, L., and Minobe, Y., (2002) Search for and analysis of single nucleotide polymorphisms (SNPs) in rice (*Oryza sativa, Oryza rufipogon*) and establishment of SNP markers, *DNA Res.* 9: 163–171.

977. Neff, B.D., (2004) Mean d² and divergence time: transformations and standardizations, *J. Hered.* 21: 165–171.

978. Nei, M., (1973) Analysis of gene diversity in subdivided populations, *Proc. Natl. Acad. Sci. U.S.A.,* 70: 3321–3323.

979. Nei, M., (1977) F-statistics and analysis of gene diversity in subdivided populations, *Ann. Hum. Genet.* 41: 225–233.

980. Nei, M., (1978) Estimation of average heterozygosity and genetic distance from a small number of individuals, *Genetics* 89: 583–590.

981. Nei, M., and Li, W.-H., (1979) Mathematical model for studying genetic variation in terms of restriction endonucleases, *Proc. Natl. Acad. Sci. U.S.A.* 76: 5269–5273.

982. Nei, M, and Kumar, S., (2000) *Molecular Evolution and Phylogenetics*, Oxford University Press, Oxford, U.K.

983. Deleted in proof.

984. Neilan, B.A., Leigh, D.A., Rapley, E., and McDonald, B.L., (1994) Microsatellite genome screening: rapid non-denaturing, non-isotopic dinucleotide repeat analysis, *BioTechniques* 17: 708–712.

985. Nelson, D.L., Ledbetter, S.A., Corbo, L., Victoria, M.F., Ramiréz-Solis, R., Webster, T.D., Ledbetter, D.H., and Caskey, C.T., (1989) *Alu* polymerase chain reaction: a method for rapid isolation of human-specific sequences from complex DNA sources, *Proc. Natl. Acad. Sci. U.S.A.* 86: 6686–6690.

986. Neu, C., Kaemmer, D., Kahl, G., Fischer, D., and Weising, K., (1999) Polymorphic microsatellite markers for the banana pathogen *Mycosphaerella fijiensis, Mol. Ecol.* 8: 523–525.

987. Newbury, H.J., and Ford-Lloyd, B.V., (1993) The use of RAPD for assessing variation in plants, *Plant Growth Regul.* 12: 43–51.

988. Newton, A.C., Allnutt, T.R., Gillies, A.C.M., Lowe, A.J., and Ennos, R.A., (1999) Molecular phylogeography, intraspecific variation and the conservation of tree species, *Trends Ecol. Evol.* 14: 140–145.

989. Nickerson, D.A., Tobe, V.O., and Taylor, S.L., (1997) PolyPhred: automating the detection and genotyping of single nucleotide substitutions using fluorescence-based resequencing, *Nucleic Acids Res.* 25: 2745–2751.

990. Nickrent, D.L., (1994) From field to film: rapid sequencing methods for field-collected plant species, *BioTechniques* 16: 470–475.

991. Nishizawa, S., Kubo, T., and Mikami, T., (2000) Variable number of tandem repeat loci in the mitochondrial genomes of beets, *Curr. Genet.* 37: 34–38.

992. Nkongolo, K.K., Klimaszewska, K., and Gratton, W.S., (1998) DNA yields and optimization of RAPD patterns using spruce embryogenic lines, seedlings, and needles, *Plant Mol. Biol. Rep.* 16: 284 (1–9).

993. Noma, K., Ohtsubo, E., and Ohtsubo, H., (1999) Non-LTR retrotransposons (LINEs) as ubiquitous components of plant genomes, *Mol. Gen. Genet.* 261: 71–79.

994. Notsu, Y., Masood, S., Nishikawa, T., Kobo, N., Akiduki, G., Nakazono, M., Hirai, A., and Kadowaki, K., (2002) The complete sequence of the rice (*Oryza sativa* L.) mitochondrial genome: frequent DNA sequence acquisition and loss during the evolution of flowering plants, *Mol. Genet. Genomics* 268: 434–445.

995. Novy, R.G., and Vorsa, N., (1996) Evidence for RAPD heteroduplex formation in cranberry: implications for pedigree and genetic-relatedness studies and a source of co-dominant RAPD markers, *Theor. Appl. Genet.* 92: 840–849.

996. Nürnberg, P., and Epplen, J.T., (1989) "Hidden partials" — a cautionary note, *Fingerprint News* 1(4): 11–12.

997. Nybom, H., (1991) Applications of DNA fingerprinting in plant breeding, in *DNA Fingerprinting: Approaches and Applications*, Burke, T., Dolf, G., Jeffreys, A.J., and Wolff, R, Eds., Birkhäuser, Basel, Switzerland, pp. 294–311.

998. Nybom, H., (1993) Applications of DNA fingerprinting in plant population studies, in *DNA Fingerprinting: State of the Science*, Pena, S.D.J., Chakraborty, R., Epplen, J.T., and Jeffreys, A.J., Eds., Birkhäuser, Basel, Switzerland, pp. 293–309.

999. Nybom, H., (2001) DNA markers for different aspects of plant breeding research and its applications, *Acta Hort.* 560: 63–67.

1000. Nybom, H., (2004) Comparison of different nuclear DNA markers for estimating intraspecific genetic diversity in plants, *Mol. Ecol.* 13: 1143–1155.

1001. Nybom, H., and Bartish, I., (2000) Effects of life history traits and sampling strategies on genetic diversity estimates obtained with RAPD markers in plants, *Persp. Plant Ecol. Evol. Syst.* 3: 93–114.

1002. Nybom, H., Werlemark, G., and Olsson, Å.M.E., (2001) Between- and within population diversity in dogrose species, *Acta Hortic.* 546: 139–144.

1003. Nybom, H., Esselink, D.G., Werlemark, G., and Vosman, B., (2004) Microsatellite DNA marker inheritance indicates preferential pairing between highly homologous genomes in polyploid and hemisexual dog-roses *Rosa* L. sect. *Caninae, Heredity* 92: 139–150.

1004. O'Hanlon, P.C., and Peakall, R., (2000) A simple method for the detection of size homoplasy among amplified fragment length polymorphism fragments, *Mol. Ecol.* 9: 815–816.

1005. O'Malley, D.M., and Whetten, R., (1998) Molecular markers and forest trees. In *DNA Markers: Protocols, Applications, and Overviews*, Caetano-Anollés, G., and Gresshoff, P.M., Eds., Wiley-VCH, New York, pp. 237–257.

1006. Oard, J.H., and Dronavalli, S., (1992) Rapid isolation of rice and maize DNA for analysis by random-primer PCR, *Plant Mol. Biol. Rep.* 10: 236–241.

1007. Ochman, H., Gerber, A.S., and Hartl, D.L., (1988) Genetic applications of an inverse polymerase chain reaction, *Genetics* 120: 621–625.

1008. Ohshima, K., Hamada, M., Terai, Y., and Okada, N., (1996) The 3'-ends of tRNA-derived short interspersed repetitive elements are derived from the 3'-ends of long interspersed repetitive elements, *Mol. Cell. Biol.* 16: 3756–3764.

1009. Olmstead, R.G., and Palmer, J.D., (1994) Chloroplast DNA systematics: a review of methods and data analysis, *Am. J. Bot.* 81: 1205–1224.

1010. Olsen, K.M., (2002) Population history of *Manihot esculenta* (Euphorbiaceae) inferred from nuclear DNA sequences, *Mol. Ecol.* 11: 901–911.

1011. Olsen, K.M., and Schaal, B.A., (1999) Evidence on the origin of cassava: phylogeography of *Manihot esculenta, Proc. Natl. Acad. Sci. U.S.A.* 96: 5586–5591.

1012. Olsen, K.M., and Schaal, B.A., (2001) Microsatellite variation in cassava (*Manihot esculenta*, Euphorbiaceae) and its wild relatives: further evidence for a Southern Amazonian origin of domestication, *Am. J. Bot.* 88: 131–142.

1013. Olsson, Å.M.E., Nybom, H., and Prentice, H.C., (2000) Relationships between Nordic dogroses (*Rosa* L. sect. *Caninae*, Rosaceae) assessed by RAPDs and elliptic Fourier analysis of leaflet shape, *Syst. Bot.* 25: 511–521.

1014. Oostermeijer, J.G.B., Luijten, S.H., and Den Nijs, J.C.M., (2003) Integrating demographic and genetic approaches in plant conservation, *Biol. Conserv.* 113: 389–398.

1015. Orgel, L.E., and Crick, F.H.C., (1980) Selfish DNA: the ultimate parasite, *Nature* 284: 604–607.

1016. Orita, M., Iwahana, H., Kanazawa, H., Hayashi, K., and Sekiya, T., (1989) Detection of polymorphisms of human DNA by gel electrophoresis as single-strand conformation polymorphisms, *Proc. Natl. Acad. Sci. U.S.A.* 86: 2766–2770.

1017. Ortí, G., Pearse, D.E., and Avise, J.C., (1997) Phylogenetic assessment of length variation at a microsatellite locus, *Proc. Natl. Acad. Sci. U.S.A.* 94: 10745–10749.

1018. Ostrander, E.A., Jong, P.M., Rine, J., and Duyk, G., (1992) Construction of small-insert genomic DNA libraries highly enriched for microsatellite repeat sequences, *Proc. Natl. Acad. Sci. U.S.A.* 89: 3419–3423.

1019. Ouborg, N.J., Piquot, Y., and Van Groenendael, J.M., (1999) Population genetics, molecular markers and the study of dispersal in plants, *J. Ecol.* 87: 551–568.

1020. Ouenzar, B., Hartmann, C., Rode, A., and Benslimane, A., (1998) Date palm DNA mini-preparation without liquid nitrogen, *Plant Mol. Biol. Rep.* 16: 263–269.

1021. Özdemir, N., Horn, R., and Friedt, W., (2002) Isolation of HMW DNA from sunflower (*Helianthus annuus* L.) for BAC cloning, *Plant Mol. Biol. Rep.* 20: 239–250.

1022. Pääbo, S., (2000) Of bears, conservation genetics, and the value of time travel, *Proc. Natl. Acad. Sci. U.S.A.* 97: 1320–1321.

1023. Pääbo, S., Irwin, D.M., and Wilson, A.C., (1990) DNA damage promotes jumping between templates during enzymatic amplification, *J. Biol. Chem.* 265: 4718–4721.

1024. Paetkau, D., (1999) Microsatellites obtained using strand extension: an enrichment protocol, *BioTechniques* 26: 690–697.

1025. Paetkau, D., and Strobeck, C., (1994) Microsatellite analysis of genetic variation in black bear population, *Mol. Ecol.* 3: 489–495.

1026. Paetkau, D., and Strobeck, C., (1995) The molecular basis and evolutionary history of a microsatellite null allele in bears, *Mol. Ecol.* 4: 519–520.

1027. Page, R.D.M., and Holmes, E.C., (1998) *Molecular Evolution: A Phylogenetic Approach*, Blackwell Science Ltd., Oxford, U.K.

1028. Paglia, G., and Morgante, M., (1998) PCR-based multiplex DNA fingerprinting techniques for the analysis of conifer genomes, *Mol. Breed.* 4: 173–177.

1029. Palacios, C., and González-Candelas, F., (1999) AFLP analysis of the critically endangered *Limonium cavanillesii* (Plumbaginaceae), *J. Hered.* 90: 485–489.

1030. Palacios, C., Kresovich, S., and González-Candelas, F., (1999) A population genetic study of the endangered plant species *Limonium dufourii* (Plumbaginaceae) based on amplified fragment length polymorphism (AFLP), *Mol. Ecol.* 8: 645–657.

1031. Palacios, G., Bustamante, S., Molina, C., Winter, P., and Kahl, G., (2002) Electro-phoretic identification of new genomic profiles with a modified selective amplification of microsatellite polymorphic loci technique based on AT/AAT polymorphic repeats, *Electrophoresis* 23: 3341–3345.

1032. Palmé, A.E., and Vendramin, G.G., (2002) Chloroplast DNA variation, postglacial recolonization and hybridization in hazel, *Corylus avellana. Mol. Ecol.* 11: 1769–1779.

1033. Palmer, J.D., (1988) Intraspecific variation and multicircularity in *Brassica* mitochondrial DNAs, *Genetics* 118: 341–351.

1034. Palmer, J.D., and Herborn, L.A., (1988) Plant mitochondrial DNA evolves rapidly in structure, but slowly in sequence, *J. Mol. Evol.* 28: 87–97.

1035. Pan, Y.B., Burner, D.M., Ehrlich, K.C., Grisham, M.P., and Wei, Q., (1997) Analysis of primer-derived, nonspecific amplification products in RAPD-PCR, *BioTechniques* 22: 1071–1077.

1036. Panaud, O., Chen, X., and McCouch, S.R., (1995) Frequency of microsatellite sequences in rice (*Oryza sativa* L.), *Genome* 38: 1170–1176.

1037. Pandey, R.N., Adams, R.P., and Flournoy, L.E., (1996) Inhibition of random amplified polymorphic DNAs (RAPDs) by plant polysaccharides, *Plant Mol. Biol. Rep.* 14: 17–22.

1038. Pandian, A., Ford, R., and Taylor, P.W.J., (2000) Transferability of sequence tagged microsatellite site (STMS) primers across four major pulses, *Plant Mol. Biol. Rep.* 18: 395a–395h.

1039. Papa, R., Attene, G., Barcaccia, G., Ohgata, A., and Konishi, T., (1998) Genetic diversity in landrace populations of *Hordeum vulgare* L. from Sardinia, Italy, as revealed by RAPDs, isozymes and morphophenological traits, *Plant Breed.* 117: 523–530.

1040. Paran, I., and Michelmore, R.W., (1993) Development of reliable PCR-based markers linked to downy mildew resistance genes in lettuce, *Theor. Appl. Genet.* 85: 985–993.

1041. Parducci, L., Szmidt, A.E., Madaghiele, A., Anzidei, M., and Vendramin, G.G., (2001) Genetic variation at chloroplast microsatellites (cpSSRs) in *Abies nebrodensis* (Lojac.) Mattei and three neighboring *Abies* species, *Theor. Appl. Genet.* 102: 733–740.

1042. Paris, M., and Carter, M., (2000) Cereal DNA: a rapid high-throughput extraction method for marker assisted selection, *Plant Mol. Biol. Rep.* 18: 357–360.

1043. Paris, M., and Jones, M.G.K., (2002) Microsatellite genotyping by primer extension and MALDI-ToF mass spectrometry, *Plant Mol. Biol. Rep.* 20: 259–263.

1044. Park, K.C., Kim, N.H., Cho, Y.S., Kang, K.H., Lee, J.K., and Kim, N.-S., (2003) Genetic variations of AA genome *Oryza* species measured by MITE-AFLP, *Theor. Appl. Genet.* 197: 203–209.

1045. Park, Y.-H., and Kohel, R.J., (1994) Effect of concentration of $MgCl_2$ on random-amplified DNA polymorphism, *BioTechniques* 16: 652–655.

1046. Parks, J.C., and Werth, C.R., (1993) A study of spatial features of clones on a population of bracken fern, *Pteridium aquilinum* (Dennstaedtiaceae), *Am. J. Bot.* 80: 537–544.

1047. Parsons, B.J., Newbury, N.J., Jackson, M.T., and Ford-Lloyd, B.V., (1999) The genetic structure and conservation of aus, aman and boro rices from Bangladesh, *Genet. Res. Crop. Evol.* 46: 587–598.

1048. Paszkowski, J, and Whitham, S., (2001) Gene silencing and DNA methylation processes, *Curr. Opin. Plant Biol.* 4: 123–129.

1049. Paterson, A.H., DeVerna, J.W., Lanini, B., and Tanksley, S.D. (1990) Fine mapping of quantitative trait loci using a selected overlapping recombinant chromosome in an interspecies cross of tomato, *Genetics* 124: 735–744.

1050. Paterson, A.H., Brubaker, C.L., and Wendel, J.F., (1993) A rapid method for extraction of cotton (*Gossypium* spec.) genomic DNA suitable for RFLP or PCR analysis, *Plant Mol. Biol. Rep.* 11: 122–127.

1051. Patzak, J., (2001) Comparison of RAPD, STS, ISSR and AFLP molecular methods used for assessment of genetic diversity in hop (*Humulus lupulus* L.), *Euphytica* 121: 9–18.

1052. Patzak, J., (2003) Assessment of somaclonal variability in hop (*Humulus lupulus* L.) *in vitro* meristem cultures and clones by molecular methods, *Euphytica* 131: 343–350.

1053. Pay, A., and Smith, M.A., (1988) A rapid method for purification of organelles for DNA isolation: self-generated percoll gradients, *Plant Cell Rep.* 7: 96–99.

1054. Peakall, R., Gilmore, S., Keys, W., Morgante, M., and Rafalski, A., (1998) Cross-species amplification of soybean (*Glycine max*) simple sequence repeats (SSRs) within the genus and other legume genera: implications for the transferability of SSRs in plants, *Mol. Biol. Evol.* 15: 1275–1287.

1055. Pearce, S.R., Harrison, G., Li, D., Heslop-Harrison, J.S., Kumar, A., and Flavell, A.J., (1996) The Ty1-*copia* group retrotransposons in *Vicia* species: copy number, sequence heterogeneity, and chromosomal localization, *Mol. Gen. Genet.* 250: 305–315.

1056. Pearce, S.R., Stuart-Rogers, C., Knox, M.R., Kumar, A., Ellis, T.H.N., and Flavell, A.J., (1999) Rapid isolation of plant Ty1-*copia* group retrotransposon LTR sequences for molecular marker studies, *Plant J.* 19: 711–717.

1057. Pearce, S.R., Knox, M., Ellis, T.H.N., Flavell, A.J., and Kumar, A., (2000) Pea Ty1-*copia* group retrotransposons: transpositional activity and use as markers to study genetic diversity in *Pisum*, *Mol. Gen. Genet.* 263: 898–907.

1058. Pehu, E., Thomas, M., Poutala, T., Karp, A., and Jones, M.G.K., (1990) Species-specific sequences in the genus *Solanum*: identification, characterization, and application to study somatic hybrids of *S. brevidens* and *S. tuberosum*, *Theor. Appl. Genet.* 80: 693–698.

1059. Pejic, I., Ajmine-Marsan, P., Morgante, M., Kozumplick, V., Castiglioni, P., Taramino, G., and Motto, M., (1998) Comparative analysis of genetic similarity among maize inbred lines detected by RFLPs, RAPDs, SSRs, and AFLPs, *Theor. Appl. Genet.* 97: 1248–1255.

1060. Pelgas, B., Isabel, N., and Bousquet, J., (2004) Efficient screening for expressed sequence tag polymorphisms (ESTPs) by DNA pool sequencing and denaturing gradient gel electrophoresis (DGGE) in spruces, *Mol. Breed.* 13: 263–279.

1061. Pemberton, J.M., Slate, J., Bancroft, D.R., and Barrett, J.A., (1995) Nonamplifying alleles at microsatellite loci: a caution for parentage and population studies, *Mol. Ecol.* 4: 249–252.

1062. Pena, S.D.J., Chakraborty, R., Epplen, J.T., and Jeffreys, A.J., Eds., (1993) *DNA Fingerprinting: State of the Science*. Birkhäuser, Basel, Switzerland.

1063. Penner, G.A., and Bezte, L.J., (1994) Increased detection of polymorphism among randomly amplified wheat DNA fragments using a modified temperature sweep gel electrophoresis (TSGE) technique, *Nucleic Acids Res.* 22: 1780–1781.

1064. Penner, G.A., Bush, A., Wise, R., Kim, W., Domier, L., Kasha, K., Laroche, A., Scoles, G., Molnar, S.J., and Fedak, G., (1993) Reproducibility of random amplified polymorphic DNA (RAPD) analysis among laboratories, *PCR Methods Appl.* 2: 341–345.

1065. Penner, G.A., Lee, S.J., Bezte, L.J., and Ugali, E., (1996) Rapid RAPD screening of plant DNA using dot blot hybridization, *Mol. Breed.* 2: 7–10.

1066. Pépin, L., Amigues, Y., Lépingle, A., Berthier, J.-L., Bensaid, A., and Vaiman, D., (1995) Sequence conservation of microsatellites between *Bos taurus* (cattle), *Capra hircus* (goat) and related species, *Heredity* 74: 53–61.

1067. Peraza-Echeverria, S., Herrera-Valencia, V.A., and James-Kay, A., (2001) Detection of DNA methylation changes in micropropagated banana plants using methylation-sensitive amplification polymorphism (MSAP), *Plant Sci.* 161: 359–367.

1068. Pereira, M.G., Lee, M., Bramel-Cox, P., Woodman, W., Doebley, J., and Whitkus, R., (1994) Construction of an RFLP map in sorghum and comparative mapping in maize, *Genome* 37: 236–243.

1069. Permingeat, H.R., Romagnoli, M.V., and Vallejos, R.H., (1998) A simple method for isolating high yield and quality DNA from cotton (*Gossypium hirsutum* L.) leaves, *Plant Mol. Biol. Rep.* 16: 89 (1–6).

1070. Perry, M.D., Davey, M.R., Power, J.B., Lowe, K.C., Bligh, H.F.J., Roach, P.S., and Jones, C., (1998) DNA isolation and AFLP genetic fingerprinting of *Theobroma cacao* (L.), *Plant Mol. Biol. Rep.* 16: 49–59.

1071. Persson, H.A., Rumpunen, K., and Möllerstedt, L.K., (2000) Identification of culinary rhubarb (*Rheum* spp.) cultivars using morphological characterization and RAPD markers, *J. Hort. Sci. Biotechnol.* 75: 684–689.

1072. Peterson, D.G., Boehm, K.S., and Stack, S.M., (1997) Isolation of milligram quantities of nuclear DNA from tomato (*Lycopersicon esculentum*), a plant containing high levels of polyphenolic compounds, *Plant Mol. Biol. Rep.* 15: 148–153.

1073. Peterson-Burch, B.D., Wright, D.A., Laten, H.M., and Voytas, D.F., (2000) Retroviruses in plants?, *Trends Genet.* 16: 151–152.

1074. Petes, T.D., Greenwell, P.W., and Dominska, M., (1997) Stabilization of microsatellite sequences by variant repeats in the yeast *Saccharomyces cerevisiae*, *Genetics* 146: 491–498.

1075. Pfaff, T., and Kahl, G., (2003) A genetic linkage map of chickpea with gene-specific markers using an interspecific population (*C. arietinum x C. reticulatum*), *Mol. Genet. Genomics* 269: 243–251.

1076. Phippen, W.B., Kresovich, S., Candelas, F.G., and McFerson, J.R., (1997) Molecular characterization can quantify and partition variation among genebank holdings: a case study with phenotypically similar accessions of *Brassica oleracea* var. *capitata* L. (cabbage) "Golden Acre", *Theor. Appl. Genet.* 94: 227–234.

1077. Pich, U., and Schubert, I., (1993) Midiprep method for isolation of DNA from plants with a high content of polyphenolics, *Nucleic Acids Res.* 21: 3328.

1078. Picoult-Newberg, L., Ideker, T.E., Pohl, M.G., Taylor, S.L., Donaldson, M.A., Nickerson, D.A., and Boyce-Jacino, M., (1999) Mining SNPs from EST databases, *Genome Res.* 9: 167–174.

1079. Pielou, E.C., (1969) *An Introduction to Mathematical Ecology*, Wiley-Interscience, New York.

1080. Piepho, H.P., (2001) Exploiting qualitative information in the analysis of dominant markers, *Theor. Appl. Genet.* 103: 462–468.

1081. Pikaart, M.J., and Villeponteau, B., (1993) Suppression of PCR amplification by high levels of RNA, *BioTechniques* 14: 33–34.

1082. Pirttilä, A.M., Hirsikorpi, M., Kämäräinen, T., Jaakola, L., and Hohtola, A., (2001) DNA isolation methods for medicinal and aromatic plants, *Plant Mol. Biol. Rep.* 19: 273a–273f.

1083. Plasterk, R.H.A., Izsvák, Z., and Ivics, Z., (1999) Resident aliens — the Tc1/*mariner* superfamily of transposable elements, *Trends Genet.* 15: 326–332.

1084. Plomion, C., Hurme, P., Frigerio, J-M., Ridolfi, M., Pot, D., Pionneau, C., Avila, C., Gallardo, F., David, H., Neutelings, G., Campbell, M., Canovas, F.M., Savolainen, O., Bodénès, C., and Kremer, A., (1999) Developing SSCP markers in two *Pinus* species, *Mol. Breed.* 5: 21–31.

1085. Pooler, M.R., (2003) Molecular genetic diversity among 12 clones of *Lagerstroemia fauriei* revealed by AFLP and RAPD markers, *HortScience* 38: 256–259.

1086. Porceddu, A., Albertini, E., Barcaccia, G., Marconi, G., Bertoli, F.B., and Veronesi, F., (2002) Development of S-SAP markers based on an LTR-like sequence from *Medicago sativa* L., *Mol. Genet. Genomics* 267: 107–114.

1087. Porebski, S., Bailey, L.G., and Baum, B.R., (1997) Modification of a CTAB DNA extraction protocol for plants containing high polysaccharide and polyphenol components, *Plant Mol. Biol. Rep.* 15: 8–15.

1088. Pornon, A., Escaravage, N., Thomas, P., and Taberlet, P., (2000) Dynamics of genotypic structure in clonal *Rhododendron ferrugineum* (Ericaceae) populations, *Mol. Ecol.* 9: 1099–1111.

1089. Porteous, L.A., Seidler, R.J., and Watrud, L.S., (1997) An improved method for purifying DNA from soil for polymerase chain reaction amplification and molecular ecology applications, *Mol. Ecol.* 6: 787–791.

1090. Posada, D., and Crandall, K.A., (2001) Intraspecific gene genealogies: trees grafting into networks, *Trends Ecol. Evol.* 16: 37–45.

1091. Posada, D., Crandall, K.A., and Templeton, A.R. (2000) GeoDis: a program for the cladistic nested clade analysis of the geographical distribution of genetic haplotypes, *Mol. Ecol.* 9: 487–488.

1092. Powell, W., Morgante, M., McDevitt, R., Vendramin, G.G., and Rafalski, J.A., (1995) Polymorphic simple sequence repeat regions in chloroplast genomes: applications to the population genetics of pines, *Proc. Natl. Acad. Sci. U.S.A.* 92: 7759–7763.

1093. Powell, W., Morgante, M., Andre, C., McNicol, J.W., Machray, G.C., Doyle, J.J., Tingey, S.V., and Rafalski, J.A., (1995) Hypervariable microsatellites provide a general source of polymorphic DNA markers for the chloroplast genome, *Curr. Biol.* 5: 1023–1029.

1094. Powell, W., Machray, G.C., and Provan, J., (1996) Polymorphism revealed by simple sequence repeats, *Trends Plant Sci.* 1: 215–222.

1095. Powell, W., Morgante, M., Andre, C., Hanafey, M., Vogel, J., Tingey, S., and Rafalski, A., (1996) The comparison of RFLP, RAPD, AFLP and SSR (microsatellite) markers for germplasm analysis, *Mol. Breed.* 2: 225–238.

1096. Powell, W., Morgante, M., Doyle, J.J., McNicol, J.W., Tingey, S.V., and Rafalski, A.J., (1996) Genepool variation in genus *Glycine* subgenus *Soja* revealed by polymorphic nuclear and chloroplast microsatellites, *Genetics* 144: 793–803.

1097. Prevost, A., and Wilkinson, M.J., (1999) A new system of comparing PCR primers applied to ISSR fingerprinting of potato cultivars, *Theor. Appl. Genet.* 98: 107–112.

1098. Primmer, C.R., and Ellegren, H., (1998) Patterns of molecular evolution in avian microsatellites, *Mol. Biol. Evol.* 15: 997–1008.

1099. Primmer, C.R., Ellegren, H., Saino, N., and Möller, A.P., (1996) Directional evolution in germline microsatellite mutations, *Nat. Genet.* 13: 391–393.

1100. Primmer, C.R., Möller, A.P., and Ellegren, H., (1996) A wide-range survey of cross-species microsatellite amplification in birds, *Mol. Ecol.* 5: 365–378.

1101. Primmer, C.R., Saino, N., Möller, A.P., and Ellegren, H., (1998) Unraveling the processes of microsatellite evolution through analysis of germ line mutations in barn swallows *Hirundo rustica*, *Mol. Biol. Evol.* 15: 1047–1054.

1102. Primmer, C.R., Borge, T., Lindell, J., and Saetre, G.-P., (2002) Single-nucleotide polymorphism characterization in species with limited available sequence information: high nucleotide diversity revealed in the avian genome, *Mol. Ecol.* 11: 603–612.

1103. Prober, S.M., and Brown, A.H.D., (1994) Conservation of the grassy white box woodlands: population genetics and fragmentation of *Eucalyptus albens*, *Conserv. Biol.* 8: 1003–1013.

1104. Prochazka, M., (1996) Microsatellite hybrid capture technique for simultaneous isolation of various STR markers, *Genome Res.* 6: 646–649.

1105. Provan, J., (2000) Novel chloroplast microsatellites reveal cytoplasmic variation in *Arabidopsis thaliana*, *Mol. Ecol.* 9: 2183–2185.

1106. Provan, J., Corbett, G., Waugh, R., McNicol, J.W., Morgante, M., and Powell, W., (1996) DNA fingerprints of rice (*Oryza sativa*) obtained from hypervariable chloroplast simple sequence repeats, *Proc. R. Soc. London* Ser. *B* 263: 1275–1281.

1107. Provan, J., Corbett, G., McNicol, J.W., and Powell, W., (1997) Chloroplast DNA variability in wild and cultivated rice (*Oryza* spp.) revealed by polymorphic chloroplast simple sequence repeats, *Genome* 40: 104–110.

1108. Provan, J., Lawrence, P., Young, G., Wright, F., Bird, R., Paglia, G., Cattonaro, F., Morgante, M., and Powell, W., (1999) Analysis of the genus *Zea* (Poaceae) using polymorphic chloroplast simple sequence repeats, *Plant Syst. Evol.* 218: 245–256.

1109. Provan, J., Powell, W., Dewar, H., Bryan, G., Machray, G.C., and Waugh, R., (1999) An extreme cytoplasmic bottleneck in the modern European cultivated potato (*Solanum tuberosum*) is not reflected in decreased levels of nuclear diversity, *Proc. R. Soc. London Ser. B* 266: 633–639.

1110. Provan, J., Russell, J.R., Booth, A., and Powell, W., (1999) Polymorphic chloroplast simple sequence repeat primers for systematic and population studies in the genus *Hordeum*, *Mol. Ecol.* 8: 505–511.

1111. Provan, J., Soranzo, N., Wilson, N.J., Goldstein, D.B., and Powell, W., (1999) A low mutation rate for chloroplast microsatellites, *Genetics* 153: 943–947.

1112. Provan, J., Soranzo, N., Wilson, N., McNicol, J., Morgante, M., and Powell, W., (1999) The use of uniparentally inherited simple sequence repeat markers in plant population studies and systematics, in *Molecular Systematics and Plant Evolution*, Hollingsworth, P., Bateman, R., and Gornall, R., Eds., Taylor and Francis, London, pp. 35–50.

1113. Provan, J., Thomas, W.T.B., Forster, B.P., and Powell, W., (1999) *Copia*-SSR: a simple marker technique which can be used on total genomic DNA, *Genome* 42: 363–366.

1114. Provan, J., Powell, W., and Hollingsworth, P.M., (2001) Chloroplast microsatellites: new tools for studies in plant ecology and evolution, *Trends Ecol. Evol.* 16: 142–147.

1115. Provan, J., Biss, P.M., McMeel, D., and Mathews, S., (2004) Universal primers for the amplification of chloroplast microsatellites in grasses, *Mol. Ecol. Notes* 4: 262–264.

1116. Purugganan, M.D., and Wessler, S.R., (1995) Transposon signatures: species-specific molecular markers that utilize a class of multiple-copy nuclear DNA, *Mol. Ecol.* 4: 265–269.

1117. Pyle, M.M., and Adams, R.P., (1989) *In situ* preservation of DNA in plant specimens, *Taxon* 38: 576–581.

1118. Qi, X., and Lindhout, P., (1997) Development of AFLP markers in barley, *Mol. Gen. Genet.* 254: 330–336.

1119. Qian, W., Ge, S., and Hong, D.Y., (2001) Genetic variation within and among populations of a wild rice *Oryza granulata* from China detected by RAPD and ISSR markers, *Theor. Appl. Genet.* 102: 440–449.

1120. Qin, L., Prins, P., Jones, J.T., Popejus, H., Smant, G., Bakker, J., and Helder, J., (2001) GenEST, a powerful bidirectional link between cDNA sequence data and gene expression profiles generated by cDNA-AFLP, *Nucleic Acids Res.* 29: 1616–1622.

1121. Quackenbush, J., (2001) Computational analysis of microarray data, *Nat. Rev. Genet.* 2: 418–427.

1122. Quint, M., Mihaljevic, R., Dussle, M., Xu, M.L., Melchinger, A.E., and Lübberstedt, T., (2002) Development of RGA-CAPS markers and genetic mapping of candidate genes for sugarcane mosaic virus resistance in maize, *Theor. Appl. Genet.* 105: 355–363.

1123. Quiros, C.F., This, P., Laudie, M., Benet, A., Chevre, A.-M., and Delseny, M., (1995) Analysis of a set of RAPD markers by hybridization and sequencing in *Brassica*: a note of caution, *Plant Cell Rep.* 14: 630–634.

1124. Rabouam, C., Comes, A.M., Bretagnolle, V., Humbert, J.-F., Periquet, G., and Bigot, Y., (1999) Features of DNA fragments obtained by random amplified polymorphic DNA (RAPD) analysis, *Mol. Ecol.* 8: 493–503.

1125. Rafalski, J.A., (1998) Random amplified polymorphic DNA (RAPD) analysis. In *DNA Markers: Protocols, Applications, and Overviews*, Caetano-Anollés, G., and Gresshoff, P., Eds., Wiley-VCH, New York, pp. 75–83.

1126. Rafalski, A., (2002) Applications of single nucleotide polymorphisms in crop genetics, *Curr. Opin. Plant Biol.* 5: 94–100.

1127. Rafalski, J.A., (2002) Novel genetic mapping tools in plants: SNPs and LD-based approaches, *Plant Sci.* 162: 329–333.
1128. Rafalski, J.A., and Tingey, S.V., (1993) Genetic diagnostics in plant breeding: RAPDs, microsatellites and machines, *Trends Genet.* 9: 275–280.
1129. Rahman, M.H., Jaquish, B., and Khasa, P.D., (2000) Optimization of PCR protocol in microsatellite analysis with silver and SYBR^R stains, *Plant Mol. Biol. Rep.* 18: 339–348.
1130. Raina, S.N., Rani, V., Kojima, T., Ogihara, Y., Singh, K.P., and Devarumath, R.M., (2001) RAPD and ISSR fingerprints as useful genetic markers for analysis of genetic diversity, varietal identification, and phylogenetic relationships in peanut (*Arachis hypogaea*) cultivars and wild species, *Genome* 44: 763–772.
1131. Rajapakse, S., Hubbard, M., Kelly, J., Abbott, A.G., and Ballard, R.E., (1992) Identification of rose cultivars by restriction fragment length polymorphism, *Sci. Hortic.* 52: 237–245.
1132. Rajesh, P.N., Sant, V.J., Gupta, V.S., Muehlbauer, F.J., and Ranjekar, P.K., (2002) Genetic relationships among annual and perennial wild species of *Cicer* using inter simple sequence repeat (ISSR) polymorphism, *Euphytica* 129: 15–23.
1133. Rajora, O.P., Rahman, M.H., Buchert, G.P., and Dancik, B.P., (2000) Microsatellite DNA analysis of genetic effects of harvesting in old-growth eastern white pine (*Pinus strobus*) in Ontario, Canada, *Mol. Ecol.* 9: 339–348.
1134. Rallo, P., Tenzer, I., Gessler, C., Baldoni, L., Dorado, G., and Martín, A., (2003) Transferability of olive microsatellite loci across the genus *Olea*, *Theor. Appl. Genet.* 107: 940–946.
1135. Ramsay, L., Macaulay, M., Cardle, L., Morgante, M., degli Ivanissevich, S., Maestri, E., Powell, W., and Waugh, R., (1999) Intimate association of microsatellite repeats with retrotransposons and other dispersed repetitive elements in barley, *Plant J.* 17: 415–425.
1136. Ramsay, L., Macaulay, M., degli Ivanissevich, S., MacLean, K., Cardle, L., Fuller, J., Edwards, K.J., Tuvesson, S., Morgante, M., Massari, A., Maestri, E., Marmiroli, N., Sjakste, T., Ganal, M., Powell, W., and Waugh, R., (2000) A simple sequence repeat-based linkage map of barley, *Genetics* 156: 1997–2005.
1137. Ramsay, M.M., and Stewart, J., (1998) Re-establishment of the lady's slipper orchid (*Cypripedium calceolus* L.) in Britain, *Bot. J. Linn. Soc.* 126: 173–181.
1138. Ramser, J., Lopez-Peralta, C., Wetzel, R., Weising, K., and Kahl, G., (1996) Genomic variation and relationships in aerial yam (*Dioscorea bulbifera* L) detected by random amplified polymorphic DNA, *Genome* 39: 17–25.
1139. Ramser, J., Weising, K., Lopez-Peralta, C., Terhalle, W., Terauchi, R., and Kahl, G., (1997) Molecular marker-based taxonomy and phylogeny of Guinea yam (*Dioscorea rotundata - D. cayenensis*), *Genome* 40: 903–915.
1140. Ramser, J., Weising, K., Chikaleke, V., and Kahl, G., (1997) Increased informativeness of RAPD analysis by detection of microsatellite motifs, *BioTechniques* 23: 285–290.
1141. Ranamukhaarachchi, D.G., Kane, M.E., Guy, C.L., and Li, Q.B., (2000) Modified AFLP technique for rapid genetic characterization in plants, *BioTechniques* 29: 858–866.
1142. Rapley, R., (1999) *The Nucleic Acid Protocols Handbook*, Humana Press, Totowa, NJ.
1143. Rassmann, K., Schlötterer, C., and Tautz, D., (1991) Isolation of simple-sequence loci for use in polymerase chain reaction-based DNA fingerprinting, *Electrophoresis* 12: 113–118.
1144. Deleted in proof.

1145. Raymond, M., and Rousset, F., (1995) GENEPOP (version 1.2): population genetics software for exact tests and ecumenicism, *J. Hered.* 86: 248–249.

1146. Reboud, X., and Zeyl, C., (1994) Organelle inheritance in plants, *Heredity* 72: 132–140.

1147. Reddy, M.P., Sarla, N., and Siddiq, E.A., (2002) Inter simple sequence repeat (ISSR) polymorphism and its application in plant breeding, *Euphytica* 128: 9–17.

1148. Regner, F., Stadbauer, A., and Eisenheld, C., (2001) Molecular markers for genotyping grapevine and for identifying clones of traditional varieties, *Acta Hortic.* 546: 331–341.

1149. Reineke, A., and Karlovsky, P., (2000) Simplified AFLP protocol: replacement of primer labeling by the incorporation of α-labeled nucleotides during PCR, *BioTechniques* 28: 622–623.

1150. Reineke, A., Karlovsky, P., and Zebitz, C.P.W., (1999) Suppression of randomly primed polymerase chain reaction products (random amplified polymorphic DNA) in heterozygous diploids, *Mol. Ecol.* 8: 1449–1455.

1151. Reiter, R.S., Williams, J.G.K., Feldmann, K.A., Rafalski, J.A., Tingey, S.V., and Scolnik, P.A., (1992) Global and local genome mapping in *Arabidopsis thaliana* by using recombinant inbred lines and random amplified polymorphic DNAs, *Proc. Natl. Acad. Sci. U.S.A.* 89: 1477–1481.

1152. Ren, N., and Timko, M.P., (2001) AFLP analysis of genetic polymorphism and evolutionary relationships among cultivated and wild *Nicotiana* species, *Genome* 44: 559–571.

1153. Rendell, S., and Ennos, R.A., (2002) Chloroplast DNA diversity in *Calluna vulgaris* (heather) populations in Europe, *Mol. Ecol.* 11: 69–78.

1154. Rether, B., Delmas, G., and Laouedj, A., (1993) Isolation of polysaccharide-free DNA from plants, *Plant Mol. Biol. Rep.* 11: 333–337.

1155. Reusch, T.B.H., Stam, W.T., and Olsen, J.L., (2000) A microsatellite-based estimation of clonal diversity and population subdivision in *Zostera marina*, a marine flowering plant, *Mol. Ecol.* 9: 127–140.

1156. Reymond, P., Weber, H., Damond, M., and Farmer, E.E., (2000) Differential gene expression in response to mechanical wounding and insect feeding in *Arabidopsis*, *Plant Cell* 12: 707–720.

1157. Reyna-López, G.E., Simpson, J., and Ruiz-Herrera, J., (1997) Differences in DNA methylation patterns are detectable during the dimorphic transition of fungi by amplification of restriction polymorphisms, *Mol. Gen. Genet.* 253: 703–710.

1158. Ribeiro, M.M., Plomion, C., Petit, R., Vendramin, G.G., and Szmidt, A.E., (2001) Variation in chloroplast simple-sequence repeats in Portuguese maritime pine (*Pinus pinaster* Ait.), *Theor. Appl. Genet.* 102: 97–103.

1159. Ribeiro, M.M., Mariette, S., Vendramin, G.G., Szmidt, A.E., Plomion, C., and Kremer, A., (2002) Comparison of genetic diversity estimates within and among populations of maritime pine using chloroplast simple-sequence repeat and amplified fragment length polymorphism data, *Mol. Ecol.* 11: 869–877.

1160. Richards, R.I., and Sutherland, G.R., (1992) Dynamic mutations: a new class of mutations causing human disease, *Cell* 70: 709–712.

1161. Richards, R.I., and Sutherland, G.R., (1994) Simple repeat DNA is not replicated simply, *Nat. Genet.* 6: 114–116.

1162. Richards, R.I., and Sutherland, G.R., (1997) Dynamic mutation: possible mechanisms and significance in human disease, *Trends Biochem. Sci.* 22: 432–436.

1163. Richardson, T., Cato, S., Ramser, J., Kahl, G., and Weising, K., (1995) Hybridization of microsatellites to RAPD: a new source of polymorphic markers, *Nucleic Acids Res.* 23: 3798–3799.

1164. Richmond, T., and Somerville, S., (2000) Chasing the dream: Plant EST microarrays, *Curr. Opin. Plant Biol.* 3: 108–116.

1165. Rico, C., Rico, I., and Hewitt, G., (1996) 470 million years of conservation of microsatellite loci among fish species, *Proc. R. Soc. London Ser. B* 263: 549–557.

1166. Riede, C.R., Fairbanks, D.J., Andersen, W.R., Kehrer, R.L., and Robison, L.R., (1994) Enhancement of RAPD analysis by restriction-endonuclease digestion of template DNA in wheat, *Plant Breed.* 113: 254–257.

1167. Riedy, M.F., Hamilton W.J., III, and Aquadro, C.F., (1992) Excess of non-parental bands in offspring from known primate pedigrees assayed using RAPD PCR, *Nucleic Acids Res.* 20: 918.

1168. Rieseberg, L.H., (1995) The role of hybridisation in evolution: old wine in new skin, *Am. J. Bot.* 82: 944–953.

1169. Rieseberg, L.H., (1996) Homology among RAPD fragments in interspecific comparisons, *Mol. Ecol.* 5: 99–105.

1170. Rieseberg, L.H., (1997) Hybrid origins of plant species, *Annu. Rev. Ecol. Syst.* 28: 359–389.

1171. Rieseberg, L.H., and Soltis, D.E., (1991) Phylogenetic consequences of cytoplasmic gene flow in plants, *Evol. Trends Plants* 5: 65–84.

1172. Rieseberg, L.H., Fossen, C.V., Desrochers, A.M., (1995) Hybrid speciation accompanied by genomic reorganization in the wild sun flower, *Nature* 375: 313–316.

1173. Rieseberg, L.H., Sinervo, B., Linder, C.R., Ungerer, M. C., and Arias, D.M., (1996) Role of gene interactions in hybrid speciation: evidence from ancient and experimental hybrids, *Science* 272: 741–745.

1174. Rigby, P.W.J., Dieckmann, M., Rhodes, C., and Berg, P., (1977) Labeling deoxyribonucleic acid to high specific activity in vitro by nick translation with DNA polymerase I, *J. Mol. Biol.* 113: 237–251.

1175. Ristaino, J.B., Groves, C.T., and Parra, G.R., (2001) PCR amplification of the Irish potato famine pathogen from historic specimens, *Nature* 411: 695–697.

1176. Rival, A., Bertrand, L., Beule, T., Combes, M.C., Trouslot, P., and Lashermes, P., (1998) Suitability of RAPD analysis for the detection of somaclonal variants in oil palm (*Elaeis guineensis* Jacq.), *Plant Breed.* 117: 73–76.

1177. Roberts, R.J., Vincze, T., Posfai, J., and Macelis, D., (2003) REBASE: restriction enzymes and methyltransferases, *Nucleic Acids Res.* 31: 418–420.

1178. Robinson, J., and Harris, S.A., (2000) A plastid DNA phylogeny of the genus *Acacia* Miller (Acaciaceae, Leguminosae), *Bot. J. Linn. Soc.* 132: 195–222.

1179. Röder, M.S., Plaschke, J., König, S.U., Börner, A., Sorrells, M.E., Tanksley, S.D., and Ganal, M.W., (1995) Abundance, variability and chromosomal location of microsatellites in wheat, *Mol. Gen. Genet.* 246: 327–333.

1180. Röder, M., Korzun, V., Gill, B.S., and Ganal, M.W., (1998) The physical mapping of microsatellite markers in wheat, *Genome* 41: 278–283.

1181. Röder, M.S., Korzun, V., Wendehake, K., Plaschke, J., Tixier, M.-H., Leroy, P., and Ganal, M., (1998) A microsatellite map of wheat, *Genetics* 149: 2007–2023.

1182. Rogers, H.J., Burns, N.A., and Parkes, H.C., (1996) Comparison of small-scale methods for the rapid extraction of plant DNA suitable for PCR analysis, *Plant Mol. Biol. Rep.* 14: 170–183.

1183. Rogers, S.O., and Bendich, A.J., (1985) Extraction of DNA from milligram amounts of fresh, herbarium and mummified plant tissues, *Plant Mol. Biol.* 5: 69–76.

1184. Rogstad, S.H., (1992) Saturated NaCl-CTAB solution as a means of field preservation of leaves for DNA analyses, *Taxon* 41: 701–708.

1185. Rogstad, S.H., (2003) Plant DNA extraction using silica, *Plant Mol. Biol. Rep.* 21: 463a–463g.

1186. Rogstad, S.H., Patton J.C., II, and Schaal, B.A., (1988) M13 repeat probe detects DNA minisatellite-like sequences in gymnosperms and angiosperms, *Proc. Natl. Acad. Sci. U.S.A.* 85: 9176–9178.

1187. Rogstad, S.H., Keane, B., Keiffer, C.H., Hebard, F., and Sisco, P., (2001) DNA extraction from plants: the use of pectinase, *Plant Mol. Biol. Rep.* 19: 353–359.

1188. Roman, B.L., Pham, V.N., Bennett, P.E., and Weinstein, B.M., (1999) Non-radioisotopic AFLP method using PCR primers fluorescently labeled with Cy™5, *BioTechniques* 26: 236–238.

1189. Rossetto, M., (2001) Sourcing of SSR markers from related plant species, in *Plant Genotyping: The DNA Fingerprinting of Plants*, Henry, R.J., Ed., CABI Publishing, Wallingford, U.K., pp. 211–224.

1190. Rousset, F., and Raymond, M., (1997) Statistical analyses of population genetic data: new tools, old concepts, *Trends Ecol. Evol.* 12: 313–317.

1191. Routman, E., and Cheverud, J., (1994) A rapid method of scoring simple sequence repeat polymorphisms with agarose gel electrophoresis, *Mamm. Genome* 5: 187–188.

1192. Roux, K.H., (1995) Optimization and troubleshooting in PCR, *PCR Method Appl.* 4: S185–S194.

1193. Rowland, L.J., and Nguyen, B., (1993) Use of polyethylene glycol for purification of DNA from leaf tissue of woody plants, *BioTechniques* 14: 735–736.

1194. Roy, J.K., Balyan, H.S., Prasad, M., and Gupta, P.K., (2002) Use of SAMPL for a study of DNA polymorphism, genetic diversity and possible gene tagging in bread wheat, *Theor. Appl. Genet.* 104: 465–472.

1195. Royle, N.J., Clarkson, R.E., Wong, Z., and Jeffreys, A.J., (1988) Clustering of hypervariable minisatellites in the proterminal regions of human autosomes, *Genomics* 3: 352–360.

1196. Rozen, S., and Skaletsky, H.J., (2000) Primer 3 on the WWW for general users and for biologist programmers. In *Bioinformatics Methods and Protocols: Methods in Molecular Biology*, Eds. Krawetz, S. and Misener, S., Humana Press, Totowa, N.J., pp. 365–386.

1197. Ruan, Y., Gilmore, J., and Conner, T. (1998) Towards *Arabidopsis* genome analysis: Monitoring expression profiles of 1400 genes using cDNA microarrays, *Plant J.* 15: 821–833.

1198. Rubinsztein, D.C., Amos, W., Leggo, J., Goodbourn, S., Jain, S., Li, S.-H., Margolis, R.L., Ross, C.A., and Ferguson-Smith, M.A., (1995) Microsatellite evolution — evidence for directionality and variation in rate between species, *Nat. Genet.* 10: 337–343.

1199. Rudi, K., Kroken, M., Dahlberg, O.J., Deggerdal, A., Jakobsen, K.S., and Larsen, F., (1997) Rapid, universal method to isolate PCR-ready DNA using magnetic beads, *BioTechniques* 22: 506–511.

1200. Rueda, J., Linacero, R., and Vasquez, A.M., (1998) Plant total DNA extraction. In *Molecular Tools for Screening Biodiversity*, Karp, A., Isaac, P.G., and Ingram, D.S., Eds., Chapman & Hall, London, pp. 10–17.

1201. Rumpunen, K., and Bartish, I.V., (2002) Comparison of differentiation estimates based on morphometric and molecular data, exemplified by various leaf shape descriptors and RAPDs in the genus *Chaenomeles* (Rosaceae), *Taxon* 51: 69–82.

1202. Russell, J.R., Fuller, J.D., Macaulay, M., Hatz, B.G., Jahoor, A., and Powell, W., (1997) Direct comparison of levels of genetic variation among barley accessions detected by RFLPs, AFLPs, SSRs and RAPDs, *Theor. Appl. Genet.* 95: 714–722.

1203. Rychlik, W., and Rhoads, R.E., (1989) A computer program for choosing optimal oligonucleotides for filter hybridization, sequencing and *in vitro* amplification of DNA, *Nucleic Acids Res.* 17: 8543–8551.

1204. Ryskov, A.P., Jincharadze, A.G., Prosnyak, M.I., Ivanov, P.L., and Limborska, S.A., (1988) M13 phage DNA as a universal marker for DNA fingerprinting of animals, plants and microorganisms, *FEBS Lett.* 233: 388–392.

1205. Saal, B., and Wricke, G., (2002) Clustering of amplified fragment length polymorphism markers in a linkage map of rye, *Plant Breed.* 121: 117–123.

1206. Saar, D.E., Polans, N.O., Sörensen, P.D., and Duvall, M.R., (2001) Angiosperm DNA contamination by endophytic fungi: detection and methods of avoidance, *Plant Mol. Biol. Rep.* 19: 249–260.

1207. Sachidanandam R., and 40 coauthors, (2001) A map of human genome sequence variation containing 1.42 million single nucleotide polymorphisms, *Nature* 409: 928–933.

1208. Saedler, H., and Gierl, A., (1996) *Transposable Elements*, Springer, Berlin.

1209. Saghai-Maroof, M.A., Soliman, K.M., Jorgensen, R.A., and Allard, R.W., (1984) Ribosomal DNA spacer-length polymorphisms in barley: Mendelian inheritance, chromosomal location, and population dynamics, *Proc. Natl. Acad. Sci. U.S.A.* 81: 8014–8018.

1210. Saghai-Maroof, M.A., Biyashev, R.M., Yang, G.P., Zhang, Q., and Allard, R.W., (1994) Extraordinarily polymorphic microsatellite DNA in barley: species diversity, chromosomal locations, and population dynamics, *Proc. Natl. Acad. Sci. U.S.A.* 91: 5466–5470.

1211. Saha, S., Karaca, M., Jenkins, J.N., Zipf, A.E., Reddy, O.U.K., and Kantety, R.V., (2003) Simple sequence repeats as useful resources to study transcribed genes of cotton, *Euphytica* 130: 355–364.

1212. Saiki, R.K., Gelfand, D.H., Stoffel, S., Scharf, S.J., Higuchi, R., Horn, G.T., Mullis, K.B., and Erlich, H.A., (1988) Primer-directed enzymatic amplification of DNA with a thermostable DNA polymerase, *Science* 239: 487–491.

1213. Saitou, N., and Nei, M., (1987) The neighbour joining method: a new method for reconstructing phylogenetic trees, *Mol. Biol. Evol.* 4: 406–425.

1214. Salama, N., Guillemin, K., McDaniel, T.K., Sherlock, G., Tompkins, L., and Falkow, S., (2000) A whole genome microarray reveals genetic diversity among *Helicobacter pylori* strains, *Proc. Natl. Acad. Sci. U.S.A.* 97: 14668–14673.

1215. Salimath, S.S., De Oliveira, A.C., Godwin, I.D., and Bennetzen, J.L., (1995) Assessment of genome origins and genetic diversity in the genus *Eleusine* with DNA markers, *Genome* 38: 757–763.

1216. Saltonstall, K., (2001) A set of primers for amplification of noncoding region of chloroplast DNA in the grasses, *Mol. Ecol. Notes* 1: 76–78.

1217. Sambrook, J., and Russell, D.W., (2001) *Molecular Cloning: A Laboratory Manual, 3rd Edition*, Cold Spring Harbor Laboratory Press, Cold Spring Harbor, NY.

1218. Sanchez de la Hoz, M.P., Dávila, J.A., Loarce, Y., and Ferrer, E., (1996) Simple sequence repeat primers used in polymerase chain reaction amplifications to study genetic diversity in barley, *Genome* 39: 112–117.

1219. Sanger, F., Nicklen, S., and Coulson, A.R., (1977) DNA sequencing with chain-terminating inhibitors, *Proc. Natl. Acad. Sci. U.S.A.* 74: 5463–5467.

1220. Sangwan, N.S., Sangwan, R.S., and Kumar, S., (1998) Isolation of genomic DNA from the antimalarial plant *Artemisia annua*, *Plant Mol. Biol. Rep.* 16: 365 (1–8).

1221. Sangwan, R.S., Yadav, U., and Sangwan, N.S., (2000) Isolation of genomic DNA from defatted oil seed residue of opium poppy (*Papaver somniferum*), *Plant Mol. Biol. Rep.* 18: 265–270.

1222. Sankar, A.A., and Moore, G.A., (2001) Evaluation of inter-simple sequence repeat analysis for mapping in *Citrus* and extension of the genetic linkage map, *Theor. Appl. Genet.* 102: 206–214.

1223. SanMiguel, P., Tikhonov, A., Jin, Y.-K., Motchoulskaya, N., Zakharov, D., Melake-Berhan, A., Springer, P.S., Edwards, K.J., Lee, M., Avramova, Z., and Bennetzen, J.L., (1996) Nested retrotransposons in the intergenic regions of the maize genome, *Science* 274: 765–768.

1224. Sarkar, G., and Sommer, S.S., (1990) Shedding light on PCR contamination, *Nature* 343: 27.

1225. Sarkar, G., Kapelner, S., and Sommer, S.S., (1990) Formamide can dramatically improve the specificity of PCR, *Nucleic Acids Res.* 18: 7465.

1226. Sato, S., Nakamura, Y., Kaneko, T., Asamizu, E., and Tabata, S., (1999) Complete structure of the chloroplast genome of *Arabidopsis thaliana*, *DNA Res.* 6: 283–290.

1227. Saunders, G.W., (1993) Gel purification of red algal genomic DNA: an inexpensive and rapid method for the isolation of polymerase chain reaction-friendly DNA, *J. Phycol.* 29: 251–254.

1228. Savolainen, V., Cuénoud, P., Spichiger, R., Martinez, M.D., Crèvecoeur, M., and Manen, J.-F., (1995) The use of herbarium specimens in DNA phylogenetics: evaluation and improvement, *Plant Syst. Evol.* 197: 87–98.

1229. Schaal, B.A., Hayworth, D.A., Olsen, K.M., Rauscher, J.T., and Smith, W.A., (1998) Phylogeographic studies in plants: problems and prospects, *Mol. Ecol.* 7: 465–474.

1230. Schadt, E.E., Monks, S.A., Drake, T.A., Lusis, A.J., Che, N., Colinayo, V., Ruff, T.G., Milligan, S.B., Lamb, J.R., Cavet, G., Linsley, P.S., Mao, M., Stoughton, R.B., and Friend, S.H., (2003) Genetics of gene expression surveyed in maize, mouse and man, *Nature* 422: 297–302.

1231. Schaffer, R., Landgraf, J., Perez-Amador, M., and Wisman, E. (2000) Monitoring genome-wide expression in plants, *Curr. Opin. Biotechnol.* 11: 162–167.

1232. Schena M., (2003) *Microarray Analysis*, Wiley-Liss, New York.

1233. Schena, M., Shalon, D., Davis, R.W., and Brown P.O., (1995) Quantitative monitoring of gene expression patterns with a complementary DNA microarray, *Science* 270: 467–470.

1234. Schierenbeck, K.A., (1994) Modified polyethylene glycol DNA extraction procedure for silica gel-dried tropical wood plants, *BioTechniques* 16: 393–394.

1235. Schierwater, B., and Ender, A., (1993) Different thermostable DNA polymerases may amplify different RAPD products, *Nucleic Acids Res.* 21: 4647–4648.

1236. Schlegel, J., Vogt, T., Münkel, K., and Rüschoff, J., (1996) DNA fingerprinting of mammalian cell lines using nonradioactive arbitrarily primed PCR (AP-PCR), *BioTechniques* 20: 178–180.

1237. Schlink, K., and Reski, R., (2002) Preparing high-quality DNA from moss *(Physcomitrella patens)*, *Plant Mol. Biol. Rep.* 20: 423a–423f.

1238. Schlötterer, C., and Tautz, D., (1992) Slippage synthesis of simple sequence DNA, *Nucleic Acids Res.* 20: 211–215.

1239. Schlötterer, C., Amos, B., and Tautz, D., (1991) Conservation of polymorphic simple sequence loci in cetacean species, *Nature* 354: 63–65.

1240. Schlötterer, C., Ritter, R., Harr, B., and Brem, G., (1998) High mutation rate of a long microsatellite allele in *Drosophila melanogaster* provides evidence for allele-specific mutation rates, *Mol. Biol. Evol.* 15: 1269–1274.

1241. Schmidt, R., (2000) Synteny: recent advances and future prospects, *Curr. Opin. Plant Biol.* 3: 97–102.

1242. Schmidt, R., (2002) Plant genome evolution: lessons from comparative genomics at the DNA level, *Plant Mol. Biol.* 48: 21–37.
1243. Schmidt, T., (1999) LINEs, SINEs and repetitive DNA: non-LTR retrotransposons in plant genomes, *Plant Mol. Biol.* 40: 903–910.
1244. Schmidt, T., and Heslop-Harison, J.S., (1996) The physical and genomic organization of microsatellites in sugar beet, *Proc. Natl. Acad. Sci. U.S.A.* 93: 8761–8765.
1245. Schmidt, T., Kubis, S., and Heslop-Harrison, J.S., (1995) Analysis and chromosomal localization of retrotransposons in sugar beet (*Beta vulgaris* L.): LINEs and Ty1-*copia*-like elements as major components of the genome, *Chromosome Res.* 3: 335–345.
1246. Schnabel, A. (1998) Parentage analysis in plants: mating systems, gene flow and relative fertilities, in *Advances in Molecular Ecology*, Carvalho, G., Ed., IOS Press, Amsterdam, pp. 173–189.
1247. Schneerman, M.C., Mwangi, J., Hobart, B., Arbuckle, J., Vaske, D.A., Register J.C., III, and Weber, D.F., (2002) The dried corncob as a source of DNA for PCR analysis, *Plant Mol Biol. Rep.* 20: 59–65.
1248. Schneider, K., Weisshaar, B., Borchardt, D.C., and Salamini, F., (2001) SNP frequency and allelic haplotype structure of *Beta vulgaris* expressed genes, *Mol. Breed.* 8: 63–74.
1249. Schoen, D.J., and Brown, A.H.D., (1993) Conservation of allelic richness in wild crop relatives is aided by assessment of genetic markers, *Proc. Natl. Acad. Sci. U.S.A.* 90: 10623–10627.
1250. Schug, M.D., Mackay, T.F.C., and Aquadro, C.F., (1997) Low mutation rates of microsatellite loci in *Drosophila melanogaster*, *Nat. Genet.* 15: 99–102.
1251. Schug, M.D., Hutter, C.M., Wetterstrand, K.A., Gaudette, M.S., Mackay, T.F.C., and Aquadro, C.F., (1998) The mutation rates of di-, tri- and tetranucleotide repeats in *Drosophila melanogaster*, *Mol. Biol. Evol.* 15: 1751–1760.
1252. Schultz, B., Wanke, U., Draeger, S., and Aust, H.-J., (1993) Endophytes from herbaceous plants and shrubs: effectiveness of surface sterilization methods, *Mycol. Res.* 97: 1447–1450.
1253. Schütze, P., Freitag, H., and Weising, K., (2003) An integrated molecular and morphological study of the subfamily Suaedoideae Ulbr. (Chenopodiaceae), *Plant Syst. Evol.* 239: 257–286.
1254. Schwarz, G., Herz, M., Huang, X.Q., Michalek, W., Jahoor, A., Wenzel, G., and Mohler, V., (2000) Application of fluorescence-based semi-automated AFLP analysis in barley and wheat, *Theor. Appl. Genet.* 100: 545–551.
1255. Schwarz, K., Hansen-Hagge, T., and Bartram, C., (1990) Improved yields of long PCR products using gene 32 protein, *Nucleic Acids Res.* 18: 1079.
1256. Schweder, M.E., Shatters, R.G., West, S.H., and Smith, R.L., (1995) Effect of transition interval between melting and annealing temperatures on RAPD analyses, *BioTechniques* 19: 38–42.
1257. Schweizer, G., Ganal, M., Ninnemann, H., and Hemleben, V., (1988) Species-specific DNA sequences for identification of somatic hybrids between *Lycopersicon esculentum* and *Solanum acaule*, *Theor. Appl. Genet.* 75: 679–684.
1258. Schwengel, D.A., Jedlicka, A.E., Nanthakumar, E.J., Weber, J.L., and Levitt, R.C., (1994) Comparison of fluorescence-based semi-automated genotyping of multiple microsatellite loci with autoradiographic techniques, *Genomics* 22: 46–54.
1259. Scott, D.L., Walker, M.D., Clarck, C.W., Prakash, C.S., and Deahl, K.L., (1998) Rapid assessment of primer combinations and recovery of AFLP products using ethidium bromide staining, *Plant Mol. Biol. Rep.* 16: 41–47.
1260. Scott, K.D., and Playford, J., (1996) DNA extraction technique for PCR in rain forest plant species, *BioTechniques* 20: 974–979.

1261. Scott, K.D., Eggler, P., Seaton, G., Rossetto, M., Ablett, E.M., Lee, L.S., and Henry, R.J., (2000) Analysis of SSRs derived from grape ESTs, *Theor. Appl. Genet.* 100: 723–726.

1262. Scott, M.P., Haymes, K.M., and Williams, S.M., (1992) Parentage analysis using RAPD PCR, *Nucleic Acids Res.* 20: 5493.

1263. Scotti, I., Magni, F., Fink, R., Powell, W., Binelli, G., and Hedley, P.E., (2000) Microsatellite repeats are not randomly distributed within Norway spruce (*Picea abies* K.) expressed sequences, *Genome* 43: 41–46.

1264. Scotti, I., Magni, F., Paglia, G.P., and Morgante, M., (2002) Trinucleotide microsatellites in Norway spruce (*Picea abies*): their features and the development of molecular markers, *Theor. Appl. Genet.* 106: 40–50.

1265. Scrimshaw, B.J., (1992) A simple nonradioactive procedure for visualization of $(dC-dA)_n$ dinucleotide repeat length polymorphisms, *BioTechniques* 13: 189.

1265a. Seaton, G., Haley, C.S., Knott, S.A., Kearsey, M., and Visscher, P.M., (2002) QTL Express: mapping quantitative trait loci in simple and complex pedigrees, *Bioinformatics* 18: 339–340.

1266. Sebastiani, F., Carnevale, S., and Vendramin, G.G., (2004) A new set of mono- and dinucleotide chloroplast microsatellites in Fagaceae, *Mol. Ecol. Notes* 4: 259–261.

1267. Sefc, K.M., Steinkellner, H., Glössl, J., Kampfer, S., and Regner, F., (1998) Reconstruction of a grapevine pedigree by microsatellite analysis, *Theor. Appl. Genet* 97: 227–231.

1268. Sefc, K.M., Lopes, M.S., Lefort, F., Botta, R., Roubelakis-Angelakis, K.A., Ibáñez, J., Pejic, I., Wagner, H.W., Glössl, J., and Steinkellner, H., (2000) Microsatellite variability in grapevine cultivars from different European regions and evaluation of assignment testing to assess the geographic origin of cultivars, *Theor. Appl. Genet.* 100: 498–505.

1269. Segarra-Moragues, J.G., and Catalan, P., (2003) Life history variation between species of the relictual genus *Borderea* (Dioscoreaceae): phylogeography, genetic diversity, and population genetic structure assessed by RAPD markers, *Biol. J. Linn. Soc.* 80: 483–498.

1270. Segarra-Moragues, J.G., Palop-Esteban, M., Gonzalez-Candelas, F., and Catalan, P., (2004) Characterisation of seven $(CTT)_n$ microsatellites loci in the Pyrenean endemic *Borderea pyrenaica* (Dioscoreaceae): remarks on ploidy level and hybrid origin assessed through allozymes and microsatellite analyses, *J. Hered.* 95: 177–183.

1271. Seki, M., Narusaka, M., Abe, H., Kasuga, M., Yamaguchi-Shinozaki, K., Carninci, P., Hayashizaki, Y., and Shinozaki K., (2001) Monitoring the expression pattern of 1300 *Arabidopsis* genes under drought and cold stresses by using a full-length cDNA microarray, *Plant Cell* 13: 61–72.

1272. Shan, X., Blake, T.K., and Talbert, L.E., (1999) Conversion of AFLP markers to sequence-specific PCR markers in barley and wheat, *Theor. Appl. Genet.* 98: 1072–1078.

1273. Sharma, A.D., Gill, P.K., and Singh, P., (2002) DNA isolation from dry and fresh samples of polysaccharide-rich plants, *Plant Mol. Biol. Rep.* 20: 415a–415f.

1274. Sharma, J.K., Gopalkrishna, V., and Das, B.C., (1992) A simple method for elimination of unspecific amplifications in polymerase chain reaction, *Nucleic Acids Res.* 20: 6117–6118.

1275. Sharma, K.K., Lavanya, M., and Anjaiah, V., (2000) A method for isolation and purification of peanut genomic DNA suitable for analytical applications, *Plant Mol. Biol. Rep.* 18: 393a–393h.

1276. Sharma, P.C., Hüttel, B., Winter, P., Kahl, G., Gardner, R.C., and Weising, K., (1995) The potential of microsatellites for hybridization- and polymerase chain reaction-based DNA fingerprinting of chickpea (*Cicer arietinum* L.) and related species, *Electrophoresis* 16: 1755–1761.

1277. Sharma, P.C., Winter, P., Bünger, T., Hüttel, B., Weigand, F., Weising, K., and Kahl, G., (1995) Abundance and polymorphism of di-, tri- and tetra-nucleotide tandem repeats in chickpea (*Cicer arietinum* L.), *Theor. Appl. Genet.* 90: 90–96.

1278. Sharma, R., John, S.J., Damgaard, D.M., and McAllister, T.A., (2003) Extraction of PCR-quality plant and microbial DNA from total rumen contents, *BioTechniques* 34: 92–97.

1279. Sharma, R., Mahla, H.R., Mohapatra, T., Bhargava, S.C., and Sharma, M.M., (2003) Isolating plant genomic DNA without liquid nitrogen, *Plant Mol. Biol. Rep.* 21: 43–50.

1280. Shashidhara, G., Hema, M.V., Koshy, B., and Farooqi, A.A., (2003) Assessment of genetic diversity and identification of core collection in sandalwood germplasm using RAPDs, *J. Hort. Sci. Biotechnol.* 78: 528–536.

1281. Sheffield, V.C., Beck, J.S., Kwitek, A.E., Sandstrom, D.W., and Stone, E.M., (1993) The sensitivity of single-strand conformation polymorphism analysis for the detection of single base substitutions, *Genomics* 16: 325–332.

1282. Shepherd, M., Cross, M., Maguire, T.L., Dieters, M.J., Williams, C.G., and Henry, R.J., (2002) Transpecific microsatellites for hard pines, *Theor. Appl. Genet.* 104: 819–827.

1283. Shepherd, M., Cross, M., Stokoe, R.L., Scott, L.J., and Jones, M.E., (2002) High-throughput DNA extraction from forest trees, *Plant Mol. Biol. Rep.* 20: 425a–425j.

1284. Sherman, J.D., and Talbert, L.E., (2002) Vernalization-induced changes of the DNA methylation pattern in winter wheat, *Genome* 45: 253–260.

1285. Sherry, S.T., Ward, M.-H., Kholodov, M., Baker, J., Phan, L., Smigielski, E.M., and Sirotkin, K. (2001) dbSNP: the NCBI database of genetic variation, *Nucleic Acids Res.* 29: 308–311.

1286. Shibata, K., Bandoh, K., Yaekashiwa, N., Matsuzaka, T., and Tamate, H.B., (2003) A simple method for isolation of microsatellites from the Japanese squirrel, *Sciurus lis*, without constructing a genomic library, *Mol. Ecol. Notes* 3: 657–658.

1287. Shimoda, N., Knapik, E.W., Ziniti, J., Sim, C., Yamada, E., Kaplan, S., Jackson, D., de Sauvage, F., Jacob, H., and Fishman, M.C., (1999) Zebrafish genetic map with 2000 microsatellite markers, *Genomics* 58: 219–232.

1288. Shoemaker, J.S., (1999) Bayesian statistics in genetics: a guide for the uninitiated, *Trends Genet.* 15: 354–358.

1289. Siebert, P.D., Chenchik, A., Kellogg, D.E., Lukyanov, K.A., and Lukyanov, S.A., (1995) An improved PCR method for walking in uncloned genomic DNA, *Nucleic Acids Res.* 23: 1087–1088.

1290. Simel, E.J., Saidak, L.R., and Tuskan, G.A., (1997) Method for extracting genomic DNA from non-germinated gymnosperm and angiosperm pollen, *BioTechniques* 22: 390–394.

1291. Simons, G., van der Lee, T., Diergaarde, P., van Daelen, R., Groenendijk, J., Frijters, A., Büschges, R., Hollricher, K., Töpsch, S., Schulze-Lefert, P., Salamini, F., Zabeau, M., and Vos, P. (1997) AFLP-based fine mapping of the *mlo* gene to a 30-kb DNA segment of the barley genome, *Genomics* 44: 61–70.

1292. Singh, A., Chaudhury, A., Srivastava, P.S., and Lakshmikumaran, M., (2002) Comparison of AFLP and SAMPL markers for assessment of intra-populational genetic variation in *Azadirachta indica* A. Juss, *Plant Sci.* 162: 17–25.

1293. Singh, B.M., and Ahuja, P.S., (1999) Isolation and PCR amplification of genomic DNA from market samples of dry tea, *Plant Mol. Biol. Rep.* 17: 171–178.

1294. Skinner, D.M., Beattie, W.G., and Blattner, F.R., (1974) The repeat sequence of a hermit crab satellite deoxyribonucleic acid is $(\text{-T-A-G-G-})_n$ x $(\text{-A-T-C-C-})_n$, *Biochemistry* 13: 3930–3937.

1295. Skroch, P.W., and Nienhuis, J., (1995) Qualitative and quantitative characterization of RAPD variation among snap bean (*Phaseolus vulgaris*) genotypes, *Theor. Appl. Genet.* 91: 1078–1085.

1296. Skroch, P.W., and Nienhuis, J., (1995) Impact of scoring error and reproducibility of RAPD data on RAPD based estimates of genetic distance, *Theor. Appl. Genet.* 91: 1086–1091.

1297. Slatkin, M., (1995) A measure of population subdivision based on microsatellite allele frequencies, *Genetics* 139: 457–462.

1298. Smeets, H.J.M., Brunner, H.G., Ropers, H.-H., and Wieringa, B., (1989) Use of variable simple sequence motifs as genetic markers: application to study of myotonic dystrophy, *Hum Genet.* 93: 245–251.

1299. Smith, L.M., Sanders, J.Z., Kaiser, R.J., Hughes, P., Dodd, C., Connell, C.R., Heiner, C., Kent, S.B.H., and Hood, L.E., (1986) Fluorescence detection in automated DNA sequence analysis, *Nature* 321: 674–679.

1300. Smith, S., Hughes, J., and Wardell-Johnson, G., (2003) High population differentiation and extensive clonality in a rare mallee eucalypt: *Eucalyptus curtisii*, *Conserv. Genet.* 4: 289–300.

1301. Smouse, P.E., Long, J.C., and Sokal, R.R., (1986) Multiple regression and correlation extensions of the Mantel test of matrix correspondence. *Syst. Zool.* 35: 627–632.

1302. Smulders, M.J.M., Bredemeijer, G., Rus-Kortekaas, W., Arens, P., and Vosman, B., (1997) Use of short microsatellites from database sequences to generate polymorphisms among *Lycopersicon esculentum* cultivars and accessions of other *Lycopersicon* species, *Theor. Appl. Genet.* 94: 264–272.

1303. Sobral, B.W.S., and Honeycutt, R.J., (1993) High output genetic mapping of polyploids using PCR-generated markers, *Theor. Appl. Genet.* 86: 105–112.

1304. Sokal, R.R., and Oden, N.L., (1978) Spatial autocorrelation in biology. 1. Methodology, *Biol. J. Linn. Soc.* 10: 199–228.

1305. Soleimani, V.D., Baum, B.R., and Johnson, D.A., (2003) Efficient validation of single nucleotide polymorphisms in plants by allele-specific PCR, with an example from barley, *Plant Mol. Biol. Rep.* 21: 281–288.

1306. Soltis, P.S., and Soltis, D.E., (1998) Molecular evolution of 18S rDNA in angiosperms: implications for character weighting in phylogenetic analysis. In *Molecular Systematics of Plants II: DNA Sequencing*, Soltis, D.E., Soltis, P.S., and Doyle, J.J., Eds., Kluwer Academic, Dordrecht, pp. 188–210.

1307. Soltis, D.E., Soltis, P.S., and Milligan, B.G., (1992) Intraspecific chloroplast DNA variation: systematic and phylogenetic implications. In *Molecular Systematics of Plants*, Soltis, P.S., Soltis, D.E., and Doyle, J.J., Eds., Chapman & Hall, New York, pp. 117–150.

1308. Soltis, D.E., Soltis, P.S., and Doyle, J.J., Eds., (1998) *Molecular Systematics of Plants II: DNA Sequencing*, Kluwer Academic, Dordrecht.

1309. Somers, D.J., Zhou, Z., Bebeli, P.J., and Gustafson, J.P., (1996) Repetitive, genome-specific probes in wheat (*Triticum aestivum* L. em Thell) amplified with minisatellite core sequences, *Theor. Appl. Genet.* 93: 982–989.

1310. Somers, D.J., Kirkpatrick, R., Moniwa, M., and Walsh, A., (2003) Mining single-nucleotide polymorphisms from hexaploid wheat ESTs, *Genome* 49: 431–437.

1311. Song, Q.J., Fickus, E.W., and Cregan, P.B., (2002) Characterization of trinucleotide SSR motifs in wheat, *Theor. Appl. Genet.* 104: 286–293.

1312. Song, W.Y., Wang, G.L., Chen, L.L., Kim, H.S., Pi, L.Y., Holsten, T., Gardner, J., Wang, B., Zhai, W.X., Zhu, L.H., Fauquet, C., and Ronald, P., (1995) A receptor kinase-like protein encoded by the rice disease resistance gene *Xa21*, *Science* 270: 1804–1806.

1313. Soranzo, N., Provan, J., and Powell, W., (1999) An example of microsatellite length variation in the mitochondrial genome of conifers, *Genome* 42: 158–161.

1314. Sork, V.L., Nason, J., Campbell, D.R., Fernandez, J.F., (1999) Landscape approaches to historical and contemporary gene flow in plants, *Trends Ecol. Evol.* 14: 219–224.

1315. Sourdille, P., Tavaud, M., Charmet, G., and Bernard, M., (2001) Transferability of wheat microsatellites to diploid Triticeae species carrying the A, B and D genomes, *Theor. Appl. Genet.* 103: 346–352.

1316. Southern, E.M., (1975) Detection of specific sequences among DNA fragments separated by gel electrophoresis, *J. Mol. Biol.* 98: 503–517.

1317. Spencer, P.B.S., Odorico, D.M., Jones, S.J., Marsh, H.D., and Miller, D.J., (1995) Highly variable microsatellites in isolated colonies of the rock-wallaby (*Petrogale assimilis*), *Mol. Ecol.* 4: 523–525.

1318. Sperisen, C., Büchler, U., Gugerli, F., Matyas, G., Geburek, T., and Vendramin, G.G., (2001) Tandem repeats in plant mitochondrial genomes: application to the analysis of population differentiation in the conifer Norway spruce, *Mol. Ecol.* 10: 257–263.

1319. Spooner, D.M., Tivang, J., Nienhuis, J., Miller, J.T., Douches, D.S., and Contreras, M.A., (1996) Comparison of four molecular markers in measuring relationships among the wild potato relatives *Solanum* section *Etuberosum* (Subgenus *Potatoe*), *Theor. Appl. Genet.* 92: 532–540.

1320. Squirrell, J., and Wolff, K., (2001) Isolation of polymorphic microsatellite loci in *Plantago major* and *P. intermedia*, *Mol. Ecol. Notes* 1: 179–181.

1321. Squirrell, J., Hollingsworth, P.M., Bateman, R.M., Dickson, J.H., Light, M.H.S., MacConaill, M., and Tebbitt, M.C., (2001) Partitioning and diversity of nuclear and organelle markers in native and introduced populations of *Epipactis helleborine* (Orchidaceae), *Am. J. Bot.* 88: 1409–1418.

1322. Squirrell, J., Hollingsworth, P.M., Woodhead, M., Russell, J., Lowe, A.J., Gibby, M., and Powell, W., (2003) How much effort is required to isolate nuclear microsatellites from plants?, *Mol. Ecol.* 12: 1339–1348.

1323. Stam, P., (1993) Construction of integrated genetic linkage maps by means of a new computer package: JOINMAP, *Plant J.* 3: 739–744.

1324. Stange, C., Prehn, D., and Arce-Johnson, P., (1998) Isolation of *Pinus radiata* genomic DNA suitable for RAPD analysis, *Plant Mol. Biol. Rep.* 16: 366 (1–8).

1325. Starman, T.W., and Abbitt, S., (1997) Evaluating genetic relationships of geranium using arbitrary signatures from amplification profiles, *HortScience* 32: 1288–1291.

1326. Staub, J., Bacher, J., and Poetter, K., (1996) Sources of potential errors in the application of random amplified polymorphic DNAs in cucumber, *HortScience* 31: 262–266.

1327. Staub, J.E., Danin-Poleg, Y., Fazio, G., Horejsi, T., Reis, N., and Katzir, N., (2000) Comparative analysis of cultivated melon groups (*Cucumis melo* L.) using random amplified polymorphic DNA and simple sequence repeat markers, *Euphytica* 115: 225–241.

1328. Steemers, F.J., Ferguson, J.A., and Walt, D.R., (2000) Screening unlabeled DNA targets with randomly ordered fiber-optic gene arrays, *Nat. Biotechnol.* 18: 91–94.

1329. Stein, D.B., (1993) Isolation and comparison of nucleic acids from land plants: nuclear and organellar genes, *Meth. Enzymol.* 224: 153–167.

1330. Steiner, J.J., Poklemba, C.J., Fjellstrom, R.G., and Elliott, L.F., (1995) A rapid one-tube genomic DNA extraction process for PCR and RAPD analyses, *Nucleic Acids Res.* 23: 2569–2570.

1331. Steinger, T., Korner, C., and Schmid, B., (1996) Long-term persistence in a changing climate: DNA analysis suggests very old ages of clones of alpine *Carex curvula*, *Oecologia* 105: 94–99.

1332. Steinkellner, H., Lexer, C., Turetschek, E., and Glössl, J., (1997) Conservation of (GA)$_n$ microsatellite loci between *Quercus* species, *Mol. Ecol.* 6: 1189–1194.

1333. Stewart, C.N., and Porter, D.M., (1995) RAPD profiling in biological conservation: an application to estimating clonal variation in rare and endangered *Iliamna* in Virginia, *Biol. Conserv.* 74: 135–142.

1334. Stewart, C.N., and Via, L.E., (1993) A rapid CTAB DNA isolation technique useful for RAPD fingerprinting and other PCR applications, *BioTechniques* 14: 748–750.

1335. Stommel, J.R., Panta, G.R., Levi, A., and Rowland, L.J., (1997) Effects of gelatin and BSA on the amplification reaction for generating RAPD, *BioTechniques* 22: 1064–1066.

1336. Stone, J., and Björklund, M., (2001) Delirious: a computer program designed to analyse molecular marker data and calculate delta and relatedness estimates with confidence, *Mol. Ecol. Notes* 1: 209–212.

1337. Storchova, H., Hrdlickova, R., Chrtek, J., Jr., Tetera, M., Fitze, D., and Fehrer, J. (2000) An improved method of DNA isolation from plants collected in the field and conserved in saturated NaCl/CTAB solution, *Taxon* 49: 79–84.

1338. Strain, S.R., and Chmielewski, J.G., (2001) ROCK: a spreadsheet-based program for the generation and analysis of random oligonucleotide primers used in PCR, *BioTechniques* 30: 1286–1291.

1339. Strand, A.E., Leebens-Mack, J., and Milligan, B.G., (1997) Nuclear DNA-based markers for plant evolutionary biology, *Mol. Ecol.* 6: 113–118.

1340. Streiff, R., Labbe, T., Bacilieri, R., Steinkellner, H., Glössl, J., and Kremer, A., (1998) Within-polulation genetic structure in *Quercus robur* L. and *Quercus petraea* (Matt.) Liebl. assessed with isozymes and microsatellites, *Mol. Ecol.* 7: 317–328.

1341. Su, X.-Z., Wu, Y., Sifri, C.D., and Wellems, T.E., (1996) Reduced extension temperatures required for PCR amplification of extremely A+T-rich DNA, *Nucleic Acids Res.* 24: 1574–1575.

1342. Suazo, A., and Hall, H.G., (1999) Modification of the AFLP protocol applied to honey bee (*Apis mellifera* L.) DNA, *BioTechniques* 26: 704–709.

1343. Sun, G.L., Díaz, O., Salomon, B., and von Bothmer, R., (1998) Microsatellite variation and its comparison with allozyme and RAPD variation in *Elymus fibrosus* (Schrenk) Tzvel. (Poaceae), *Hereditas* 129: 275–282.

1344. Sun, G.L., Díaz, O., Salomon, B., and von Bothmer, R., (1999) Genetic diversity in *Elymus caninus* as revealed by isozyme, RAPD, and microsatellite markers, *Genome* 42: 420–431.

1345. Sunnucks, P., Wilson, A.C.C., Beheregaray, L.B., Zenger, K., French, J., and Taylor, A.C., (2000) SSCP is not so difficult: the application and utility of single-stranded conformation polymorphism in evolutionary biology and molecular ecology, *Mol. Ecol.* 9: 1699–1710.

1346. Suoniemi, A., Tanskanen, J., and Schulman, A.H., (1998) *Gypsy*-like retrotransposons are widespread in the plant kingdom, *Plant J.* 13: 699–705.

1347. Sutherland, G.R., and Richards, R.I., (1995) Simple tandem DNA repeats and human genetic disease, *Proc. Natl. Acad. Sci. U.S.A.* 92: 3636–3641.

1348. Sutherland, G.R., Baker, E., and Richards, R.I., (1998) Fragile sites still breaking, *Trends Genet.* 14: 501–506.

1349. Suyama, Y., Obayashi, K., and Hayashi, I., (2000) Clonal structure in a dwarf bamboo (*Sasa senanensis*) population inferred from amplified fragment length polymorphism (AFLP) fingerprints, *Mol. Ecol.* 9: 901–906.

1350. Swallow, D.M., Gendler, S., Griffiths, B., Corney, G., Taylor-Papadimitriou, J., and Bramwell, M.E, (1987) The human tumour-associated epithelial mucins are coded by an expressed hypervariable gene locus PUM, *Nature* 328: 82–84.

1351. Swensen, S.M., Allan, G.J., Howe, M., Elisens, W.J., Junak, S.A., and Rieseberg,
       L.H., (1995) Genetic analysis of the endangered island endemic *Malacothamnus
       fasciculatus* (Nutt.) Greene var. *nesioticus* (Rob.) Kearn. (Malvaceae). *Conserv. Biol.*
       9: 404–415.

1352. Swofford, D.L., Olsen, G.J., Waddell, P.J., and Hillis, D.M., (1996) Phylogenetic
       inference. In *Molecular Systematics, 2nd Edition*, Hillis, D.M., Moritz, C., and Mable,
       B.K., Eds., Sinauer Associates, Sunderland, MA, pp. 407–514.

1353. Syamkumar, S., Lowarence, B., and Sasikumar, B., (2003) Isolation and amplification
       of DNA from rhizomes of turmeric and ginger, *Plant Mol. Biol. Rep.* 21: 171a–171e.

1354. Sydes, M.A., and Peakall, R., (1998) Extensive clonality in the endangered shrub
       Haloragodendron lucasii (Haloragaceae) revealed by allozymes and RAPD, *Mol.
       Ecol.* 7: 87–93.

1355. Sytsma, K.J., Givnish, T.J., Smith, J.F., and Hahn, W.J., (1993) Collection and storage
       of land plant samples for macromolecular comparisons, *Meth. Enzymol.* 224: 23–37.

1356. Syvanen, A.C., (2001) Genotyping single nucleotide polymorphisms, *Nat. Rev. Genet.*
       2: 930–942.

1357. Szmidt, A.E., Nilsson, M.C., Briceno, E., Zackrisson, O., and Wang, X.R., (2002)
       Establishment and genetic structure of *Empetrum hermaphroditum* populations in
       northern Sweden, *J. Veg. Sci.* 13: 627–634.

1358. Taberlet, P., Gielly, L., Pautou, G., and Bouvet, J., (1991) Universal primers for
       amplification of three non-coding regions of chloroplast DNA, *Plant Mol. Biol.* 17:
       1105–1109.

1359. Taberlet, P., Fumagalli, L., Wust-Saucy, A.-G., and Cosson, J.-F., (1998) Comparative
       phylogeography and postglacial colonization routes in Europe, *Mol. Ecol.* 7: 453–464.

1360. Taguchi, G., (1986) *Introduction to Quality Engineering*, Asian Productivity Organi-
       zation, Unipub, New York.

1361. Tai, T.H., and Tanksley, S.D., (1990) A rapid and inexpensive method for isolation
       of total DNA from dehydrated plant tissue, *Plant Mol. Biol. Rep.* 8: 297–303.

1362. Tang, S., Yu, J.-K., Slabaugh, M.B., Shintani, D.K., and Knapp, S.J., (2002) Simple
       sequence repeat map of the sunflower genome, *Theor. Appl. Genet.* 105: 1124–1136.

1363. Tang, S., Kishore, V.K., and Knapp, S.J., (2003) PCR-multiplexes for a genome-wide
       framework of simple sequence repeat marker loci in cultivated sunflower, *Theor. Appl.
       Genet.* 107: 6–19.

1364. Tanksley, S.D., Young, N.D., Paterson, A.H., and Bonierbale, M.W., (1989) RFLP
       mapping in plant breeding: new tools for an old science, *Bio/Technology* 7: 257–264.

1365. Tanksley, S.D., Ganal, M.W., and Martin, G.B., (1995) Chromosomal landing: a
       paradigm for map-based gene cloning in plants with large genomes, *Trends Genet.*
       11: 63–68.

1366. Taramino, G., and Tingey, S., (1996) Simple sequence repeats for germplasm analysis
       and mapping in maize, *Genome* 39: 277–287.

1367. Tautz, D., (1989) Hypervariability of simple sequences as a general source for poly-
       morphic DNA markers, *Nucleic Acids Res.* 17: 6463–6471.

1368. Tautz, D., (1993) Notes on the definition and nomenclature of tandemly repetitive
       DNA sequences. In *DNA Fingerprinting: State of the Science*, Pena, S.D.J.,
       Chakraborty, R., Epplen, J.T., and Jeffreys, A.J., Eds., Birkhäuser, Basel, Switzerland,
       pp. 21–28.

1369. Tautz, D., and Renz, M., (1984) Simple sequences are ubiquitous repetitive compo-
       nents of eukaryotic genomes, *Nucleic Acids Res.* 12: 4127–4138.

1370. Tautz, D., and Schlötterer, C., (1994) Simple sequences, *Curr. Opin. Genet. Dev.* 4:
       832–837.

1371. Tautz, D., Trick, M., and Dover, G.A., (1986) Cryptic simplicity in DNA is a major source of genetic variation, *Nature* 322: 652–656.

1372. Taylor, J.S., and Breden, F., (2000) Slipped-strand mispairing at noncontiguous repeats in *Poecilia reticulata*: a model for minisatellite birth, *Genetics* 155: 1313–1330.

1373. Taylor, J.S., Durkin, J.M.H., and Breden, F., (1999) The death of a microsatellite: a phylogenetic perspective on microsatellite interruptions, *Mol. Biol. Evol.* 16: 567–572.

1374. Taylor, J.W., and Swann, E.C., (1994) DNA from herbarium specimens. In *Ancient DNA*, Hermann, B., and Hummel, S., Eds., Springer, New York, pp. 166–181.

1375. Taylor, P.W.J., Geijskes, J.R., Ko, H.-L., Fraser, T.A., Henry, R.J., and Birch, R.G., (1995) Sensitivity of random amplified polymorphic DNA analysis to detect genetic change in sugarcane during tissue culture, *Theor. Appl. Genet.* 90: 1169–1173.

1376. Tegelström. H., (1992) Detection of mitochondrial DNA fragments. In *Molecular Genetic Analysis of Populations: A Practical Approach*, Hoelzel, A.R., Ed., IRL Press, Oxford, U.K., pp. 89–114.

1377. Tel-Zur, N., Abbo, S., Myslabodski, D., and Mizrahi, Y., (1999) Modified CTAB procedure for DNA isolation from epiphytic cacti of the genera *Hylocereus* and *Selenicereus* (Cactaceae), *Plant Mol. Biol. Rep.* 17: 249–254.

1378. Temnykh, S., Park, W.D., Ayres, N., Cartinhour, S., Hauck, N., Lipovich, L., Cho, Y.G., Ishii, T., and McCouch, S.R., (2000) Mapping and genome organization of microsatellite sequences in rice (*Oryza sativa* L.), *Theor. Appl. Genet.* 100: 697–712.

1379. Temnykh, S., DeClerck, G., Lukashova, A., Lipovich, L., Cartinhour, S., and McCouch, S., (2001) Computational and experimental analysis of microsatellites in rice (*Oryza sativa* L.): frequency, length variation, transposon associations, and genetic marker potential, *Genome Res.* 11: 1441–1452.

1380. Templeton, A.R., (1998) Nested clade analysis of phylogeographic data: testing hypotheses about gene flow and population history, *Mol. Ecol.* 7: 381–397.

1381. Templeton, A.R., (2004) Statistical phylogeography: methods of evaluating and minimizing inference errors, *Mol. Ecol.* 13: 789–809.

1382. Templeton, A.R., Routman, E., and Phillips, C.A., (1995) Separating population structure from population history: a cladistic analysis of the geographical distribution of mitochondrial DNA haplotypes in the tiger salamander, *Ambystoma tigrinum*, *Genetics* 14: 767–782.

1383. Tenaillon, M.I., Sawkins, M.C., Long, A.D., Gaut, R.L., Doebley, J.F., and Gaut, B.S., (2001) Patterns of DNA sequence polymorphism along chromosome 1 of maize (*Zea mays* ssp. *mays* L.), *Proc. Natl. Acad. Sci. U.S.A.* 98: 9161–9166.

1384. Tessier, C., David, J., This, P., Boursiquot, J.M., and Charrier, A., (1999) Optimization of the choice of molecular markers for varietal identification in *Vitis vinifera* L., *Theor. Appl. Genet.* 98: 171–177.

1385. The *Arabidopsis* Genome Initiative, (2000) Analysis of the genome sequence of the flowering plant *Arabidopsis thaliana*, *Nature* 408: 796–815.

1386. The *C. elegans* Sequencing Consortium, (1998) Genome sequence of the nematode *C. elegans*: a platform for investigating biology, *Science* 282: 2012–2018.

1387. The International HapMap Consortium, (2003) The International HapMap Project, *Nature* 426: 789–796.

1388. The International Human Genome Sequencing Consortium, (2001) Initial sequencing and analysis of the human genome, *Nature* 409: 860–921.

1389. The Mouse Genome Sequencing Consortium, (2002) Initial sequencing and comparative analysis of the mouse genome, *Nature* 420: 520–562.

1390. The Rat Genome Sequencing Project Consortium, (2004) Genome sequence of the brown Norway rat yields insights into mammalian evolution, *Nature* 428: 493–521.

1391. Thein, S.L., and Wallace, B., (1986) The use of synthetic oligonucleotides as specific hybridization probes in the diagnosis of genetic disorders, in *Human Genetic Diseases — A Practical Approach*, Davies, K.E., Ed., IRL Press, Oxford, U.K., pp. 33–50.

1392. Thiel, T., Michalek, W., Varshney, R.K., and Graner, A., (2003) Exploiting EST databases for the development and characterization of gene-derived SSR-markers in barley (*Hordeum vulgare* L.), *Theor. Appl. Genet.* 106: 411–422.

1393. Thomas, M.R., and Scott, N.S., (1993) Microsatellite repeats in grapevine reveal DNA polymorphisms when analysed as sequence-tagged sites, *Theor. Appl. Genet.* 86: 985–990.

1394. Thomas, M.R., Cain, P., and Scott, N.S., (1994) DNA typing of grapevines: a universal methodology and database for describing cultivars and evaluating genetic relatedness, *Plant Mol. Biol.* 25: 939–949.

1395. Thompson, J.D., Gibson, T.J., Plewniak, F., Jeanmougin, F., and Higgins, D.G., (1997) The CLUSTAL X windows interface: flexible strategies for multiple sequence alignment aided by quality analysis tools, *Nucleic Acids Res.* 25: 4876–4882.

1396. Thomson, D., and Henry, R., (1993) Use of DNA from dry leaves for PCR and RAPD analysis, *Plant Mol. Biol. Rep.* 11: 202–206.

1397. Thomson, D., and Henry, R., (1995) Single-step protocol for preparation of plant tissue for analysis by PCR, *BioTechniques* 19: 394–400.

1398. Thormann, C.E., Ferreira, M.E., Camargo, L.E.A., Tivang, J.G., and Osborn, T.C., (1994) Comparison of RFLP and RAPD markers to estimating genetic relationships within and among cruciferous species, *Theor. Appl. Genet.* 88: 973–980.

1399. Thrall, P.H., and Young, A., (2000) AUTOTET: a program for analysis of autotetraploid genotypic data, *J. Hered.* 91: 348–349.

1400. Thuillet, A.-C., Bru, D., David, J., Roumet, P., Santoni, S., Sourdille, P., and Bataillon, T., (2002) Direct estimation of mutation rate for 10 microsatellite loci in durum wheat, *Triticum turgidum* (L.) Thell. ssp. *durum*, *Mol. Biol. Evol.* 19: 122–125.

1401. Tillib, S.V., and Mirzabekov, A.D., (2001) Advances in the analysis of DNA sequence variations using oligonucleotide microchip technology, *Curr. Opin. Biotechnol.* 12: 53–58.

1402. Tingey, S.V., and del Tufo, J.P., (1993) Genetic analysis with random amplified polymorphic DNA markers, *Plant Physiol.* 101: 349–352.

1403. Tivang, J., Skroch, P.W., Nienhuis, J., and de Vos, N., (1996) Randomly amplified polymorphic DNA (RAPD) variation among and within artichoke (*Cynara scolymus* L.) cultivars and breeding populations, *J. Am. Soc. Hortic. Sci.* 121: 783–788.

1404. Todokoro, S., Terauchi, R., and Kawano, S., (1995) Microsatellite polymorphism in natural populations of *Arabidopsis thaliana* in Japan, *Jpn. J. Genet.* 70: 543–554.

1405. Tohme, J., Gonzalez, D.O., Beebe, S., and Duque, M.C., (1996) AFLP analysis of gene pools of a wild bean core collection, *Crop Sci.* 36: 1375–1384.

1406. Tollefsrud, M.M., Bachmann, K., Jakobsen, K.S., and Brochmann, C., (1998) Glacial survival does not matter — II: RAPD phylogeography of Nordic *Saxifraga cespitosa*, *Mol. Ecol.* 7: 1219–1232.

1407. Tomiuk, J., and Loeschke, V., (2003) Comments on: evolutionary and statistical properties of three genetic distances (Kalinowski, 2002), *Mol. Ecol.* 12: 2275–2277.

1408. Tomiuk, J., Guldbrandtsen, B., and Loeschke, V., (1998) Population differentiation through mutation and drift — a comparison of genetic identity measures, *Genetica* 102/103: 545–558.

1409. Tomkins, J.P., Wood, T.C., Barnes, L.S., Westman, A., and Wing, R.A., (2001) Evaluation of genetic variation in the daylily (*Hemerocallis* spp.) using AFLP markers, *Theor. Appl. Genet.* 102: 489–496.

1410. Torres, E., Iriondo, J.M., Escudero, A., and Pérez, C., (2003) Analysis of within-population spatial genetic structure in *Antirrhinum microphyllum* (Scrophulariaceae), *Am. J. Bot.* 90: 1688–1695.

1411. Tosto, D.S., and Hopp, H.E., (2000) Suitability of AFLP markers for the study of genomic relationships within the *Oxalis tuberosa* alliance, *Plant Syst. Evol.* 223: 201–209.

1412. Toth, G., Gaspari, Z., and Jurka, J., (2000) Microsatellites in different eukaryotic genomes: survey and analysis, *Genome Res.* 10: 967–981.

1413. Tourmente, S., Deragon, J.M., Lafleuriel, J., Tutois, S., Cuvillier, C., Espagnol, M.C., and Picard, G., (1994) Characterization of minisatellites in *Arabidopsis thaliana* with sequence similarity to the human minisatellite core sequence, *Nucleic Acids Res.* 22: 3317–3321.

1414. Tourmente, S., Lazreg, A., Lafleuriel, J., Cuvillier, C., Espagnol, M.C., and Picard, G., (1998) Identification of new minisatellites loci in *Arabidopsis thaliana*, *J. Exp. Bot.* 49: 21–25.

1415. Tragoonrung, S., Kanazin, V., Hayes, P.M., and Blake, T.K., (1992) Sequence-tagged-site-facilitated PCR for barley genome mapping, *Theor. Appl. Genet.* 84: 1002–1008.

1416. Trepicchio, W.L., and Krontiris, T.G., (1992) Members of the *rel/NF-kB* family of transcriptional regulatory proteins bind the HRAS1 minisatellite DNA sequence, *Nucleic Acids Res.* 20: 2427–2434.

1417. Trepicchio, W.L., and Krontiris, T.G., (1993) IGH minisatellite suppression of USF-binding-site- and Eu-mediated transcriptional activation of the adenovirus late promoter, *Nucleic Acids Res.* 21: 977–985.

1418. Triboush, S.O., Danilenko, N.G., and Davydenko, O.G., (1998) A method for isolation of chloroplast DNA and mitochondrial DNA from sunflower, *Plant Mol. Biol. Rep.* 16: 183–189.

1419. Trigiano, R.N., Scott, M.C., and Caetano-Anollés, G., (1998) Genetic signatures from amplification profiles characterize DNA mutation in somatic and radiation-induced sports of chrysanthemum, *J. Am. Soc. Hortic. Sci.* 123: 642–646.

1420. Triglia, T., Peterson, M.G., and Kemp, D.J., (1988) A procedure for *in vitro* amplification of DNA segments that lie outside the boundaries of known sequences, *Nucleic Acids Res.* 16: 8186.

1421. Tsao, S.G.S., Brunk, C.F., and Pearlman, R.E., (1983) Hybridization of nucleic acids directly in agarose gels, *Anal. Biochem.* 131: 365–372.

1422. Tsumura, Y., Ohba, K., and Strauss, S.H., (1996) Diversity and inheritance of inter-simple sequence repeat polymorphisms in Douglas-fir (*Pseudotsuga menziesii*) and sugi (*Cryptomeria japonica*), *Theor. Appl. Genet.* 92: 40–45.

1423. Tulsieram, L.K., Glaubitz, J.C., Kiss, G., and Carlson, J.E., (1992) Single tree genetic linkage mapping in conifers using haploid DNA from megagametophytes, *Bio/Technology* 10: 686–690.

1424. Turcotte, K., Srinivasan, S., and Bureau, T., (2001) Survey of transposable elements from rice genomic sequences, *Plant J.* 25: 169–179.

1425. Turpeinen, T., Tenhola, T., Manninen, O., Nevo, O., and Nissilä, E., (2001) Microsatellite diversity associated with ecological factors in *Hordeum spontaneum* populations in Israel, *Mol. Ecol.* 10: 1577–1592.

1426. Turri, M.G., Cuin, K.A., and Porter, A.C., (1995) Characterisation of a novel mini-satellite that provides multiple splice donor sites in an interferon-induced transcript, *Nucleic Acids Res.* 23: 1854–1861.

1427. Tyrka, M., Dziadczyk, P., and Hortynski, J.A., (2002) Simplified AFLP procedure as a tool for identification of strawberry cultivars and advanced breeding lines, *Euphytica* 125: 273–280.

1428. Ude, G., Pillay, M., Nwakanma, D., and Tenkouano, A., (2002) Analysis of genetic diversity and sectional relationships in *Musa* using AFLP markers, *Theor. Appl. Genet.* 104: 1239–1245.

1429. Udupa, S.M., and Baum, M., (2001) High mutation rate and mutational bias at $(TAA)_n$ microsatellite loci in chickpea (*Cicer arietinum* L.), *Mol. Genet. Genomics* 265: 1097–1103.

1430. Ueno, S., Yoshimaru, H., Tomaru, N., and Yamamoto, S., (1999) Development and characterization of microsatellite markers in *Camellia japonica* L., *Mol. Ecol.* 8: 335–336.

1431. Ueno, S., Tsumura, Y., and Washitani, I., (2003) Cost-effective method to synthesize fluorescently labeled DNA size standards using cloned AFLP fragments, *BioTechniques* 34: 1146–1148.

1432. Unkles, S.E., Duncan, J.M., and Kinghorn, J.R., (1992) Zinc fingerprinting for *Phytophthora* species: ZIF markers, *Curr. Genet.* 22: 317–318.

1433. Unseld, M., Marienfeld, J.R., Brandt, P., and Brennicke, A., (1997) The mitochondrial genome of *Arabidopsis thaliana* contains 57 genes in 366,924 nucleotides, *Nat. Genet.* 15: 57–61.

1434. Uptmoor, R., Wenzel, W., Friedt, W., Donaldson, G., Ayisi, K., and Ordon, F., (2003) Comparative analysis on the genetic relatedness of *Sorghum bicolor* accessions from Southern Africa by RAPDs, AFLPs and SSRs, *Theor. Appl. Genet.* 106: 1316–1325.

1435. Urquhart, A., Oldroyd, N.J., Kimpton, C.P., and Gill, P., (1995) Highly discriminating heptaplex short tandem repeat PCR system for forensic identification, *BioTechniques* 18: 116–121.

1436. Vaillancourt, R.E., and Jackson, H.D., (2000) A chloroplast DNA hypervariable region in eucalpyts, *Theor. Appl. Genet.* 101: 473–477.

1437. Valdes, A.M., Slatkin, M., and Freimer, N.B., (1993) Allele frequencies at microsatellite loci: the stepwise mutation model revisited, *Genetics* 133: 737–749.

1438. Van Belkum, A., Scherer, S., Van Alphen, L., and Verbrugh, H., (1998) Short-sequence DNA repeats in prokaryotic genomes, *Microbiol. Mol. Biol. Rev.* 62: 275–293.

1439. Van Daelen, R.A.J., Jonkers, J.J., and Zabel, P., (1989) Preparation of megabase-sized tomato DNA and separation of large restriction fragments by field inversion gel electrophoresis, *Plant Mol. Biol.* 12: 341–352.

1440. Van De Casteele, T., Galbusera, P., and Matthysen, E., (2001) A comparison of microsatellite-based pairwise relatedness estimators, *Mol. Ecol.* 10: 1539–1549.

1441. Van de Wouw, M., Maxted, N., Chabane, K., and Ford-Lloyd, B.V., (2001) Molecular taxonomy of *Vicia* ser. *Vicia* based on amplified fragment length polymorphisms, *Plant Syst. Evol.* 229: 91–105.

1442. Van den Berg, R.G., Bryan, G.J., Del Rio, A., and Spooner, D.M., (2002) Reduction of species in the wild potato *Solanum* section *Petota* series *Longipedicellata*: AFLP, RAPD and chloroplast SSR data, *Theor. Appl. Genet.* 105: 1109–1114.

1443. Van den Broeck, D., Maes, T., Sauer, M., Zetho, J., De Keukeleire, P., D'Hauw, M., Van Montagu, M., and Gerats, T. (1998) Transposon display identifies individual transposable elements in high copy number lines, *Plant J.* 13: 121–129.

1444. Van der Hulst, R.G.M., Mes, T.H.M., den Nijs, J.C.M., and Bachmann, K., (2000) Amplified fragment length polymorphism (AFLP) markers reveal that population structure of triploid dandelions (*Taraxacum officinale*) exhibits both clonality and recombination, *Mol. Ecol.* 9: 1–8.

1445. Van der Hulst, R.G.M., Mes, T.H.M., Falque, M., Stam, P., Den Nijs, J.C.M., and Bachmann, K., (2003) Genetic structure of a population sample of apomictic dandelion, *Heredity* 90: 326–335.

1446. Van der Velde, M, Van de Zande, L., and Bijlsma, R., (2001) Genetic structure of *Polytrichum formosum* in relation to the breeding system as revealed by microsatellites, *J. Evol. Biol.* 14: 288–295.

1447. Van der Wurff, A.W.G., Chan, Y.L., Van Straalen, N.M., and Schouten, J., (2000) TE-AFLP: combining rapidity and robustness in DNA fingerprinting, *Nucleic Acids Res.* 28: e105.

1448. Van der Wurff, A.W.G., Isaaks, J.A., Ernsting, G., and Van Straalen, N.M., (2003) Population structures in the soil invertebrate *Orchesella cincta*, as revealed by microsatellite and TE-AFLP markers, *Mol. Ecol.* 12: 1349–1359.

1449. Van Droogenbroeck, B., Breyne, P., Goetghebeur, P., Romeijn-Peeters, E., Kyndt, T., and Gheysen, G., (2002) AFLP analysis of genetic relationships among papaya and its wild relatives (Caricaceae) from Ecuador, *Theor. Appl. Genet.* 105: 289–297.

1450. Van Eeuwijk, F.A., and Baril, C.P., (2001) Conceptual and statistical issues related to the use of molecular markers for distinctness and essential derivation, *Acta Hortic.* 546: 35–53.

1451. Van Ham, R.C.H.J., 't Hart, H., Mes, T.H.M., and Sandbrink, J.M., (1994) Molecular evolution of noncoding regions of the chloroplast genome in the Crassulaceae and related species, *Curr. Genet.* 25: 558–566.

1452. Van Leven, F., (1991) The trouble with PCR machines: fill up the empty spaces!, *Trends Genet.* 7: 142.

1453. Van Raamsdonk, L.W.D., Vrielink-van Ginkel, M., and Kik, C., (2000) Phylogeny reconstruction and hybrid analysis in *Allium* subgenus *Rhizirideum*, *Theor. Appl. Genet.* 100: 1000–1009.

1454. Van Treuren, R., Kuittinen, H., Kärkkäinen, K., Baena-Gonzalez, E., and Savolainen, O., (1997) Evolution of microsatellites in *Arabis petraea* and *Arabis lyrata*, outcrossing relatives of *Arabidopsis thaliana*, *Mol. Biol. Evol.* 14: 220–229.

1455. Varadarajan, G.S., and Prakash, C.S., (1991) A rapid and efficient method for the extraction of total DNA from the sweet potato and its related species, *Plant Mol. Biol. Rep.* 9: 6–12.

1456. Vassart, G., Georges, M., Monsieur, R., Brocas, H., Lequarre, A.S., and Christophe, D., (1987) A sequence in M13 phage detects hypervariable minisatellites in human and animal DNA, *Science* 235: 683–684.

1457. Vazquez-Marrufo, G., Vazquez-Garciduenas, M.S., Gómez-Luna, B.E., and Olalde-Portugal, V., (2002) DNA isolation from forest soil suitable for PCR assays of fungal and plant rRNA genes, *Plant Mol. Biol. Rep.* 20: 379–390.

1458. Vekemans, X., and Jacquemart, A.L., (1997) Perspectives on the use of molecular markers in plant population biology, *Belg. J. Bot.* 129: 91–100.

1459. Vekemans, X., Beauwens, T., Lemaire, M., and Roldán-Ruiz, I., (2002) Data from amplified fragment length polymorphism (AFLP) markers show indication of size homoplasy and of a relationship between degree of homoplasy and fragment size, *Mol. Ecol.* 11: 139–152.

1460. Velculescu, V.E., Zhang, I., Vogelstein, B., and Kinzler, K.W., (1995) Serial analysis of gene expression, *Science* 270: 484–487.

1461. Vendramin, G.G., and Ziegenhagen, B., (1997) Characterisation and inheritance of polymorphic plastid microsatellites in *Abies, Genome* 40: 857–864.

1462. Vendramin, G.G., Lelli, L., Rossi, P., and Morgante, M., (1996) A set of primers for the amplification of 20 chloroplast microsatellites in Pinaceae, *Mol. Ecol.* 5: 595–598.

1463. Vendramin, G.G., Anzidei, M., Madaghiele, A., and Bucci, G., (1998) Distribution of genetic diversity in *Pinus pinaster* Ait. as revealed by chloroplast microsatellites, *Theor. Appl. Genet.* 97: 456–463.

1464. Vendramin, G.G., Deen, B., Petit, R.J., Anzidei, M., Madaghiele, A., and Ziegenhagen, B., (1999) High level of variation at *Abies alba* chloroplast microsatellite loci in Europe, *Mol. Ecol.* 8: 1117–1126.

1465. Vergnaud, G., and Denoeud, F., (2000) Minisatellites: mutability and genome architecture, *Genome Res.* 10: 899–907.

1466. Vergnaud, G., Mariat, D., Apiou, F., Aurias, A., Lathrop, M., and Lauthier, V., (1991) The use of synthetic tandem repeats to isolate new VNTR loci: cloning of a human hypermutable sequence, *Genomics* 11: 135–144.

1467. Viard, F., El-Kassaby, Y.A., and Ritland, K., (2001) Diversity and genetic structure in populations of *Pseudotsuga menziesii* (Pinaceae) at chloroplast microsatellite loci, *Genome* 44: 336–344.

1468. Viard, F., Bernard, J., and Desplanque, B., (2002) Crop-weed interactions in the *Beta vulgaris* complex at a local scale: allelic diversity and gene flow within sugar beet fields, *Theor. Appl. Genet.* 104: 688–697.

1469. Vigouroux, Y., Jaqueth, J.S., Matsuoka, Y., Smith, O.S., Beavis, W.D., Smith, J.S.C., and Doebley, J., (2002) Rate and pattern of mutation at microsatellite loci in maize, *Mol. Biol. Evol.* 19: 1251–1260.

1470. Vilain, C., Libert, F., Venet, D., Costagliola, S., and Vassart, G., (2003) Small amplified RNA-SAGE: an alternative approach to study transcriptome from limiting amount of mRNA, *Nucleic Acids Res.* 31: e24.

1471. Virk, P.S., Newbury, H.J., Jackson, M.T., and Ford-Lloyd, B.V., (1995) The identification of duplicate accessions within a rice germplasm collection using RAPD analysis, *Theor. Appl. Genet.* 90: 1049–1055.

1472. Virk, P.S., Pooni, H.S., Syed, N.H., and Kearsey, M.J., (1999) Fast and reliable genotype validation using microsatellite markers in *Arabidopsis thaliana, Theor. Appl. Genet.* 98: 462–464.

1473. Virk, P.S., Zhu, J., Newbury, H.J., Bryan, G.J., Jackson, M.T., and Ford-Lloyd, B.V., (2000) Effectiveness of different classes of molecular marker for classifying and revealing variation in rice (*Oryza sativa*) germplasm, *Euphytica* 112: 275–284.

1474. Vlácilová, K., Ohri, D., Vrána J., Cihaliková, J., Kubaláková, M., Kahl, G., and Dolezel, J., (2002) Development of flow cytogenetics and physical genome mapping in chickpea (*Cicer arietinum* L.), *Chromosome Res.* 10: 695–706.

1475. Vogel, J.M., and Scolnik, P.A., (1998) Direct amplification from microsatellites: detection of simple sequence repeat-based polymorphisms without cloning. In *DNA Markers: Protocols, Applications, and Overviews*, Caetano-Anollés, G., and Gresshoff, P.M., Eds., Wiley-VCH, New York, pp. 133–150.

1476. Vogel, M., Bänfer, G., Moog, U., and Weising, K., (2003) Development and characterization of chloroplast microsatellite markers in *Macaranga* (Euphorbiaceae), *Genome* 46: 845–857.

1477. Vogelstein, B., and Gillespie, D., (1979) Preparative and analytical purification of DNA from agarose, *Proc. Natl. Acad. Sci. U.S.A.* 76: 615–619.

1478. Vogl, C., Karhu, A., Moran, G., and Savolainen, O., (2002) High resolution analysis of mating systems: inbreeding in natural populations of *Pinus radiata*, *J. Evol. Biol.* 15: 433–439.

1479. Von Post, R., Von Post, L., Dayteg, C., Nilsson, M., Forster, B.P., and Tuvesson, S., (2003) A high-throughput DNA extraction method for barley seed, *Euphytica* 130: 255–260.

1480. Vos, P., and Kuiper, M., (1998) AFLP analysis. In *DNA Markers: Protocols, Applications, and Overviews*, Caetano-Anollés, G., and Gresshoff, P.M., Eds., Wiley-VCH, New York, pp. 115–131.

1481. Vos, P., Hogers, R., Bleeker, M., Reijans, M., Van de Lee, T., Hornes, M., Frijters, A., Pot, J., Peleman, J., Kuiper, M., and Zabeau, M., (1995) AFLP: a new technique for DNA fingerprinting, *Nucleic Acids Res.* 23: 4407–4414.

1482. Vosman, B., and Arens, P., (1997) Molecular characterization of GATA/GACA microsatellite repeats in tomato, *Genome* 40: 25–33.

1483. Voytas, D.F., Cummings, M.P., Konieczny, A., Ausubel, F.M., and Rodermel, S.R., (1992) *copia*-like retrotransposons are ubiquitous among plants, *Proc. Natl. Acad. Sci. U.S.A.* 89: 7124–7128.

1484. Vrieling, K., Peters, J., and Sandbrink, H., (1997) Amplified fragment length polymorphisms (AFLPs) detected with non-radioactive digoxigenin labelled primers in three plant species, *Plant Mol. Biol. Rep.* 15: 255–262.

1485. Wachira, F., Tanaka, J., and Takeda, Y., (2001) Genetic variation and differentiation in tea (*Camellia sinensis*) germplasm revealed by RAPD and AFLP variation, *J. Hortic. Sci. Biotechnol.* 76: 557–563.

1486. Wahls, W.P., and Moore, P.D., (1998) Recombination hotspot activity of hypervariable minisatellite DNA requires minisatellite DNA binding proteins, *Somat. Cell Mol. Genet.* 24: 41–51.

1487. Wahls, W.P., Swenson, G., and Moore, P.D., (1991) Two hypervariable minisatellite DNA binding proteins, *Nucleic Acids Res.* 19: 3269–3274.

1488. Waits, L.P., Luikart, G., and Taberlet, P., (2001) Estimating the probability of identity among genotypes in natural populations: cautions and guidelines, *Mol. Ecol.* 10: 249–256.

1489. Waldbieser, G.C., (1995) PCR-based identification of AT-rich tri- and tetranucleotide repeat loci in an enriched plasmid library, *BioTechniques* 19: 742–744.

1490. Walsh, P.S., Metzger, D.A., and Higuchi, R., (1991) Chelex^R 100 as a medium for simple extraction of DNA for PCR-based typing from forensic material, *BioTechniques* 10: 506–513.

1491. Walsh, P.S., Fildes, N.J., and Reynolds, R., (1996) Sequence analysis and characterization of stutter products at the tetranucleotide repeat locus vWA, *Nucleic Acids Res.* 24: 2807–2812.

1492. Walter, R., and Epperson, B.K., (2001) Geographic pattern of genetic variation in *Pinus resinosa*: area of greatest diversity is not the origin of postglacial populations, *Mol. Ecol.* 10: 103–111.

1493. Wan, C.-Y., and Wilkins, T.A., (1993) Spermidine facilitates PCR amplification of target DNA, *PCR Methods Appl.* 3: 208–210.

1494. Wang D.G., and 24 coauthors, (1998) Large-scale identification, mapping and genotyping of single nucleotide polymorphisms in the human genome, *Science* 280: 1077–1082.

1495. Wang, G., Mahalingam, R., and Knap, H.T., (1998) (C-A) and (G-A) anchored simple sequence repeats (ASSRs) generated polymorphism in soybean, *Glycine max* (L.) Merr., *Theor. Appl. Genet.* 96: 1086–1096.

1496. Wang, G.-L., Wing, R.A., and Paterson, A.H., (1993) PCR amplification from single seeds, facilitating DNA marker-assisted breeding, *Nucleic Acids Res.* 21: 2527.

1497. Wang, H., Qi, M., and Cutler, A.J., (1993) A simple method of preparing plant samples for PCR, *Nucleic Acids Res.* 21: 4153–4154.

1498. Wang, X.-D., Wang, Z.-P., and Zou, Y.-P., (1996) An improved procedure for the isolation of nuclear DNA from leaves of wild grapevine dried with silica gel, *Plant Mol. Biol. Rep.* 14: 369–373.

1499. Wang, Z., Weber, J.L., Zhong, G., and Tanksley, S.D., (1994) Survey of plant short tandem DNA repeats, *Theor. Appl. Genet.* 88: 1–6.

1500. Warburton, M.L., and Bliss, F. A., (1996) Genetic diversity in peach (*Prunus persica* (L.) Batch) revealed by randomly amplified polymorphic DNA (RAPD) markers and compared to inbreeding coefficients, *J. Am. Soc. Hortic. Sci.* 121: 1012–1019.

1501. Warburton, M.L., Becerra-Velasquez, V.L., Goffreda, J.C., and Bliss, F.A., (1996) Utility of RAPD markers in identifying genetic linkages to genes of economic interest in peach, *Theor. Appl. Genet.* 93: 920–925.

1502. Wartell, R.M., Hosseini, S.H., and Moran, C.P., (1990) Detecting base pair substitutions in DNA fragments by temperature gradient gel electrophoresis, *Nucleic Acids Res.* 18: 2699–2705.

1503. Warude, D., Chavan, P., Joshi, K., and Patwardhan, B., (2003) DNA isolation from fresh and dry plant samples with highly acidic tissue extracts, *Plant Mol. Biol. Rep.* 21: 467a–467f.

1504. Waser, P.M., and Strobeck, C., (1998) Genetic signatures of interpopulation dispersal, *Trends Ecol. Evol.* 13: 43–44.

1505. Watson, J.C., and Thompson, W.F., (1986) Purification and restriction endonuclease analysis of plant nuclear DNA, *Meth. Enzymol.* 118: 57–75.

1506. Wattier, R., Engel, C.R., Saumitou-Laprade, P., and Vakero, M., (1998) Short allele dominance as a source of heterozygote deficiency at microsatellite loci: experimental evidence at the dinucleotide locus Gv1CT in *Gracilaria gracilis* (Rhodophyta), *Mol. Ecol.* 7: 1569–1573

1507. Deleted in proof.

1508. Wattier, R.A., Prodöhl, P.A., and Maggs, C.A., (2000) DNA isolation protocol for red seaweed (Rhodophyta), *Plant Mol. Biol. Rep.* 18: 275–281.

1509. Waugh, R., and Powell, W., (1992) Using RAPD markers for crop improvement, *Trends Biotechnol.* 10: 186–191.

1510. Waugh, R., McLean, K., Flavell, A.J., Pearce, S.R., Kumar, A., Thomas, B.B.T., and Powell, W., (1997) Genetic distribution of *Bare*-1-like retrotransposable elements in the barley genome revealed by sequence-specific amplification polymorphisms (S-SAP), *Mol. Gen. Genet.* 253: 687–694.

1511. Webb, D.M., and Knapp, S.J., (1990) DNA extraction from a previously recalcitrant plant genus, *Plant Mol. Biol. Rep.* 8: 180–185.

1512. Weber, J.L., (1990) Informativeness of human (dC-dA)$_n$ x (dG-dT)$_n$ polymorphisms, *Genomics* 7: 524–530.

1513. Weber, J.L., and May, P.E., (1989) Abundant class of human DNA polymorphisms which can be typed using the polymerase chain reaction, *Am. J. Hum. Genet.* 44: 388–396.

1514. Weber, J.L., and Myers, E.W., (1997) Human whole-genome shotgun sequencing, *Genome Res.* 7: 401–409.

1515. Weeks, D.P., Beerman, N., and Griffith, O.M., (1986) A small-scale five-hour procedure for isolating multiple samples of CsCl-purified DNA: application to isolations from mammalian, insect, higher plant, algal, yeast, and bacterial sources, *Anal. Biochem.* 152: 376–385.

1516. Weining, S., and Langridge, P., (1991) Identification and mapping of polymorphisms in cereals based on the polymerase chain reaction, *Theor. Appl. Genet.* 82: 209–216.

1517. Weir B.S., and Cockerham, C.C., (1984) Estimating F-statistics for the analysis of population structure, *Evolution* 38: 1358–1370.

1518. Weir, B.S., (1996) *Genetic Data Analysis II*, Sinauer Associates, Sunderland, Mass.

1519. Weising, K., and Kahl, G., (1998) Hybridization-based microsatellite fingerprinting of plants and fungi, In *DNA Markers: Protocols, Applications, and Overviews*, Caetano-Anollés, G., and Gresshoff, P.M., Eds., Wiley-VCH, New York, pp. 27–54.

1520. Weising, K., and Gardner, R.C., (1999) A set of conserved PCR primers for the analysis of simple sequence repeat polymorphisms in chloroplast genomes of dicotyledonous angiosperms, *Genome* 42: 9–19.

1521. Weising, K., Weigand, F., Driesel, A., Kahl, G., Zischler, H., and Epplen, J.T., (1989) Polymorphic simple GATA/GACA repeats in plant genomes, *Nucleic Acids Res.* 17: 10128.

1522. Weising, K., Beyermann, B., Ramser, J., and Kahl, G., (1991) Plant DNA fingerprinting with radioactive and digoxigenated oligonucleotide probes complementary to simple repetitive DNA sequences, *Electrophoresis* 12: 159–169.

1523. Weising, K., Kaemmer, D., Weigand, F., Epplen, J.T., and Kahl, G., (1992) Oligonucleotide fingerprinting reveals various probe-dependent levels of informativeness in chickpea (*Cicer arietinum*), *Genome* 35: 436–442.

1524. Weising, K., Atkinson, R.G., and Gardner, R.C., (1995) Genomic fingerprinting by microsatellite-primed PCR: a critical evaluation, *PCR Methods Appl.* 4: 249–255.

1525. Weising, K., Fung, R.W.M., Keeling, D.J., Atkinson, R.G., and Gardner, R.C., (1996) Characterisation of microsatellites from *Actinidia chinensis*, *Mol. Breed.* 2: 117–131.

1526. Weising, K., Winter, P., Hüttel, B., and Kahl, G., (1998) Microsatellite markers for molecular breeding, *J. Crop Prod.* 1: 113–143.

1527. Welsh, J., and McClelland, M., (1990) Fingerprinting genomes using PCR with arbitrary primers, *Nucleic Acids Res.* 18: 7213–7218.

1528. Welsh, J., and McClelland, M., (1991) Genomic fingerprints produced by PCR with consensus tRNA gene primers, *Nucleic Acids Res.* 19: 861–866.

1529. Welsh, J., and McClelland, M., (1991) Genomic fingerprinting using arbitrarily primed PCR and a matrix of pairwise combinations of primers, *Nucleic Acids Res.* 19: 5275–5279.

1530. Welsh, J., Chada, K., Dalal, S.S., Cheng, R., Ralph, D., and McClelland, M., (1992) Arbitrarily primed PCR fingerprinting of RNA, *Nucleic Acids Res.* 20: 4965–4970.

1531. Wenz, H.-M., Robertson, J.M., Menchen, S., Oaks, F., Demorest, D.M., Scheibler, D., Rosenblum, B.B., Wike, C., Gilbert, D.A., and Efcavitch, J.W., (1998) High-precision genotyping by denaturing capillary electrophoresis, *Genome Res.* 8: 69–80.

1532. Wessler, S.R., Bureau, T., and White, S.E., (1995) LTR-retrotransposons and MITEs: important players in the evolution of plant genomes, *Curr. Opin. Genet. Dev.* 5: 814–821.

1533. Westman, A.L., and Kresovich, S., (1998) The potential for cross-taxa simple-sequence repeat (SSR) amplification between *Arabidopsis thaliana* L. and crop brassicas, *Theor. Appl. Genet.* 96: 272–281.

1534. Westneat, D.F., Noon, W.A., Reeve, H.K., and Aquadro, C.F., (1988) Improved hybridization conditions for DNA "fingerprints" probed with M13, *Nucleic Acids Res.* 16: 4161.

1535. White, E., Sahota, R., and Edes, S., (2002) Rapid microsatellite analysis using discontinuous polyacrylamide gel electrophoresis, *Genome* 45: 1107–1109.

1536. White, G.W., and Powell, W., (1997) Isolation and characterization of microsatellite loci in *Swietenia humilis* (Meliaceae): an endangered tropical hardwood species, *Mol. Ecol.* 6: 851–860.

1537. White, G.M., Boshier, D.H., and Powell, W., (1999) Genetic variation within a fragmented population of *Swietenia humilis* Zucc., *Mol. Ecol.* 8: 1899–1909.

1538. White, H.W., and Kusukawa, N., (1997) Agarose-based system for separation of short tandem repeat loci. *BioTechniques* 22: 976–980.

1539. White, M.B., Carvalho, M., Derse, D., O'Brien, S.J., and Dean, M., (1992) Detecting single-base substitutions as heteroduplex polymorphisms, *Genomics* 12: 301–306.

1540. Whitlock, M.C., and McCauley, D.E., (1999) Indirect measures of gene flow and migration: $F_{ST} \neq 1/(4Nm + 1)$, *Heredity* 82: 117–125.

1541. Whitton, J., Rieseberg, L.H., and Ungerer, M.C., (1997) Microsatellite loci are not conserved across the Asteraceae, *Mol. Biol. Evol.* 14: 204–209.

1542. Wiesner, I., and Wiesnerová, D., (2003) Effect of resolving medium and staining procedure on inter-simple-sequence-repeat (ISSR) patterns in cultivated flax germplasm, *Genet. Res. Crop Evol.* 50: 849–853.

1543. Wilkie, S.E., Isaac, P.G., and Slater, R.J., (1993) Random amplified polymorphic DNA (RAPD) markers for genetic analysis in *Allium*, *Theor. Appl. Genet.* 86: 497–504.

1544. Wilkinson, M.J., Sweet, J., and Poppy, G.M., (2003) Risk assessment of GM plants: avoiding gridlock?, *Trends Plant Sci.* 8: 208–212.

1545. Williams, C.E., and Ronald, P.C., (1994) PCR template-DNA isolated quickly from monocot and dicot leaves without tissue homogenization, *Nucleic Acids Res.* 22: 1917–1918.

1546. Williams, J.G.K., Kubelik, A.R., Livak, K.J., Rafalski, J.A., and Tingey, S.V., (1990) DNA polymorphisms amplified by arbitrary primers are useful as genetic markers, *Nucleic Acids Res.* 18: 6231–6235.

1547. Williams, J.G.K., Hanafey, M.K., Rafalski, J.A., and Tingey, S.V., (1993) Genetic analysis using random amplified polymorphic markers, *Meth. Enzymol.* 218: 704–740.

1548. Williams, N.M.V., Pande, N., Nair, S., Mohan, M., and Bennett, J., (1991) Restriction fragment length polymorphism analysis of polymerase chain reaction products amplified from mapped loci of rice (*Oryza sativa* L.) genomic DNA, *Theor. Appl. Genet.* 82: 489–498.

1549. Willmitzer, L., and Wagner, K.G., (1981) The isolation of nuclei from tissue-cultured plant cells, *Exp. Cell Res.* 135: 69–77.

1550. Wilson, A.J., and Chourey, P.S., (1984) A rapid inexpensive method for the isolation of restrictable mitochondrial DNA from various plant sources, *Plant Cell Rep.* 3: 237–239.

1551. Winberg, B.C., Zhou, Z., Dallas, J.F., McIntyre, C.L., and Gustafson, J.P., (1993) Characterization of minisatellite sequences from *Oryza sativa*, *Genome* 36: 978–983.

1552. Winfield, M.O., Arnold, G.M., Cooper, F., Le Ray, M., White, J., Karp, A., and Edwards, K.J., (1998) A study of genetic diversity in *Populus nigra* subsp. *betulifolia* in the Upper Severn area of the U.K. using AFLP markers, *Mol. Ecol.* 7: 3–10.

1553. Wing, R.A., Rastogi, V.K., Zhang, H.-B., Paterson, A.H., and Tanksley, S.D., (1993) An improved method of plant megabase DNA isolation in agarose microbeads suitable for physical mapping and YAC cloning, *Plant J.* 4: 893–898.

1554. Winter, P., and Kahl, G., (1995) Molecular marker technologies for plant improvement, *World J. Microbiol. Biotechnol.* 11: 438–448.

1555. Winter, P., Benko-Iseppon, A.-M., Hüttel, B., Ratnaparkhe, M., Tullu, A., Sonnante, G., Pfaff, T., Tekeoglu, M., Santra, D., Sant, V.J., Rajesh, P.N., Kahl, G., and Muehlbauer, F.J., (2000) A linkage map of the chickpea (*Cicer arietinum L.*) genome based on recombinant inbred lines from a *C. arietinum* x *C. reticulatum* cross: localization of resistance genes for *Fusarium* wilt races 4 and 5, *Theor. Appl. Genet.* 101: 1155–1163.

1556. Witsenboer, H., Vogel, J., and Michelmore, R.W., (1997) Identification, genetic localization, and allelic diversity of selectively amplified microsatellite polymorphic loci in lettuce and wild relatives (*Lactuca* spp.), *Genome* 40: 923–936.

1557. Wöhl, T., Brecht, M., Lottspeich, F., and Ammer, H., (1995) The use of genomic DNA probes for in-gel hybridization, *Electrophoresis* 16: 739–741.

1558. Wolfe, A.D., and Liston, A., (1998) Contributions of PCR-based methods to plant systematics and evolutionary biology. In *Molecular Systematics of Plants II. DNA Sequencing*, Soltis, D.E., Soltis, P.S., and Doyle, J.J., Eds., Kluwer, Dordrecht, pp. 43–86.

1559. Wolfe, K.H., Li, H.-H., and Sharp, P.M., (1987) Rates of nucleotide substitution vary greatly among plant mitochondrial, chloroplast and nuclear DNAs, *Proc. Natl. Acad. Sci. U.S.A.* 84: 9054–9058.

1560. Wolff, K., (1991) Analysis of allozyme variability in three *Plantago* species and a comparison to morphological variability, *Theor. Appl. Genet.* 81: 119–126.

1561. Wolff, K., Schoen, E.D., and Peters-Van Rijn, J., (1993) Optimizing the generation of random amplified polymorphic DNAs in chrysanthemum, *Theor. Appl. Genet.* 86: 1033–1037.

1562. Wolff, K., Schaal, B.A., and Rogstad, S.H., (1994) Population and species variation of minisatellite DNA in *Plantago*, *Theor. Appl. Genet.* 87: 733–740.

1563. Wolff, K., Zietkiewicz, E., and Hofstra, H., (1995) Identification of chrysanthemum cultivars and stability of fingerprint patterns, *Theor. Appl. Genet.* 91: 439–447.

1564. Wolff, K., El-Akkad, S., and Abbott, R.J., (1997) Population substructure in *Alkanna orientalis* (Boraginaceae) in the Sinai Desert, in relation to its pollinator behaviour, *Mol. Ecol.* 6: 365–372.

1565. Wong, K.C., and Sun, M., (1999) Reproductive biology and conservation genetics of *Goodyera procera* (Orchidaceae), *Am. J. Bot.* 86: 1406–1413.

1566. Wong, Z., Wilson, V., Jeffreys, A.J., and Thein, S.L., (1986) Cloning a selected fragment from a human DNA "fingerprint": isolation of an extremely polymorphic minisatellite, *Nucleic Acids Res.* 14: 4605–4616.

1567. Woo, S.-S., Rastogi, V.K., Zhang, H.-B., Paterson, A.H., Schertz, K.F., and Wing, R.A., (1995) Isolation of megabase-size DNA from *Sorghum* and applications for physical mapping and bacterial and yeast artificial chromosome library construction, *Plant Mol. Biol. Rep.* 13: 82–94.

1568. Woodhead, M., Russell, J., Squirrell, J., Hollingsworth, P.M., Cardle, L., Ramsay, L., Gibby, M., and Powell, W., (2003) Development of EST-SSRs from the alpine ladyfern, *Athyrium distentifolium*, *Mol. Ecol. Notes* 3: 287–290.

1569. Wright, D.A., Ke, N., Smalle, J., Hauge, B.M., Goodman, H.M., and Voytas, D.F., (1996) Multiple Non-LTR retrotransposons in the genome of *Arabidopsis thaliana*, *Genetics* 142: 569–578.

1570. Wright, J.M., (1994) Mutation at VNTRs: are minisatellites the evolutionary progeny of microsatellites? *Genome* 37: 345–346.

1571. Wright, S., (1931) Evolution in Mendelian populations, *Genetics* 16: 97–159.

1572. Wright, S., (1965) The interpretation of population structure by F-statistics with special regard to system of mating, *Evolution* 19: 395–420.

1573. Wu, J., Krutovskii, K.V., and Strauss, S.H., (1999) Nuclear DNA diversity, population differentiation, and phylogenetic relationships in the California closed-cone pines based on RAPD and allozyme markers, *Genome* 42: 893–908.

1574. Wu, K., Jones, R., Danneberger, L., and Scolnik, P.A., (1994) Detection of microsatellite polymorphisms without cloning, *Nucleic Acids Res.* 22: 3257–3258.

1575. Wu, K.-S., and Tanksley, S.D., (1993) Abundance, polymorphism and genetic mapping of microsatellites in rice, *Mol. Gen. Genet.* 241: 225–235.

1576. Wulff, E.G., Torres, S., and Vigil, E.G., (2002) Protocol for DNA extraction from potato tubers, *Plant Mol. Biol. Rep.* 20: 187a–187e.

1577. Wyman, A.R., and White, R., (1980) A highly polymorphic locus in human DNA, *Proc. Natl. Acad. Sci. U.S.A.* 77: 6754–6758.

1578. Xia, X., and Zie, Z., (2001) DAMBE: software package for data analysis in molecular biology and evolution, *J. Hered.* 92: 371–373.

1579. Xie, Z.W., Lu, Y.Q., Ge, S., Hong, D.Y., and Li, F.Z., (2001) Clonality in wild rice (*Oryza rufipogon*, Poaceae) and its implications for conservation management, *Am. J. Bot.* 88: 1058–1064.

1580. Xiong, L.Z., Xu, C.G., Saghai-Maroof, M.A., and Zhang, Q., (1999) Patterns of cytosine methylation in an elite rice hybrid and its parental lines, detected by a methylation-sensitive amplification polymorphism technique, *Mol. Gen. Genet.* 261: 439–446.

1581. Xu, C.Y., Zhang, W.J., Fu, C.Z., and Lu, B.R., (2003) Genetic diversity of alligator weed in China by RAPD analysis, *Biodivers. Conserv.* 12: 637–645.

1582. Xu, M., Huaracha, E., and Korban, S.S., (2001) Development of sequence-characterized amplified regions (SCARs) from amplified fragment length polymorphism (AFLP) markers tightly linked to the *Vf* gene in apple, *Genome* 44: 63–70.

1583. Xu, M., Li, X., and Korban, S.S., (2000) AFLP-based detection of DNA methylation, *Plant Mol. Biol. Rep.* 18: 361–368.

1584. Yamazaki, H., Nomoto, S., Mishima, Y., and Kominami, R., (1992) A 35-Kda protein binding to a cytosine-rich strand of hypervariable minisatellite DNA, *J. Biol. Chem.* 267: 12311–12316.

1585. Yang, H., Sweetingham, M.W., Cowling, W.A., and Smith, P.M.C., (2001) DNA fingerprinting based on microsatellite-anchored fragment length polymorphisms, and isolation of sequence-specific PCR markers in lupin (*Lupinus angustifolius* L.), *Mol. Breed.* 7: 203–209.

1586. Yang, H., Shankar, M., Buirchell, B.J., Sweetingham, M.W., Caminero, C., and Smith, P.M.C., (2002) Development of molecular markers using MFLP linked to a gene conferring resistance to *Diaporthe toxica* in narrow-leafed lupin (*Lupinus angustifolius* L.), *Theor. Appl. Genet.* 195: 265–270.

1587. Yang, Y.-W., Lai, K.-N., Tai, P.-Y., and Li, W.-H., (1999) Rates of nucleotide substitution in angiosperm mitochondrial DNA sequences and dates of divergence between *Brassica* and other angiosperm lineages, *J. Mol. Evol.* 48: 597–604.

1588. Yap, P.H., and O'Ghee, J., (1992) Nonisotopic SSCP detection in PCR products by ethidium bromide staining, *Trends Genet.* 8: 49.

1589. Yasui, Y., Nasuda, S., Matsuoka, Y., and Kawahara, T., (2001) The *Au* family, a novel short interspersed element (SINE) from *Aegilops umbellulata*, *Theor. Appl. Genet.* 102: 463–470.

1590. Ye, G.-N., Hemmat, M., Lodhi, M.A., Weeden, N.F., and Reisch, B.I., (1996) Long primers for RAPD mapping and fingerprinting of grape and pear, *BioTechniques* 20: 368–371.

1591. Yeh, F.C., Chong, D.K.X., and Yan, R.C., (1995) RAPD variation within and among natural populations of trembling aspen (Michx.) from Alberta, *J. Hered.* 86: 454–460.

1592. Yoon, C.-S., and Glawe, D.A., (1993) Pretreatment with RNase to improve PCR amplification of DNA using 10-mer primers, *BioTechniques* 14: 908–910.

1593. Yoshimura, S., Yoshimura, A., and Iwata, N., (1992) Simple and rapid PCR method by using crude extracts from rice seedlings, *Jpn. J. Breed.* 42: 669–674.

1594. Yoshimura, S., Yoshimura, A., Iwata, N., McCouch, S., Abenes, M.L., Baraoidan, M.R., Mew, T., and Nelson, R.J. (1995) Tagging and combining bacterial blight resistance genes in rice using RAPD and RFLP markers, *Mol. Breed.* 1: 375–387.

1595. Yoshioka, Y., Matsumoto, S., Kojima, S., Ohshima, K., Okada, N., and Machida, Y., (1993) Molecular characterization of a short interspersed repetitive element from tobacco that exhibits sequence homology to specific tRNAs, *Proc. Natl. Acad. Sci. U.S.A.* 90: 6562–6566.

1596. Young, A., Boyle, T., and Brown, T., (1996) The population genetic consequences of habitat fragmentation for plants, *Trends Ecol. Evol.* 11: 413–418.

1597. Young, E.T., Sloan, J.S., and Van Riper, K., (2000) Trinucleotide repeats are clustered in regulatory genes in *Saccharomyces cerevisiae*, *Genetics* 154: 1053–1068.

1598. Young, N.D. (1994) Constructing a plant genetic linkage map with DNA markers. In *DNA-Based Markers in Plants*, Phillips, R.L., and Vasil, I.K., Eds., Kluwer Academic, Dordrecht, pp. 39–57.

1599. Young, N.D., (2000) The genetic architecture of resistance, *Curr. Opin. Plant Biol.* 3: 285–290.

1600. Young, N.D., Zamir, D., Ganal, M.W., and Tanksley, S.D., (1988) Use of isogenic lines and simultaneous probing to identify DNA markers tightly linked to the *Tm-2a* gene in tomato, *Genetics* 120: 579–585.

1601. Young, W.P., Schupp, J.M., and Keim, P., (1999) DNA methylation and AFLP marker distribution in the soybean genome, *Theor. Appl. Genet.* 99: 785–790.

1602. Yu, J., and 99 co-authors, (2002) A draft sequence of the rice genome (*Oryza sativa* L. ssp. *indica*), *Science* 296: 79–92.

1603. Yu, K., and Pauls, K.P., (1992) Optimization of the PCR program for RAPD analysis, *Nucleic Acids Res.* 20: 2606.

1604. Yu, Y.G., Buss, G.R., and Saghai-Maroof, M.A., (1996) Isolation of a superfamily of candidate disease-resistance genes in soybean based on a conserved nucleotide-binding site, *Proc. Natl. Acad. Sci. U.S.A.* 93: 11751–11756.

1605. Zabeau, M., and Vos, P., (1993) Selective restriction fragment amplification: a general method for DNA fingerprinting, European Patent Application EP 0534858.

1606. Zane, L., Bargelloni, L., and Patarnello, T., (2002) Strategies for microsatellite isolation: a review, *Mol. Ecol.* 11: 1–16.

1607. Zawko, G., Krauss, S.L., Dixon, K.W., and Sivasithamparam, K., (2001) Conservation genetics of the rare and endangered *Leucopogon obtectus* (Ericaceae), *Mol. Ecol.* 10: 2389–2396.

1608. Zhang, D.P., Huaman, Z., Rodriguez, F., Rossel, G., and Ghislain, M., (2001) Identifying duplicates in sweet potato (*Ipomoea batatas* (L.) Lam) cultivars using RAPD, *Acta Hortic.* 546: 535–541.

1609. Zhang, H.-B., Zhao, X., Ding, X., Paterson, A.H., and Wing, R.A., (1995) Preparation of megabase-size DNA from plant nuclei, *Plant J.* 7: 175–184.

1610. Zhang, L.-B., Comes, H.P., and Kadereit, J.W., (2001) Phylogeny and quaternary history of the European montane/alpine endemic *Soldanella* (Primulaceae) based on ITS and AFLP variation, *Am. J. Bot.* 88: 2331–2345.

1611. Zhang, Q., Arbuckle, J., and Wessler, S.R., (2000) Recent, extensive, and preferential insertion of members of the miniature inverted-repeat transposable element family *Heartbreaker* into genic regions of maize, *Proc. Natl. Acad. Sci. U.S.A.* 97: 1160–1165.

1612. Zhang, W., Gianibelli, M.C., Ma, W., Rampling, L., and Gale, K.R., (2003) Identification of SNPs and development of allele-specific PCR markers for γ-gliadin alleles in *Triticum aestivum*, *Theor. Appl. Genet.* 107: 130–138.

1613. Zhang, W., Wendel, J.F., and Clark, L.G., (1997) Bamboozled again! Inadvertant isolation of fungal rDNA sequences from bamboos (Poaceae: Bambusoideae), *Mol. Phylogenet. Evol.* 8: 205–217.

1614. Zhao, X., Zhang, H.-B., Wing, R.A., and Paterson, A.H., (1994) A simple method for isolation of megabase DNA from cotton, *Plant Mol. Biol. Rep.* 12: 110–115.

1615. Zhivotovsky, L.A., (1999) Estimating population structure in diploids with multilocus dominant DNA markers, *Mol. Ecol.* 8: 907–914.

1616. Zhou, Y., Bui, T., Auckland, L.D., and Williams, C.G., (2001) Undermethylated DNA as a source of microsatellites from a conifer genome, *Genome* 45: 91–99.

1617. Zhou, Z., Bebeli, P.J., Somers, D.J., and Gustafson, J.P., (1997) Direct amplification of minisatellite-region DNA with VNTR core sequences in the genus *Oryza, Theor. Appl. Genet.* 95: 942–949.

1618. Zhu, H., Qu, F., and Zhu, L.-H., (1993) Isolation of genomic DNA from plants, fungi and bacteria using benzyl chloride, *Nucleic Acids Res.* 21: 5279–5280.

1619. Ziegenhagen, B., Guillemaut, P., and Scholz, F., (1993) A procedure for mini-preparations of genomic DNA from needles of silver fir (*Abies alba* Mill.), *Plant Mol. Biol. Rep.* 11: 117–121.

1620. Ziegle, J.S., Su, Y., Corcoran, K.P., Nie, L., Mayrand, P.E., Hoff, L.B., McBride, L.J., Kronick, M.N., and Diehl, S.R., (1992) Application of automated DNA sizing technology for genotyping microsatellite loci, *Genomics* 14: 1026–1031.

1621. Zietkiewicz, E., Rafalski, A., and Labuda, D., (1994) Genome fingerprinting by simple sequence repeat (SSR)-anchored polymerase chain reaction amplification, *Genomics* 20: 176–183.

1622. Zong, X.X., Kaga, A., Tomooka, N., Wang, C.W., Han, O.K., and Vaughan, D., (2003) The genetic diversity of the *Vigna angularis* complex in Asia, *Genome* 46: 647–658.

1623. Zuliani, G., and Hobbs, H.H., (1990) A high frequency of length polymorphisms in repeated sequences adjacent to *Alu* sequences, *Am. J. Hum. Genet.* 46: 963–969.

Index

H

I

S